READINGS IN THE PHILOSOPHY OF SCIENCE

READINGS IN THE PHILOSOPHY OF SCIENCE
From Positivism to Postmodernism

Theodore Schick, Jr.
Muhlenberg College

Mayfield Publishing Company
Mountain View, California
London • Toronto

Library of Congress Cataloging-in-Publication Data

Schick, Theodore.
    Readings in the philosophy of science : from positivism to postmodernism /
Theodore Schick, Jr.
       p.   cm.
    Includes bibliographical references.
    ISBN 0-7674-0277-4
    1. Science—Philosophy.   I. Title.
Q175.S3513 1999
501—dc21
                                          99-39029
                                           CIP

Manufactured in the United States of America

10   9   8   7   6   5   4   3   2

Mayfield Publishing Company
1280 Villa Street
Mountain View, California, 94041

Sponsoring editor, Kenneth King; production, Publication Services; manuscript editor, Dave Mason; design manager, Susan Breitbard; text and cover designer, Linda M. Robertson; manufacturing manager, Randy Hurst. The text was set in 9.5/12 Plantin Light by Publication Services and printed on 50# Finch Opaque by R. R. Donnelley & Sons Co., Inc.

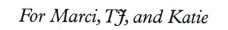

For Marci, TJ, and Katie

PREFACE

The philosophy of science has become too diverse and complex a discipline to be adequately represented in one anthology. Consequently, most anthologies provide a smorgasbord of articles on a variety of topics in the hope of giving students the flavor of the discipline. What is often lost in this approach is a sense of the historical and logical development of the ideas in the discipline. This anthology addresses that problem by tracing the development of thinking on key issues from their classical formulation to the present. Subsequent articles often clarify or critique preceding ones. In this way, students get a sense of how philosophical theories are developed in response to one another.

Logical positivism was one of the most influential philosophical movements in the twentieth century. Not only did it redefine traditional philosophical problems, but it also affected how scientists thought about their work. Contemporary philosophy of science is by and large a reaction to the claims of the logical positivists. An understanding of current issues in the philosophy of science, then, requires a knowledge of their logical positivist underpinnings. This anthology contains a number of classic articles by logical positivists as well as contemporary critiques of them.

The critique of logical positivism paved the way for the work of Thomas Kuhn and Paul Feyerabend. Their rejection of the tenets of logical positivism called into question the rationality and objectivity of science and ushered in the new fields of sociology of knowledge and feminist epistemology. These "postmodernist" approaches to science and knowledge serve as the battleground for what is known as the "science wars." The present anthology examines these approaches by providing a balanced mix of articles, both pro and con.

The new philosophy of science has also resurrected the old debate about the nature and scope of science and religion. Feyerabend claims that science is a religion because it rests on articles of faith that cannot be rationally justified. The present anthology devotes an entire section to the relationship between science and religion and includes opposing views from some of today's best-known scientists and science writers, including Richard Dawkins and Martin Gardner. The most visible clash between science and religion is the creation/evolution controversy. This topic is explored in a number of sections of this anthology. What constitutes science is one of the central problems in the philosophy of science. The creation/evolution controversy serves as a touchstone against which various solutions to this problem can be evaluated.

A number of people contributed to this book with a thoughtful reading of the manuscript. Thanks are due to Frank Fair, Sam Houston State University; Eric Kraemer, University of Wisconsin at La Crosse; Marc Lange, University of Washington; Joseph E. Martire, Southwest Missouri State University; Tim McGrew, Western Michigan University; Lenny Moss, University of Notre Dame; Mark Perlman, West Oregon University; and Nicholas Power, University of West Florida.

CONTENTS

PART 6: SCIENCE AND OBJECTIVITY: THE SCIENCE WARS 198

PART 7: REALISM AND ANTIREALISM: DOES SCIENCE REVEAL REALITY? 256

READINGS IN THE PHILOSOPHY OF SCIENCE

GENERAL INTRODUCTION

Every inquiry makes certain assumptions about what it's trying to find and how best to look for it. Science is no exception. The philosophy of science critically examines the assumptions underlying scientific inquiry in an attempt to understand its nature and evaluate its results. It is commonly assumed, for example, that scientific inquiry gives us objective knowledge of the world by providing explanations based on laws or theories that have been confirmed through observation or experiment. To evaluate this claim, the philosophy of science asks such questions as: What is a scientific theory? What distinguishes scientific theories from nonscientific ones? How are scientific theories confirmed? What is a scientific law? What is a scientific explanation? How are different sciences related to one another? Are all sciences reducible to physics? What is the relationship between theory and observation? Is what we observe affected by the theory we accept? If so, can scientific inquiry be objective? Are scientific theories invented or discovered? If they are invented, what guarantee do we have that they correspond to reality? What is the relation between science and religion? Can everything be explained in natural terms, or must we appeal to the supernatural? By answering these questions, we not only gain a better understanding of science, but we also gain a better understanding of the scope and limits of human knowledge.

Philosophy is not the only discipline that studies science. History, psychology, and sociology have also made a study of science. But their concern is with the action of scientists, not with the underlying assumptions of science. They attempt to determine what caused certain scientists or groups of scientists to behave as they did, not whether the assumptions underlying their behavior are justified. A historian may want to know what effect the Inquisition had on the development of science. A psychologist may want to know why James Watson (one of the discoverers of DNA) treated the research of Rosaline Franklin the way he did. A sociologist may want to know why the Chinese never developed a science comparable to that of the West. But none of these disciplines attempts to analyze the conceptual underpinnings of science in the way that philosophy does.

What distinguishes good science from bad science is not a question that can be answered by science. We can't use a particular model of science to determine whether that model is adequate, because in using it we would have already assumed

1

its adequacy. Any judgment concerning the acceptability of a particular mode of inquiry must appeal to general principles of knowledge. Only philosophy provides the conceptual resources needed to make such a judgment.

Because of its interest in the nature of knowledge and reality, philosophy has long been concerned with the nature of science, supposedly the most reliable source of knowledge about reality. Philosophical inquiries into the scientific enterprise have not only elucidated it; they have also changed it. For over 2000 years science was conducted using an Aristotelian model: to understand something was to know its purpose. Philosophical critiques of this notion led to the scientific revolution of the sixteenth century. Scientists now seek to understand things by identifying their causes.

positivism & postmodernism

In the twentieth century, our understanding of the nature of science has undergone almost as radical a transformation as it did in the sixteenth. New developments in logic led a number of thinkers to believe that the notions of confirmation, explanation, and reduction could be expressed in purely formal terms. Known as logical positivists, they developed the most detailed and precise philosophy of science to date. The logical positivists were empiricists: they believed that the only source of knowledge about the world is sense experience. They believed that sense experience is the same for everybody and that it can be expressed in a language free of any theoretical bias. This, they thought, is what makes science objective. Both of these positions have been called into question, and so has the objectivity of science. Many "postmodernists" do not believe that science provides us with objective knowledge. For them, science is merely one way of making sense of the world, and not necessarily the best. To help the reader understand how someone could hold this view, this anthology presents the classic critiques of logical positivism that led up to it.

The text is divided into nine parts. The first deals with the problem of demarcation: what distinguishes science from nonscience? The second deals with the nature of scientific inference: what is the relationship between theories and their data? The third deals with the nature of scientific theories: what is a scientific explanation? The fourth deals with the question of the unity of science: are all sciences reducible to physics? The fifth deals with the nature of observation: is all observation theory laden? The sixth deals with the question of scientific objectivity: is scientific inquiry free of any bias? The seventh deals with the status of theoretical entities: do theoretical entities such as electrons exist? The eighth deals with the relationship between science and religion: is science a religion? Finally, the ninth part deals with some contemporary issues in the philosophy of physics, the philosophy of biology, and the philosophy of psychology.

PART 1

SCIENCE AND NONSCIENCE
Defining the Boundary

Science is considered by many to be the royal road to the truth. If you want to acquire knowledge about the world, they say, you should either consult a scientist or engage in scientific inquiry. But many different disciplines claim to be scientific, and not all of them are. So to keep from falling into error, we need some way of telling legitimate scientific claims from illegitimate ones. What makes a claim scientific? What distinguishes real science from pseudoscience? This is the problem of demarcation. Nowhere is this question more pressing than in the creation-versus-evolution controversy.

Scientific creationism as propounded by the Institute for Creation Research holds (1) that the universe, energy, and life were created from nothing relatively recently (around 6,000 to 10,000 years ago), (2) that living things could not have developed from a single organism through mutation and natural selection, (3) that there is very little variation among members of the same species, (4) that humans did not develop from the apes, and (5) that the earth's geology can be explained by the occurrence of various catastrophes, including a worldwide flood. This account of the creation of the universe and its inhabitants is derived primarily from the Bible's Book of Genesis.

Promoting religion in the public schools, however, violates the establishment clause of the First Amendment, which reads, "Congress shall make no law respecting an establishment of religion." Consequently, the courts have consistently found laws requiring the teaching of creationism to be unconstitutional. Our concern, however, is not with the constitutionality of the teaching of creationism but with its status as a scientific theory. We want to know whether creationism really is as good a theory as evolution.

A. J. Ayer presents the logical positivist's view of the distinction between scientific and nonscientific claims. Scientific claims are empirically verifiable: their truth can be established through observation. Other claims, such as "The absolute enters into, but is itself incapable of, evolution and progress," are not empirically verifiable, and hence are nonscientific. Because there is no way to verify such claims, they do not say anything that can be considered true or false. And because these claims have no truth value, they are cognitively meaningless—they do not convey any information. They may excite certain emotions in us, but they don't tell us anything about the world. The view that only verifiable claims are meaningful is a fundamental principle of logical positivism and is known as "the verifiability theory of meaning."

Karl Popper does not consider verifiability to be the mark of the scientific because verifying theories is too easy. Some theories, such as those of Freud, Adler, and Marx, have been verified many times over and yet do not tell us anything about the world. To be informative, a theory should rule something out; it should "be incompatible with certain results of observation." For example, the statement "Either it's raining or it isn't" tells us nothing about the weather because it's consistent with all possible observations. The same goes for the theories of Freud, Adler, and Marx. There is no conceivable situation that would provide grounds for rejecting these theories. They can account for every eventuality in their respective domains. Because these theories are not falsifiable, Popper claims that they are not scientific.

Thomas Kuhn argues that what's unique about science is its process, not its products. The way that practitioners conduct their inquiry is what distinguishes science from pseudoscience. Popper judges the status of a field by its fruits; if it produces claims that are falsifiable, it is scientific. Kuhn objects that this criterion is too lax because, by its lights, astrology would be a science. Not only has astrology produced many claims that are falsifiable, but many of those claims have turned out to be false. Astrology is not a science because its practitioners do not engage in the sort of puzzle solving that is characteristic of normal science. According to Kuhn, scientific inquiry is guided by a paradigm or model that indicates what problems are worth investigating and how one should go about investigating them. Astrology is a pseudoscience because it is not governed by a paradigm; there are no agreed-upon criteria that astrologers use to determine what an astrological problem is or how one should go about solving it.

Imre Lakatos agrees with Kuhn that scientific inquiry must be guided by a paradigm. But he disagrees with Kuhn's assertion that the choice of paradigms is not governed by rules or reason. For if that were true, we would have to conclude that "scientific revolution is irrational; a matter of mob psychology." Lakatos proposes what he calls a "sophisticated falsificationist theory," which maintains that one theory has been falsified if and only if another theory has been found that explains everything explained by the former theory and also predicts new facts that would have been difficult or impossible to predict from the old theory.

The boundary between science and nonscience became a legal issue after some state legislatures passed laws mandating the teaching of creationism in the classroom. In his ruling regarding the Arkansas statute, Judge Overton identified the following essential characteristics of science: "(1) It is guided by natural law; (2) it has to be explanatory by reference to natural law; (3) it is testable against the empirical world; (4) its conclusions are tentative, i.e., are not necessarily the final word; and (5) it is falsifiable" (*McLean v. Arkansas Board of Education,* 50 L.W. 2412 (1982)). Overton rejected the claim that creationism is science on the grounds that it fails to meet these criteria.

Larry Laudan finds Judge Overton's decision suspect because creationism does meet some of these criteria, and those it doesn't meet are not essential to science. Laudan claims that creationism is both testable and falsifiable because it makes a number of claims whose falsity can be established through observation, such as the earth is 6,000 years old, there was a worldwide flood, and dinosaurs and men walked the earth at the same time. What's wrong with creationism is not that it isn't falsifiable, but that its claims have been tested and have turned out to be false. What's more, contrary to what Ruse, the author of the next reading, would have us believe, many scientific theories, such as Darwinian evolution, were not initially guided by natural law or explanatory by reference to natural law. In Laudan's view, Judge Overton made the right decision but for the wrong reasons.

Michael Ruse testified at the Arkansas trial and offered the criteria that Laudan attacks. He acknowledges that Darwin's theory of evolution may not have been guided by natural law or explained by reference to natural law when it was introduced in the nineteenth century. But he claims that our concept of science has evolved since then. What we admit as science should be consistent with our current conception of it, which in the case of Darwinian evolution does include those criteria. As for testability, falsifiability, and revisability, Ruse claims that creationism meets none of these criteria because there are no possible circumstances under which the creationists would give up any of the central doctrines. They believe that the Bible is the word of God, and if the evidence seems to conflict with the holy word, it must be mistaken.

1

A. J. AYER

The Elimination of Metaphysics

The traditional disputes of philosophers are, for the most part, as unwarranted as they are unfruitful. The surest way to end them is to establish beyond question what should be the purpose and method of a philosophical inquiry. And this is by no means so difficult a task as the history of philosophy would lead one to suppose. For if there are any questions which science leaves it to philosophy to answer, a straightforward process of elimination must lead to their discovery.

We may begin by criticizing the metaphysical thesis that philosophy affords us knowledge of a reality transcending the world of science and common sense. Later on, when we come to define metaphysics and account for its existence, we shall find that it is possible to be a metaphysician without believing in a transcendent reality; for we shall see that many metaphysical utterances are due to the commission of logical errors, rather than to a conscious desire on the part of their authors to go beyond the limits of experience. But it is convenient for us to take the case of those who believe that it is possible to have knowledge of a transcendent reality as a starting-point for our discussion. The arguments which we use to refute them will subsequently be found to apply to the whole of metaphysics.

One way of attacking a metaphysician who claimed to have knowledge of a reality which transcended the phenomenal world would be to inquire from what premises his propositions were deduced. Must he not begin, as other men do, with the evidence of his senses? And if so, what valid process of reasoning can possibly lead him to the conception of a transcendent reality? Surely from empirical premises nothing whatsoever concerning the properties, or even the existence, of anything super-empirical can legitimately be inferred. But this objection would be met by a denial on the part of the metaphysician that his assertions were ultimately based on the evidence of his senses. He would say that he was endowed with a faculty of intellectual intuition which enabled him to know facts that could not be known through sense-experience. And even if it could be shown that he was relying on empirical premises, and that his venture into a nonempirical world was therefore logically unjustified, it would not follow that the assertions which he made concerning this nonempirical world could not be true. For the fact that a conclusion does not follow from its putative premise is not sufficient to show that it is false. Consequently one cannot overthrow a system of transcendent metaphysics merely by criticizing the way in which it comes into being. What is required is rather a criticism of the nature of the actual statements which comprise it. And this is the line of argument which we shall, in fact, pursue. For we shall maintain that no statement which refers to a "reality" transcending the limits of all possible sense-experience can possibly have any literal significance; from which it must follow that the labors of those who have striven to describe such a reality have all been devoted to the production of nonsense.

It may be suggested that this is a proposition which has already been proved by Kant. But although Kant also condemned transcendent metaphysics, he did so on different grounds. For he said that the human understanding was so constituted that it lost itself in contradictions when it ventured

Language, Truth, and Logic (New York: Dover, 1952), pp. 33–45. Reprinted by permission of the publisher.

out beyond the limits of possible experience and attempted to deal with things in themselves. And thus he made the impossibility of a transcendent metaphysic not, as we do, a matter of logic, but a matter of fact. He asserted, not that our minds could not conceivably have had the power of penetrating beyond the phenomenal world, but merely that they were in fact devoid of it. And this leads the critic to ask how, if it is possible to know only what lies within the bounds of sense-experience, the author can be justified in asserting that real things do exist beyond, and how he can tell what are the boundaries beyond which the human understanding may not venture, unless he succeeds in passing them himself. As Wittgenstein says, "in order to draw a limit to thinking, we should have to think both sides of this limit,"[1] a truth to which Bradley gives a special twist in maintaining that the man who is ready to prove that metaphysics is impossible is a brother metaphysician with a rival theory of his own.[2]

Whatever force these objections may have against the Kantian doctrine, they have none whatsoever against the thesis that I am about to set forth. It cannot here be said that the author is himself overstepping the barrier he maintains to be impassable. For the fruitlessness of attempting to transcend the limits of possible sense-experience will be deduced, not from a psychological hypothesis concerning the actual constitution of the human mind, but from the rule which determines the literal significance of language. Our charge against the metaphysician is not that he attempts to employ the understanding in a field where it cannot profitably venture, but that he produces sentences which fail to conform to the conditions under which alone a sentence can be literally significant. Nor are we ourselves obliged to talk nonsense in order to show that all sentences of a certain type are necessarily devoid of literal significance. We need only formulate the criterion which enables us to test whether a sentence expresses a genuine proposition about a matter of fact, and then point out that the sentences under consideration fail to satisfy it. And this we shall now proceed to do. We shall first of all formulate the criterion in somewhat vague terms, and then give the explanations which are necessary to render it precise.

The criterion which we use to test the genuineness of apparent statements of fact is the criterion of verifiability. We say that a sentence is factually significant to any given person, if, and only if, he knows how to verify the proposition which it purports to express—that is, if he knows what observations would lead him, under certain conditions, to accept the proposition as being true, or reject it as being false. If, on the other hand, the putative proposition is of such a character that the assumption of its truth, or falsehood, is consistent with any assumption whatsoever concerning the nature of his future experience, then, as far as he is concerned, it is, if not a tautology, a mere pseudo-proposition. The sentence expressing it may be emotionally significant to him; but it is not literally significant. And with regard to questions the procedure is the same. We inquire in every case what observations would lead us to answer the question, one way or the other; and, if none can be discovered, we must conclude that the sentence under consideration does not, as far as we are concerned, express a genuine question, however strongly its grammatical appearance may suggest that it does.

As the adoption of this procedure is an essential factor in the argument of this book, it needs to be examined in detail.

In the first place, it is necessary to draw a distinction between practical verifiability, and verifiability in principle. Plainly we all understand, in many cases believe, propositions which we have not in fact taken steps to verify. Many of these are propositions which we could verify if we took enough trouble. But there remain a number of significant propositions, concerning matters of fact, which we could not verify even if we chose; simply because we lack the practical means of placing ourselves in the situation where the relevant observations could be made. A simple and familiar example of such a proposition is the proposition that there are mountains on the farther side of the moon.[3] No rocket has yet been invented which would enable me to go and look at the farther side of the moon, so that I am unable to decide the matter by actual observation. But I do know what observations would decide it for me, if, as is theoretically conceivable, I were once in a position to make them. And therefore I say that the proposition is verifiable in principle, if not in practice, and is accordingly significant. On the other hand, such a metaphysical pseudo-proposition as "the Absolute enters into, but is itself incapable of, evolution and progress,"[4] is not even in principle verifiable. For one cannot conceive of an observation which would enable one to determine whether the Absolute did, or did not, enter into evolution and progress. Of course it is possible that the author of such a remark is using English words in a way in which they are not commonly used by English-speaking people, and that he does, in fact, intend to assert something which could be empirically verified. But until he makes us understand how the proposition that he wishes to express would be verified, he fails to communicate anything to us. And if he admits, as I think the author of the remark in question would have admitted, that his words were not intended to express either a tautology or a proposition which was capable, at least in principle, of being verified, then it follows that he has made an utterance which has no literal significance even for himself.

A further distinction which we must make is the distinction between the "strong" and the "weak" sense of the term "verifiable." A proposition is said to be verifiable, in the strong sense of the term, if, and only if, its truth could be conclusively established in experience. But it is verifiable, in the weak sense, if it is possible for experience to render it probable. In which sense are we using the term when we say that a putative proposition is genuine only if it is verifiable?

It seems to me that if we adopt conclusive verifiability as our criterion of significance, as some positivists have proposed,[5] our argument will prove too much. Consider, for example, the case of general propositions of law—such propositions, namely, as "arsenic is poisonous"; "all men are mortal"; "a body tends to expand when it is heated." It is of the very nature of these propositions that their truth cannot be established with certainty by any finite series of observations. But if it is recognized that such general propositions of law are designed to cover an infinite number of cases, then it must be admitted that they cannot, even in principle, be verified conclusively. And then, if we adopt conclusive verifia-

bility as our criterion of significance, we are logically obliged to treat these general propositions of law in the same fashion as we treat the statements of the metaphysician.

In face of this difficulty, some positivists[6] have adopted the heroic course of saying that these general propositions are indeed pieces of nonsense, albeit an essentially important type of nonsense. But here the introduction of the term "important" is simply an attempt to hedge. It serves only to mark the authors' recognition that their view is somewhat too paradoxical, without in any way removing the paradox. Besides, the difficulty is not confined to the case of general propositions of law, though it is there revealed most plainly. It is hardly less obvious in the case of propositions about the remote past. For it must surely be admitted that, however strong the evidence in favor of historical statements may be, their truth can never become more than highly probable. And to maintain that they also constituted an important, or unimportant, type of nonsense would be unplausible, to say the very least. Indeed, it will be our contention that no proposition, other than a tautology, can possibly be anything more than a probable hypothesis. And if this is correct, the principle that a sentence can be factually significant only if it expresses what is conclusively verifiable is self-stultifying as a criterion of significance. For it leads to the conclusion that it is impossible to make a significant statement of fact at all.

NOTES

1. *Tractatus Logico-Philosophicus*, Preface.
2. Bradley, *Appearance and Reality*, 2nd ed., p. 1.
3. This example has been used by Professor Schlick to illustrate the same point.
4. A remark taken at random from *Appearance and Reality*, by F. H. Bradley.
5. E.g., M. Schlick, "Positivismus und Realismus," *Erkenntnis*, Vol. I, 1930. F. Waismann, "Logische Analyse des Warscheinlichkeitsbegriffs," *Erkenntnis*, Vol. I, 1930.
6. E.g., M. Schlick, "Die Kausalität in der gegenwärtigen Physik," *Naturwissenschaft*, Vol. 19, 1931.

2

KARL R. POPPER

Science: Conjectures and Refutations

When I received the list of participants in this course and realized that I had been asked to speak to philosophical colleagues I thought, after some hesitation and consultation, that you would probably prefer me to speak about those problems which interest me most, and about those developments with which I am most intimately acquainted. I therefore decided to do what I have never done before: to give you a report on my own work in the philosophy of science, since the autumn of 1919 when I first began to grapple with the problem, *'When should a theory be ranked as scientific?'* or *'Is there a criterion for the scientific character or status of a theory?'*

The problem which troubled me at the time was neither, 'When is a theory true?' nor, 'When is a theory acceptable?' My problem was different. I *wished to distinguish between science and pseudo-science;* knowing very well that science often errs, and that pseudo-science may happen to stumble on the truth.

I knew, of course, the most widely accepted answer to my problem: that science is distinguished from pseudo-science—or from 'metaphysics'—by its *empirical method,* which is essentially *inductive,* proceeding from observation or experiment. But this did not satisfy me. On the contrary, I often formulated my problem as one of distinguishing between a genuinely empirical method and a non-empirical or even a pseudo-empirical method—that is to say, a method which, although it appeals to observation and experiment, nevertheless does not come up to scientific standards. The latter method may be exemplified by astrology, with its stupendous mass of empirical evidence based on observation—on horoscopes and on biographies.

But as it was not the example of astrology which led me to my problem I should perhaps briefly describe the atmosphere in which my problem arose and the examples by which it was stimulated. After the collapse of the Austrian Empire there had been a revolution in Austria: the air was full of revolutionary slogans and ideas, and new and often wild theories. Among the theories which interested me Einstein's theory of relativity was no doubt by far the most important. Three others were Marx's theory of history, Freud's psycho-analysis, and Alfred Adler's so-called 'individual psychology'.

There was a lot of popular nonsense talked about these theories, and especially about relativity (as still happens even today), but I was fortunate in those who introduced me to the study of this theory. We all—the small circle of students to which I belonged—were thrilled with the result of Eddington's eclipse observations which in 1919 brought the first important confirmation of Einstein's theory of gravitation. It was a great experience for us, and one which had a lasting influence on my intellectual development.

The three other theories I have mentioned were also widely discussed among students at that time. I myself happened to come into personal contact with Alfred Adler, and even to cooperate with him in his social work among the children and young people in the working-class districts of Vienna where he had established social guidance clinics.

It was during the summer of 1919 that I began to feel more and more dissatisfied with these three theories—the Marxist theory of history, psycho-analysis,

Conjectures and Refutations: The Growth of Scientific Knowledge (London: Routledge and Kegan Paul, 1963), pp. 33–39. Reprinted with permission.

and individual psychology; and I began to feel dubious about their claims to scientific status. My problem perhaps first took the simple form, 'What is wrong with Marxism, psycho-analysis, and individual psychology? Why are they so different from physical theories, from Newton's theory, and especially from the theory of relativity?'

To make this contrast clear I should explain that few of us at the time would have said that we believed in the *truth* of Einstein's theory of gravitation. This shows that it was not my doubting the *truth* of those other three theories which bothered me, but something else. Yet neither was it that I merely felt mathematical physics to be more *exact* than the sociological or psychological type of theory. Thus what worried me was neither the problem of truth, at that stage at least, nor the problem of exactness or measurability. It was rather that I felt that these other three theories, though posing as sciences, had in fact more in common with primitive myths than with science; that they resembled astrology rather than astronomy.

I found that those of my friends who were admirers of Marx, Freud, and Adler, were impressed by a number of points common to these theories, and especially by their apparent *explanatory power*. These theories appeared to be able to explain practically everything that happened within the fields to which they referred. The study of any of them seemed to have the effect of an intellectual conversion or revelation, opening your eyes to a new truth hidden from those not yet initiated. Once your eyes were thus opened you saw confirming instances everywhere: the world was full of *verifications* of the theory. Whatever happened always confirmed it. Thus its truth appeared manifest; and unbelievers were clearly people who did not want to see the manifest truth; who refused to see it, either because it was against their class interest, or because of their repressions which were still 'un-analyzed' and crying aloud for treatment.

The most characteristic element in this situation seemed to me the incessant stream of confirmations, of observations which 'verified' the theories in question; and this point was constantly emphasized by their adherents. A Marxist could not open a newspaper without finding on every page confirming evidence for his interpretation of history; not only in the news, but also in its presentation—which revealed the class bias of the paper—and especially of course in what the paper did *not* say. The Freudian analysts emphasized that their theories were constantly verified by their 'clinical observations'. As for Adler, I was much impressed by a personal experience. Once, in 1919, I reported to him a case which to me did not seem particularly Adlerian, but which he found no difficulty in analyzing in terms of his theory of inferiority feelings, although he had not even seen the child. Slightly shocked, I asked him how he could be so sure. 'Because of my thousandfold experience,' he replied; whereupon I could not help saying: 'And with this new case, I suppose, your experience has become thousand-and-one-fold.'

What I had in mind was that his previous observations may not have been much sounder than this new one; that each in its turn had been interpreted in the light of 'previous experience', and at the same time counted as additional confirmation. What, I asked myself, did it confirm? No more than that a case could be interpreted in the light of the theory. But this meant very little, I reflected, since every conceivable case could be interpreted in the light of Adler's theory, or equally of Freud's. I may illustrate this by two very different examples of human behavior: that of a man who pushes a child into the water with the intention of drowning it; and that of a man who sacrifices his life in an attempt to save the child. Each of these two cases can be explained with equal ease in Freudian and in Adlerian terms. According to Freud the first man suffered from repression (say, of some component of his Oedipus complex), while the second man had achieved sublimation. According to Adler the first man suffered from feelings of inferiority (producing perhaps the need to prove to himself that he dared to commit some crime), and so did the second man (whose need was to prove to himself that he dared to rescue the child). I could not think of any human behavior which could not be interpreted in terms of either theory. It was precisely this fact—that they always fitted, that they were always confirmed—which in the eyes of their admirers constituted the strongest argument in favor of these theories. It began to dawn on me that this apparent strength was in fact their weakness.

With Einstein's theory the situation was strikingly different. Take one typical instance—Einstein's prediction, just then confirmed by the findings of Eddington's expedition. Einstein's gravitational theory had led to the result that light must be attracted by heavy bodies (such as the sun), precisely as material bodies were attracted. As a consequence it could be calculated that light from a distant fixed star whose apparent position was close to the sun would reach the earth from such a direction that the star would seem to be slightly shifted away from the sun; or, in other words, that stars close to the sun would look as if they had moved a little away from the sun, and from one another. This is a thing which cannot normally be observed since such stars are rendered invisible in daytime by the sun's overwhelming brightness; but during an eclipse it is possible to take photographs of them. If the same constellation is photographed at night one can measure the distances on the two photographs, and check the predicted effect.

Now the impressive thing about this case is the *risk* involved in a prediction of this kind. If observation shows that the predicted effect is definitely absent, then the theory is simply refuted. The theory is *incompatible with certain possible results of observation*—in fact with results which everybody before Einstein would have expected.[1] This is quite different from the situation I have previously described, when it turned out that the theories in question were compatible with the most divergent human behavior, so that it was practically impossible to describe any human behavior that might not be claimed to be a verification of these theories.

These considerations led me in the winter of 1919–20 to conclusions which I may now reformulate as follows.

1. It is easy to obtain confirmations, or verifications, for nearly every theory—if we look for confirmations.

2. Confirmations should count only if they are the result of *risky predictions;* that is to say, if, unenlightened by the theory in question, we should have expected an event which was incompatible with the theory—an event which would have refuted the theory.

3. Every 'good' scientific theory is a prohibition: it forbids certain things to happen. The more a theory forbids, the better it is.

4. A theory which is not refutable by any conceivable event is non-scientific. Irrefutability is not a virtue of a theory (as people often think) but a vice.

5. Every genuine *test* of a theory is an attempt to falsify it, or to refute it. Testability is falsifiability; but there are degrees of testability: some theories are more testable, more exposed to refutation, than others; they take, as it were, greater risks.

6. Confirming evidence should not count *except when it is the result of a genuine test of the theory;* and this means that it can be presented as a serious but unsuccessful attempt to falsify the theory. (I now speak in such cases of 'corroborating evidence'.)

7. Some genuinely testable theories, when found to be false, are still upheld by their admirers—for example by introducing *ad hoc* some auxiliary assumption, or by reinterpreting the theory *ad hoc* in such a way that it escapes refutation. Such a procedure is always possible, but it rescues the theory from refutation only at the price of destroying, or at least lowering, its scientific status. (I later described such a rescuing operation as a *'conventionalist twist'* or a *'conventionalist stratagem'.*)

One can sum up all this by saying that *the criterion of the scientific status of a theory is its falsifiability, or refutability, or testability.*

II

I may perhaps exemplify this with the help of the various theories so far mentioned. Einstein's theory of gravitation clearly satisfied the criterion of falsifiability. Even if our measuring instruments at the time did not allow us to pronounce on the results of the tests with complete assurance, there was clearly a possibility of refuting the theory.

Astrology did not pass the test. Astrologers were greatly impressed, and misled, by what they believed to be confirming evidence—so much so that they were quite unimpressed by any unfavorable evidence. Moreover, by making their interpretations and prophecies sufficiently vague they were able to explain away anything that might have been a refutation of the theory had the theory and the prophecies been more precise. In order to escape falsification they destroyed the testability of their theory. It is a typical soothsayer's trick to predict things so vaguely that the predictions can hardly fail: that they become irrefutable.

The Marxist theory of history, in spite of the serious efforts of some of its founders and followers, ultimately adopted this soothsaying practice. In some of its earlier formulations (for example in Marx's analysis of the character of the 'coming social revolution') their predictions were testable, and in fact falsified.[2] Yet instead of accepting the refutations the followers of Marx reinterpreted both the theory and the evidence in order to make them agree. In this way they rescued the theory from refutation; but they did so at the price of adopting a device which made it irrefutable. They thus gave a 'conventionalist twist' to the theory; and by this stratagem they destroyed its much advertised claim to scientific status.

The two psycho-analytic theories were in a different class. They were simply non-testable, irrefutable. There was no conceivable human behavior which could contradict them. This does not mean that Freud and Adler were not seeing certain things correctly: I personally do not doubt that much of what they say is of considerable importance, and may well play its part one day in a psychological science which is testable. But it does mean that those 'clinical observations' which analysts naïvely believe confirm their theory cannot do this any more than the daily confirmations which astrologers find in their practice.[3] And as for Freud's epic of the Ego, the Super-ego, and the Id, no substantially stronger claim to scientific status can be made for it than for Homer's collected stories from Olympus. These theories describe some facts, but in the manner of myths. They contain most interesting psychological suggestions, but not in a testable form.

At the same time I realized that such myths may be developed, and become testable; that historically speaking all—or very nearly all—scientific theories originate from myths, and that a myth may contain important anticipations of scientific theories. Examples are Empedocles' theory of evolution by trial and error, or Parmenides' myth of the unchanging block universe in which nothing ever happens and which, if we add another dimension, becomes Einstein's block universe (in which, too, nothing ever happens, since everything is, four-dimensionally speaking, determined and laid down from the beginning). I thus felt that if a theory is found to be non-scientific, or 'metaphysical' (as we might say), it is not thereby found to be unimportant, or insignificant, or 'meaningless', or 'nonsensical'.[4] But it cannot claim to be backed by empirical evidence in the scientific sense—although it may easily be, in some genetic sense, the 'result of observation'.

(There were a great many other theories of this pre-scientific or pseudo-scientific character, some of them, unfortunately, as influential as the Marxist interpretation of history; for example, the racialist interpretation of history—another of those impressive and all-explanatory theories which act upon weak minds like revelations.)

Thus the problem which I tried to solve by proposing the criterion of falsifiability was neither a problem of meaningfulness or significance, nor a problem of truth or acceptability. It was the problem of drawing a line (as well as this can be done) between the statements, or systems of statements, of the empirical sciences, and all other statements—whether they are of a religious or of a metaphysical character, or simply pseudo-scientific. Years later—it must have been in 1928 or 1929—I called this first problem of mine the *'problem of demarcation'*. The criterion of falsifiability is a solution to this problem of demarcation, for it says that statements or systems of statements, in order to be ranked as scientific, must be capable of conflicting with possible, or conceivable, observations.

NOTES

1. This is a slight oversimplification, for about half of the Einstein effect may be derived from the classical theory, provided we assume a ballistic theory of light.
2. See, for example, my *Open Society and Its Enemies*, ch. 15, section iii, and notes 13–14.
3. 'Clinical observations', like all other observations, are *interpretations in the light of theories;* and for this

reason alone they are apt to seem to support those theories in the light of which they were interpreted. But real support can be obtained only from observations undertaken as tests (by 'attempted refutations'); and for this purpose *criteria of refutation* have to be laid down beforehand: it must be agreed which observable situations, if actually observed, mean that the theory is refuted. But what kind of clinical responses would refute to the satisfaction of the analyst not merely a particular analytic diagnosis but psycho-analysis itself? And have such criteria ever been discussed or agreed upon by analysts? Is there not, on the contrary, a whole family of analytic concepts, such as 'ambivalence' (I do not suggest that there is no such thing as ambivalence), which would make it difficult, if not impossible, to agree upon such criteria? Moreover, how much headway has been made in investigating the question of the extent to which the (conscious or unconscious) expectations and theories held by the analyst influence the 'clinical responses' of the patient? (To say nothing about the conscious attempts to influence the patient by proposing interpretations to him, etc.) Years ago I introduced the term *'Oedipus effect'* to describe the influence of a theory or expectation or prediction *upon the event which it predicts* or describes: it will be remembered that the causal chain leading to Oedipus' parricide was started by the oracle's prediction

of this event. This is a characteristic and recurrent theme of such myths, but one which seems to have failed to attract the interest of the analysts, perhaps not accidentally. (The problem of confirmatory dreams suggested by the analyst is discussed by Freud, for example in *Gesammelte Schriften*, III, 1925, where he says on p. 315: 'If anybody asserts that most of the dreams which can be utilized in an analysis . . . owe their origin to [the analyst's] suggestion, then no objection can be made from the point of view of analytic theory. Yet there is nothing in this fact', he surprisingly adds, 'which would detract from the reliability of our results.')

4. The case of astrology, nowadays a typical pseudo-science, may illustrate this point. It was attacked, by Aristotelians and other rationalists, down to Newton's day, for the wrong reason—for its now accepted assertion that the planets had an 'influence' upon terrestrial ('sublunar') events. In fact Newton's theory of gravity, and especially the lunar theory of the tides, was historically speaking an offspring of astrological lore. Newton, it seems, was most reluctant to adopt a theory which came from the same stable as for example the theory that 'influenza' epidemics are due to an astral 'influence'. And Galileo, no doubt for the same reason, actually rejected the lunar theory of the tides; and his misgivings about Kepler may easily be explained by his misgivings about astrology.

3

THOMAS S. KUHN

Logic of Discovery or Psychology of Research?

My object in these pages is to juxtapose the view of scientific development outlined in my book, *The Structure of Scientific Revolutions,* with the better-

Criticism and the Growth of Knowledge, Imre Lakatos and Alan Musgrave, eds. (New York: Cambridge University Press, 1970), pp. 1–23. Reprinted with permission of the publisher.

known views of our chairman, Sir Karl Popper.[1] Ordinarily I should decline such an undertaking, for I am not so sanguine as Sir Karl about the utility of confrontations. Besides, I have admired his work for too long to turn critic easily at this date. Nevertheless, I am persuaded that for this occasion the attempt must be made. Even before my book was published two and a half years ago, I had begun to discover special and often puzzling characteristics of the relation between my views and his. That relation and the divergent reactions I have encountered to it suggest that a disciplined comparison of the two may produce peculiar enlightenment. Let me say why I think this could occur.

On almost all the occasions when we turn explicitly to the same problems, Sir Karl's view of science and my own are very nearly identical.[2] We are both concerned with the dynamic process by which scientific knowledge is acquired rather than with the logical structure of the products of scientific research. Given that concern, both of us emphasize, as legitimate data, the facts and also the spirit of actual scientific life, and both of us turn often to history to find them. From this pool of shared data, we draw many of the same conclusions. Both of us reject the view that science progresses by accretion; both emphasize instead the revolutionary process by which an older theory is rejected and replaced by an incompatible new one[3]; and both deeply underscore the role played in this process by the older theory's occasional failure to meet challenges posed by logic, experiment, or observation. Finally, Sir Karl and I are united in opposition to a number of classical positivism's most characteristic theses. We both emphasize, for example, the intimate and inevitable entanglement of scientific observation with scientific theory; we are correspondingly skeptical of efforts to produce any neutral observation language; and we both insist that scientists may properly aim to invent theories that *explain* observed phenomena and that do so in terms of *real* objects, whatever the latter phrase may mean.

That list, though it by no means exhausts the issues about which Sir Karl and I agree,[4] is already extensive enough to place us in the same minority among contemporary philosophers of science. Presumably that is why Sir Karl's followers have with some regularity provided my most sympathetic philosophical audience, one for which I continue to be grateful. But my gratitude is not unmixed. The same agreement that evokes the sympathy of this group too often misdirects its interest. Apparently Sir Karl's followers can often read much of my book as chapters from a late (and, for some, a drastic) revision of his classic, *The Logic of Scientific Discovery*.

One of them asks whether the view of science outlined in my *Scientific Revolutions* has not long been common knowledge. A second, more charitably, isolates my originality as the demonstration that discoveries-of-fact have a life cycle very like that displayed by innovations-of-theory. Still others express general pleasure in the book but will discuss only the two comparatively secondary issues about which my disagreement with Sir Karl is most nearly explicit: my emphasis on the importance of deep commitment to tradition and my discontent with the implications of the term 'falsification'. All these men, in short, read my book through a quite special pair of spectacles. The view through those spectacles is not wrong—my agreement with Sir Karl is real and substantial. Yet readers outside of the Popperian circle almost invariably fail even to notice that the agreement exists, and it is these readers who most often recognize (not necessarily with sympathy) what seem to me the central issues. I conclude that a gestalt switch divides readers of my book into two or more groups. What one of these sees as striking parallelism is virtually invisible to the others. The desire to understand how this can be so motivates the present comparison of my view with Sir Karl's.

The comparison must not, however, be a mere point by point juxtaposition. What demands attention is not so much the peripheral area in which our occasional secondary disagreements are to be isolated but the central region in which we appear to agree. Sir Karl and I do appeal to the same data; to an uncommon extent we are seeing the same lines on the same paper; asked about those lines and those data, we often give virtually identical responses, or at least responses that inevitably seem identical in the isolation enforced by the question-and-answer mode. Nevertheless, experiences like those mentioned above convince me that our intentions are often quite different when we say the same things. Though the lines are the same, the figures which emerge from them are not. That is why I call what separates us a gestalt switch rather than a disagreement and also why I am at once perplexed and intrigued about how best to explore the separation. How am I to persuade Sir Karl, who knows everything I know about scientific development and who has somewhere or other said it, that what he calls a duck can be seen as a rabbit? How am I to show him what it would be like to wear my spectacles when he has already learned to look at everything I can point to through his own?

In this situation a change in strategy is called for, and the following suggests itself. Reading over once more a number of Sir Karl's principal books and essays, I encounter again a series of recurrent phrases which, though I understand them and do not quite disagree, are locutions that I could never have used in the same places. Undoubtedly they are most often intended as metaphors applied rhetorically to situations for which Sir Karl has elsewhere provided unexceptionable descriptions. Nevertheless, for present purposes these metaphors, which strike me as patently inappropriate, may prove more useful than straightforward descriptions. They may, that is, be symptomatic of contextual differences that a careful literal expression hides. If that is so, then these locutions may function not as the lines-on-paper but as the rabbit-ear, the shawl, or the ribbon-at-the-throat which one isolates when teaching a friend to transform his way of seeing a gestalt diagram. That, at least, is my hope for them. I have four such differences of locutions in mind and shall treat them *seriatim.* [Serially]

Among the most fundamental issues on which Sir Karl and I agree is our insistence that an analysis of the development of scientific knowledge must take account of the way science has actually been practiced. That being so, a few of his recurrent generalizations startle me. One of these provides the opening sentences of the first chapter of the *Logic of Scientific Discovery:* 'A scientist', writes Sir Karl, 'whether theorist or experimenter, puts forward statements, or systems of statements, and tests them step by step. In the field of the empirical sciences, more particularly, he constructs hypotheses, or systems of theories, and tests them against experience by observation and experiment.'[5] The statement is virtually a cliché, yet in application it presents three problems. It is ambiguous in its failure to specify which of two sorts of 'statements' or 'theories' are being tested. That ambiguity can, it is true, be eliminated by reference to other passages in Sir Karl's writings, but the generalization that results is historically mistaken. Furthermore, the mistake proves important, for the unambiguous form of the description misses just that characteristic of scientific practice which most nearly distinguishes the sciences from other creative pursuits.

There is one sort of 'statement' or 'hypothesis' that scientists do repeatedly subject to systematic test. I have in mind statements of an individual's best guesses about the proper way to connect his own research problem with the corpus of accepted scientific knowledge. He may, for example, conjecture that a given chemical unknown contains the salt of a rare earth, that the obesity of his experimental rats is due to a specified component in their diet, or that a newly discovered spectral pattern is to be understood as an effect of nuclear spin. In each case, the next steps in his research are intended to try out or test the conjecture or hypothesis. If it passes enough or stringent enough tests, the scientist has made a discovery or has at least resolved the puzzle he had been set. If not, he must either abandon the puzzle entirely or attempt to solve it with the aid of some other hypothesis. Many research problems, though by no means all, take this form. Tests of this sort are a standard component of what I have elsewhere labeled 'normal science' or 'normal research', an enterprise which accounts for the overwhelming majority of the work done in basic science. In no usual sense, however, are such tests directed to current theory. On the contrary, when engaged with a normal research problem, the scientist must *premise* current theory as the rules of his game. His object is to solve a puzzle, preferably one at which others have failed, and current theory is required to define that puzzle and to guarantee that, given sufficient brilliance, it can be solved.[6] Of course the practitioner of such an enterprise must often test the conjectural puzzle solution that his ingenuity suggests. But only his personal conjecture is tested. If it fails the test, only his own ability, not the corpus of current science, is impugned. In short, though tests occur frequently in normal science, these tests are of a peculiar sort, for in the final analysis it is the individual scientist rather than current theory which is tested.

This is not, however, the sort of test Sir Karl has in mind. He is above all concerned with the procedures through which science grows, and he is convinced that 'growth' occurs not primarily by accretion but by the revolutionary overthrow of an accepted theory and its replacement by a better one.[7] (The subsumption under 'growth' of 'repeated overthrow'

Subsumption: placement in a more comprehensive category/general principle
accretion: gradual growth through inclusion
Seriation: Serially, in order

is itself a linguistic oddity whose *raison d'être* may become more visible as we proceed.) Taking this view, the tests which Sir Karl emphasizes are those which were performed to explore the limitations of accepted theory or to subject a current theory to maximum strain. Among his favorite examples, all of them startling and destructive in their outcome, are Lavoisier's experiments on calcination, the eclipse expedition of 1919, and the recent experiments on parity conservation.[8] All, of course, are classic tests, but in using them to characterize scientific activity Sir Karl misses something terribly important about them. Episodes like these are very rare in the development of science. When they occur, they are generally called forth either by a prior crisis in the relevant field (Lavoisier's experiments or Lee and Yang's[9]) or by the existence of a theory which competes with the existing canons of research (Einstein's general relativity). These are, however, aspects of or occasions for what I have elsewhere called 'extraordinary research', an enterprise in which scientists do display very many of the characteristics Sir Karl emphasizes, but one which, at least in the past, has arisen only intermittently and under quite special circumstances in any scientific speciality.[10]

I suggest then that Sir Karl has characterized the entire scientific enterprise in terms that apply only to its occasional revolutionary parts. His emphasis is natural and common: the exploits of a Copernicus or Einstein make better reading than those of a Brahe or Lorentz; Sir Karl would not be the first if he mistook what I call normal science for an intrinsically uninteresting enterprise. Nevertheless, neither science nor the development of knowledge is likely to be understood if research is viewed exclusively through the revolutions it occasionally produces. For example, though testing of basic commitments occurs only in extraordinary science, it is normal science that discloses both the points to test and the manner of testing. Or again, it is for the normal, not the extraordinary practice of science that professionals are trained; if they are nevertheless eminently successful in displacing and replacing the theories on which normal practice depends, that is an oddity which must be explained. Finally, and this is for now my main point, a careful look at the scientific enterprise suggests that it is normal science, in which Sir Karl's sort of testing does not occur, rather than extraordi-

nary science which most nearly distinguishes science from other enterprises. If a demarcation criterion exists (we must not, I think, seek a sharp or decisive one), it may lie just in that part of science which Sir Karl ignores.

In one of his most evocative essays, Sir Karl traces the origin of 'the tradition of critical discussion [which] represents the only practicable way of expanding our knowledge' to the Greek philosophers between Thales and Plato, the men who, as he sees it, encouraged critical discussion both between schools and within individual schools.[11] The accompanying description of Presocratic discourse is most apt, but what is described does not at all resemble science. Rather it is the tradition of claims, counterclaims, and debates over fundamentals which, except perhaps during the Middle Ages, have characterized philosophy and much of social science ever since. Already by the Hellenistic period mathematics, astronomy, statics, and the geometric parts of optics had abandoned this mode of discourse in favor of puzzle solving. Other sciences, in increasing numbers, have undergone the same transition since. In a sense, to turn Sir Karl's view on its head, it is precisely the abandonment of critical discourse that marks the transition to a science. Once a field has made that transition, critical discourse recurs only at moments of crisis when the bases of the field are again in jeopardy.[12] Only when they must choose between competing theories do scientists behave like philosophers. That, I think, is why Sir Karl's brilliant description of the reasons for the choice between metaphysical systems so closely resembles my description of the reasons for choosing between scientific theories.[13] In neither choice, as I shall shortly try to show, can testing play a quite decisive role.

There is, however, good reason why testing has seemed to do so, and in exploring it Sir Karl's duck may at last become my rabbit. No puzzle-solving enterprise can exist unless its practitioners share criteria which, for that group and for that time, determine when a particular puzzle has been solved. The same criteria necessarily determine failure to achieve a solution, and anyone who chooses may view that failure as the failure of a theory to pass a test. Normally, as I have already insisted, it is not viewed that way. Only the practitioner is blamed, not his tools. But under the special circumstances which

raison d'être = reason for going; justification for existing induce a crisis in the profession (e.g., gross failure, or repeated failure by the most brilliant professionals) the group's opinion may change. A failure that had previously been personal may then come to seem the failure of a theory under test. Thereafter, because the test arose from a puzzle and thus carried settled criteria of solution, it proves both more severe and harder to evade than the tests available within a tradition whose normal mode is critical discourse rather than puzzle solving.

In a sense, therefore, severity of test-criteria is simply one side of the coin whose other face is a puzzle-solving tradition. That is why Sir Karl's line of demarcation and my own so frequently coincide. That coincidence is, however, only in their *outcome;* the *process* of applying them is very different, and it isolates distinct aspects of the activity about which the decision—science or non-science—is to be made. Examining the vexing cases, for example, psychoanalysis or Marxist historiography, for which Sir Karl tells us his criterion was initially designed,[14] I concur that they cannot now properly be labeled 'science'. But I reach that conclusion by a route far surer and more direct than his. One brief example may suggest that of the two criteria, testing and puzzle solving, the latter is at once the less equivocal and the more fundamental.

To avoid irrelevant contemporary controversies, I consider astrology rather than, say, psychoanalysis. Astrology is Sir Karl's most frequently cited example of a 'pseudo-science'.[15] He says: 'By making their interpretations and prophecies sufficiently vague they [astrologers] were able to explain away anything that might have been a refutation of the theory had the theory and the prophecies been more precise. In order to escape falsification they destroyed the testability of the theory.'[16] Those generalizations catch something of the spirit of the astrological enterprise. But taken at all literally, as they must be if they are to provide a demarcation criterion, they are impossible to support. The history of astrology during the centuries when it was intellectually reputable records many predictions that categorically failed.[17] Not even astrology's most convinced and vehement exponents doubted the recurrence of such failures. Astrology cannot be barred from the sciences because of the form in which its predictions were cast.

Nor can it be barred because of the way its practitioners explained failure. Astrologers pointed out, for example, that, unlike general predictions about, say, an individual's propensities or a natural calamity, the forecast of an individual's future was an immensely complex task, demanding the utmost skill, and extremely sensitive to minor errors in relevant data. The configuration of the stars and eight planets was constantly changing; the astronomical tables used to compute the configuration at an individual's birth were notoriously imperfect; few men knew the instant of their birth with the requisite precision.[18] No wonder, then, that forecasts often failed. Only after astrology itself became implausible did these arguments come to seem question-begging.[19] Similar arguments are regularly used today when explaining, for example, failures in medicine or meteorology. In times of trouble they are also deployed in the exact sciences, fields like physics, chemistry, and astronomy.[20] There was nothing unscientific about the astrologer's explanation of failure.

Nevertheless, astrology was not a science. Instead it was a craft, one of the practical arts, with close resemblances to engineering, meteorology, and medicine as these fields were practiced until little more than a century ago. The parallels to an older medicine and to contemporary psychoanalysis are, I think, particularly close. In each of these fields shared theory was adequate only to establish the plausibility of the discipline and to provide a rationale for the various craft-rules which governed practice. These rules had proved their use in the past, but no practitioner supposed they were sufficient to prevent recurrent failure. A more articulated theory and more powerful rules were desired, but it would have been absurd to abandon a plausible and badly needed discipline with a tradition of limited success simply because these desiderata were not yet at hand. In their absence, however, neither the astrologer nor the doctor could do research. Though they had rules to apply, they had no puzzles to solve and therefore no science to practice.[21]

Compare the situations of the astronomer and the astrologer. If an astronomer's prediction failed and his calculations checked, he could hope to set the situation right. Perhaps the data were at fault: old observations could be re-examined and new measurements made, tasks which posed a host of calculational and

desiderata — something necessary or desirable —

instrumental puzzles. Or perhaps theory needed adjustment, either by the manipulation of epicycles, eccentrics, equants, etc., or by more fundamental reforms of astronomical technique. For more than a millennium these were the theoretical and mathematical puzzles around which, together with their instrumental counterparts, the astronomical research tradition was constituted. The astrologer, by contrast, had no such puzzles. The occurrence of failures could be explained, but particular failures did not give rise to research puzzles, for no man, however skilled, could make use of them in a constructive attempt to revise the astrological tradition. There were too many possible sources of difficulty, most of them beyond the astrologer's knowledge, control, or responsibility. Individual failures were correspondingly uninformative, and they did not reflect on the competence of the prognosticator in the eyes of his professional compeers.[22] Though astronomy and astrology were regularly practiced by the same people, including Ptolemy, Kepler, and Tycho Brahe, there was never an astrological equivalent of the puzzle-solving astronomical tradition. And without puzzles, able first to challenge and then to attest the ingenuity of the individual practitioner, astrology could not have become a science even if the stars had, in fact, controlled human destiny.

In short, though astrologers made testable predictions and recognized that these predictions sometimes failed, they did not and could not engage in the sorts of activities that normally characterize all recognized sciences. Sir Karl is right to exclude astrology from the sciences, but his over-concentration on science's occasional revolutions prevents his seeing the surest reason for doing so.

That fact, in turn, may explain another oddity of Sir Karl's historiography. Though he repeatedly underlines the role of tests in the replacement of scientific theories, he is also constrained to recognize that many theories, for example the Ptolemaic, were replaced before they had in fact been tested.[23] On some occasions, at least, tests are not requisite to the revolutions through which science advances. But that is not true of puzzles. Though the theories Sir Karl cites had not been put to the test before their displacement, none of these was replaced before it had ceased adequately to support a puzzle-solving tradition. The state of astronomy was a scandal in the early sixteenth century. Most astronomers nevertheless felt that normal adjustments of a basically Ptolemaic model would set the situation right. In this sense the theory had not failed a test. But a few astronomers, Copernicus among them, felt that the difficulties must lie in the Ptolemaic approach itself rather than in the particular versions of Ptolemaic theory so far developed, and the results of that conviction are already recorded. The situation is typical.[24] With or without tests, a puzzle-solving tradition can prepare the way for its own displacement. To rely on testing as the mark of a science is to miss what scientists mostly do and, with it, the most characteristic feature of their enterprise.

NOTES

1. For purposes of the following discussion I have reviewed Sir Karl Popper's [1959], his [1963], and his [1957]. I have also occasionally referred to his original [1935] and his [1945]. My own [1962] provides a more extended account of many of the issues discussed below.

2. More than coincidence is presumably responsible for this extensive overlap. Though I had read none of Sir Karl's work before the appearance in 1959 of the English translation of his [1935] (by which time my book was in draft), I had repeatedly heard a number of his main ideas discussed. In particular, I had heard him discuss some of them as William James Lecturer at Harvard in the spring of 1950. These circumstances do not permit me to specify an intellectual debt to Sir Karl, but there must be one.

3. Elsewhere I use the term 'paradigm' rather than 'theory' to denote what is rejected and replaced during scientific revolutions. Some reasons for the change of term will emerge below.

4. Underlining one additional area of agreement about which there has been much misunderstanding may further highlight what I take to be the real differences between Sir Karl's views and mine. We both insist that adherence to a tradition has an essential role in scientific development. He has written, for example, 'Quantitatively and qualitatively by far the most important source of our knowledge—apart from inborn knowledge—is tradition' (Popper [1963], p. 27). Even more to the point, as early as 1948 Sir Karl wrote, 'I do not think that we could ever free ourselves entirely from the bonds of tradition. The so-called freeing is really only a change from one tradition to another' ([1963], p. 122).

5. Popper [1959], p. 27.

6. For an extended discussion of normal science, the activity which practitioners are trained to carry on, see my [1962], pp. 23–42, and 135–42. It is important to notice that when I describe the scientist as a puzzle solver and Sir Karl describes him as a problem solver (e.g., in his [1963], pp. 67, 222), the similarity of our terms disguises a fundamental divergence. Sir Karl writes (the italics are his), 'Admittedly, our expectations, and thus our theories, may precede, historically, even our problems. *Yet science starts only with problems.* Problems crop up especially when we are disappointed in our expectations, or when our theories involve us in difficulties, in contradictions'. I use the term 'puzzle' in order to emphasize that the difficulties which *ordinarily* confront even the very best scientists are, like crossword puzzles or chess puzzles, challenges only to his ingenuity. *He* is in difficulty, not current theory. My point is almost the converse of Sir Karl's.

7. Cf. Popper [1963], pp. 129, 215, and 221, for particularly forceful statements of this position.

8. For example, Popper [1963], p. 220.

9. For the work on calcination see, Guerlac [1961]. For the background of the parity experiments see, Hafner and Presswood [1965].

10. The point is argued at length in my [1962], pp. 52–97.

11. Popper [1963], Chapter 5, especially pp. 148–52.

12. Though I was not then seeking a demarcation criterion, just these points are argued at length in my [1962], pp. 10–22 and 87–90.

13. Cf. Popper [1963], pp. 192–200, with my [1962], pp. 143–58.

14. Popper [1963], p. 34.

15. The index to Popper [1963] has eight entries under the heading 'astrology as a typical pseudo science'.

16. Popper [1963], p. 37.

17. For examples see, Thorndike [1923–58], 5, pp. 225 ff., 6, pp. 71, 101, 114.

18. For reiterated explanations of failure see, *ibid.* I, pp. 11 and 514 f.; 4, 368; 5, 279.

19. A perceptive account of some reasons for astrology's loss of plausibility is included in Stahlman [1956]. For an explanation of astrology's previous appeal see, Thorndike [1955].

20. Cf. my [1962], pp. 66–76.

21. This formulation suggests that Sir Karl's criterion of demarcation might be saved by a minor restatement entirely in keeping with his apparent intent. For a field to be a science its conclusions must be *logically derivable* from *shared premises*. On this view astrology is to be barred not because its forecasts were not testable but because only the most general and least testable ones could be derived from accepted theory. Since any field that did satisfy this condition might support a puzzle-solving tradition, the suggestion is clearly helpful. It comes close to supplying a sufficient condition for a field's being a science. But in this form, at least, it is not even quite a sufficient condition, and it is surely not a necessary one. It would, for example, admit surveying and navigation as sciences, and it would bar taxonomy, historical geology, and the theory of evolution. The conclusions of a science may be both precise and binding without being fully derivable by logic from accepted premises. Cf. my [1962], pp. 35–51.

22. This is not to suggest that astrologers did not criticize each other. On the contrary, like practitioners of philosophy and some social sciences, they belonged to a variety of different schools, and the inter-school strife was sometimes bitter. But these debates ordinarily revolved about the *implausibility* of the particular theory employed by one or another school. Failures of individual predictions played very little role. Compare Thorndike [1923–58], 5, p. 233.

23. Cf. Popper [1963], p. 246.

24. Cf. my [1962], pp. 77–87.

REFERENCES

Guerlac [1961]: *Lavoisier—The Crucial Year,* 1961.

Hafner and Presswood [1965]: "Strong Interference and Weak Interactions," *Science,* 149, pp. 503–10.

Kuhn [1962]: *The Structure of Scientific Revolutions,* 1962.

Popper [1935]: *Logik der Forschung,* 1935.

Popper [1945]: *The Open Society and Its Enemies,* 2 vols., 1945.

Popper [1957]: *The Poverty of Historicism,* 1957.

Popper [1959]: *Logic of Scientific Discovery,* 1959.

Popper [1963]: *Conjectures and Refutations,* 1963.

Stahlman [1956]: "Astrology in Colonial America: An Extended Query," *William and Mary Quarterly,* 13, pp. 551–63.

Thorndike [1923–58]: *A History of Magic and Experimental Science,* 8 vols, 1923–58.

Thorndike [1955]: "The True Place of Astrology in the History of Science," *Isis,* 46, pp. 273–8.

4

IMRE LAKATOS

Falsification and the Methodology
of Scientific Research Programs

1. SCIENCE: REASON OR RELIGION?

For centuries knowledge meant proven knowledge—proven either by the power of the intellect or by the evidence of the senses. Wisdom and intellectual integrity demanded that one must desist from unproven utterances and minimize, even in thought, the gap between speculation and established knowledge. The proving power of the intellect or the senses was questioned by the skeptics more than two thousand years ago; but they were browbeaten into confusion by the glory of Newtonian physics. Einstein's results again turned the tables and now very few philosophers or scientists still think that scientific knowledge is, or can be, proven knowledge. But few realize that with this the whole classical structure of intellectual values falls in ruins and has to be replaced: one cannot simply water down the ideal of proven truth—as some logical empiricists do—to the ideal of 'probable truth'[1] or—as some sociologists of knowledge do—to 'truth by [changing] consensus'.[2]

Popper's distinction lies primarily in his having grasped the full implications of the collapse of the best-corroborated scientific theory of all times: Newtonian mechanics and the Newtonian theory of gravitation. In his view virtue lies not in caution in avoiding errors, but in ruthlessness in eliminating them. Boldness in conjectures on the one hand and austerity in refutations on the other: this is Popper's recipe. Intel-lectual honesty does not consist in trying to entrench, or establish one's position by proving (or 'probabilifying') it—intellectual honesty consists rather in specifying precisely the conditions under which one is willing to give up one's position. Committed Marxists and Freudians refuse to specify such conditions: this is the hallmark of their intellectual dishonesty. *Belief* may be a regrettably unavoidable biological weakness to be kept under the control of criticism: but *commitment* is for Popper an outright crime.

Kuhn thinks otherwise. He too rejects the idea that science grows by accumulation of eternal truths.[3] He too takes his main inspiration from Einstein's overthrow of Newtonian physics. His main problem too is *scientific revolution*. But while according to Popper science is 'revolution in permanence', and criticism the heart of the scientific enterprise, according to Kuhn revolution is exceptional and, indeed, extra-scientific, and criticism is, in 'normal' times, anathema. Indeed for Kuhn the transition from criticism to commitment marks the point where progress—and 'normal' science—begins. For him the idea that on 'refutation' one can demand the rejection, the elimination of a theory, is 'naive' falsificationism. Criticism of the dominant theory and proposals of new theories are only allowed in the rare moments of 'crisis'. This last Kuhnian thesis has been widely criticized and I shall not discuss it. My concern is rather that Kuhn, having recognized the failure both of justificationism and falsificationism in providing rational accounts of scientific growth, seems now to fall back on irrationalism.

For Popper scientific change is rational or at least rationally reconstructible and falls in the realm of

Criticism and the Growth of Knowledge, Imre Lakatos and Alan Musgrave, eds. (New York: Cambridge University Press, 1970), pp. 91–195. Reprinted with permission from the publisher.

anathema - one that is greatly reviled or shunned

the *logic of discovery*. For Kuhn scientific change—from one 'paradigm' to another—is a mystical conversion which is not and cannot be governed by rules of reason and which falls totally within the realm of the *(social) psychology of discovery*. Scientific change is a kind of religious change.

The clash between Popper and Kuhn is not about a mere technical point in epistemology. It concerns our central intellectual values, and has implications not only for theoretical physics but also for the underdeveloped social sciences and even for moral and political philosophy. If even in science there is no other way of judging a theory but by assessing the number, faith and vocal energy of its supporters, then this must be even more so in the social sciences: truth lies in power. Thus Kuhn's position would vindicate, no doubt, unintentionally, the basic political *credo* of contemporary religious maniacs ('student revolutionaries').

In this paper I shall first show that in Popper's logic of scientific discovery two different positions are conflated. Kuhn understands only one of these, 'naive falsificationism' (I prefer the term 'naive methodological falsificationism'); I think that his criticism of it is correct, and I shall even strengthen it. But Kuhn does not understand a more sophisticated position the rationality of which is not based on 'naive' falsificationism. I shall try to explain—and further strengthen—this stronger Popperian position which, I think, may escape Kuhn's strictures and present scientific revolutions as constituting rational progress rather than as religious conversions.

2. FALLIBILISM VERSUS FALSIFICATIONISM

Sophisticated versus naive methodological falsificationism. Progressive and degenerating problemshifts.

Sophisticated falsificationism differs from naive falsificationism both in its rules of *acceptance* (or 'demarcation criterion') and its rules of *falsification* or elimination. For the naive falsificationist any theory which can be interpreted as experimentally falsifiable, is 'acceptable' or 'scientific'. For the sophisticated falsificationist a theory is 'acceptable' or 'scientific' only if it has corroborated excess empirical content over its predecessor (or rival), that is, only if it leads to the discovery of novel facts. This condition can be analyzed into two clauses: that the new theory has excess empirical content ('*acceptability*'₁) and that some of this excess content is verified ('*acceptability*'₂). The first clause can be checked instantly by *a priori* logical analysis; the second can be checked only empirically and this may take an indefinite time.

Again, for the naive falsificationist a theory is *falsified* by a '(fortified) observational' statement which conflicts with it (or rather, which he decides to interpret as conflicting with it). The sophisticated falsificationist regards a scientific theory T as falsified if and only if another theory T' has been proposed with the following characteristics: (1) T' has excess empirical content over T: that is, it predicts *novel* facts, that is, facts improbable in the light of, or even forbidden, by T;[4] (2) T' explains the previous success of T, that is, all the unrefuted content of T is contained (within the limits of observational error) in the content of T'; and (3) some of the excess content of T' is corroborated.[5] . . .

Let us take a series of theories, T_1, T_2, T_3, . . . where each subsequent theory results from adding auxiliary clauses to (or from semantical reinterpretations of) the previous theory in order to accommodate some anomaly, each theory having at least as much content as the unrefuted content of its predecessor. Let us say that such a series of theories is *theoretically progressive (or 'constitutes a theoretically progressive problemshift')* if each new theory has some excess empirical content over its predecessor, that is, if it predicts some novel, hitherto unexpected fact. Let us say that a theoretically progressive series of theories is also *empirically progressive (or 'constitutes an empirically progressive problemshift')* if some of this excess empirical content is also corroborated, that is, if each new theory leads us to the actual discovery of some *new fact*.[6] Finally, let us call a problemshift *progressive* if it is both theoretically and empirically progressive, and *degenerating* if it is not.[7] We '*accept*' problemshifts as 'scientific' only if they are at least theoretically progressive; if they are not, we '*reject*' them as 'pseudoscientific'. Progress is measured by the degree to which a problemshift is progressive, by the degree to which the series of theories leads us to the discovery of novel facts. We regard a theory in the series 'falsified' when it is superseded by a theory with higher corroborated content.

This demarcation between progressive and degenerating problemshifts sheds new light on the appraisal of *scientific—or, rather, progressive—explanations.* If we put forward a theory to resolve a contradiction between a previous theory and a counterexample in such a way that the new theory, instead of offering a content-increasing (scientific) *explanation,* only offers a content-decreasing (linguistic) *reinterpretation,* the contradiction is resolved in a merely semantical, unscientific way. *A given fact is explained scientifically only if a new fact is also explained with it.*[8]

Sophisticated falsificationism thus shifts the problem of how to appraise *theories* to the problem of how to appraise *series of theories.* Not an isolated *theory,* but only a series of theories can be said to be scientific or unscientific: to apply the term 'scientific' to one *single* theory is a category mistake.[9]

The time-honored empirical criterion for a satisfactory theory was agreement with the observed facts. Our empirical criterion for a series of theories is that it should produce new facts. *The idea of growth and the concept of empirical character are soldered into one.*

This revised form of methodological falsificationism has many new features. First, it denies that 'in the case of a scientific theory, our decision depends upon the results of experiments. If these confirm the theory, we may accept it until we find a better one. If they contradict the theory, we reject it.'[10] It denies that 'what ultimately decides the fate of a theory is the result of a test, *i.e.,* an agreement about basic statements.'[11] Contrary to naive falsificationism, *no experiment, experimental report, observation statement or well-corroborated low-level falsifying hypothesis alone can lead to falsification. There is no falsification before the emergence of a better theory.*[12] But then the distinctively negative character of naive falsificationism vanishes; criticism becomes more difficult, and also positive, constructive. But, of course, if falsification depends on the emergence of better theories, on the invention of theories which anticipate new facts, then falsification is *not* simply a relation between a theory and the empirical basis, but a multiple relation between competing theories, the original 'empirical basis', and the empirical growth resulting from the competition. Falsification can thus be said to have a *'historical character'.*[13] Moreover, some of the theories which bring about falsification are frequently proposed *after* the 'counterevidence'. This may sound paradoxical for people indoctrinated with naive falsificationism. Indeed, this epistemological theory of the relation between theory and experiment differs sharply from the epistemological theory of naive falsificationism. The very term 'counterevidence' has to be abandoned in the sense that no experimental result must be interpreted directly as 'counterevidence'. If we still want to retain this time-honored term, we have to redefine it like this: 'counterevidence to T_1' is a corroborating instance to T_2 which is either inconsistent with or independent of T_1 (with the *proviso* that T_2 is a theory which satisfactorily explains the empirical success of T_1). This shows that *'crucial counterevidence'*—or *'crucial experiments'*—can be recognized as such among the scores of anomalies only *with hindsight,* in the light of some superceding theory.[14]

Thus the crucial element in falsification is whether the *new theory* offers any novel, excess information compared with its predecessor and whether some of this excess information is corroborated. Justificationists valued 'confirming' instances of a theory; naive falsificationists stressed 'refuting' instances; for the methodological falsificationists it is the—rather rare—corroborating instances of the *excess* information which are the crucial ones; these receive all the attention. We are no longer interested in the thousands of trivial verifying instances nor in the hundreds of readily available anomalies: the few crucial *excess-verifying instances* are decisive.[15] This consideration rehabilitates—and reinterprets—the old proverb: *Exemplum docet, exempla obscurant.*

'Falsification' in the sense of naive falsificationism (corroborated counterevidence) is not a *sufficient* condition for eliminating a specific theory: in spite of hundreds of known anomalies we do not regard it as falsified (that is, eliminated) until we have a better one.[16] Nor is 'falsification' in the naive sense *necessary* for falsification in the sophisticated sense: a progressive problemshift does not have to be interspersed with 'refutations.' Science can grow without any 'refutations' leading the way. Naive falsificationists suggest a linear growth of science, in

[handwritten margin notes:]
Something old theory does not explain does like same thing

Theory is scientific if it explains it explains empirical theoretical the sense that theories are followed by powerful refutations which eliminate them; these refutations in turn are followed by new theories.[17] It is perfectly *possible* that theories be put forward 'progressively' in such a rapid succession that the 'refutation' of the *n*-th appears only as the corroboration of the *n* + 1-th. The problem fever of science is raised by proliferation of rival theories rather than counterexamples or anomalies.

This shows that the slogan of *proliferation of theories* is much more important for sophisticated than for naive falsificationism. For the naive falsificationist science grows through repeated experimental overthrow of theories; new rival theories proposed before such 'overthrows' may speed up growth but are not absolutely necessary;[18] constant proliferation of theories is optional but not mandatory. For the sophisticated falsificationist proliferation of theories cannot wait until the accepted theories are 'refuted' (or until their protagonists get into a Kuhnian crisis of confidence).[19] While naive falsificationism stresses 'the urgency of replacing a *falsified* hypothesis by a better one',[20] sophisticated falsificationism stresses the urgency of replacing *any* hypothesis by a better one. Falsification cannot 'compel the theorist to search for a better theory',[21] simply because falsification cannot precede the better theory.

THE POPPERIAN VERSUS THE KUHNIAN RESEARCH PROGRAM

Let us now sum up the Kuhn–Popper controversy.

We have shown that Kuhn is right in objecting to naive falsificationism, and also in stressing the *continuity* of scientific growth, the *tenacity* of some scientific theories. But Kuhn is wrong in thinking that by discarding naive falsificationism he has discarded thereby all brands of falsificationism. Kuhn objects to the entire Popperian research program, and he excludes *any* possibility of a rational reconstruction of the growth of science. In a succinct comparison of Hume, Carnap and Popper, Watkins points out that the growth of science is inductive and irrational according to Hume, inductive and irrational according to Carnap, non-inductive and rational according to Popper.[22] But Watkins's comparison can be ex-

tended by adding that it is non-inductive and irrational according to Kuhn. *In Kuhn's view there can be no logic, but only psychology of discovery.*[23] For instance, in Kuhn's conception, anomalies, inconsistencies *always* abound in science, but in 'normal' periods the dominant paradigm secures a pattern of growth which is eventually overthrown by a 'crisis'. 'Crisis' is a psychological concept; it is a contagious panic. Then a new 'paradigm' emerges, incommensurable with its predecessor. There are no rational standards for their comparison. Each paradigm contains its own standards. The crisis sweeps away not only the old theories and rules but also the standards which made us respect them. The new paradigm brings a totally new rationality. There are no superparadigmatic standards. The change is a bandwagon effect. Thus *in Kuhn's view scientific revolution is irrational, a matter for mob psychology.*

The reduction of philosophy of science to psychology of science did not start with Kuhn. An earlier wave of 'psychologism' followed the breakdown of justificationism. For many, justificationism represented the only possible form of rationality: the end of justificationism meant the end of rationality. The collapse of the thesis that scientific theories are provable, that the progress of science is cumulative, made justificationists panic. If 'to discover is to prove', but nothing is provable, then there can be no discoveries, only discovery-claims. Thus disappointed justificationists—ex-justificationists—thought that the elaboration of rational standards was a hopeless enterprise and that all one can do is to study—and imitate—the Scientific Mind, as it is exemplified in famous scientists. After the collapse of Newtonian physics, Popper elaborated new, non-justificationist critical standards. Now some of those who had already learned of the collapse of justificationist rationality now learned, mostly by hearsay, of Popper's colorful slogans which suggested naive falsificationism. Finding them untenable, they identified the collapse of naive falsificationism with the end of rationality itself. The elaboration of rational standards was again regarded as a hopeless enterprise; the best one can do is to study, they thought once again, the Scientific Mind.[24] Critical philosophy was to be replaced by what Polanyi called a 'post-critical' philosophy. But the Kuhnian research program contains a new feature: we have to study not the mind of the individual scientist but the mind of the Scientific Community. Individual psychology is now replaced by social psychology; imitation of the great scientists by submission to the collective wisdom of the community.

But Kuhn overlooked Popper's sophisticated falsificationism and the research program he initiated. Popper replaced the central problem of classical rationality, *the old problem of foundations*, with *the new problem of fallible-critical growth*, and started to elaborate objective standards of this growth. In this paper I have tried to develop his program a step further. I think this small development is sufficient to escape Kuhn's strictures.[25]

NOTES

1. The main contemporary protagonist of the ideal of 'probable truth' is Rudolf Carnap. For the historical background and a criticism of this position, cf. Lakatos [1968a].

2. The main contemporary protagonists of the ideal of 'truth by consensus' are Polanyi and Kuhn. For the historical background and a criticism of this position, cf. Musgrave [1969a], Musgrave [1969b] and Lakatos [1970].

3. Indeed he introduces his [1962] by arguing against the 'development-by-accumulation' idea of scientific growth. But his intellectual debt is to Koyré rather than to Popper. Koyré showed that positivism gives bad guidance to the historian of science, for the history of physics can only be understood in the context of a succession of 'metaphysical' research programs. Thus scientific changes are connected with vast cataclysmic metaphysical revolutions. Kuhn develops this message of Burtt and Koyré and the vast success of his book was partly due to his hard-hitting, direct criticism of justificationist historiography—which created a sensation among ordinary scientists and historians of science whom Burtt's, Koyré's (or Popper's) message has not yet reached. But, unfortunately, his message had some authoritarian and irrationalist overtones.

4. I use 'prediction' in a wide sense that includes 'postdiction'.

5. *For a detailed discussion of these acceptance and rejection rules and for references to Popper's work, cf. my [1968a], pp. 375–90.*

6. If I already know P_1: 'Swan A is white', P_ω: 'All swans are white' represents no progress, because it may only lead to the discovery of such further

similar facts as P_2: 'Swan B is white'. So-called 'empirical generalizations' constitute no progress. A *new* fact must be improbable or even impossible in the light of previous knowledge.

7. The appropriateness of the term 'problemshift' for a series of theories rather than of problems may be questioned. I chose it partly because I have not found a more appropriate alternative—'theoryshift' sounds dreadful—partly because theories are always problematical, they never solve all the problems they have set out to solve.

8. Indeed, in the original manuscript of my [1968a] I wrote: 'A theory without excess corroboration has no excess explanatory power; *therefore, according to Popper, it does not represent growth and therefore it is not "scientific"; therefore, we should say, it has no explanatory power*' (p. 386). I cut out the italicized half of the sentence under pressure from my colleagues who thought it sounded too eccentric. I regret it now.

9. Popper's conflation of 'theories' and 'series of theories' prevented him from getting the basic ideas of sophisticated falsificationism across more successfully. His ambiguous usage led to such confusing formulations as 'Marxism [as the core of a series of theories or of a "research program"] is irrefutable' and, at the same time, 'Marxism [as a particular conjunction of this core and some specified auxiliary hypotheses, initial conditions and a *ceteris paribus* clause] has been refuted.' (Cf. Popper [1963].)

   Of course, there is nothing wrong in saying that an isolated, single theory is 'scientific' if it represents an advance on its predecessor, as long as one clearly realizes that in this formulation we appraise the theory as the outcome of—and in the context of—a certain historical development.

10. Popper [1945], vol. II, p. 233. Popper's more sophisticated attitude surfaces in the remark that 'concrete and practical consequences can be *more* directly tested by experiment' (*ibid.* my italics).

11. Popper [1934], section 30.

12. 'In most cases we have, before falsifying a hypothesis, another one up our sleeves' (Popper [1959a], p. 87, footnote 1). But, as our argument shows, we *must* have one. Or, as Feyerabend put it: 'The best criticism is provided by those theories which can replace the rivals they have removed' ([1965], p. 227). He notes that in *some* cases 'alternatives will be quite indispensable for the purpose of refutation' (*ibid.* p. 254). But according to our argument *refutation without an alternative shows nothing but the poverty of our imagination in providing a rescue hypothesis.*

13. Cf. my [1968a], pp. 387 ff.

14. In the distorting mirror of naive falsificationism, new theories which replace old refuted ones, are themselves born unrefuted. Therefore they do not believe that there is a relevant difference between anomalies and crucial counterevidence. For them, anomaly is a dishonest euphemism for counterevidence. But in actual history new theories are born refuted: they inherit many anomalies of the old theory. Moreover, frequently it is *only* the new theory which dramatically predicts that fact which will function as crucial counterevidence against its predecessor, while the 'old' anomalies may well stay on as 'new' anomalies.

15. *Sophisticated falsificationism adumbrates a new theory of learning.*

16. It is clear that the theory T' may have excess corroborated empirical content over another theory T even if both T and T' are refuted. Empirical content has nothing to do with truth or falsity. Corroborated contents can also be compared irrespective of the refuted content. Thus we may see the rationality of the elimination of Newton's theory in favor of Einstein's, even though Einstein's theory may be said to have been born—like Newton's—'refuted'. We have only to remember that 'qualitative confirmation' is a euphemism for 'quantitative disconfirmation'. (Cf. my [1968a], pp. 384–6.)

17. Cf. Popper [1934], section 85, p. 279 of the 1959 English translation.

18. It is true that a certain type of *proliferation of rival theories* is allowed to play an accidental heuristic role in falsification. In many cases falsification heuristically 'depends on [the condition] that sufficiently many and sufficiently different theories are offered' (Popper [1940]). For instance, we may have a theory T which is apparently unrefuted. But it may happen that a new theory T', inconsistent with T, is proposed which equally fits the available facts: the differences are smaller than the range of observational error. In such cases the inconsistency prods us into improving our 'experimental techniques', and thus refining the 'empirical basis' so that either T or T' (or, incidentally, both) can be falsified: 'We need [a] new theory in order to find out where the old theory was deficient' (Popper [1963], p. 246). But the role of this proliferation is *accidental* in the sense that, once the empirical basis is refined, the fight is between this refined empirical basis and the theory T under test; the rival theory T' acted only as a *catalyst*.

19. Also cf. Feyerabend [1965], pp. 254–5.

20. Popper [1959a], p. 87, footnote 1.

21. Popper [1934], section 30.

22. Watkins [1968], p. 281.

23. Kuhn [1965]. But this position is already implicit in his [1962].

24. Incidentally, just as some earlier ex-justificationists led the wave of skeptical irrationalism, so now some ex-falsificationists lead the *new* wave of skeptical irrationalism and anarchism. This is best exemplified in Feyerabend [1970].

25. Indeed, as I had already mentioned, *my concept of a 'research program' may be construed as an objective, 'third world' reconstruction of Kuhn's socio-psychological concept of 'paradigm'*: thus the Kuhnian 'Gestalt-switch' can be performed without removing one's Popperian spectacles.

(I have not dealt with Kuhn's and Feyerabend's claim that theories cannot be eliminated on any *objective* grounds because of the 'incommensurability' of rival theories. Incommensurable theories are neither inconsistent with each other, nor comparable for content. But we can *make* them, by a dictionary, inconsistent and their content comparable. If we want to eliminate a program, we need some methodological determination. This determination is the heart of methodological falsificationism; for instance, no result of statistical sampling is ever inconsistent with a statistical theory unless we *make them* inconsistent with the help of Popperian rejection rules.)

REFERENCES

Feyerabend [1965]: 'Reply to Criticism', in Cohen and Wartofsky (eds.): *Boston Studies in the Philosophy of Science*, II, pp. 223–61.

Feyerabend [1970]: 'Against Method', *Minnesota Studies for the Philosophy of Science*, 4, 1970.

Kuhn [1962]: *The Structure of Scientific Revolutions*, 1962.

Kuhn [1965]: 'Logic of Discovery or Psychology of Research', *Criticism and the Growth of Knowledge*, 1970, pp. 1–23.

Lakatos [1968a]: 'Changes in the Problem of Inductive Logic', in Lakatos (ed.): *The Problem of Inductive Logic*, 1968, pp. 315–417.

Lakatos [1968b]: 'Criticism and the Methodology of Scientific Research Programmes', in *Proceedings of the Aristotelian Society*, 69, pp. 149–86.

Lakatos [1970]: *The Changing Logic of Scientific Discovery*, 1970.

Musgrave [1969a]: *Impersonal Knowledge*, Ph.D. Thesis, University of London, 1969.

Musgrave [1969*b*]: Review of Ziman's 'Public Knowledge: An Essay Concerning the Social Dimensions of Science', in *The British Journal for the Philosophy of Science*, 20, pp. 92–4.

Popper [1934]: *Logik der Forschung*, 1935 (expanded English edition: Popper [1959*a*]).

Popper [1940]: 'What is Dialectic?' *Mind*, N.S. 49, pp. 403–26; reprinted in Popper [1963], pp. 312–35.

Popper [1959*a*]: *The Logic of Scientific Discovery*, 1959.

Popper [1959*b*]: 'Testability and "ad-Hocness" of the Contraction Hypothesis', *British Journal for the Philosophy of Science*, 10, p. 50.

Popper [1963]: *Conjectures and Refutations*, 1963.

Watkins [1968]: 'Hume, Carnap and Popper', in Lakatos (ed.): *The Problem of Inductive Logic*, 1968, pp. 271–82.

5

LARRY LAUDAN

Science at the Bar–Causes for Concern

In the wake of the decision in the Arkansas Creationism trial *(McLean v. Arkansas)*,[1] the friends of science are apt to be relishing the outcome. The creationists quite clearly made a botch of their case and there can be little doubt that the Arkansas decision may, at least for a time, blunt legislative pressure to enact similar laws in other states. Once the dust has settled, however, the trial in general and Judge William R. Overton's ruling in particular may come back to haunt us; for, although the verdict itself is probably to be commended, it was reached for all the wrong reasons and by a chain of argument which is hopelessly suspect. Indeed, the ruling rests on a host of misrepresentations of what science is and how it works.

The heart of Judge Overton's Opinion is a formulation of "the essential characteristics of science." These characteristics serve as touchstones for contrasting evolutionary theory with Creationism; they lead Judge Overton ultimately to the claim, specious in its own right, that since Creationism is not

"science," it must be religion. The Opinion offers five essential properties that demarcate scientific knowledge from other things: "(1) It is guided by natural law; (2) it has to be explanatory by reference to natural law; (3) it is testable against the empirical world; (4) its conclusions are tentative, i.e., are not necessarily the final word; and (5) it is falsifiable."

These fall naturally into families: properties (1) and (2) have to do with lawlikeness and explanatory ability; the other three properties have to do with the fallibility and testability of scientific claims. I shall deal with the second set of issues first, because it is there that the most egregious errors of fact and judgment are to be found.

At various key points in the Opinion, Creationism is charged with being untestable, dogmatic (and thus non-tentative), and unfalsifiable. All three charges are of dubious merit. For instance, to make the interlinked claims that Creationism is neither falsifiable nor testable is to assert that Creationism makes no empirical assertions whatever. That is surely false. Creationists make a wide range of testable assertions about empirical matters of fact. Thus, as Judge Overton himself grants (apparently without seeing its im-

Science, Technology, & Human Values 7 (1982), pp. 16–19. Copyright © 1982 Sage Publications, Inc. Reprinted by permission of the publisher.

plications), the creationists say that the earth is of very recent origin (say 6,000 to 20,000 years old); they argue that most of the geological features of the earth's surface are diluvial in character (i.e., products of the postulated Noachian deluge); they are committed to a large number of factual historical claims with which the Old Testament is replete; they assert the limited variability of species. They are committed to the view that, since animals and man were created at the same time, the human fossil record must be paleontologically co-extensive with the record of lower animals. It is fair to say that no one has shown how to reconcile such claims with the available evidence—evidence which speaks persuasively to a long earth history, among other things.

In brief, these claims are testable, they have been tested, and they have failed those tests. Unfortunately, the logic of the Opinion's analysis precludes saying any of the above. By arguing that the tenets of Creationism are neither testable nor falsifiable, Judge Overton (like those scientists who similarly charge Creationism with being untestable) deprives science of its strongest argument against Creationism. Indeed, if any doctrine in the history of science has ever been falsified, it is the set of claims associated with "creation-science." Asserting that Creationism makes no empirical claims plays directly, if inadvertently, into the hands of the creationists by immunizing their ideology from empirical confrontation. The correct way to combat Creationism is to confute the empirical claims it does make, not to pretend that it makes no such claims at all.

It is true, of course, that some tenets of Creationism are not testable in isolation (e.g., the claim that man emerged by a direct supernatural act of creation). But that scarcely makes Creationism "unscientific." It is now widely acknowledged that many scientific claims are not testable in isolation, but only when embedded in a larger system of statements, some of whose consequences can be submitted to test.

Judge Overton's third worry about Creationism centers on the issue of revisability. Over and over again, he finds Creationism and its advocates "unscientific" because they have "refuse[d] to change it regardless of the evidence developed during the course of the[ir] investigation." In point of fact, the charge is mistaken. If the claims of modern-day creationists

are compared with those of their nineteenth-century counterparts, significant shifts in orientation and assertion are evident. One of the most visible opponents of Creationism, Stephen Gould, concedes that creationists have modified their views about the amount of variability allowed at the level of species change. Creationists do, in short, change their minds from time to time. Doubtless they would credit these shifts to their efforts to adjust their views to newly emerging evidence, in what they imagine to be a scientifically respectable way.

Perhaps what Judge Overton had in mind was the fact that some of Creationism's core assumptions (e.g., that there was a Noachian flood, that man did not evolve from lower animals, or that God created the world) seem closed off from any serious modification. But historical and sociological researches on science strongly suggest that the scientists of any epoch likewise regard some of their beliefs as so fundamental as not to be open to repudiation or negotiation. Would Newton, for instance, have been tentative about the claim that there were forces in the world? Are quantum mechanicians willing to contemplate giving up the uncertainty relation? Are physicists willing to specify circumstances under which they would give up energy conservation? Numerous historians and philosophers of science (e.g., Kuhn, Mitroff, Feyerabend, Lakatos) have documented the existence of a certain degree of dogmatism about core commitments in scientific research and have argued that such dogmatism plays a constructive role in promoting the aims of science. I am not denying that there may be subtle but important differences between the dogmatism of scientists and that exhibited by many creationists; but one does not even begin to get at those differences by pretending that science is characterized by an uncompromising open-mindedness.

Even worse, the *ad hominem* charge of dogmatism against Creationism egregiously confuses doctrines with the proponents of those doctrines. Since no law mandates that creationists should be invited into the classroom, it is quite irrelevant whether they themselves are close-minded. The Arkansas statute proposed that Creationism be taught, not that creationists should teach it. What counts is the epistemic status of Creationism, not the cognitive idiosyncrasies of the creationists. Because many of the theses of Creationism are testable, the mind set of creationists has no bearing in law or in fact on the merits of Creationism.

What about the other pair of essential characteristics which the *McLean* Opinion cites, namely, that science is a matter of natural law and explainable by natural law? I find the formulation in the Opinion to be rather fuzzy; but the general idea appears to be that it is inappropriate and unscientific to postulate the existence of any process or fact which cannot be explained in terms of some known scientific laws—for instance, the creationists' assertion that there are outer limits to the change of species "cannot be explained by natural law." Earlier in the Opinion, Judge Overton also writes, "there is no scientific explanation for these limits which is guided by natural law," and thus concludes that such limits are unscientific. Still later, remarking on the hypothesis of the Noachian flood, he says: "A worldwide flood as an explanation of the world's geology is not the product of natural law, nor can its occurrence be explained by natural law." Quite how Judge Overton knows that a worldwide flood "cannot" be explained by the laws of science is left opaque; and even if we did not know how to reduce a universal flood to the familiar laws of physics, this requirements is an altogether inappropriate standard for ascertaining whether a claim is scientific. For centuries scientists have recognized a difference between establishing the existence of a phenomenon and explaining that phenomenon in a lawlike way. Our ultimate goal, no doubt, is to do both. But to suggest, as the *McLean* Opinion does repeatedly, that an existence claim (e.g., there was a worldwide flood) is unscientific until we have found the laws on which the alleged phenomenon depends is simply outrageous. Galileo and Newton took themselves to have established the existence of gravitational phenomena, long before anyone was able to give a causal or explanatory account of gravitation. Darwin took himself to have established the existence of natural selection almost a half-century before geneticists were able to lay out the laws of heredity on which natural selection depended. If we took the *McLean* Opinion criterion seriously, we should have to say that Newton and Darwin were unscientific; and, to take an example from our own time, it would follow that plate tectonics is unscientific because we have not yet identified the laws of physics and chemistry which account for the dynamics of crustal motion.

The real objection to such creationist claims as that of the (relative) invariability of species is not that such invariability has not been explained by scientific laws, but rather that the evidence for invariability is less robust than the evidence for its contrary, variability. But to say as much requires renunciation of the Opinion's other charge—to wit, that Creationism is not testable.

I could continue with this tale of woeful fallacies in the Arkansas ruling, but that is hardly necessary. What is worrisome is that the Opinion's line of reasoning—which neatly coincides with the predominant tactic among scientists who have entered the public fray on this issue—leaves many loopholes for the creationists to exploit. As numerous authors have shown, the requirements of testability, revisability, and falsifiability are exceedingly *weak* requirements. Leaving aside the fact that (as I pointed out above) it can be argued that Creationism already satisfies these requirements, it would be easy for a creationist to say the following: "I will abandon my views if we find a living specimen of a species intermediate between man and apes." It is, of course, extremely unlikely that such an individual will be discovered. But, in that statement the creationist would satisfy, in one fell swoop, all the formal requirements of testability, falsifiability, and revisability. If we set very weak standards for scientific status—and, let there be no mistake, I believe that all of the Opinion's last three criteria fall in this category—then it will be quite simple for Creationism to qualify as "scientific."

Rather than taking on the creationists obliquely in wholesale fashion by suggesting that what they are doing is "unscientific" *tout court* (which is doubly silly because few authors can even agree on what makes an activity scientific), we should confront their claims directly and in piecemeal fashion by asking what evidence and arguments can be marshaled for and against each of them. The core issue is not whether Creationism satisfies some undemanding and highly controversial definitions of what is scientific; the real question is whether the existing evidence provides stronger arguments for evolutionary theory than for Creationism. Once that question is settled, we will know what belongs in the classroom and what does not. Debating the scientific status of Creationism (especially when "science" is construed in such an unfortunate manner) is a red herring that diverts attention away from the issues that should concern us.

Some defenders of the scientific orthodoxy will probably say that my reservations are just nit-picking ones, and that—at least to a first order of approximation—Judge Overton has correctly identified what is fishy about Creationism. The apologists for science, such as the editor of *The Skeptical Inquirer,* have already objected to those who criticize this whitewash of science "on arcane, semantic grounds . . . [drawn] from the most remote reaches of the academic philosophy of science."[2] But let us be clear about what is at stake. In setting out in the *McLean* Opinion to characterize the "essential" nature of science, Judge Overton was explicitly venturing into philosophical terrain. His *obiter dicta* are about as remote from well-founded opinion in the philosophy of science as Creationism is from respectable geology. It simply will not do for the defenders of science to invoke philosophy of science when it suits them (e.g., their much-loved principle of falsifiability comes directly from the philosopher Karl Popper) and to dismiss it as "arcane" and "remote" when it does not. However noble the motivation, bad philosophy makes for bad law.

The victory in the Arkansas case was hollow, for it was achieved only at the expense of perpet-uating and canonizing a false stereotype of what science is and how it works. If it goes unchallenged by the scientific community, it will raise grave doubts about that community's intellectual integrity. No one familiar with the issues can really believe that anything important was settled through anachronistic efforts to revive a variety of discredited criteria for distinguishing between the scientific and the non-scientific. Fifty years ago, Clarence Darrow asked, *à propos* the Scopes trial, "Isn't it difficult to realize that a trial of this kind is possible in the twentieth century in the United States of America?" We can raise that question anew, with the added irony that, this time, the pro-science forces are defending a philosophy of science which is, in its way, every bit as outmoded as the "science" of the creationists.

NOTES

1. *McLean v. Arkansas Board of Education,* 529 F. Supp. 1255 (E.D. Ark. 1982). For the text of the law, the decision, and essays by participants in the trial, see 7 *Science, Technology, & Human Values* 40 (Summer 1982), and also Marcel LaFollette, *Creationism, Science and the Law* (The MIT Press, 1983).
2. "The Creationist Threat: Science Finally Awakens." VI *The Skeptical Inquirer* 3 (Spring 1982): 2–5.

6

MICHAEL RUSE

Pro Judice

As always, my friend Larry Laudan writes in an entertaining and provocative manner, but, in his complaint against Judge William Overton's ruling in *McLean v. Arkansas,*[1] Laudan is hopelessly wide of the mark. Laudan's outrage centers on the criteria for the demarcation of science which Judge Overton adopted, and the judge's conclusion that, evaluated by these criteria, creation-science fails as science. I shall respond directly to this concern—after making three preliminary remarks.

First, although Judge Overton does not need defense from me or anyone else, as one who participated in the Arkansas trial, I must go on record as saying that I was enormously impressed by his handling of the case. His written judgment is a first-class piece of reasoning. With cause, many have criticized the State of Arkansas for passing the "Creation-Science Act," but we should not ignore that, to the state's credit, Judge Overton was born, raised, and educated in Arkansas.

Second, Judge Overton, like everyone else, was fully aware that proof that something is not science is not the same as proof that it is religion. The issue of what constitutes science arose because the creationists claim that their ideas qualify as genuine science rather than as fundamentalist religion. The attorneys developing the American Civil Liberties Union (ACLU) case believed it important to show that creation-science is not genuine science. Of course, this demonstration does raise the question of what creation-science really is. The plaintiffs claimed that creation-science always was

Science, Technology, & Human Values 7 (1982), pp. 19–23. Copyright © 1982 Sage Publications, Inc. Reprinted by permission of the publisher.

(and still is) religion. The plaintiffs' lawyers went beyond the negative argument (against science) to make the positive case (for religion). They provided considerable evidence for the religious nature of creation-science, including such things as the creationists' explicit reliance on the Bible in their various writings. Such arguments seem about as strong as one could wish, and they were duly noted by Judge Overton and used in support of his ruling. It seems a little unfair, in the context, therefore, to accuse him of "specious" argumentation. He did not adopt the naïve dichotomy of "science or religion but nothing else."

Third, whatever the merits of the plaintiffs' case, the kinds of conclusions and strategies apparently favored by Laudan are simply not strong enough for legal purposes. His strategy would require arguing that creation-science is weak science and therefore ought not to be taught:

> The core issue is not whether Creationism satisfies some undemanding and highly controversial definitions of what is scientific; the real question is whether the existing evidence provides stronger arguments for evolutionary theory than for Creationism. Once that question is settled, we will know what belongs in the classroom and what does not.[2]

Unfortunately, the U.S. Constitution does not bar the teaching of weak science. What it bars (through the Establishment Clause of the First Amendment) is the teaching of religion. The plaintiffs' tactic was to show that creation-science is less than weak or bad science. It is not science at all.

Turning now to the main issue, I see three questions that must be addressed. Using the five criteria listed by Judge Overton, can one distinguish science from non-science? Assuming a positive answer to the first question, does creation-science fail as genuine science when it is judged by these criteria? And, assuming a positive answer to the second, does the Opinion in *McLean* make this case?

The first question has certainly tied philosophers of science in knots in recent years. Simple criteria that supposedly give a clear answer to every case—for example, Karl Popper's single stipulation of falsifiability[3]—will not do. Nevertheless, although there may be many grey areas, white does seem to be white and black does seem to be black. Less metaphorically, something like psychoanalytic theory may or may not be science, but there do appear to be clear-cut cases of real science and of real non-science. For instance, an explanation of the fact that my son has blue eyes, given that both parents have blue eyes, done in terms of dominant and recessive genes and with an appeal to Mendel's first law, is scientific. The Catholic doctrine of transubstantiation (i.e., that in the Mass the bread and wine turn into the body and blood of Christ) is not scientific.

Furthermore, the five cited criteria of demarcation do a good job of distinguishing the Mendelian example from the Catholic example. Law and explanation through law come into the first example. They do not enter the second. We can test the first example, rejecting it if necessary. In this sense, it is tentative, in that something empirical might change our minds. The case of transubstantiation is different. God may have His own laws, but neither scientist nor priest can tell us about those which turn bread and wine into flesh and blood. There is no explanation through law. No empirical evidence is pertinent to the miracle. Nor would the believer be swayed by any empirical facts. Microscopic examination of the Host is considered irrelevant. In this sense, the doctrine is certainly not tentative.

One pair of examples certainly do not make for a definitive case, but at least they do suggest that Judge Overton's criteria are not quite as irrelevant as Laudan's critique implies. What about the types of objections (to the criteria) that Laudan does or could make? As far as the use of law is concerned, he might complain that scientists themselves have certainly not always been that particular about reference to

law. For instance, consider the following claim by Charles Lyell in his *Principles of Geology* (1830/3): "We are not, however, contending that a real departure from the antecedent course of physical events cannot be traced in the introduction of man."[4] All scholars agree that in this statement Lyell was going beyond law. The coming of man required special divine intervention. Yet, surely the *Principles* as a whole qualify as a contribution to science.

Two replies are open: either one agrees that the case of Lyell shows that science has sometimes mingled law with non-law; or one argues that Lyell (and others) mingled science and non-science (specifically, religion at this point). My inclination is to argue the latter. Insofar as Lyell acted as scientist, he appealed only to law. A century and a half ago, people were not as conscientious as today about separating science and religion. However, even if one argues the former alternative—that some science has allowed place for non-lawbound events—this hardly makes Laudan's case. Science, like most human cultural phenomena, has evolved. What was allowable in the early nineteenth century is not necessarily allowable in the late twentieth century. Specifically, science today does not break with law. And this is what counts for us. We want criteria of science for today, not for yesterday. (Before I am accused of making my case by fiat, let me challenge Laudan to find one point within the modern geological theory of plate tectonics where appeal is made to miracles, that is, to breaks with law. Of course, saying that science appeals to law is not asserting that we know all of the laws. But, who said that we did? Not Judge Overton in his Opinion.)

What about the criterion of tentativeness, which involves a willingness to test and reject if necessary? Laudan objects that real science is hardly all that tentative: "[H]istorical and sociological researches on science strongly suggest that the scientists of any epoch likewise regard some of their beliefs as so fundamental as not to be open to repudiation or negotiation."[5]

It cannot be denied that scientists do sometimes—frequently—hang on to their views, even if not everything meshes precisely with the real world. Nevertheless, such tenacity can be exaggerated. Scientists, even Newtonians, have been known to change their minds. Although I would not want to say that the empirical evidence is all-decisive, it plays a major role in such mind changes. As an example, consider a major revolution of our own time, namely that which occurred in geology. When I was an undergraduate in 1960, students were taught that continents do not move. Ten years later, they were told that they do move. Where is the dogmatism here? Furthermore, it was the new empirical evidence—e.g., about the nature of the sea-bed—which persuaded geologists. In short, although science may not be as open-minded as Karl Popper thinks it is, it is not as close-minded as, say, Thomas Kuhn[6] thinks it is.

Let me move on to the second and third questions, the status of creation-science and Judge Overton's treatment of the problem. The slightest acquaintance with the creation-science literature and Creationism movement shows that creation-science fails abysmally as science. Consider the following passage, written by one of the leading creationists, Duane T. Gish, in *Evolution: The Fossils Say No!*:

> *CREATION.* By creation we mean the bringing into being by a supernatural Creator of the basic kinds of plants and animals by the process of sudden, or fiat, creation.
>
> We do not know how the Creator created, what processes He used, *for He used processes which are not now operating anywhere in the natural universe.* This is why we refer to creation as Special Creation. We cannot discover by scientific investigations anything about the creative processes used by the Creator.[7]

The following similar passage was written by Henry M. Morris, who is considered to be the founder of the creation-science movement:

> . . . it is . . . quite impossible to determine anything about Creation through a study of present processes, because present processes are not created in character. If man wishes to know anything about Creation (the time of Creation, the duration of Creation, the order of Creation, the methods of Creation, or anything else) his sole source of true information is that of divine revelation. God was there when it happened. We were not there . . . therefore, we are completely limited to what God has seen fit to tell us, and

this information is in His written Word. This is our textbook on the science of Creation![8]

By their own words, therefore, creation-scientists admit that they appeal to phenomena not covered or explicable by any laws that humans can grasp as laws. It is not simply that the pertinent laws are not yet known. Creative processes stand outside law as humans know it (or could know it) on Earth—at least there is no way that scientists can know Mendel's laws through observation and experiment. Even if God did use His own laws, they are necessarily veiled from us forever in this life, because Genesis says nothing of them.

Furthermore, there is nothing tentative or empirically checkable about the central claims of creation-science. Creationists admit as much when they join the Creation Research Society (the leading organization of the movement). As a condition of membership applicants must sign a document specifying that they now believe and will continue to believe:

> (1) The Bible is the written Word of God, and because we believe it to be inspired throughout, all of its assertions are historically and scientifically true in all of the original autographs. To the student of nature, this means that the account of origins in Genesis is a factual presentation of simple historical truths. (2) All basic types of living things, including man, were made by direct creative acts of God during Creation Week as described in Genesis. Whatever biological changes have occurred since Creation have accomplished only changes within the original created kinds. (3) The great Flood described in Genesis, commonly referred to as the Noachian Deluge, was an historical event, worldwide in its extent and effect. (4) Finally, we are an organization of Christian men of science, who accept Jesus Christ as our Lord and Savior. The account of the special creation of Adam and Eve as one man and one woman, and their subsequent fall into sin, is the basis for our belief in the necessity of a Savior for all mankind. Therefore, salvation can come only thru accepting Jesus Christ as our Savior.[9]

It is difficult to imagine evolutionists signing a comparable statement, that they will never deviate from the literal text of Charles Darwin's *On the Origin of Species.* The non-scientific nature of creation-science is evident for all to see, as is also its religious nature. Moreover, the quotes I have used above were all used by Judge Overton, in the *McLean* Opinion, to make exactly the points I have just made. Creation-science is not genuine science, and Judge Overton showed this.

Finally, what about Laudan's claim that some parts of creation-science (e.g., claims about the Flood) are falsifiable and that other parts (e.g., about the originally created "kinds") are revisable? Such parts are not falsifiable or revisable in a way indicative of genuine science. Creation-science is not like physics, which exists as part of humanity's common cultural heritage and domain. It exists solely in the imaginations and writing of a relatively small group of people. Their publications (and stated intentions) show that, for example, there is no way they will relinquish belief in the Flood, whatever the evidence.[10] In this sense, their doctrines are truly unfalsifiable.

Furthermore, any revisions are not genuine revisions, but exploitations of the gross ambiguities in the creationists' own position. In the matter of origins, for example, some elasticity could be perceived in the creationist position, given the conflicting claims that the possibility of (degenerative) change within the originally created "kinds." Unfortunately, any open-mindedness soon proves illusory, for creationists have no real idea about what God is supposed to have created in the beginning, except that man was a separate species. They rely solely on the Book of Genesis:

And God said, Let the waters bring forth abundantly the moving creature that hath life, and the fowl that may fly above the earth in the open firmament of heaven.

And God created great whales, and every living creature that moveth, which the waters brought forth abundantly, after their kind, and every winged fowl after his kind: and God saw that it was good.

And God blessed them, saying Be fruitful, and multiply, and fill the waters in the seas, and let fowl multiply in the earth.

And the evening and the morning were the fifth day.

And God said, Let the earth bring forth the living creature after his kind, cattle, and creeping thing, and beast of the earth after his kind: and it was so.

And God made the beast of the earth after his kind, and cattle after their kind, and everything that creepeth upon the earth after his kind: and God saw that it was good.[11]

But the *definition* of "kind," what it really is, leaves creationists as mystified as it does evolutionists. For example, creationist Duane Gish makes this statement on the subject:

[W]e have defined a basic kind as including all of those variants which have been derived from a single stock. . . . We cannot always be sure, however, what constitutes a separate kind. The division into kinds is easier the more the divergence observed. It is obvious, for example, that among invertebrates the protozoa, sponges, jellyfish, worms, snails, trilobites, lobsters, and bees are all different kinds. Among the vertebrates, the fishes, amphibians, reptiles, birds, and mammals are obviously different basic kinds.

Among the reptiles, the turtles, crocodiles, dinosaurs, pterosaurs (flying reptiles), and ichthyosaurs (aquatic reptiles) would be placed in different kinds. Each one of these major groups of reptiles could be further subdivided into the basic kinds within each.

Within the mammalian class, duck-billed platypus, bats, hedgehogs, rats, rabbits, dogs, cats, lemurs, monkeys, apes, and men are easily assignable to different basic kinds. Among the apes, the gibbons, orangutans, chimpanzees, and gorillas would each be included in a different basic kind.[12]

Apparently, a "kind" can be anything from humans (one species) to trilobites (literally thousands of species). The term is flabby to the point of inconsistency. Because humans are mammals, if one claims (as creationists do) that evolution can occur within but not across kinds, then humans could have evolved from common mammalian stock—but because humans themselves are kinds such evolution is impossible.

In brief, there is no true resemblance between the creationists' treatment of their concept of "kind" and the openness expected of scientists. Nothing can be said in favor of creation-science or its inventors. Overton's judgment emerges unscathed by Laudan's complaints.

NOTES

1. For the text of Judge Overton's Opinion, see *Science, Technology, & Human Values* 40 (Summer 1982): 28–42; and Marcel LaFollette, *Creationism, Science, and the Law* (The MIT Press, 1983).
2. Larry Laudan, "Commentary: Science at the Bar—Causes for Concern," *Science, Technology, & Human Values* 7 (1982): 19–23.
3. Karl Popper, *The Logic of Scientific Discovery* (London: Hutchinson, 1959).
4. Charles Lyell, *Principles of Geology*, Volume I (London: John Murray, 1830), p. 162.
5. Laudan, *op. cit.,* p. 17.
6. Thomas Kuhn, *The Structure of Scientific Revolutions* (Chicago, Ill.: University of Chicago Press, 1962).
7. Duane Gish, *Evolution: The Fossils Say No!,* 3rd edition (San Diego, Calif.: Creation-Life Publishers, 1979), p. 40 (his italics).
8. Henry M. Morris, *Studies in the Bible and Science* (Philadelphia, Penn.: Presbyterian and Reformed Publishing Company, 1966), p. 114.
9. Application form for the Creation Research Society, reprinted in Plaintiffs' trial briefs, *McLean v. Arkansas* (1981).
10. See, for instance, Henry M. Morris, *Scientific Creationism* (San Diego, Calif.: Creation-Life Publishers, 1974; and my own detailed discussion in Michael Ruse, *Darwinism Defended: A Guide to the Evolution Controversies* (Reading, Mass.: Addison-Wesley, 1982).
11. Genesis, Book I, Verses 20–25.
12. Gish, *op. cit.,* pp. 34–35.

PART 1 SUGGESTIONS FOR FURTHER READING

Ayer, A. J., ed. *Logical Positivism.* New York: Free Press, 1959.

Campbell, N. *What Is Science?* New York: Dover Publications, 1952.

Churchland, Paul M. "Karl Popper's Philosophy of Science." *Canadian Journal of Philosophy* 5 (1975): 145–156.

Kitcher, Philip. *Abusing Science.* Cambridge, MA: MIT Press, 1982.

Kuhn, T. S. *The Structure of Scientific Revolutions.* 2d ed. Chicago: University of Chicago Press, 1971.

Lakatos, Imre, and Alan Musgrave, eds. *Criticism and the Growth of Knowledge.* New York: Cambridge University Press, 1970.

Nagel, E. *The Structure of Science.* New York: Harcourt, Brace and World, 1961.

Popper, K. *Conjectures and Refutations.* New York: Harper & Row, 1968.

———. *The Logic of Scientific Discovery.* New York: Harper & Row, 1968.

Toulmin, S. *Foresight and Understanding.* London: Hutchinson, 1961.

PART 2

INDUCTION AND CONFIRMATION
The Nature of Scientific Inference

Scientific theories are based on empirical evidence. The data they try to explain have been obtained through observation and experiment, not wishful thinking or divine revelation. But what sort of relationship must exist between data and theory in order for the data to support the theory? When are we justified in believing that a theory is true?

Two means of evidential support have traditionally been recognized by philosophers of science: deduction and induction. A theory is deductively supported by its evidence if it logically follows from that evidence. A theory logically follows from its evidence just in case it's impossible for the evidence to be true and the theory false. For example, the statement "Some birds are animals" logically follows from the statement "Some animals are birds." If the latter statement is true, the former has to be true. Because the truth of the latter statement guarantees the truth of the former, deductive support is said to be "truth preserving." Any theory that can be deduced from true evidence must itself be true.

Most scientific theories cannot be deduced from their evidence. Some can be induced from their evidence, however. If every raven that has ever been observed has been found to be black, we may use induction to arrive at the conclusion that, probably, every raven that ever will be observed will be black. Inductive inference does not guarantee the truth of its conclusion, however, for no matter how many ravens we observe, there is always the chance that there is a nonblack one that we did not observe. Nonetheless, induction is able to establish the truth of certain statements with a high degree of probability. The following problem arises, however: if induction does not guarantee the truth of its conclusions, how can science give us knowledge about the world? This is the problem that David Hume addresses.

Enumerative induction has the form "Every A that has been observed has been found to be F. Therefore, every A that ever will be observed will be found to be F." Hume realized that this form of inference assumes that the future will resemble the past. But what justifies our believing that? We can't provide a deductive argument for the claim that the future will resemble the past because there is no more fundamental claim from which it logically follows. Nor can we provide an inductive argument for that claim because such an argument would be circular: it would assume what it is trying to prove. So there appears to be no way to justify induction. The belief that the future will resemble the past is not something we arrived at through a process of inference. Rather, it is a bias that is built into our thinking.

The scientific method is often thought to consist of four steps: (1) Observe and record all the relevant facts, (2) analyze and classify those facts, (3) use induction to derive generalizations from the facts, and (4) test the generalizations. Carl Hempel calls this the "narrow inductivist" conception of scientific inquiry and objects to it on the grounds that scientists do not and cannot follow it. Facts cannot be observed or analyzed in the absence of a hypothesis, and induction is rarely used to generate hypotheses. It can be used to generate certain simple hypotheses, such as "Whenever copper is heated, it expands"; but it can't be used to generate more complex hypotheses, such as "Matter is composed of atoms," because when that hypothesis was introduced, atoms had not been observed. Scientific hypotheses often postulate the existence of unobserved entities to explain the observed. The conclusion of an enumerative inductive inference, however, can't make reference to things not covered by the data. So enumerative induction cannot generate hypotheses that contain novel concepts or ideas.

If scientific hypotheses are not derived from the data, how are they arrived at? Hempel claims that they are invented to account for the data. Since most hypotheses go beyond their data in various ways, there can be no set of rules for generating them. Hypotheses simply represent one's best guess about the way things are. Hempel claims that the scientific method is the "method of hypotheses," or what has come to be known as the "hypothetico-deductive method." This method consists of three steps: (1) Invent a hypothesis, (2) deduce a test implication from the hypothesis, and (3) perform the test. A test implication is a statement that should be true if the hypothesis in question is true. This statement can usually be expressed as a conditional statement saying that if certain conditions are realized, then certain things should be observed. Scientists test their hypotheses by creating those conditions in the laboratory or by locating them in the field and determining whether what they find is what the hypothesis predicted. If things are as the hypothesis says they should be, then the hypothesis has been confirmed. This does not establish the truth of the hypothesis, but the more tests it has successfully passed, the more likely it is to be true.

According to the hypothetico-deductive method, the successful test of a hypothesis has the following form:

If H, then P.

P.

Therefore, (probably) H.

where H stands for the hypothesis being tested, and P stands for the test implication. From a logical point of view, however, this inference is suspect because it commits the fallacy of affirming the consequent. To see this, consider this inference: If it rained, the streets are wet. The streets are wet. Therefore, it rained. The conclusion of this argument doesn't logically follow from the premises because it's possible for the premises to be true and the conclusion false. For example, the streets could have become wet because a water main broke, a spring water truck tipped over, or the street washer came by. Even the claim that it probably rained is problematic because until we know the relative likelihood of the other possibilities, we are not justified in claiming that it probably rained. (In the desert, one of these other possibilities might be more likely.)

Because of the difficulty of assessing the relative probabilities of various possibilities, Karl Popper claims that inductivism, in either the narrow or the wide sense, is untenable. Instead he offers a view that he calls "deductivism." In this view, the job of the scientist is not to confirm hypotheses, but to refute them. He agrees with Hempel that hypotheses are invented rather than discovered. But he disagrees with Hempel's claim that hypotheses are made more likely by successful tests. If a hypothesis successfully passes a number of tests, the best we can say is that it has been "corroborated."

If a theory fails to pass a test, however, we can reject it out of hand. This rejection is justified because it is the result of a valid inference. An unsuccessful test has the following form:

If H, then P.

Not P.

Therefore, not H.

This form of argument—known as denying the consequent—is not suspect because the conclusion logically follows from the premises. To see this, consider again the rain example. If it rained, the streets are wet. The streets are not wet. Therefore it didn't rain. In this case, it's impossible for the premises to be true and the conclusion false. In Popper's deductivist conception of science, induction plays no role. Thus Popper's conception avoids the problem of induction.

Popper's deductivism depends on a critical assumption: hypotheses can be tested in isolation from other beliefs we hold. But Pierre Duhem convincingly argues that hypotheses have testable consequences only in the context of certain background assumptions. These background assumptions provide information about the objects under study as well as the apparatus used to study them. If a test is unsuccessful, we can always save the hypothesis by rejecting one or more of the background assumptions. So it appears that hypotheses can neither be conclusively verified nor conclusively falsified.

For any set of data, it is possible in principle to construct an infinite number of explanations to account for those data. For example, think of all the different lines that can be drawn through a set of data points on a graph. So when we ask the question "Why did this happen?" it may be difficult to know how to go about answering it. But if we ask the question "Why did this happen rather than that?" we've narrowed the field of possible answers and provided a focus for our inquiry. Peter Lipton claims that the method of inference to the best contrastive explanation more accurately reflects the actual practice of scientists than either Hempel's or Popper's model.

7

DAVID HUME

The Problem of Induction

When it is asked, *What is the nature of all our reasonings concerning matter of fact?* the proper answer seems to be, that they are founded on the relation of cause and effect. When again it is asked, *What is the foundation of all our reasonings and conclusions concerning that relation?* it may be replied in one word, Experience. But if we still carry on our sifting humor, and ask, *What is the foundation of all conclusions from experience?* this implies a new question, which may be of more difficult solution and explication. Philosophers, that give themselves airs of superior wisdom and sufficiency, have a hard task when they encounter persons of inquisitive dispositions, who push them from every corner to which they retreat, and who are sure at last to bring them to some dangerous dilemma. The best expedient to prevent this confusion, is to be modest in our pretensions; and even to discover the difficulty ourselves before it is objected to us. By this means, we may make a kind of merit of our very ignorance.

I shall content myself, in this section, with an easy task, and shall pretend only to give a negative answer to the question here proposed. I say then, that, even after we have experience of the operations of cause and effect, our conclusions from that experience are *not* founded on reasoning, or any process of the understanding. This answer we must endeavor both to explain and to defend.

It must certainly be allowed, that nature has kept us at a great distance from all her secrets, and has afforded us only the knowledge of a few superficial qualities of objects; while she conceals from us those powers and principles on which the influence of those objects entirely depends. Our senses inform us of the color, weight, and consistence of bread; but neither sense nor reason can ever inform us of those qualities which fit it for the nourishment and support of a human body. Sight or feeling conveys an idea of the actual motion of bodies; but as to that wonderful force or power, which would carry on a moving body for ever in a continued change of place, and which bodies never lose but by communicating it to others; of this we cannot form the most distant conception. But notwithstanding this ignorance of natural powers[1] and principles, we always presume, when we see like sensible qualities, that they have like secret powers, and expect that effects, similar to those which we have experienced, will follow from them. If a body of like color and consistence with that bread, which we have formerly eaten, be presented to us, we make no scruple of repeating the experiment, and foresee, with certainty, like nourishment and support. Now this is a process of the mind or thought, of which I would willingly know the foundation. It is allowed on all hands that there is no known connection between the sensible qualities and the secret powers; and consequently, that the mind is not led to form such a conclusion concerning their constant and regular conjunction, by anything which it knows of their nature. As to past *Experience,* it can be allowed to give *direct* and *certain* information of those precise objects only, and that precise period of time, which fell under its cognizance: but why this experience should be extended to future times, and to other objects, which for aught we know, may be only in appearance similar; this is the main question on which I would insist. The bread, which I formerly ate, nourished me; that is, a body of such sensible qualities was, at that time, endued with such secret powers: but does it follow, that other bread must also nourish me at another time, and that like sensible qualities must always be attended with

An Inquiry Concerning Human Understanding, Section IV, Part II.

like secret powers? The consequence seems nowise necessary. At least, it must be acknowledged that there is here a consequence drawn by the mind; that there is a certain step taken; a process of thought, and an inference, which wants to be explained. These two propositions are far from being the same, *I have found that such an object has always been attended with such an effect*, and *I foresee, that other objects, which are, in appearance, similar, will be attended with similar effects.* I shall allow, if you please, that the one proposition may justly be inferred from the other: I know, in fact, that it always is inferred. But if you insist that the inference is made by a chain of reasoning, I desire you to produce that reasoning. The connection between these propositions is not intuitive. There is required a medium, which may enable the mind to draw such an inference, if indeed it be drawn by reasoning and argument. What that medium is, I must confess, passes my comprehension; and it is incumbent on those to produce it, who assert that it really exists, and is the origin of all our conclusions concerning matter of fact.

This negative argument must certainly, in process of time, become altogether convincing, if many penetrating and able philosophers shall turn their inquiries this way and no one be ever able to discover any connecting proposition or intermediate step, which supports the understanding in this conclusion. But as the question is yet new, every reader may not trust so far to his own penetration, as to conclude, because an argument escapes his inquiry, that therefore it does not really exist. For this reason it may be requisite to venture upon a more difficult task; and enumerating all the branches of human knowledge, endeavor to show that none of them can afford such an argument.

All reasonings may be divided into two kinds, namely, demonstrative reasoning, or that concerning relations of ideas, and moral reasoning, or that concerning matter of fact and existence. That there are no demonstrative arguments in the case seems evident; since it implies no contradiction that the course of nature may change, and that an object, seemingly like those which we have experienced, may be attended with different or contrary effects. May I not clearly and distinctly conceive that a body, falling from the clouds, and which, in all other respects, resembles snow, has yet the taste of salt or feeling of fire? Is there any more intelligible proposition than to affirm, that all the trees will flourish in December and January, and decay in May and June? Now whatever is intelligible, and can be distinctly conceived, implies no contradiction, and can never be proved false by any demonstrative argument or abstract reasoning *à priori*.

If we be, therefore, engaged by arguments to put trust in past experience, and make it the standard of our future judgment, these arguments must be probable only, or such as regard matter of fact and real existence, according to the division above mentioned. But that there is no argument of this kind, must appear, if our explication of that species of reasoning be admitted as solid and satisfactory. We have said that all arguments concerning existence are founded on the relation of cause and effect; that our knowledge of that relation is derived entirely from experience; and that all our experimental conclusions proceed upon the supposition that the future will be conformable to the past. To endeavor, therefore, the proof of this last supposition by probable arguments, or arguments regarding existence, must be evidently going in a circle, and taking that for granted, which is the very point in question.

In reality, all arguments from experience are founded on the similarity which we discover among natural objects, and by which we are induced to expect effects similar to those which we have found to follow from such objects. And though none but a fool or madman will ever pretend to dispute the authority of experience, or to reject that great guide of human life, it may surely be allowed a philosopher to have so much curiosity at least as to examine the principle of human nature, which gives this mighty authority to experience, and makes us draw advantage from that similarity which nature has placed among different objects. From causes which appear *similar* we expect similar effects. This is the sum of all our experimental conclusions. Now it seems evident that, if this conclusion were formed by reason, it would be as perfect at first, and upon one instance, as after ever so long a course of experience. But the case is far otherwise. Nothing so like as eggs; yet no one, on account of this appearing similarity, expects the same taste and relish in all of them. It is only after a long course of uniform experiments in any kind, that we attain a firm reliance and security with regard to a particular event. Now where is that process of reasoning which, from one instance, draws a conclusion, so different from that which it infers from a hundred instances that are nowise different from that single one? This question I propose as much for the sake of information, as with an intention of raising difficulties. I cannot find, I cannot imagine any such reasoning. But I keep my mind still open to instruction, if any one will vouchsafe to bestow it on me.

Should it be said that, from a number of uniform experiments, we *infer* a connection between the sensible qualities and the secret powers; this, I must confess, seems the same difficulty, couched in different terms. The question still recurs, on what process of argument this *inference* is founded? Where is the medium, the interposing ideas, which join propositions so very wide of each other? It is confessed that the color, consistence, and other sensible qualities of bread appear not, of themselves, to have any connection with the secret powers of nourishment and support. For otherwise we could infer these secret powers from the first appearance of these sensible qualities, without the aid of experience; contrary to the sentiment of all philosophers, and contrary to plain matter of fact. Here, then, is our natural state of ignorance with regard to the powers and influence of all objects. How is this remedied by experience? It only shows us a number of uniform effects, resulting from certain objects, and teaches us that those particular objects, at that particular time, were endowed with such powers and forces. When a new object, endowed with similar sensible qualities, is produced, we expect similar powers and forces, and look for a like effect. From a body of like color and consistence with bread we expect like nourishment and support. But this surely is a step or progress of the mind, which wants to be explained. When a man says, *I have found, in all past instances, such sensible qualities conjoined with such secret powers:* And when he says, *Similar sensible qualities will always be conjoined with similar secret powers,* he is not guilty of a tautology, nor are these propositions in any respect the same. You say that the one proposition is an inference from the other. But you must confess that the inference is not intuitive; neither is it demonstrative: Of what nature is it, then? To say it is experimental, is begging the question. For all inferences from experience suppose, as their foundation, that the future will resemble the past, and that similar powers will be conjoined with similar sensible qualities. If there be any suspicion that the course of nature may change, and that the past may be no rule for the future, all experience becomes useless, and can give rise to no inference or conclusion. It is impossible, therefore, that

any arguments from experience can prove this resemblance of the past to the future; since all these arguments are founded on the supposition of that resemblance. Let the course of things be allowed hitherto ever so regular; that alone, without some new argument or inference, proves not that, for the future, it will continue so. In vain do you pretend to have learned the nature of bodies from your past experience. Their secret nature, and consequently all their effects and influence, may change, without any change in their sensible qualities. This happens sometimes, and with regard to some objects: Why may it not happen always, and with regard to all objects? What logic, what process of argument secures you against this supposition? My practice, you say, refutes my doubts. But you mistake the purport of my question. As an agent, I am quite satisfied in the point; but as a philosopher, who has some share of curiosity, I will not say skepticism, I want to learn the foundation of this inference. No reading, no inquiry has yet been able to remove my difficulty, or give me satisfaction in a matter of such importance. Can I do better than propose the difficulty to be public, even though, perhaps, I have small hopes of obtaining a solution? We shall at least, by this means, be sensible of our ignorance, if we do not augment our knowledge.

I must confess that a man is guilty of unpardonable arrogance who concludes, because an argument has escaped his own investigation, that therefore it does not really exist. I must also confess that, though all the learned, for several ages, should have employed themselves in fruitless search upon any subject, it may still, perhaps, be rash to conclude positively that the subject must, therefore, pass all human comprehension. Even though we examine all the sources of our knowledge, and conclude them unfit for such a subject, there may still remain a suspicion, that the enumeration is not complete, or the examination not accurate. But with regard to the present subject, there are some considerations which seem to remove all this accusation of arrogance or suspicion of mistake.

It is certain that the most ignorant and stupid peasants—nay infants, nay even brute beasts—improve by experience, and learn the qualities of natural objects, by observing the effects which result from them. When a child has felt the sensation of pain from touching the flame of a candle, he will be careful not to put his hand near any candle; but will expect a similar effect from a cause which is similar in its sensible qualities and appearance. If you assert, therefore, that the understanding of the child is led into this conclusion by any process of argument or ratiocination, I may justly require you to produce that argument; nor have you any pretense to refuse so equitable a demand. You cannot say that the argument is abstruse, and may possibly escape your inquiry; since you confess that it is obvious to the capacity of a mere infant. If you hesitate, therefore, a moment, or if, after reflection, you produce any intricate or profound argument, you, in a manner, give up the question, and confess that it is not reasoning which engages us to suppose the past resembling the future, and to expect similar effects from causes which are, to

appearance, similar. This is the proposition which I intended to enforce in the present section. If I be right, I pretend not to have made any mighty discovery. And if I be wrong, I must acknowledge myself to be indeed a very backward scholar; since I cannot now discover an argument which, it seems, was perfectly familiar to me long before I was out of my cradle.

NOTE

1. The word, Power, is here used in a loose and popular sense. The more accurate explication of it would give additional evidence to this argument.

8

CARL HEMPEL

The Role of Induction in Scientific Inquiry

As a simple illustration of some important aspects of scientific inquiry let us consider Semmelweis' work on childbed fever. Ignaz Semmelweis, a physician of Hungarian birth, did this work during the years from 1844 to 1848 at the Vienna General Hospital. As a member of the medical staff of the First Maternity Division in the hospital, Semmelweis was distressed to find that a large proportion of the women who were delivered of their babies in that division contracted a serious and often fatal illness known as puerperal fever or childbed fever. In 1844, as many as 260 out of 3,157 mothers in the First Division, or 8.2 percent, died of the disease; for 1845, the death rate was 6.8 percent, and for 1846, it was 11.4 percent. These fig-

ures were all the more alarming because in the adjacent Second Maternity Division of the same hospital, which accommodated almost as many women as the First, the death toll from childbed fever was much lower: 2.3, 2.0, and 2.7 percent for the same years. In a book that he wrote later on the causation and the prevention of childbed fever, Semmelweis describes his efforts to resolve the dreadful puzzle.[1]

He began by considering various explanations that were current at the time; some of these he rejected out of hand as incompatible with well-established facts; others he subjected to specific tests.

One widely accepted view attributed the ravages of puerperal fever to "epidemic influences," which were vaguely described as "atmospheric-cosmic-telluric changes" spreading over whole districts and causing childbed fever in women in confinement. But how, Semmelweis reasons, could such influences have

Philosophy of Natural Science (Upper Saddle River, NJ: Prentice-Hall, 1996), pp. 11–18. © 1966 Prentice-Hall, Inc. Reprinted by permission of the publisher.

plagued the First Division for years and yet spared the Second? And how could this view be reconciled with the fact that while the fever was raging in the hospital, hardly a case occurred in the city of Vienna or in its surroundings: a genuine epidemic, such as cholera, would not be so selective. Finally, Semmelweis notes that some of the women admitted to the First Division, living far from the hospital, had been overcome by labor on their way and had given birth in the street: yet despite these adverse conditions, the death rate from childbed fever among these cases of "street birth" was lower than the average for the First Division.

On another view, overcrowding was a cause of mortality in the First Division. But Semmelweis points out that in fact the crowding was heavier in the Second Division, partly as a result of the desperate efforts of patients to avoid assignment to the notorious First Division. He also rejects two similar conjectures that were current, by noting that there were no differences between the two Divisions in regard to diet or general care of the patients.

In 1846, a commission that had been appointed to investigate the matter attributed the prevalence of illness in the First Division to injuries resulting from rough examination by the medical students, all of whom received their obstetrical training in the First Division. Semmelweis notes in refutation of this view that (a) the injuries resulting naturally from the process of birth are much more extensive than those that might be caused by rough examination; (b) the midwives who received their training in the Second Division examined their patients in much the same manner but without the same ill effects; (c) when, in response to the commission's report, the number of medical students was halved and their examinations of the women were reduced to a minimum, the mortality, after a brief decline, rose to higher levels than ever before.

Various psychological explanations were attempted. One of them noted that the First Division was so arranged that a priest bearing the last sacrament to a dying woman had to pass through five wards before reaching the sickroom beyond: the appearance of the priest, preceded by an attendant ringing a bell, was held to have a terrifying and debilitating effect upon the patients in the wards and thus to make them more likely victims of childbed fever. In the Second Division, this adverse factor

was absent, since the priest had direct access to the sickroom. Semmelweis decided to test this conjecture. He persuaded the priest to come by a roundabout route and without ringing of the bell, in order to reach the sick chamber silently and unobserved. But the mortality in the First Division did not decrease.

A new idea was suggested to Semmelweis by the observation that in the First Division the women were delivered lying on their backs; in the Second Division, on their sides. Though he thought it unlikely, he decided "like a drowning man clutching at a straw", to test whether this difference in procedure was significant. He introduced the use of the lateral position in the First Division, but again, the mortality remained unaffected.

At last, early in 1847, an accident gave Semmelweis the decisive clue for his solution of the problem. A colleague of his, Kolletschka, received a puncture wound in the finger, from the scalpel of a student with whom he was performing an autopsy, and died after an agonizing illness during which he displayed the same symptoms that Semmelweis had observed in the victims of childbed fever. Although the role of microorganisms in such infections had not yet been recognized at the time, Semmelweis realized that "cadaveric matter" which the student's scalpel had introduced into Kolletschka's blood stream had caused his colleague's fatal illness. And the similarities between the course of Kolletschka's disease and that of the women in his clinic led Semmelweis to the conclusion that his patients had died of the same kind of blood poisoning: he, his colleagues, and the medical students had been the carriers of the infectious material, for he and his associates used to come to the wards directly from performing dissections in the autopsy room, and examine the women in labor after only superficially washing their hands, which often retained a characteristic foul odor.

Again, Semmelweis put his idea to a test. He reasoned that if he were right, then childbed fever could be prevented by chemically destroying the infectious material adhering to the hands. He therefore issued an order requiring all medical students to wash their hands in a solution of chlorinated lime before making an examination. The mortality from childbed fever promptly began to decrease, and for the year

1848 it fell to 1.27 percent in the First Division, compared to 1.33 in the Second.

In further support of his idea, or of his *hypothesis,* as we will also say, Semmelweis notes that it accounts for the fact that the mortality in the Second Division consistently was so much lower: the patients there were attended by midwives, whose training did not include anatomical instruction by dissection of cadavers.

The hypothesis also explained the lower mortality among "street births": women who arrived with babies in arms were rarely examined after admission and thus had a better chance of escaping infection.

Similarly, the hypothesis accounted for the fact that the victims of childbed fever among the newborn babies were all among those whose mothers had contracted the disease during labor; for then the infection could be transmitted to the baby before birth, through the common bloodstream of mother and child, whereas this was impossible when the mother remained healthy.

Further clinical experiences soon led Semmelweis to broaden his hypothesis. On one occasion, for example, he and his associates, having carefully disinfected their hands, examined first a woman in labor who was suffering from a festering cervical cancer; then they proceeded to examine twelve other women in the same room, after only routine washing without renewed disinfection. Eleven of the twelve patients died of puerperal fever. Semmelweis concluded that childbed fever can be caused not only by cadaveric material, but also by "putrid matter derived from living organisms."

We have seen how, in his search for the cause of childbed fever, Semmelweis examined various hypotheses that had been suggested as possible answers. How such hypotheses are arrived at in the first place is an intriguing question which we will consider later. First, however, let us examine how a hypothesis, once proposed, is tested.

Sometimes, the procedure is quite direct. Consider the conjectures that differences in crowding, or in diet, or in general care account for the difference in mortality between the two divisions. As Semmelweis points out, these conflict with readily observable facts. There are no such differences between the divisions; the hypotheses are therefore rejected as false.

But usually the test will be less simple and straightforward. Take the hypothesis attributing the high mortality in the First Division to the dread evoked by the appearance of the priest with his attendant. The intensity of that dread, and especially its effect upon childbed fever, are not as directly ascertainable as are differences in crowding or in diet, and Semmelweis uses an indirect method of testing. He asks himself: Are there any readily observable effects that should occur if the hypothesis were true? And he reasons: If the hypothesis were true, *then* an appropriate change in the priest's procedure should be followed by a decline in fatalities. He checks this implication by a simple experiment and finds it false, and he therefore rejects the hypothesis.

Similarly, to test his conjecture about the position of the women during delivery, he reasons: If this conjecture should be true, *then* adoption of the lateral position in the First Division will reduce the mortality. Again, the implication is shown false by his experiment, and the conjecture is discarded.

In the last two cases, the test is based on an argument to the effect that *if* the contemplated hypothesis, say H, is true, *then* certain observable events (e.g., decline in mortality) should occur under specified circumstances (e.g., if the priest refrains from walking through the wards, or if the women are delivered in lateral position); or briefly, if H is true, then so is I, where I is a statement describing the observable occurrences to be expected. For convenience, let us say that I is inferred from, or implied by, H; and let us call I a *test implication of the hypothesis H.* (We will later give a more accurate description of the relation between I and H.)

In our last two examples, experiments show the test implication to be false, and the hypothesis is accordingly rejected. The reasoning that leads to the rejection may be schematized as follows:

If H is true, then so is I.

a] <u>But (as the evidence shows) I is not true.</u>

H is not true.

Any argument of this form, called *modus tollens* in logic,[2] is deductively valid; that is, if its premises (the sentences above the horizontal line) are true, then its conclusion (the sentence below the horizontal line) is unfailingly true as well. Hence, if the premises of (*a*) are properly established, the hypothesis H that is being tested must indeed be rejected.

Next, let us consider the case where observation or experiment bears out the test implication I. From his hypothesis that childbed fever is blood poisoning produced by cadaveric matter, Semmelweis infers that suitable antiseptic measures will reduce fatalities from the disease. This time, experiment shows the test implication to be true. But this favorable outcome does not conclusively prove the hypothesis true, for the underlying argument would have the form

> If H is true, then so is I.
>
> b]   (As the evidence shows) I is true.
> _____
>
> H is true.

And this mode of reasoning, which is referred to as the *fallacy of affirming the consequent,* is deductively invalid, that is, its conclusion may be false even if its premises are true.[3] This is in fact illustrated by Semmelweis' own experience. The initial version of his account of childbed fever as a form of blood poisoning presented infection with cadaveric matter essentially as the one and only source of the disease; and he was right in reasoning that if this hypothesis should be true, then destruction of cadaveric particles by antiseptic washing should reduce the mortality. Furthermore, his experiment did show the test implication to be true. Hence, in this case, the premises of (*b*) were both true. Yet, his hypothesis was false, for as he later discovered, putrid material from living organisms, too, could produce childbed fever.

Thus, the favorable outcome of a test, i.e., the fact that a test implication inferred from a hypothesis is found to be true, does not prove the hypothesis to be true. Even if many implications of a hypothesis have been borne out by careful tests, the hypothesis may still be false. The following argument still commits the fallacy of affirming the consequent:

> If H is true, then so are $I_1, I_2, \ldots, I_n$.
>
> c]   (As the evidence shows) $I_1, I_2, \ldots, I_n$, are all true.
> _____
>
> H is true.

This, too, can be illustrated by reference to Semmelweis' final hypothesis in its first version. As we noted earlier, his hypothesis also yields the test implications that among cases of street births admitted to the First Division, mortality from puerperal fever should be below the average for the Division, and that infants of mothers who escape the illness do not contract childbed fever; and these implications, too, were borne out by the evidence—even though the first version of the final hypothesis was false.

But the observation that a favorable outcome of however many tests does not afford conclusive proof for a hypothesis should not lead us to think that if we have subjected a hypothesis to a number of tests and all of them have had a favorable outcome, we are no better off than if we had not tested the hypothesis at all. For each of our tests might conceivably have had an unfavorable outcome and might have led to the rejection of the hypothesis. A set of favorable results obtained by testing different test implications, $I_1, I_2, \ldots, I_n$, of a hypothesis, shows that as far as these particular implications are concerned, the hypothesis has been borne out; and while this result does not afford a complete proof of the hypothesis, it provides at least some support, some partial corroboration or confirmation for it. The extent of this support will depend on various aspects of the hypothesis and of the test data.

The idea that in scientific inquiry, inductive inference from antecedently collected data leads to appropriate general principles is clearly embodied in the following account of how a scientist would ideally proceed:

> If we try to imagine how a mind of superhuman power and reach, but normal so far as the logical processes of its thought are concerned, . . . would use the scientific method, the process would be as follows: First, all facts would be observed and recorded, *without selection* or *a priori* guess as to their relative importance. Secondly, the observed and recorded facts would be analyzed, compared, and classified, *without hypothesis or postulates* other than those necessarily involved in the logic of thought. Third, from this analysis of the facts generalizations would be inductively drawn as to the relations, classificatory or causal, between them. Fourth, further research would be deductive as well as inductive, employing inferences from previously established generalizations.[4]

This passage distinguishes four stages in an ideal scientific inquiry: (1) observation and recording of all facts, (2) analysis and classification of these facts, (3) inductive derivation of generalizations from them, and (4) further testing of the generalizations. The first two of these stages are specifically assumed not to make use of any guesses or hypotheses as to how the observed facts might be interconnected; this restriction seems to have been imposed in the belief that such preconceived ideas would introduce a bias and would jeopardize the scientific objectivity of the investigation.

But the view expressed in the quoted passage—I will call it *the narrow inductivist conception of scientific inquiry*—is untenable, for several reasons. A brief survey of these can serve to amplify and to supplement our earlier remarks on scientific procedure.

First, a scientific investigation as here envisaged could never get off the ground. Even its first phase could never be carried out, for a collection of *all* the facts would have to await the end of the world, so to speak; and even all the facts *up to now* cannot be collected, since there are an infinite number and variety of them. Are we to examine, for example, all the grains of sand in all the deserts and on all the beaches, and are we to record their shapes, their weights, their chemical composition, their distances from each other, their constantly changing temperature, and their equally changing distance from the center of the moon? Are we to record the floating thoughts that cross our minds in the tedious process? The shapes of the clouds overhead, the changing color of the sky? The construction and the trade name of our writing equipment? Our own life histories and those of our fellow investigators? All these, and untold other things, are, after all, among "all the facts up to now".

Perhaps, then, all that should be required in the first phase is that all the *relevant* facts be collected. But relevant to what? Though the author does not mention this, let us suppose that the inquiry is concerned with a specified *problem*. Should we not then begin by collecting all the facts—or better, all available data—relevant to that problem? This notion still makes no clear sense. Semmelweis sought to solve one specific problem, yet he collected quite different kinds of data at different stages of his in-

quiry. And rightly so; for what particular sorts of data it is reasonable to collect is not determined by the problem under study, but by a tentative answer to it that the investigator entertains in the form of a conjecture or hypothesis. Given the conjecture that mortality from childbed fever was increased by the terrifying appearance of the priest and his attendant with the death bell, it was relevant to collect data on the consequences of having the priest change his routine; but it would have been totally irrelevant to check what would happen if doctors and students disinfected their hands before examining their patients. With respect to Semmelweis' eventual contamination hypothesis, data of the latter kind were clearly relevant, and those of the former kind totally irrelevant.

Empirical "facts" or findings, therefore, can be qualified as logically relevant or irrelevant only in reference to a given hypothesis, but not in reference to a given problem.

Suppose now that a hypothesis H has been advanced as a tentative answer to a research problem: what kinds of data would be relevant to H? Our earlier examples suggest an answer: A finding is relevant to H if either its occurrence or its nonoccurrence can be inferred from H. Take Torricelli's hypothesis, for example. As we saw, Pascal inferred from it that the mercury column in a barometer should grow shorter if the barometer were carried up a mountain. Therefore, any finding to the effect that this did indeed happen in a particular case is relevant to the hypothesis; but so would be the finding that the length of the mercury column had remained unchanged or that it had decreased and then increased during the ascent, for such findings would refute Pascal's test implication and would thus disconfirm Torricelli's hypothesis. Data of the former kind may be called positively, or favorably, relevant to the hypothesis; those of the latter kind negatively, or unfavorably, relevant.

In sum, the maxim that data should be gathered without guidance by antecedent hypotheses about the connections among the facts under study is self-defeating, and it is certainly not followed in scientific inquiry. On the contrary, tentative hypotheses are needed to give direction to a scientific investigation. Such hypotheses determine, among other things, what data should be collected at a given point in a scientific investigation.

It is of interest to note that social scientists trying to check a hypothesis by reference to the vast store of facts recorded by the U.S. Bureau of the Census, or by other data-gathering organizations, sometimes find to their disappointment that the values of some variable that plays a central role in the hypothesis have nowhere been systematically recorded. This remark is not, of course, intended as a criticism of data gathering: those engaged in the process no doubt try to select facts that might prove relevant to future hypotheses; the observation is simply meant to illustrate the impossibility of collecting "all the relevant data" without knowledge of the hypotheses to which the data are to have relevance.

The second stage envisaged in our quoted passage is open to similar criticism. A set of empirical "facts" can be analyzed and classified in many different ways, most of which will be unilluminating for the purposes of a given inquiry. Semmelweis could have classified the women in the maternity wards according to criteria such as age, place of residence, marital status, dietary habits, and so forth; but information on these would have provided no clue to a patient's prospects of becoming a victim of childbed fever. What Semmelweis sought were criteria that would be significantly connected with those prospects; and for this purpose, as he eventually found, it was illuminating to single out those women who were attended by medical personnel with contaminated hands; for it was with this characteristic, or with the corresponding class of patients, that high mortality from childbed fever was associated.

Thus, if a particular way of analyzing and classifying empirical findings is to lead to an explanation of the phenomena concerned, then it must be based on hypotheses about how those phenomena are connected; without such hypotheses, analysis and classification are blind.

Our critical reflections on the first two stages of inquiry as envisaged in the quoted passage also undercut the notion that hypotheses are introduced only in the third stage, by inductive inference from antecedently collected data. But some further remarks on the subject should be added here.

Induction is sometimes conceived as a method that leads, by means of mechanically applicable rules, from observed facts to corresponding general principles. In this case, the rules of inductive inference would provide effective canons of scientific discovery; induction would be a mechanical procedure analogous to the familiar routine for the multiplication of integers, which leads, in a finite number of predetermined and mechanically performable steps, to the corresponding product. Actually, however, no such general and mechanical induction procedure is available at present; otherwise, the much studied problem of the causation of cancer, for example, would hardly have remained unsolved to this day. Nor can the discovery of such a procedure ever be expected. For—to mention one reason—scientific hypotheses and theories are usually couched in terms that do not occur at all in the description of the empirical findings on which they rest, and which they serve to explain. For example, theories about the atomic and subatomic structure of matter contain terms such as 'atom', 'electron', 'proton', 'neutron', 'psi-function', etc.; yet they are based on laboratory findings about the spectra of various gases, tracks in cloud and bubble chambers, quantitative aspects of chemical reactions, and so forth—all of which can be described without the use of those "theoretical terms". Induction rules of the kind here envisaged would therefore have to provide a mechanical routine for constructing, on the basis of the given data, a hypothesis or theory stated in terms of some quite novel concepts, which are nowhere used in the description of the data themselves. Surely, no general mechanical rule of procedure can be expected to achieve this. Could there be a general rule, for example, which, when applied to the data available to Galileo concerning the limited effectiveness of suction pumps, would, by a mechanical routine, produce a hypothesis based on the concept of a sea of air?

To be sure, mechanical procedures for inductively "inferring" a hypothesis on the basis of given data may be specifiable for situations of special, and relatively simple, kinds. For example, if the length of a copper rod has been measured at several different temperatures, the resulting pairs of associated values for temperature and length may be represented by points in a plane coordinate system, and a curve may be drawn through them in accordance with some particular rule of curve fitting. The curve then graphically represents a general quantitative hypothesis that expresses the length of the rod as a specific function of its temperature. But note that this hypothesis contains no novel terms; it is expressible in terms of the concepts of temperature and length, which are used also in describing the data. Moreover, the choice of "associated" values of temperature and length as data already presupposes a guiding hypothesis; namely, that with each value of the temperature, exactly one value of the length of the copper rod is associated, so that its length is indeed a function of its temperature alone. The mechanical curve-fitting routine then serves only to select a particular function as the appropriate one. This point is important; for suppose that instead of a copper rod, we examine a body of nitrogen gas enclosed in a cylindrical container with a movable piston as a lid, and that we measure its volume at several different temperatures. If we were to use this procedure in an effort to obtain from our data a *general* hypothesis representing the volume of the gas as a function of its temperature, we would fail, because the volume of a gas is a function both of its temperature and of the pressure exerted upon it, so that at the same temperature, the given gas may assume different volumes.

Thus, even in these simple cases, the mechanical procedures for the construction of a hypothesis do only part of the job, for they presuppose an antecedent, less specific hypothesis (i.e., that a certain physical variable is a function of one single other variable), which is not obtainable by the same procedure.

There are, then, no generally applicable "rules of induction", by which hypotheses or theories can be mechanically derived or inferred from empirical data. The transition from data to theory requires creative imagination. Scientific hypotheses and theories are not *derived* from observed facts, but *invented* in order to account for them. They constitute guesses at the connections that might obtain between the phenomena under study, at uniformities and patterns that might underlie their occurrence. "Happy guesses"[5] of this kind require great ingenuity, especially if they involve a radical departure from current modes of scientific thinking, as did, for example, the theory of relativity and quantum theory. The inventive effort required in scientific research will benefit from a thorough familiarity with current knowledge in the field. A complete novice will hardly make an important scientific discovery, for the ideas that may occur to him are likely to duplicate what has been tried before or to run afoul of well-established facts or theories of which he is not aware.

Nevertheless, the ways in which fruitful scientific guesses are arrived at are very different from any process of systematic inference. The chemist Kekulé, for example, tells us that he had long been trying unsuccessfully to devise a structural formula for the benzene molecule when, one evening in 1865, he found a solution to his problem while he was dozing in front of his fireplace. Gazing into the flames, he seemed to see atoms dancing in snakelike arrays. Suddenly, one of the snakes formed a ring by seizing hold of its own tail and then whirled mockingly before him. Kekulé awoke in a flash: he had hit upon the now famous and familiar idea of representing the molecular structure of benzene by a hexagonal ring. He spent the rest of the night working out the consequences of this hypothesis.[6]

This last remark contains an important reminder concerning the objectivity of science. In his endeavor to find a solution to his problem, the scientist may give free rein to his imagination, and the course of his creative thinking may be influenced even by scientifically questionable notions. Kepler's study of planetary motion, for example, was inspired by his interest in a mystical doctrine about numbers and a passion to demonstrate the music of the spheres. Yet, scientific objectivity is safeguarded by the principle that while hypotheses and theories may be freely invented and *proposed* in science, they can be *accepted* into the body of scientific knowledge only if they pass critical scrutiny, which includes in particular the checking of suitable test implications by careful observation or experiment.

Interestingly, imagination and free invention play a similarly important role in those disciplines whose results are validated exclusively by deductive reasoning; for example, in mathematics. For the rules of deductive inference do not afford mechanical rules of discovery, either. As illustrated by our statement of *modus tollens* above, those rules are usually expressed in the form of general schemata, any instance of which is a deductively valid argument. If premises of the specified kind are given, such a schema does indeed specify a way of proceeding to a logical consequence. But for any set of premises that may be given, the rules of deductive inference specify an infinity of validly deducible conclusions.

Take, for example, one simple rule represented by the following schema:

$$\frac{p}{p \text{ or } q}$$

It tells us, in effect, that from the proposition that p is the case, it follows that p or q is the case, where p and q may be any propositions whatever. The word 'or' is here understood in the "nonexclusive" sense, so that 'p or q' is tantamount to 'either p or q or both p and q'. Clearly, if the premise of an argument of this type is true, then so must be the conclusion; hence, any argument of the specified form is valid. But this one rule alone entitles us to infer infinitely many different consequences from any one premise. Thus, from 'the Moon has no atmosphere', it authorizes us to infer any statement of the form 'The Moon has no atmosphere, or q', where for 'q' we may write any statement whatsoever, no matter whether it is true or false; for example, 'the Moon's atmosphere is very thin', 'the Moon is uninhabited', 'gold is denser than silver', 'silver is denser than gold', and so forth. (It is interesting and not difficult to prove that infinitely many different statements can be formed in English; each of these may be put in the place of the variable 'q'.) Other rules of deductive inference add, of course, to the variety of statements derivable from one premise or set of premises. Hence, if we are given a set of statements as premises, the rules of deduction give no direction to our inferential procedures. They do not single out one statement as "the" conclusion to be derived from our premises, nor do they tell us how to obtain interesting or systematically important conclusions; they provide no mechanical routine, for example, for deriving significant mathematical theorems from given postulates. The discovery of important, fruitful mathematical theorems, like the discovery of important, fruitful theories in empirical science, requires inventive ingenuity; it calls for imaginative, insightful guessing. But again, the interests of scientific objectivity are safeguarded by the demand for an *objective validation* of such conjectures. In mathematics, this means *proof* by deductive derivation from axioms. And when a mathematical proposition has been proposed as a conjecture, its proof or disproof still requires inventiveness and ingenuity, often of a very high caliber; for the rules of deductive inference do not even provide a general mechanical procedure for constructing proofs or disproofs. Their systematic role is rather the modest one of serving as *criteria of soundness for arguments* offered as proofs: an argument will constitute a valid mathematical proof if it proceeds from the axioms to the proposed theorem by a chain of inferential steps each of which is valid according to one of the rules of deductive inference. And to check whether a given argument is a valid proof in this sense is indeed a purely mechanical task.

Scientific knowledge, as we have seen, is not arrived at by applying some inductive inference procedure to antecedently collected data, but rather by what is often called "the method of hypothesis", i.e., by inventing hypotheses as tentative answers to a problem under study, and then subjecting these to empirical test. It will be part of such test to see whether the hypothesis is borne out by whatever relevant findings may have been gathered before its formulation; an acceptable hypothesis will have to fit the available relevant data. Another part of the test will consist in deriving new test implications from the hypothesis and checking these by suitable observations or experiments. As we noted earlier, even extensive testing with entirely favorable results does not establish a hypothesis conclusively, but provides only more or less strong support for it. Hence, while scientific inquiry is certainly not inductive in the narrow sense we have examined in some detail, it may be said to be *inductive in a wider sense,* inasmuch as it involves the acceptance of hypotheses on the basis of data that afford no deductively conclusive evidence for it, but lend it only more or less strong "inductive support", or confirmation. And any "rules of induction" will have to be conceived, in analogy to the rules of deduction, as canons of validation rather than of discovery. Far from generating a hypothesis that accounts for given empirical findings, such rules will presuppose that both the empirical data forming the "premises" of the "inductive argument" and a tentative hypothesis forming its "conclusion" are *given.* The rules of induction would then state criteria for the soundness of the argument. According to some theories of induction, the rules would determine the strength of the support that the data lend to the hypothesis, and they might express such support in terms of probabilities.

NOTES

1. The story of Semmelweis' work and of the difficulties he encountered forms a fascinating page in the history of medicine. A detailed account, which includes translations and paraphrases of large portions of Semmelweis' writings, is given in W. J. Sinclair, *Semmelweis: His Life and His Doctrine* (Manchester, England: Manchester University Press, 1909). Brief quoted phrases in this chapter are taken from this work. The highlights of Semmelweis' career are recounted in the first chapter of P. de Kruif, *Men Against Death* (New York: Harcourt, Brace & World, Inc., 1932).

2. For details, see another volume in this series: W. Salmon, *Logic,* pp., 24–25.

3. See Salmon, *Logic,* pp. 27–29.

4. A. B. Wolfe, "Functional Economics," in *The Trend of Economics,* ed. R. G. Tugwell (New York: Alfred A. Knopf, Inc., 1924), p. 450 (italics are quoted).

5. This characterization was given already by William Whewell in his work *The Philosophy of the Inductive*

Sciences, 2nd ed. (London: John W. Parker, 1847); II, 41. Whewell also speaks of "invention" as "part of induction" (p. 46). In the same vein, K. Popper refers to scientific hypotheses and theories as "conjectures"; see, for example, the essay "Science: Conjectures and Refutations" in his book, *Conjectures and Refutations* (New York and London: Basic Books, 1962). Indeed, A. B. Wolfe, whose narrowly inductivist conception of ideal scientific procedure was quoted earlier, stresses that "the limited human mind" has to use "a greatly modified procedure", requiring scientific imagination and the selection of data on the basis of some "working hypothesis" (p. 450 of the essay cited in note 4).

6. Cf. the quotations from Kekulé's own report in A. Findlay, *A Hundred Years of Chemistry,* 2nd ed. (London: Gerald Duckworth & Co., 1948), p. 37; and W. I. B., Beveridge, *The Art of Scientific Investigation,* 3rd ed. (London: William Heinemann, Ltd., 1957), p. 56.

9

KARL POPPER

The Problem of Induction

A scientist, whether theorist or experimenter, puts forward statements, or systems of statements, and tests them step by step. In the field of the empirical sciences, more particularly, he constructs hypotheses, or systems of theories, and tests them against experience by observation and experiment.

I suggest that it is the task of the logic of scientific discovery, or the logic of knowledge, to give a logical analysis of this procedure; that is, to analyze the method of the empirical sciences.

But what are these 'methods of the empirical sciences'? And what do we call 'empirical science'?

THE PROBLEM OF INDUCTION

According to a widely accepted view—to be opposed in this book—the empirical sciences can be characterized by the fact that they use '*inductive methods*', as they are called. According to this view, the logic of scientific discovery would be identical

The Logic of Discovery (London: Hutchinson, 1959), pp. 27–33. Reprinted with permission.

with inductive logic, *i.e.*, with the logical analysis of these inductive methods.

It is usual to call an inference 'inductive' if it passes from *singular statements* (sometimes also called 'particular' statements), such as accounts of the results of observations or experiments, to *universal statements*, such as hypotheses or theories.

Now it is far from obvious, from a logical point of view, that we are justified in inferring universal statements from singular ones, no matter how numerous; for any conclusion drawn in this way may always turn out to be false: no matter how many instances of white swans we may have observed, this does not justify the conclusion that *all* swans are white.

The question whether inductive inferences are justified, or under what conditions, is known as *the problem of induction*.

The problem of induction may also be formulated as the question of how to establish the truth of universal statements which are based on experience, such as the hypotheses and theoretical systems of the empirical sciences. For many people believe that the truth of these universal statements is *'known by experience'*; yet it is clear that an account of an experience—of an observation or the result of an experiment—can in the first place be only a singular statement and not a universal one. Accordingly, people who say of a universal statement that we know its truth from experience usually mean that the truth of this universal statement can somehow be reduced to the truth of singular ones, and that these singular ones are known by experience to be true; which amounts to saying that the universal statement is based on inductive inference. Thus to ask whether there are natural laws known to be true appears to be only another way of asking whether inductive inferences are logically justified.

Yet if we want to find a way of justifying inductive inferences, we must first of all try to establish a *principle of induction*. A principle of induction would be a statement with the help of which we could put inductive inferences into a logically acceptable form. In the eyes of the upholders of inductive logic, a principle of induction is of supreme importance for scientific method: ' . . . this principle', says Reichenbach, 'determines the truth of scientific theories. To eliminate it from science would mean nothing less than to deprive science of the power to decide the truth or falsity of its theories. Without it,

clearly, science would no longer have the right to distinguish its theories from the fanciful and arbitrary creations of the poet's mind.'[1]

Now this principle of induction cannot be a purely logical truth like a tautology or an analytic statement. Indeed, if there were such a thing as a purely logical principle of induction, there would be no problem of induction; for in this case, all inductive inferences would have to be regarded as purely logical or tautological transformations, just like inferences in deductive logic. Thus the principle of induction must be a synthetic statement; that is, a statement whose negation is not self-contradictory but logically possible. So the question arises why such a principle should be accepted at all, and how we can justify its acceptance on rational grounds.

Some who believe in inductive logic are anxious to point out, with Reichenbach, that 'the principle of induction is unreservedly accepted by the whole of science and that no man can seriously doubt this principle in everyday life either'.[2] Yet even supposing this were the case—for after all, 'the whole of science' might err—I should still contend that a principle of induction is superfluous, and that it must lead to logical inconsistencies.

That inconsistencies may easily arise in connection with the principle of induction should have been clear from the work of Hume; also, that they can be avoided, if at all, only with difficulty. For the principle of induction must be a universal statement in its turn. Thus if we try to regard its truth as known from experience, then the very same problems which occasioned its introduction will arise all over again. To justify it, we should have to employ inductive inferences; and to justify these we should have to assume an inductive principle of a higher order; and so on. Thus the attempt to base the principle of induction on experience breaks down, since it must lead to an infinite regress.

Kant tried to force his way out of this difficulty by taking the principle of induction (which he formulated as the 'principle of universal causation') to be '*a priori* valid'. But I do not think that his ingenious attempt to provide an *a priori* justification for synthetic statement was successful.

My own view is that the various difficulties of inductive logic here sketched are insurmountable. So also, I fear, are those inherent in the doctrine, so widely current today, that inductive inference, although not 'strictly valid', *can attain some degree of 'reliability' or of 'probability'*. According to this doctrine, inductive inferences are 'probable inferences'.[3] 'We have described', says Reichenbach, 'the principle of induction as the means whereby science decides upon truth. To be more exact, we should say that it serves to decide upon probability. For it is not given to science to reach either truth or falsity . . . but scientific statements can only attain continuous degrees of probability whose unattainable upper and lower limits are truth and falsity'.[4]

At this stage I can disregard the fact that the believers in inductive logic entertain an idea of probability that I shall later reject as highly unsuitable for their own purposes. I can do so because the difficulties mentioned are not even touched by an appeal to probability. For if a certain degree of probability is to be assigned to statements based on inductive inference, then this will have to be justified by invoking a new principle of induction, appropriately modified. And this new principle in its turn will have to be justified, and so on. Nothing is gained, moreover, if the principle of induction, in its turn, is taken not as 'true' but only as 'probable'. In short, like every other form of inductive logic, the logic of probable inference, or 'probability logic', leads either to an infinite regress, or to the doctrine of *apriorism*.

The theory to be developed in the following pages stands directly opposed to all attempts to operate with the ideas of inductive logic. It might be described as the theory of *the deductive method of testing*, or as the view that a hypothesis can only be empirically *tested*—and only *after* it has been advanced.

Before I can elaborate this view (which might be called 'deductivism', in contrast to 'inductivism'[5]) I must first make clear the distinction between the *psychology of knowledge* which deals with empirical facts, and the *logic of knowledge* which is concerned only with logical relations. For the belief in inductive logic is largely due to a confusion of psychological problems with epistemological ones. It may be worth noticing, by the way, that this confusion spells trouble not only for the logic of knowledge but for its psychology as well.

ELIMINATION OF PSYCHOLOGISM

I said above that the work of the scientist consists in putting forward and testing theories.

The initial stage, the act of conceiving or inventing a theory, seems to me neither to call for logical analysis nor to be susceptible of it. The question how it happens that a new idea occurs to a man—whether it is a musical theme, a dramatic conflict, or a scientific theory—may be of great interest to empirical psychology; but it is irrelevant to the logical analysis of scientific knowledge. This latter is concerned not with *questions of fact* (Kant's *quid facti?*), but only with questions of *justification or validity* (Kant's *quid juris?*). Its questions are of the following kind. Can a statement be justified? And if so, how? Is it testable? Is it logically dependent on certain other statements? Or does it perhaps contradict them? In order that a statement may be logically examined in this way, it must already have been presented to us. Someone must have formulated it, and submitted it to logical examination.

Accordingly I shall distinguish sharply between the process of conceiving a new idea, and the methods and results of examining it logically. As to the task of the logic of knowledge—in contradistinction to the psychology of knowledge—I shall proceed on the assumption that it consists solely in investigating the methods employed in those systematic tests to which every new idea must be subjected if it is to be seriously entertained.

Some might object that it would be more to the purpose to regard it as the business of epistemology to produce what has been called a *'rational reconstruction'* of the steps that have led the scientist to a discovery—to the finding of some new truth. But the question is: what, precisely, do we want to reconstruct? If it is the processes involved in the stimulation and release of an inspiration which are to be reconstructed, then I should refuse to take it as the task of the logic of knowledge. Such processes are the concern of empirical psychology but hardly of logic. It is another matter if we want to reconstruct rationally the *subsequent tests* whereby the inspiration may be discovered to be a discovery, or become known to be knowledge. In so far as the scientist critically judges, alters, or rejects his own inspiration we may, if we like, regard the methodological analysis undertaken here as a kind of 'rational reconstruction' of the corresponding thought-processes. But this reconstruction would not describe these processes as they actually happen: it can give only a logical skeleton of the procedure of testing. Still, this is perhaps all that is meant by those who speak of a 'rational reconstruction' of the ways in which we gain knowledge.

It so happens that my arguments in this book are quite independent of this problem. However, my view of the matter, for what it is worth, is that there is no such thing as a logical method of having new ideas, or a logical reconstruction of this process. My view may be expressed by saying that every discovery contains 'an irrational element', or 'a creative intuition', in Bergson's sense. In a similar way Einstein speaks of the 'search for those highly universal laws . . . from which a picture of the world can be obtained by pure deduction. There is no logical path', he says, 'leading to these . . . laws. They can only be reached by intuition, based upon something like an intellectual love (*Einfühlung*') of the objects of experience'.[6]

DEDUCTIVE TESTING OF THEORIES

According to the view that will be put forward here, the method of critically testing theories, and selecting them according to the results of tests, always proceeds on the following lines. From a new idea, put up tentatively, and not yet justified in any way—an anticipation, a hypothesis, a theoretical system, or what you will—conclusions are drawn by means of logical deduction. These conclusions are then compared with one another and with other relevant statements, so as to find what logical relations (such as equivalence, derivability, compatibility, or incompatibility) exist between them.

We may if we like distinguish four different lines along which the testing of a theory could be carried out. First there is the logical comparison of the conclusions among themselves, by which the internal consistency of the system is tested. Secondly, there is the investigation of the logical form of the theory, with the object of determining whether it has the character of an empirical or scientific theory, or whether it is, for example, tautological. Thirdly, there is the comparison with other theories, chiefly

with the aim of determining whether the theory would constitute a scientific advance should it survive our various tests. And finally, there is the testing of the theory by way of empirical applications of the conclusions which can be derived from it.

The purpose of this last kind of test is to find out how far the new consequences of the theory—whatever may be new in what it asserts—stand up to the demands of practice, whether raised by purely scientific experiments, or by practical technological applications. Here too the procedure of testing turns out to be deductive. With the help of other statements, previously accepted, certain singular statements—which we may call 'predictions'—are deduced from the theory; especially predictions that are easily testable or applicable. From among these statements, those are selected which are not derivable from the current theory, and more especially those which the current theory contradicts. Next we seek a decision as regards these (and other) derived statements by comparing them with the results of practical applications and experiments. If this decision is positive, that is, if the singular conclusions turn out to be acceptable, or *verified*, then the theory has, for the time being, passed its test: we have found no reason to discard it. But if the decision is negative, or in other words, if the conclusions have been *falsified*, then their falsification also falsifies the theory from which they were logically deduced.

It should be noticed that a positive decision can only temporarily support the theory, for subsequent negative decisions may always overthrow it. So long as a theory withstands detailed and severe tests and is not superseded by another theory in the course of scientific progress, we may say that it has 'proved its mettle' or that it is *'corroborated'*.

Nothing resembling inductive logic appears in the procedure here outlined. I never assume that we can argue from the truth of singular statements to the truth of theories. I never assume that by force of 'verified' conclusions, theories can be established as 'true', or even as merely 'probable'.

NOTES

1. H. Reichenbach, *Erkenntnis* **1**, 1930, p. 186 (*cf.* also p. 64 f.).

2. Reichenbach *ibid.,* p. 67.

3. *Cf.* J. M. Keynes, *A Treatise on Probability* (1921); O. Külpe, *Vorlesungen über Logic* (ed. by Selz, 1923); Reichenbach (who uses the term 'probability implications'), *Axiomatik der Wahrscheinlichkeitrechnung, Mathem. Zeitschr.* 34 (1932); and in many other places.

4. Reichenbach, *Erkenntnis 1,* 1930, p. 186.

5. Liebig (in *Induktion und Deduktion,* 1865) was probably the first to reject the inductive method from the standpoint of natural science; his attack is directed against Bacon. Duhem (in *La Théorie physique, son objet et sa structure,* 1906; English translation by P. P. Wiener: *The Aim and Structure of Physical Theory,* Princeton, 1954) held pronounced deductivist views. (But there are also inductivist views to be found in Duhem's book, for example in the third chapter, Part One, where we are told that only experiment, induction, and generalization have produced Descartes's law of refraction; *cf.* the English translation, p. 34.) See also V. Kraft, *Die Grundformen der Wissenschaftlichen Methoden,* 1925; and Carnap, *Erkenntnis 2,* 1932, p. 440.

6. Address on Max Planck's 60th birthday. The passage quoted begins with the words, 'The supreme task of the physicist is to search for those highly universal laws . . . ,' etc. (quoted from A. Einstein, *Mein Weltbild,* 1934, p. 168; English translation by A. Harris: *The World as I See It,* 1935, p. 125). Similar ideas are found earlier in Liebig, *op. cit.; cf.* also Mach, *Principien der Wärmelehre* (1896), p. 443 *ff.* The German word *'Einfühlung'* is difficult to translate. Harris translates: 'sympathetic understanding of experience'.

verified for the time being → based on experience → induction

10

PIERRE DUHEM

Physical Theory and Experiment

The physicist who carries out an experiment, or gives a report of one, implicitly recognizes the accuracy of a whole group of theories. Let us accept this principle and see what consequences we may deduce from it when we seek to estimate the role and logical import of a physical experiment.

In order to avoid any confusion we shall distinguish two sorts of experiments: experiments of *application,* which we shall first just mention, and experiments of *testing,* which will be our chief concern.

The Aim and Structure of Physical Theory, Philip P. Wiener, trans. (Princeton, NJ: Princeton University Press, 1954), pp. 183–190. © 1954, renewed 1982 Princeton University Press. Reprinted by permission of the publisher.

You are confronted with a problem in physics to be solved practically; in order to produce a certain effect you wish to make use of knowledge acquired by physicists; you wish to light an incandescent bulb; accepted theories indicate to you the means for solving the problem; but to make use of these means you have to secure certain information; you ought, I suppose, to determine the electromotive force of the battery of generators at your disposal; you measure this electromotive force: that is what I call an experiment of application. This experiment does not aim at discovering whether accepted theories are accurate or not; it merely intends to draw on these theories. In order to carry it out, you make use of instruments that these same theories legitimize; there is nothing to shock logic in this procedure.

But experiments of application are not the only ones the physicist has to perform; only with their aid can science aid practice, but it is not through them that science creates and develops itself; besides experiments of application, we have experiments of testing.

A physicist disputes a certain law; he calls into doubt a certain theoretical point. How will he justify these doubts? How will he demonstrate the inaccuracy of the law? From the proposition under indictment he will derive the prediction of an experimental fact; he will bring into existence the conditions under which this fact should be produced; if the predicted fact is not produced, the proposition which served as the basis of the prediction will be irremediably condemned.

F. E. Neumann assumed that in a ray of polarized light the vibration is parallel to the plane of polarization, and many physicists have doubted this proposition. How did O. Wiener undertake to transform this doubt into a certainty in order to condemn Neumann's proposition? He deduced from this proposition the following consequence: If we cause a light beam reflected at 45° from a plate of glass to interfere with the incident beam polarized perpendicularly to the plane of incidence, there ought to appear alternately dark and light interference bands parallel to the reflecting surface; he brought about the conditions under which these bands should have been produced and showed that the predicted phenomenon did not appear, from which he concluded that Neumann's proposition is false, viz., that in a polarized ray of light the vibration is not parallel to the plane of polarization.

Such a mode of demonstration seems as convincing and as irrefutable as the proof by reduction to absurdity customary among mathematicians; moreover, this demonstration is copied from the reduction to absurdity, experimental contradiction playing the same role in one as logical contradiction plays in the other.

Indeed, the demonstrative value of experimental method is far from being so rigorous or absolute: the conditions under which it functions are much more complicated than is supposed in what we have just said; the evaluation of results is much more delicate and subject to caution.

A physicist decides to demonstrate the inaccuracy of a proposition; in order to deduce from this proposition the prediction of a phenomenon and institute the experiment which is to show whether this phenomenon is or is not produced, in order to interpret the results of this experiment and establish that the predicted phenomenon is not produced, he does not confine himself to making use of the phenomenon, whose nonproduction is to cut off debate, does not derive from the proposition challenged if taken by itself, but from the proposition at issue joined to that whole group of theories; if the predicted phenomenon is not produced, not only is the proposition questioned at fault, but so is the whole theoretical scaffolding used by the physicist. The only thing the experiment teaches us is that among the propositions used to predict the phenomenon and to establish whether it would be produced, there is at least one error; but where this error lies is just what it does not tell us. The physicist may declare that this error is contained in exactly the proposition he wishes to refute, but is he sure it is not in another proposition? If he is, he accepts implicitly the accuracy of all the other propositions he has used, and the validity of his conclusion is as great as the validity of his confidence.

Let us take as an example the experiment imagined by Zenker and carried out by O. Wiener. In order to predict the formation of bands in certain circumstances and to show that these did not appear, Wiener did not make use merely of the famous proposition of F. E. Neumann, the proposition which he wished to refute; he did not merely admit that in a polarized ray vibrations are parallel to the plane of polarization; but he used, besides this, propositions, laws, and hypotheses constituting the optics commonly accepted: he admitted that light consists in simple periodic vibrations, that these vibrations are normal to the light ray, that at each point the mean kinetic energy of the vibratory motion is a measure of the intensity of light, that the more or less complete attack of the gelatine coating on a photographic plate indicates the various degrees of this intensity. By joining these propositions, and many others that would take too long to enumerate, to Neumann's proposition, Wiener was able to formulate a forecast and establish that the experiment belied it. If he attributed this solely to Neumann's proposition, if it alone bears the responsibility for the error this negative result has put in evidence, then Wiener was taking all the other propositions he invoked as beyond doubt. But this assurance is not imposed as a matter of logical necessity; nothing stops us from taking Neumann's proposition as accurate and shifting the weight of the experimental contradiction to some other proposition of the commonly accepted optics; as H. Poincaré has shown, we can very easily rescue Neumann's hypothesis from the grip of Wiener's experiment on the condition that we abandon in exchange the hypothesis which takes the mean kinetic energy as the measure of the light intensity; we may, without being contradicted by the experiment, let the vibration be parallel to the plane of polarization, provided that we measure the light intensity by the mean potential energy of the medium deforming the vibratory motion.

These principles are so important that it will be useful to apply them to another example; again we choose an experiment regarded as one of the most decisive ones in optics.

We know that Newton conceived the emission theory for optical phenomena. The emission theory supposes light to be formed of extremely thin projectiles, thrown out with very great speed by the sun and other sources of light; these projectiles penetrate all transparent bodies; on account of the various parts of the media through which they move, they undergo attractions and repulsions; when the distance separating the acting particles is very small these actions are very powerful, and they vanish when the masses between which they act are appreciably far from each other. These essential hypotheses joined to several others, which we pass over without mention, lead to the formulation of a complete theory of reflection and refraction of light; in particular, they imply the following proposition: The index of refraction of light passing from one medium into another is equal to the velocity of the light projectile within the medium it penetrates, divided by the velocity of the same projectile in the medium it leaves behind.

This is the proposition that Arago chose in order to show that the theory of emission is in contradiction with the facts. From this proposition a second follows: Light travels faster in water than in air. Now Arago had indicated an appropriate procedure for comparing the velocity of light in air with the velocity of light in water; the procedure, it is true, was inapplicable, but Foucault modified the experiment in such a way that it could be carried out; he found that the light was propagated less rapidly in water than in air. We may conclude from this, with Foucault, that the system of emission is incompatible with the facts.

I say the *system* of emission and not the *hypothesis* of emission; in fact, what the experiment declares stained with error is the whole group of propositions accepted by Newton, and after him by Laplace and Biot, that is, the whole theory from which we deduce the relation between the index of refraction and the velocity of light in various media. But in condemning this system as a whole by declaring it stained with error, the experiment does not tell us where the error lies. Is it in the fundamental hypothesis that light consists in projectiles thrown out with great speed by luminous bodies? Is it in some other assumption concerning the actions experienced by light corpuscles due to the media through which they move? We know nothing about that. It would be rash to believe, as Arago seems to have thought, that Foucault's experiment condemns once and for all the very hypothesis of emission, i.e., the assimilation of a ray of light to a swarm of projectiles. If physicists had attached some value to this task, they would undoubtedly have succeeded in founding on this assumption a system of optics that would agree with Foucault's experiment.

In sum, the physicist can never subject an isolated hypothesis to experimental test, but only a whole group of hypotheses; when the experiment is in disagreement with his predictions, what he learns is that at least one of the hypotheses constituting this group is unacceptable and ought to be modified; but the experiment does not designate which one should be changed.

We have gone a long way from the conception of the experimental method arbitrarily held by persons unfamiliar with its actual functioning. People generally think that each one of the hypotheses employed in physics can be taken in isolation, checked by experiment, and then, when many varied tests have established its validity, given a definitive place in the system of physics. In reality, this is not the case. Physics is not a machine which lets itself be taken apart; we cannot try each piece in isolation and, in order to adjust it, wait until its solidity has been carefully checked. Physical science is a system that must be taken as a whole; it is an organism in which one part cannot be made to function except when the parts that are most remote from it are called into play, some more so than others, but all to some degree. If something goes wrong, if some discomfort is felt in the functioning of the organism, the physicist will have to ferret out through its effect on the entire system which organ needs to be remedied or modified without the possibility of isolating this organ and examining it apart. The watchmaker to whom you give a watch that has stopped separates all the wheelworks and examines them one by one until he finds the part that is defective or broken. The doctor to whom a patient appears cannot dissect him in order to establish his diagnosis; he has to guess the seat and cause of the ailment solely by inspecting disorders affecting the whole body. Now, the physicist concerned with remedying a limping theory resembles the doctor and not the watchmaker.

A "CRUCIAL EXPERIMENT" IS IMPOSSIBLE IN PHYSICS

Let us press this point further, for we are touching on one of the essential features of experimental method, as it is employed in physics.

Reduction to absurdity seems to be merely a means of refutation, but it may become a method of demonstration: in order to demonstrate the truth of a proposition it suffices to corner anyone who would admit the contradictory of the given proposition into admitting an absurd consequence. We know to what extent the Greek geometers drew heavily on this mode of demonstration.

Those who assimilate experimental contradiction to reduction to absurdity imagine that in physics we may use a line of argument similar to the one Euclid employed so frequently in geometry. Do you wish to obtain from a group of phenomena a theoretically certain and indisputable explanation? Enumerate all the hypotheses that can be made to account for this group of phenomena; then, by experimental contradiction eliminate all except one; the latter will no longer be a hypothesis, but will become a certainty.

Suppose, for instance, we are confronted with only two hypotheses. Seek experimental conditions such that one of the hypotheses forecasts the production of one phenomenon and the other the production of quite a different effect; bring these conditions into existence and observe what happens; depending on whether you observe the first or the second of the predicted phenomena, you will condemn the second or the first hypothesis; the hypothesis not condemned will be henceforth indisputable; debate will be cut off, and a new truth will be acquired by science. Such is the experimental test that the author of the *Novum Organum* called the "*fact of the cross,* borrowing the expression from the crosses which at an intersection indicate the various roads."

We are confronted with two hypotheses concerning the nature of light; for Newton, Laplace, or Biot light consisted of projectiles hurled with extreme speed, but for Huygens, Young, or Fresnel light consisted of vibrations whose waves are propagated within an ether. These are the only two possible hypotheses as far as once can see: either the motion is carried away by the body it excites and remains attached to it, or else it passes from one body to another. Let us pursue the first hypothesis; it declares that light travels more quickly in water than in air; but if we follow the second, it declares that light travels more quickly in air than in water. Let us set up Foucault's apparatus; we set into motion the turning mirror; we see two luminous spots formed before us, one colorless, the other greenish. If the greenish band is to the left of the colorless one, it means that light travels faster in water than in air, and that the hypothesis of vibrating waves is false. If, on the contrary, the greenish band is to the right of the colorless one, that means that light travels faster in air than in water, and that the hypothesis of emissions is condemned. We look through the magnifying glass used to examine the two luminous spots, and we notice that the greenish spot is to the right of the colorless one; the debate is over; light is not a body, but a vibratory wave motion propagated by the ether; the emission hypothesis has had its day; the wave hypothesis has been put beyond doubt, and the crucial experiment has made it a new article of the scientific credo.

What we have said in the foregoing paragraph shows how mistaken we should be to attribute to Foucault's experiment so simple a meaning and so decisive an importance; for it is not between two hypotheses, the emission and wave hypotheses, that Foucault's experiment judges trenchantly; it decides rather between two sets of theories each of which has to be taken as a whole, i.e., between two entire systems, Newton's optics and Huygens' optics.

But let us admit for a moment that in each of these systems everything is compelled to be necessary by strict logic, except a single hypothesis; consequently, let us admit that the facts, in condemning one of the two systems, condemn once and for all the single doubtful assumption it contains. Does it follow that we can find in the "crucial experiment" and irrefutable procedure for transforming one of the two hypotheses before us into a demonstrated truth? Between two contradictory theorems of geometry there is no room for a third judgment; if one is false, the other is necessarily true. Do two hypotheses in physics ever constitute such a strict dilemma? Shall we ever dare to assert that no other hypothesis is imaginable? Light may be a swarm of projectiles, or it may be a vibratory motion whose waves are propagated in a medium; is it forbidden to

be anything else at all? Arago undoubtedly thought so when he formulated this incisive alternative: Does light move more quickly in water than in air? "Light is a body. If the contrary is the case, then light is a wave." But it would be difficult for us to take such a decisive stand; Maxwell, in fact, showed that we might just as well attribute light to a periodical electrical disturbance that is propagated within a dielectric medium.

Unlike the reduction to absurdity employed by geometers, experimental contradiction does not have the power to transform a physical hypothesis into an indisputable truth; in order to confer this power on it, it would be necessary to enumerate completely the various hypotheses which may cover a determinate group of phenomena; but the physicist is never sure he has exhausted all the imaginable assumptions. The truth of a physical theory is not decided by heads or tails.

CV's can't be eliminated

11

PETER LIPTON

Contrastive Inference

Inference to the Best Contrastive Explanation . . . is better than simple hypothetico-deductivism. It marks an improvement both where the deductive model is too strict, neglecting evidential relevance in cases where there is no appropriate deductive connection between hypothesis and data, and where it is too lenient, registering support where there is none to be had. Inference to the Best Explanation does better in the first case because, as

the analysis of contrastive explanation shows, explanatory causes need not be sufficient for their effects, so the fact that a hypothesis would explain a contrast may provide some reason to believe the hypothesis, even though the hypothesis does not entail the data. It does better in the second case because, while some contrapositive instances do support a hypothesis, not all do, and the requirement of shared antecedents helps to determine which do and which do not. The structural similarity between the Method of Difference and contrastive explanation that I will exploit in these chapters will also eventually raise the question of why Inference

---

Inference to the Best Explanation (London: Routledge, 1993), pp. 78–98. Reprinted by permission of the publisher.

to the Best Explanation is an improvement on Mill's methods. . . .

To develop these arguments and, more generally, to show just how inferences to contrastive explanations work, it is useful to consider a simple but actual scientific example in some detail. The example I have chosen is Ignaz Semmelweis's research from 1844 to 1848 on childbed fever, taken from Hempel's well-known and characteristically clear discussion. Semmelweis wanted to find the cause of this often fatal disease, which was contracted by many of the women who gave birth in the Viennese hospital in which he did his research. His central datum was that a much higher percentage of the women in the First Maternity Division of the hospital contracted the disease than in the adjacent Second Division, and he sought to explain this difference. The hypotheses he considered fell into three types. In the first were hypotheses that did not mark differences between the divisions, and so were rejected. Thus, the theory of 'epidemic influences' descending over entire districts did not explain why more women should die in one division than another; nor did it explain why the mortality among Viennese women who gave birth at home or on the way to the hospital was lower than in the First Division. Similarly, the hypotheses that the fever was caused by overcrowding, by diet, or by general care were rejected because these factors did not mark a difference between the divisions.

One striking difference between the two divisions was that medical students only used the First Division for their obstetrical training, while midwives received their training in the Second Division. This suggested the hypothesis that the high rate of fever in the First Division was caused by injuries due to rough examination by the medical students. Semmelweis rejected the rough examination hypothesis on the grounds that midwives performed their examinations in more or less the same way, and that the injuries due to childbirth were in any case greater than those due to rough examination.

The second type of hypotheses included those that did mark a difference between the divisions, but where eliminating the difference in putative cause did not affect the difference in mortality. A priest delivering the last sacrament to a dying woman had to pass through the First Division to get to the sickroom where dying women were kept, but not through the Second Division. This suggested that the psychological influence of seeing the priest might explain the difference, but Semmelweis ruled this out by arranging for the priest not to be seen by the women in the First Division either and finding that this did not affect the mortality rates. Again, women in the First Division were delivered lying on their backs, while women in the Second delivered on their sides but, when Semmelweis arranged for all women to deliver on their sides, the mortality remained the same.

The last type of hypothesis that Semmelweis considered is one that marked a difference between the divisions, and where eliminating this difference also eliminated the difference in mortality. Kolletschka, one of Semmelweis's colleagues, received a puncture wound in his finger during an autopsy, and died from an illness with symptoms like those of childbed fever. This led Semmelweis to infer that Kolletschka's death was due to the 'cadaveric matter' that the wound introduced into his bloodstream, and Semmelweis then hypothesized that the same explanation might account for the deaths in the First Division, since medical students performed their examinations directly after performing autopsies, and midwives did not perform autopsies at all. Similarly, the cadaveric hypothesis would explain why women who delivered outside the hospital had a lower mortality from childbed fever, since they were not examined. Semmelweis had the medical students disinfect their hands before examination, and the mortality rate in the First Division went down to the same low level as that in the Second Division. Here at last was a difference that made a difference, and Semmelweis inferred the cadaveric hypothesis.

Hempel's case study is a gold mine for inferences to the best contrastive explanation. Let us begin by considering Semmelweis's strategy for each of the three groups of hypotheses, those of no difference, of irrelevant differences, and of relevant differences. His rejection of the hypotheses in the first group—epidemic influences, overcrowding, general care, diet, and rough examination—show how Inference to the Best Explanation can account for negative evidence. These hypotheses are rejected on the grounds that, though they are compatible with the evidence, they would not explain the contrast between the divisions. Epidemic influences, for example, still might possibly be part of the causal history of the deaths in the First Division, say because the presence of these influences is a necessary condition for any case of childbed fever. And nobody who endorsed the epidemic hypothesis would have claimed that the influences were sufficient for the fever, since it was common knowledge that not all mothers in the district contracted childbed fever. Still, Semmelweis took the fact that the hypotheses in the first group would not explain the contrast between the divisions or the contrast between the First Division and mothers who gave birth outside the hospital to be evidence against them.

Semmelweis also used a complementary technique for discrediting the explanations in the first group that is naturally described in terms of Inference to the Best Explanation, when he argued against the epidemic hypothesis on the grounds that the mortality rate for births outside the hospital was lower than in the First Division. What he had done was to change the foil, and point out that the hypothesis also fails to explain this new contrast. It explains neither why mothers get fever in the First Division rather than in the Second, nor why mothers get fever in the First Division rather than outside the hospital. Similarly, when Semmelweis argued against the rough examination hypothesis on the grounds that childbirth is rougher on the mother than any examination, he pointed out not only that it fails to explain why there is fever in the First Division rather than in the Second, but also why there is fever in the First Division rather than among other mothers generally. New foils provide new evidence, in these cases additional evidence against the putative explanations.

The mere fact that the hypotheses in the first group did not explain some evidence cannot, however, account for Semmelweis's negative judgment. No hypothesis explains every observation, and most evidence that is not explained by a hypothesis is simply irrelevant to it. But Semmelweis's observation that the hypotheses do not explain the contrast in mortality between the divisions seems to count against those hypotheses in a way that, say, the observation that those hypotheses would not explain why the women in the First Division were wealthier than those in the Second Division (if they were)

would not. Of course, since Semmelweis was interested in reducing the incidence of childbed fever, he was naturally more interested in an explanation of the contrast in mortality than in an explanation of the contrast in wealth, but this does not show why the failure of the hypotheses to explain the first contrast counts against them. This poses a general puzzle for Inference to the Best Explanation: how can that account distinguish negative evidence from irrelevant evidence, when the evidence is logically consistent with the hypothesis?

One straightforward mechanism is rival support. In some cases, evidence counts against one hypothesis by improving the explanatory power of a competitor. The fact that the mortality in the First Division went down when the medical students disinfected their hands before examination supports the cadaveric matter hypothesis, and so indirectly counts against all the hypotheses inconsistent with it that cannot explain this contrast. But this mechanism of disconfirming an explanation by supporting a rival does not seem to account for Semmelweis's rejection of the hypotheses in the first group, since at that stage of his inquiry he had not yet produced an alternative account.

Part of the answer to this puzzle about the difference in the epistemic relevance of a contrast in mortality and a contrast in wealth is that the rejected hypotheses would have enjoyed some support from the fact of mortality but not from the fact of wealth. The epidemic hypothesis, for example, was not Semmelweis's invention, but a popular explanation at the time of his research. Its acceptance presumably depended on the fact that it seemed to provide an explanation, if a weak one, for the non-contrastive observations of the occurrence of childbed fever. In the absence of a stronger and competing explanation, this support might have seemed good enough to justify the inference. But by pointing out that the hypothesis does not explain the contrast between the divisions, Semmelweis undermines this support. On the other hand, the epidemic hypothesis never explained, and so was never supported by observations about, the wealth of the victims of childbed fever, so its failure to explain why the women in the First Division were wealthier than those in the Second Division would not take away any support it had hitherto enjoyed.

On this view, the observation that the hypotheses in the first group do not explain the contrast in mortality and the observation that they do not explain the contrast in wealth are alike in that they both show that these data do not support the hypothesis. The difference in impact only appears when we take into account that only evidence about mortality had been supposed to support the hypothesis, so only in this case is there a net loss of support. This view seems to me to be correct as far as it goes, but it leaves a difficult question. Why, exactly, does the failure to explain the contrast in mortality undermine prior support for hypotheses in the first group? Those hypotheses would still give some sort of explanation for the cases of the fever in the hospital, even if they would not explain the contrast between the divisions. Consider a different example. Suppose that we had two wards of patients who suffered from syphilis and discovered that many more of them in one ward contracted paresis than in the other. The hypothesis that syphilis is a necessary cause of paresis would not explain this contrast, but this would not, I think, lead us to abandon the hypothesis on the grounds that its support had been undermined. Instead, we would continue to accept it and look for some further and complementary explanation for the difference between the wards, say in terms of a difference in the treatments provided. Why, then, is Semmelweis's case any different?

The difference must lie in the relative weakness of the initial evidence in support of the hypotheses in the first group. If the only evidence in favor of the epidemic hypothesis is the presence of childbed fever, the contrast in mortality does undermine the hypothesis, because it suggests that the correct explanation of the contrast will show that epidemic influences have nothing to do with fever. If, on the other hand, the epidemic hypothesis would also explain why there were outbreaks of fever at some times rather than others, or in some hospitals rather than others, even though these cases seemed similar in all other plausibly relevant respects, then we would be inclined to hold on to that hypothesis and look for a complementary explanation of the contrast between the divisions. In the case of syphilis and paresis, we presumably have extensive evidence that there are no known cases of paresis not preceded by syphilis. The syphilis hypothesis not only would explain why those with paresis have it, but

also the many contrasts between people with paresis and those without it. This leads us to say that the correct explanation of the contrast between the wards is more likely to complement the syphilis hypothesis than to replace it.

If this account is along the right lines, then the strength of the disconfirmation provided by the failure to explain a contrast depends on how likely it seems that the correct explanation of the contrast will pre-empt the original hypothesis. This explains our different reaction to the wealth case. We may have no idea why the women in the First Division are wealthier than those in the Second, but it seems most unlikely that the reason for this will pre-empt the hypotheses of the first group. When we judge that pre-emption is likely, we are in effect betting that the best explanation of the contrast will either contradict the original hypothesis or show it to be unnecessary, and so that the evidence that originally supported it will instead support a competitor. So the mechanism here turns out to be an attenuated version of disconfirmation by rival support after all. The inability of the hypotheses in the first group to explain the contrast between the divisions and the contrast between the First Division and births outside the hospital disconfirms those hypotheses because, although the contrastive data do not yet support a competing explanation, since none has yet been formulated, Semmelweis judged that the best explanation of those contrasts would turn out to be a competing rather than a complementary account. This judgment can itself be construed as an overarching inference to the best explanation. If we reject the hypotheses in the first group because they fail to explain the contrasts, this is because we regard the conjecture that the hypotheses are wrong to be a better explanation of the failures than that they are merely incomplete. Judgments of this sort are speculative, and we may in the end find ourselves inferring an explanation of the contrasts that is compatible with the hypotheses in the first group, but insofar as we do take their explanatory failures to count against them, I think it must be because we do make these judgments.

On this view, given a hypothesis about the etiology of a fact, and faced with the failure of that hypothesis to explain a contrast between that fact and a similar foil, the scientist must choose between the overarching explanations that the failure is due to incompleteness and that it is due to incorrectness. Semmelweis's rejections of the hypotheses in the first group are examples of choosing the incorrectness explanation. It is further corroboration of the claim that these choices must be made that we cannot make sense of Semmelweis's research without supposing that he also sometimes inferred incompleteness. For while the cadaveric hypothesis had conspicuous success in explaining the contrast between the divisions, it failed to explain other contrasts that formed part of Semmelweis's evidence. For example, it did not explain why some women in the Second Division contracted childbed fever while others in that division did not, since none of the midwives who performed the deliveries in that division performed autopsies. Similarly, the cadaveric hypothesis did not explain why some women who had street births on their way to the hospital contracted the fever, since those women were rarely examined by either medics or midwives after they arrived. Consequently, if we take it that Semmelweis nevertheless had good reason to believe that infection by cadaveric matter was a cause of childbed fever, it can only be because he reasonably inferred that the best explanation of these explanatory failures was that the cadaveric hypothesis was incomplete, not the only cause of the fever, rather than that it was incorrect. These cases also show that we cannot in general avoid the speculative judgment by waiting until we actually produce an explanation for all the known relevant contrasts, since in many cases this would postpone inference indefinitely.

Let us turn now to the two hypotheses of the second group, concerning the priest and delivery position. Unlike the hypotheses of the first group, these did mark differences between the divisions and so might explain the contrast in mortality. The priest bearing the last sacrament only passed through the First Division, and only in that division did mothers deliver on their backs. Since these factors were under Semmelweis's control, he tested these hypotheses in the obvious way, by seeing whether the contrast in mortality between the divisions remained when these differences were eliminated. Since that contrast remained, even when the priest was removed from the scene and when the mothers in both divisions were delivered on their sides, these hypotheses could no longer be held to explain the original contrast.

This technique of testing a putative cause by seeing whether the effect remains when it is removed is widely employed. Semmelweis could have used it even without the contrast between the divisions, and it is worth seeing how a contrastive analysis could account for this. Suppose that all the mothers in the hospital were delivered on their backs, and Semmelweis tested the hypothesis that this delivery position is a cause of childbed fever by switching positions. He might have done this for only some of the women, using the remainder as a control. In this case, the two groups would have provided a potential contrast. If a smaller percentage of the women who were delivered on their sides contracted childbed fever, the delivery hypothesis would have explained and so been supported by this contrast. And even if Semmelweis had switched all the mothers, he would have had a potential diachronic (before and after) contrast, by comparing the incidence of fever before and after the switch. In either case, a contrast would have supported the explanatory inference. In fact, however, these procedures would not have produced a contrast, since delivery position is irrelevant to childbed fever. This absence of contrast would not disprove the delivery hypothesis. Delivering on the back might still be a cause of fever, though there might be some obscure alternate cause that came into play when the delivery position was switched. But the absence of the contrast certainly would disconfirm the delivery hypothesis. The reason for this is the same as in the case of the epidemic hypothesis: the likeliness of a better, pre-emptive explanation. Even if Semmelweis did not have an alternative explanation for the cases of fever, there must be another explanation in the cases of side delivery, and it is likely that this explanation will show that back delivery is irrelevant even when it occurs. As in the case of the hypotheses in the first group, when we take an explanatory failure to count against a hypothesis, even when we do not have an alternative explanation, this is because we infer that the falsity of the hypothesis is a better explanation for its explanatory failure than its incompleteness.

This leaves us with Semmelweis's final hypothesis, that the difference in mortality is explained by the cadaveric matter that the medical students introduced into the First Division. Here too we have an overarching explanation in play. Semmelweis had already conjectured that the difference in mortality was somehow explained by the fact that mothers were attended by medical students in the First Division, and by midwives in the Second Division. This had initially suggested the hypothesis that the rough examinations given by the medical students was the cause, but this neither explained the contrast between the divisions nor the contrast between the mothers in the First Division and mothers generally, who suffered more from labor and childbirth than from any examination. The cadaveric hypothesis was another attempt to explain the difference between the divisions under the overarching hypothesis that the contrast is due to the difference between medical students and midwives. In addition to explaining the difference between divisions, this hypothesis would explain Kolletschka's illness, as well as the difference between the First Division and births outside the hospital.

Finally, Semmelweis tested this explanation by eliminating the cadaveric matter with disinfectant and finding that this eliminated the difference in the mortality between the divisions. This too can be seen as the inference to a contrastive explanation for a new contrast, where now the difference that is explained is not the simple difference in mortality between the divisions, but the diachronic contrast between the initial presence of that difference and its subsequent absence. The best explanation of the fact that removing the cadaveric matter was followed by the elimination of the difference in mortality was that the cadaveric matter was responsible for that difference. By construing Semmelweis's evidence as a diachronic contrast, we bring out the important point that the comparative data have a special probative force that we would miss if we simply treated them as two separate confirmations of Semmelweis's hypothesis.

Semmelweis's research into the causes of childbed fever brings out many of the virtues of Inference to the Best Explanation when that account is tied to a model of contrastive explanation. In particular, it shows how explanatory considerations focus and direct inquiry. Semmelweis's work shows how the strategy of considering potential contrastive explanations focuses inquiry, even when the ultimate goal is not simply an explanation. Semmelweis's primary interest was to eliminate or at least reduce the cases of childbed fever, but he nevertheless posed an explanatory question: Why do women contract childbed fever? His initial ignorance was such, however, that simply asking why those with the fever have it did not generate a useful set of hypotheses. So he focused his inquiry by asking contrastive why-questions. His choice of the Second Division as foil was natural because it provided a case where the effect is absent yet the causal histories are very similar. By asking why the contrast obtains, Semmelweis focused his search for explanatory hypotheses on the remaining differences. This strategy is widely applicable. If we want to find out why some phenomenon occurs, the class of possible causes is often too big for the process of Inference to the Best Explanation to get a handle on. If, however, we are lucky or clever enough to find or produce a contrast where fact and foil have similar histories, most potential explanations are immediately 'cancelled out' and we have a manageable and directed research program. The contrast will be particularly useful if, as in Semmelweis's case, in addition to meeting the requirement of shared history, it is also a contrast that various available hypotheses will not explain. Usually, this will still leave more than one hypothesis in the field, but then further observation and experiment may produce new contrasts that leave only one explanation. This shows how the interest relativity of explanation is at the service of inference. By tailoring his explanatory interests (and his observational and experimental procedures) to contrasts that would help to discriminate between competing hypotheses, Semmelweis was able to judge which hypothesis would provide the best overall explanation of the wide variety of contrasts (and absences of contrast) he observed, and so to judge which hypothesis he ought to infer. Semmelweis's inferential interests determined his explanatory interests, and the best explanation then determined his inference.

EXPLANATION AND DEDUCTION

Semmelweis's research is a simple and striking illustration of inferences to the best explanation in action, and of the way they exploit contrastive data. It is also

Hempel's paradigm of the hypothetico-deductive method. So this case is particularly well suited for a comparison of the virtues of Inference to the Best Explanation and the deductive model. The example, I suggest, shows that Inference to the Best Explanation is better than hypothetico-deductivism.

Consider first the context of discovery. Semmelweis's use of contrasts and prior differences to help generate a list of candidate hypotheses illustrates one of the ways Inference to the Best Explanation elucidates the context of discovery, a central feature of our inductive practice neglected by the hypothetico-deductive model. The main reason for this neglect is easy to see. Hypothetico-deductivists emphasize the hopelessness of 'narrow inductivism', the view that scientists ought to proceed by first gathering all the relevant data without theoretical preconception and then using some inductive algorithm to infer from those data to the hypothesis they best support. Scientists never have all the relevant data, they often cannot tell whether or not a datum is relevant without theoretical guidance, and there is no general algorithm that could take them from data to a hypothesis that refers to entities and processes not mentioned in the data. The hypothetico-deductive alternative is that, while scientists never have all the data, they can at least determine relevance if the hypothesis comes first. Given a conjectural hypothesis, they know to look for data that either can be deduced from it or would contradict it. The cost of this account is that we are left in the dark about the source of the hypotheses themselves. According to Hempel, scientists need to be familiar with the current state of research, and the hypotheses they generate should be consistent with the available evidence but, in the end, generating good hypotheses is a matter of 'happy guesses'.

The hypothetico-deductivist must be right in claiming that there are no universally shared mechanical rules that generate a unique hypothesis from any given pool of data since, among other things, different scientists generate different hypotheses, even when they are working with the same data. Nevertheless, this 'narrow hypothetico-deductivist' conception of inquiry badly distorts the process of scientific invention. Most hypotheses consistent with the data are non-starters, and the use of contrastive evidence and explanatory inference is one way the field is narrowed. In advance of an explanation for some effect, we know to look for a foil with a similar history. If we find one, this sharply constrains the class of hypotheses that are worth testing. A reasonable conjecture must provide a potential explanation of the contrast, and most hypotheses that are consistent with the data will not provide this. (For hypotheses that traffic in unobservables, the restriction to potential contrastive explanations still leaves a lot of play. . . .)

The slogan 'Inference to the Best Explanation' may itself bring to mind an excessively passive picture of scientific inquiry, suggesting perhaps that we simply infer whatever seems the best explanation of the data we happen to have. But the Semmelweis example shows that the account, properly construed, allows for the feedback between the processes of hypothesis formation and data acquisition that characterizes actual inquiry. Contrastive data suggest explanatory hypotheses, and these hypotheses in turn suggest manipulations and controlled experiments that may reveal new contrasts that help to determine which of the candidates is the best explanation. This is one of the reasons the subjunctive element in Inference to the Best Explanation is important. By considering what sort of explanation the hypothesis would provide, if it were true, we assess not only how good an explanation it would be, but also what as yet unobserved contrasts it would explain, and this directs future observation and experiment. Semmelweis's research also shows that Inference to the Best Explanation is well suited to describe the role of overarching hypotheses in directing inquiry. Semmelweis's path to his cadaveric hypothesis is guided by his prior conjecture that the contrast in mortalities between the divisions is somehow due to the fact that deliveries are performed by medical students in the First Division, but by midwives in the Second Division. He then searches for ways of fleshing out this explanation and for the data that would test various proposals. Again, I have suggested that we can understand Semmelweis's rejection of the priest and the birth position hypotheses in terms of an inference to a negative explanation. The best explanation for the observed fact that eliminating these differences between the divisions did not affect mortality is that the mortality had a different cause. In both cases, the intermediate explanations focus research, either by marking the causal region within which the final explanation is likely to be found, or by showing that a certain region is unlikely to include the cause Semmelweis is looking for.

The hypothetico-deductive model emphasizes the priority of theory as a guide to observation and experiment, at the cost of neglecting the sources of theory. I want now to argue that the model also fails to give a good account of the way scientists decide which observations and experiments are worth making. According to the deductive model, scientists should check the observable consequences of their theoretical conjectures, or of their theoretical systems, consisting of the conjunction of theories and suitable auxiliary statements. As we will see below, this account is too restrictive, since there are relevant data not entailed by the theoretical system. It is also too permissive, since most consequences are not worth checking. Any hypothesis entails the disjunction of itself and any observational claim whatever, but establishing the truth of such a disjunction by checking the observable disjunct rarely bears on the truth of the hypothesis. The contrastive account of Inference to the Best Explanation is more informative, since it suggests Semmelweis's strategy of looking for observable contrasts that distinguish one causal hypothesis from competing explanations.

Even if we take both Semmelweis's hypotheses and his data as given, the hypothetico-deductive model gives a relatively poor account of their relevance to each other. This is particularly clear in the case of negative evidence. According to the deductive model, evidence disconfirms a hypothesis just in case the evidence either contradicts the hypothesis or contradicts the conjunction of the hypothesis and suitable auxiliary statements. None of the hypotheses Semmelweis rejects contradicts his data outright. For example, the epidemic hypothesis does not contradict the observed contrast in mortality between the divisions. Proponents of the epidemic hypothesis would have acknowledged that, like any other epidemic, not everyone who is exposed to the influence succumbs to the fever. They realized that not all mothers contract childbed fever, but rightly held that this did not refute their hypothesis, which was that the epidemic influence was a cause of the fever

in those mothers that did contract it. So the hypothesis does not entail that the mortality in the two divisions is the same. Similarly, the delivery position hypothesis does not entail that the mortality in the two divisions is different when the birth positions are different; nor does it entail that the mortality will be the same when the positions are the same. Even if back delivery is a cause of childbed fever, the mortality in the Second Division could have been as high as in the First, because the fever might have had other causes there. Similarly, the possibility of additional causes shows that back delivery could be a cause of fever even though the mortality in the First Division is lower than in the Second Division when all the mothers deliver on their sides. The situation is the same for all the other hypotheses Semmelweis rejects.

What does Hempel say about this? He finds a logical conflict, but in the wrong place. According to him, the hypotheses that appealed to overcrowding, diet, or general care were rejected because the claims that the difference in mortality between the divisions was due to such differences 'conflict with readily observable facts'. The claim that, for example, the difference in mortality is due to a difference in diet is incompatible with the observation that there is no difference in diet. These are clearly cases of logical incompatibility, but they are not the ones Hempel needs: the claims that are incompatible with observation are not the general hypotheses Semmelweis rejects. Like the cadaveric hypothesis he eventually accepts, the hypotheses of overcrowding, diet, and care are surely general conjectures about causes of childbed fever, not specific claims about the differences between the divisions. But the hypotheses that overcrowding, diet, or general care is a cause of childbed fever are logically compatible with everything Semmelweis observes.

The hypothetico-deductivist must claim that hypotheses are rejected because, although they are compatible with the data, each of them, when conjoined with suitable auxiliary statements, is not. But what could such statements be? Each hypothesis must have a set of auxiliaries that allows the deduction that the mortality in the divisions is the same, and this contradicts the data. The auxiliaries need not be known to be true, but they need to be specified. This, however, cannot be done. The proponent of the epidemic hypothesis, for example, does not know what additional factors determine just who succumbs to the influence, so he cannot say how the divisions must be similar in order for it to follow that the mortality should be the same. Similarly, Semmelweis knew from the beginning that back delivery cannot be necessary for childbed fever, since there are cases of fever in the Second Division where all the women are delivered on their sides, but he cannot specify what all the other relevant factors ought to be. The best the hypothetico-deductivist can do, then, is to rely on *ceteris paribus* auxiliaries. If fever is caused by epidemic influence, or by back delivery, and everything else 'is equal', the mortality in the divisions ought to be the same. This, however, does not provide a useful analysis of the situation. Any proponent of the rejected hypotheses will reasonably claim that precisely what the contrast between the divisions shows is that not everything is equal. This shows that there is more to be said about the etiology of childbed fever, but it does not show why we should reject any of the hypotheses that Semmelweis does reject. Semmelweis's observations show that none of these hypotheses would explain the contrasts, but they do not show that the hypotheses are false, on hypothetico-deductive grounds.

There is another objection to the *ceteris paribus* approach, and indeed to any other scheme that would generate the auxiliaries the deductive model requires to account for negative evidence. It would disprove too much. Recall that the cadaveric hypothesis does not itself explain all the relevant contrasts, such as why some women in the Second Division contracted childbed fever while others in that division did not, or why some women who had 'street births' on their way to the hospital contracted the fever while others did not. If the other hypotheses were rejected because they, along with *ceteris paribus* clauses, entailed that there ought to be no difference in mortality between the divisions, then the model does not help us to understand why similar clauses did not lead Semmelweis to reject the cadaveric hypothesis as well.

From the point of view of Inference to the Best Explanation, we can see that there are several general and related reasons why the hypothetico-deductive model does not give a good description of the way causal hypotheses are disconfirmed by con-

trastive data. The most important is that the model does not account for the negative impact of explanatory failure. Semmelweis rejected hypotheses because they failed to explain contrasts, not because they were logically incompatible with them. Even on a deductive-nomological account of explanation, the failure to explain is not tantamount to a contradiction. In order to register the negative impact of these failures, the hypothetico-deductive model must place them on the Procrustean bed of logical incompatibility, which requires auxiliary statements that are not used by scientists and not usually available even if they were wanted.

Second, the hypothetico-deductive model misconstrues the nature of explanatory failure, in the case of contrastive explanations. . . . [T]o explain a contrast is not to deduce the conjunction of the fact and the negation of the foil, but to find some causal difference. The hypotheses Semmelweis rejects do not fail to explain because they do not entail the contrast between the divisions: the cadaveric hypothesis does not entail this either. They fail because they do not mark a difference between the divisions, either initially or after manipulation.

Third, the model does not reflect the need, in the case of explanatory failure, to judge whether this is due to incompleteness or error. In the model, this decision becomes one of whether we should reject the hypothesis or the auxiliaries in a case where their conjunction contradicts the evidence. This, however, is not the decision Semmelweis had to make. When he had all the mothers in both divisions deliver on their sides, and found that this did not affect the contrast in mortality, he did not have to choose between saying that the hypothesis that delivery position is a cause of fever is false and saying that the claim that everything else was equal is false. After his experiment, he knew that not everything else was equal, but this left him with the question of whether he ought to reject the delivery hypothesis or just judge it to be incomplete.

The failures of the hypothetico-deductive model to capture the force of disconfirmation through explanatory failure also clearly count against Karl Popper's account of theory testing through falsification. Although he is wrong to suppose that we can give an adequate account of science without relying on some notion of positive inductive support,

Popper is right to suppose that much scientific research consists in the attempt to select from among competing conjectures by disconfirming all but one of them. Popper's mistake here is to hold that disconfirmation works exclusively through refutation. As the Semmelweis example shows, scientists also reject theories as false because, while they are not refuted by the evidence, they fail to explain the salient contrasts. Moreover, if my account of the way this sort of negative evidence operates is along the right lines, this is a form of disconfirmation that Popper's account cannot be modified to capture without abandoning his central proscription on positive support, since it requires that we make a positive judgment about whether the explanatory failure is more likely to be due to incompleteness or error, a judgment that depends on inductive considerations.

The hypothetico-deductive model appears to do a better job of accounting for Semmelweis's main positive argument for his cadaveric hypothesis, that disinfection eliminated the contrast in mortality between the divisions. Suppose we take it that the cadaveric hypothesis says that infection with cadaveric matter is a necessary cause of childbed fever, that everyone who contracts the fever was so infected. In this case, the hypothesis entails that where there is no infection, there is no fever. This, along with plausible auxiliaries about the influence of the disinfectant, entails that there should be no fever in the First Division after disinfection. But this analysis does not do justice to the experiment, for three reasons. First of all, the claim that cadaveric infection is strictly necessary for fever, which is needed for the deduction, is not strictly a tenable form of the cadaveric hypothesis, since Semmelweis knew of some cases of fever, such as those in the Second Division and those among street births, where there was no cadaveric infection. Similarly, given that disinfection is completely effective, this version of the hypothesis entails that there should be no cases of fever in the First Division after disinfection, which is not what Semmelweis observed. What he found was rather that the mortality in the First Division went down to the same low level (just over one percent) as in the Second Division. As Hempel himself observes, Semmelweis eventually went on to 'broaden' his hypothesis, by allowing that childbed fever could also

be caused by 'putrid matter derived from living organisms'. But if this is to count as broadening the hypothesis, rather than rejecting it, the original cadaveric hypothesis cannot have been that cadaveric infection is a necessary cause of the fever.

The second reason the deductive analysis of the disinfection experiment does not do justice to it is that the analysis does not bring out the special probative force of the contrastive experiment. Even if we suppose that cadaveric infection is necessary for fever, the hypothesis does not entail the *change* in mortality, but only that there should be no fever where there is disinfection, since it does not entail that there should be fever where there is no disinfection. But it is precisely this contrast that makes the experiment persuasive. What the hypothetico-deductivist could say here, I suppose, is that the change is entailed if we use the observed prior mortality as a premise in the argument. If cadaveric infection is necessary for fever, and if there was fever and infection but then the infection is removed, it follows that the fever will disappear as well. Even this, however, leaves out an essential feature of the experiment, which was the knowledge that, apart from disinfection, all the antecedents of the diachronic fact and foil were held constant. Finally, what makes the cadaveric experiment so telling is not only that it provides evidence that is well explained by the cadaveric hypothesis, but that the evidence simultaneously disconfirms the competitors. None of the other hypotheses can explain the temporal difference, since they all appeal to factors that were unchanged in this experiment. As we have seen in our discussion of negative evidence, however, the deductive model does not account for this process of disconfirmation through explanatory failure, and so it does not account for the way the evidence makes the cadaveric hypothesis the best explanation by simultaneously strengthening it and weakening its rivals.

I conclude that Inference to the Best Explanation, linked up to an account of contrastive explanation that provides an alternative to the deductive-nomological model, is an improvement over the hypothetico deductive model in its account of the context of discovery, the determination of relevant evidence, the nature of disconfirmation, and the special positive support that certain contrastive experiments provide.

In particular, Inference to the Best Explanation is an improvement because it allows for evidential relevance in the absence of plausible deductive connections, since contrastive explanations need not entail what they explain. If Inference to the Best Explanation is to come out as a suitable replacement for the hypothetico-deductive model, however, it is important to see that it does not conflict with the obvious fact that scientific research is shot through with deductive inferences. To deny that all scientific explanations can be cast in deductive form is not to deny that some of them can, or that deduction often plays an essential role in those that cannot be so cast. Semmelweis certainly relied on deductive inferences, many of them elementary. For example, he needed to use deductive calculations to determine the relative frequencies of fever mortalities for the two divisions and for street births.

Moreover, in many cases of causal scientific explanation, deduction is required to see whether a putative cause would explain a particular contrast. One reason for this is that an effect may be due to many causes, some of which are already known, and calculation is required to determine whether an additional putative cause would explain the residual effect. Consider, for example, the inference from the perturbation in the orbit of Uranus to the existence of Neptune. In order to determine whether Neptune would explain this perturbation, Adams and Leverrier had first to calculate the influence of the Sun and known planets on Uranus, in order to work out what the perturbation was, and then had to do further calculations to determine what sort of planet and orbit would account for it. As Mill points out, this 'Method of Residues' is an elaboration of the Method of Difference where the negative instance is 'not the direct result of observation and experiment, but has been arrived at by deduction'. Through deduction, Adams and Leverrier determined that Neptune would explain why Uranus had a perturbed orbit rather than the one it would have had if only the Sun and known planets were influencing its motion. This example also illustrates other roles for deduction, since calculation was required to solve Newton's equations even for the sole influence of the Sun, and to go from the subsequent observations of Neptune to Neptune's mass and orbit. This particular inference to the best

contrastive explanation would not have been possible without deduction.

Let us return now to the two challenges for Inference to the Best Explanation, that it mark an improvement over the hypothetico-deductive model, and that it tell us more than that inductive inference is often Inference to the Likeliest Cause. I have argued that the Semmelweis case shows that Inference to the Best Explanation passes the first test. It helps to show how the account passes the second test, by illustrating some of the ways explanatory considerations guide inference and judgments of likeliness. Although Semmelweis's overriding interest was in control rather than in understanding, he focused his inquiry by asking a contrastive explanatory question. Faced with the brute fact that many women were dying of childbed fever, and the many competing explanations for this, Semmelweis did not simply consider which explanation seemed the most plausible. Instead, he followed an organized research program based on evidential contrasts. By means of a combination of conjecture, observation, and manipulation, Semmelweis tried to show that the cadaveric hypothesis is the only available hypothesis that adequately explains his central contrast in mortality between the divisions. This entire process is governed by explanatory considerations that are not simply reducible to independent judgments of likeliness. By asking why the mortality in the two divisions was different, Semmelweis was able to generate a pool of candidate hypotheses, which he then evaluated by appeal to what they could and could not explain; his experimental procedure was governed by the need to find contrasts that would distinguish between them on explanatory grounds. When Semmelweis inferred the cadaveric hypothesis, it was not simply that what turned out to be the likeliest hypothesis also seemed the best explanation: he judged that the likeliest cause of most of the cases of childbed fever in his hospital was infection by cadaveric matter *because* this was the best explanation of his evidence.

The picture of Inference to the Best Explanation that has emerged from the example of Semmelweis's research is, I think, somewhat different from the one that the slogan initially suggests in two important respects. The slogan calls to mind a fairly passive process, where we take whatever data happen to be to hand and infer an explanation, and where the central judgment we must make in this process is which of a battery of explanations of the same data would, if true, provide the loveliest explanation. But as the example shows, Inference to the Best Explanation supports a picture of research that is at once more active and realistic, where explanatory considerations guide the program of observation and experiment, as well as of conjecture. The upshot of this program is an inference to the loveliest explanation but the technique is eliminative. Through the use of judiciously chosen experiments, Semmelweis determined the loveliest explanation by a process of manipulation and elimination that left only a single explanation of the salient contrasts. In effect, he converted the question of the loveliest explanation into the question of the only explanation of various contrasts. Research programs that make this conversion are common in science, and it is one of the merits of Inference to the Best Explanation that it elucidates this strategy. And it is because Semmelweis successfully pursued it that we have been able to say something substantial about how explanatory considerations can be a guide to inference without getting bogged down in the daunting question of comparative loveliness where two hypotheses do both explain the same data. At the same time, this question cannot be avoided in a full assessment of Inference to the Best Explanation, since scientists are not always as fortunate as Semmelweis was in finding contrasts that discriminate between all the competitors.

PART 2 SUGGESTIONS FOR FURTHER READING

Achinstein, Peter, ed. *The Concept of Evidence*. New York: Oxford University Press, 1983.

Barker, Stephen F. *Induction and Hypothesis: A Study of the Logic of Confirmation*. Ithaca: Cornell University Press, 1957.

Foster, Marguerite, and Michael Martin. *Probability, Confirmation, and Simplicity*. New York: Odyssey Press, 1966.

Fumerton, Richard A. "Induction and Reasoning to the Best Explanation." *Philosophy of Science* 47 (1980): 589–600.

Glymour, Clark. "Hypothetico-deductivism Is Hopeless." *Philosophy of Science* 47 (1980): 322–325.

Goodman, Nelson. *Fact, Fiction, and Forecast*. New York: Bobbs-Merrill, 1965.

Harman, Gilbert. "The Inference to the Best Explanation." *Philosophical Review* 74 (1965): 88–95.

Rappaport, Steven. "Inference to the Best Explanation: Is It Really Different from Mill's Methods?" *Philosophy of Science* 63 (1996): 65–80.

Salmon, Wesley C. *The Foundations of Scientific Inference*. University of Pittsburgh Press, 1967.

Scheffler, Israel. *The Anatomy of Inquiry*. New York: Knopf, 1969.

PART 3

LAWS AND EXPLANATION
The Nature of Scientific Theories

Explanation and prediction have long been considered the most fundamental goals of scientific inquiry. Scientists not only want to know why certain phenomena occur; they also want to be able to predict their occurrence. Being able to predict a natural phenomenon can help us control it. And, as we have seen, when a theory leads to successful predictions, there is reason to believe that it is true. Achieving these goals has traditionally been thought to involve the construction of a special sort of deductive argument. The logical positivists followed this tradition but improved on it by specifying in detail the form such an argument must take.

The selection from Carl Hempel provides the logical positivist view of the nature of explanation. According to Hempel, a scientific explanation must meet two requirements: explanatory relevance and testability. The information contained in a scientific explanation must provide a reason for believing that the phenomenon in question occurred (or, in the case of prediction, will occur), and the explanation itself must be empirically verifiable. Scientific explanations meet these requirements by using statements referring to initial conditions and statements referring to natural laws. These two types of statements together deductively imply a statement describing the phenomenon. For example, one could explain why a metal rod expanded when heated by citing the fact that the rod is made of copper and the law that all copper expands when heated. Because Hempel thought that explaining something involves deducing a description of it from a natural law, his theory has become known as the deductive-nomological model of explanation (*nomos* is Greek for "law").

Natural laws, according to Hempel, are universal statements that say, in effect, that whenever conditions of kind F are realized, conditions of kind G are realized. Not all such statements are natural laws, however. As an example, Hempel cites the statement "All rocks in this box contain iron." Although this is a universal statement, it is not a natural law because it's not the case that if a rock were in the box, it would contain iron. Natural laws differ from ordinary universal generalizations in that they apply not only to what is the case, but also to what would be the case if certain conditions were met. This is what allows us to make predictions from them. For example, from the natural law "All metals expand when heated," we can predict that if a piece of copper were heated, it would expand. From the statement "All rocks in this box contain iron," however, we can't predict that if a rock were in this box, it would contain iron. One thing that distinguishes natural laws from other universal generalizations, then, is that they support "contrary-to-fact conditionals."

In Hempel's view, the laws used in scientific explanations do not have to express a causal relationship. They can also express mathematical relationships between various quantities. Consider, for example, Newton's law that force equals mass times acceleration ($F = ma$). This law could be used to predict the force with which a projectile will hit a wall. But it could also be used to retrodict how massive a projectile was that hit a wall. But even though such retrodiction meets the requirements of a deductive-nomological explanation, it doesn't really explain why the projectile had the mass that it did. Presumably such an explanation would refer to the processes that brought the projectile into being. One problem with Hempel's account, then, is that it doesn't recognize the temporal asymmetry of explanation: although the present or future can be explained in terms of what happened in the past, the past cannot be explained in terms of what is happening now or what will happen in the future.

Wesley Salmon argues that the deductive-nomological model of explanation not only fails to deal with the asymmetry problem, but it also fails to provide an adequate account of statistical explanation. Some events only occur a certain percentage of the time. Some drugs, for example, may cure a disease only 90 percent of the time. We can explain why someone who took the drug was cured by appealing to a statistical law such as "90 percent of the people who take this drug are cured." On the deductive-nomological model, however, such laws can explain only phenomena that have better than a 50-50 chance of occurring under the specified conditions. Many phenomena that stand in need of an explanation, however, occur much less frequently than that.

Consider, for example, the relationship between smoking and lung cancer. Those who smoke have less than a 50 percent chance of developing lung cancer. Yet we can explain a smoker's getting lung cancer in terms of smoking because smokers are much more likely to get lung cancer than nonsmokers. From the fact that people smoke, we cannot predict that they will get lung cancer. Nevertheless, we can use the fact that they smoke to explain their lung cancer because there is a statistically relevant connection between smoking and lung cancer.

According to Salmon, explaining a phenomenon does not require deducing it from a law. All we need to do is identify its cause. In many cases, he claims, this can be done through a process of "screening off." Just as we can find a source of light by moving a screen around until it cuts off the beam, so we can find the cause of a phenomenon by altering various factors until the effect is no longer produced. In other cases, this can be done by analyzing the interaction among various causal processes. Salmon claims that his causal account of explanation is superior to Hempel's because, among other reasons, it does not face the asymmetry problem (an effect cannot precede its cause), and it can explain events whose probability is less than 50 percent.

One problem in finding a scientific explanation is that there are many laws, causes, and statistical relationships surrounding any phenomenon. Most of these are related to the phenomenon but would be poor explanations. Consider the bank robber whose explanation of "Why do you rob banks?" is "Because that's where the money is." True enough, but not all that enlightening. So how are we to pick the most relevant factor to arrive at a good explanation?

According to Bas van Fraassen, an explanation is an answer to a why-question. To answer a why-question, however, we have to understand its context. When we ask why something is the case, we want to know why this rather than that. For example, when we ask, "Why is the sky blue?" we want to know why it's blue rather than green or orange or some other color. These other possibilities constitute what van Fraassen calls the "contrast class" of a why-question.

Why-questions are answered by providing certain sorts of information. But the information is not the same in every case. The information we desire depends on the context of our inquiry. In one context, the phenomenon might be explained by a law, in another by citing a cause, and in still another by citing statistical correlations. Because explanation is relative to context, van Fraassen believes that the notion of scientific explanation cannot be analyzed in terms of any one particular relation between theory and fact, such as causation.

Philip Kitcher believes that an adequate theory of explanation helps us evaluate the claim that science is giving us an increasingly better understanding of the world, and it enables us to comprehend past disputes in science. Van Fraassen's theory, however, does neither. It does not explain, for example, why scientists chose Darwin's theory over creationism long before it made any successful predictions. Kitcher proposes that by construing explanation as unification, both goals of a theory of explanation can be met. In his view, to understand something is to see it as part of a pattern. The more that pattern encompasses—the more it systematizes and unifies our knowledge—the more understanding it produces.

Theories unify our knowledge by providing us with certain patterns of argument that can be used to construct explanations. Darwin's theory, for example, tells us that the physical features of an organism can be explained by appeal to natural selection. Although Darwin didn't identify any natural laws from which descriptions of biological phenomena could be derived, he provided a pattern of argument that promised to explain many different biological phenomena. The success of this pattern of argument explains not only its explanatory power, but also its adoption by the scientific community.

Social science has not achieved the same sort of explanatory success as natural science. Some attribute this to the fact that social science can't offer the same sort of explanations as natural science. Merilee Salmon examines three different accounts of explanation in the social sciences: interpretativism, nomological skepticism, and critical theory. Interpretativism claims that human behavior cannot be explained in terms of laws because there are no empirical laws connecting reasons with actions. Nomological skepticism admits that reasons can be causes, but considers it doubtful that laws connecting reasons with actions can be found because human behavior is too complex. Critical theory opposes explanations in terms of laws because such explanations undercut the belief in free will. Salmon does not find any of these accounts entirely convincing. She suggests that social phenomena can be explained in the same way that natural phenomena are explained.

12

CARL G. HEMPEL

Laws and Their Role in Scientific Explanations

That man has long and persistently been concerned to achieve some understanding of the enormously diverse, often perplexing, and sometimes threatening occurrences in the world around him is shown by the manifold myths and metaphors he has devised in an effort to account for the very existence of the world and of himself, for life and death, for the motions of the heavenly bodies, for the regular sequence of day and night, for the changing seasons, for thunder and lightning, sunshine and rain. Some of these explanatory ideas are based on anthropomorphic conceptions of the forces of nature, others invoke hidden powers or agents, still others refer to God's inscrutable plans or to fate.

Accounts of this kind undeniably may give the questioner a sense of having attained some understanding; they may resolve his perplexity and in this sense "answer" his question. But however satisfactory these answers may be psychologically, they are not adequate for the purposes of science, which, after all, is concerned to develop a conception of the world that has a clear, logical bearing on our experience and is thus capable of objective test. Scientific explanations must, for this reason, meet two systematic requirements, which will be called the requirement of explanatory relevance and the requirement of testability.

The astronomer Francesco Sizi offered the following argument to show why, contrary to what his contemporary, Galileo, claimed to have seen through his telescope, there could be no satellites circling around Jupiter:

Philosophy of Natural Science (Upper Saddle River, NJ: Prentice-Hall, 1966), pp. 47–67. © 1966 Prentice-Hall, Inc. Reprinted by permission of the publisher.

There are seven windows in the head, two nostrils, two ears, two eyes and a mouth; so in the heavens there are two favorable stars, two unpropitious, two luminaries, and Mercury alone undecided and indifferent. From which and many other similar phenomena of nature such as the seven metals, etc., which it were tedious to enumerate, we gather that the number of planets is necessarily seven. . . . Moreover, the satellites are invisible to the naked eye and therefore can have no influence on the earth and therefore would be useless and therefore do not exist.[1]

The crucial defect of this argument is evident: the "facts" it adduces, even if accepted without question, are entirely irrelevant to the point at issue; they do not afford the slightest reason for the assumption that Jupiter has no satellites; the claim of relevance suggested by the barrage of words like 'therefore', 'it follows', and 'necessarily' is entirely spurious.

• Consider by contrast the physical explanation of a rainbow. It shows that the phenomenon comes about as a result of the reflection and refraction of the white light of the sun in spherical droplets of water such as those that occur in a cloud. By reference to the relevant optical laws, this account shows that the appearance of a rainbow is to be expected whenever a spray or mist of water droplets is illuminated by a strong white light behind the observer. Thus, even if we happened never to have seen a rainbow, the explanatory information provided by the physical account would constitute good grounds for expecting or believing that a rainbow will appear under the specified circumstances. We will refer to this characteristic by saying that the physical explanation meets the *requirement of ex-*

planatory relevance: the explanatory information adduced affords good grounds for believing that the phenomenon to be explained did, or does, indeed occur. This condition must be met if we are to be entitled to say: "That explains it—the phenomenon in question was indeed to be expected under the circumstances!"

The requirement represents a necessary condition for an adequate explanation, but not a sufficient one. For example, a large body of data showing a red-shift in the spectra of distant galaxies provides strong grounds for believing *that* those galaxies recede from our local one at enormous speeds, yet it does not explain *why*.

To introduce the second basic requirement for scientific explanations, let us consider the conception of gravitational attraction as manifesting a natural tendency akin to love. This conception has no test implications whatever. Hence, no empirical finding could possibly bear it out or disconfirm it. Being thus devoid of empirical content, the conception surely affords no grounds for expecting the characteristic phenomena of gravitational attraction: it lacks objective explanatory power. Similar comments apply to explanations in terms of an inscrutable fate: to invoke such an idea is not to achieve an especially profound insight, but to give up the attempt at explanation altogether. By contrast, the statements on which the physical explanation of a rainbow is based do have various test implications; these concern, for example, the conditions under which a rainbow will be seen in the sky, and the order of the colors in it; the appearance of rainbow phenomena in the spray of a wave breaking on the rocks and in the mist of a lawn sprinkler; and so forth. These examples illustrate a second condition for scientific explanations, which we will call the *requirement of testability:* the statements constituting a scientific explanation must be capable of empirical test.

It has already been suggested that since the conception of gravitation in terms of an underlying universal affinity has no test implications, it can have no explanatory power: it cannot provide grounds for expecting that universal gravitation will occur, nor that gravitational attraction will show such and such characteristic features; for if it

did imply such consequences either deductively or even in a weaker, inductive-probabilistic sense, then it would be testable by reference to those consequences. As this example shows, the two requirements just considered are interrelated: a proposed explanation that meets the requirement of relevance also meets the requirement of testability. (The converse clearly does not hold.)

Now let us see what forms scientific explanations take, and how they meet the two basic requirements.

Consider . . . Périer's finding in the Puy-de-Dôme experiment, that the length of the mercury column in a Torricelli barometer decreased with increasing altitude. Torricelli's and Pascal's ideas on atmospheric pressure provided an explanation for this phenomenon; somewhat pedantically, it can be spelled out as follows:

a. At any location, the pressure that the mercury column in the closed branch of the Torricelli apparatus exerts upon the mercury below equals the pressure exerted on the surface of the mercury in the open vessel by the column of air above it.

b. The pressures exerted by the columns of mercury and of air are proportional to their weights; and the shorter the columns, the smaller their weights.

c. As Périer carried the apparatus to the top of the mountain, the column of air above the open vessel became steadily shorter.

d. (Therefore,) the mercury column in the closed vessel grew steadily shorter during the ascent.

Thus formulated, the explanation is an argument to the effect that the phenomenon to be explained, as described by the sentence (*d*), is just what is to be expected in view of the explanatory facts cited in (*a*), (*b*), and (*c*); and that, indeed, (*d*) follows deductively from the explanatory statements. The latter are of two kinds; (*a*) and (*b*) have the character of general laws expressing uniform empirical connections; whereas (*c*) describes certain particular facts. Thus, the shortening of the mercury column is here explained by showing that it occurred in accordance with certain laws of nature, as a result of certain particular circumstances. The explanation fits the phenomenon to be explained into a pattern of uniformities and shows that its occurrence was to be expected, given the specified laws and the pertinent particular circumstances.

The phenomenon to be accounted for by an explanation will henceforth also be referred to as the *explanandum phenomenon;* the sentence describing it, as the *explanandum sentence.* When the context shows which is meant, either of them will simply be called the explanandum. The sentences specifying the explanatory information—(*a*), (*b*), (*c*) in our example—will be called the *explanans sentences;* jointly they will be said to form the *explanans.*

As a second example, consider the explanation of a characteristic of image formation by reflection in a spherical mirror; namely, that generally $1/u + 1/v = 2/r$, where u and v are the distances of object-point and image-point from the mirror, and r is the mirror's radius of curvature. In geometrical optics, this uniformity is explained with the help of the basic law of reflection in a plane mirror, by treating the reflection of a beam of light at any one point of a spherical mirror as a case of reflection in a plane tangential to the spherical surface. The resulting explanation can be formulated as a deductive argument whose conclusion is the explanandum sentence, and whose premises include the basic laws of reflection and of rectilinear propagation, as well as the statement that the surface of the mirror forms a segment of a sphere.[2]

A similar argument, whose premises again include the law for reflection in a plane mirror, offers an explanation of why the light of a small light source placed at the focus of a paraboloidal mirror is reflected in a beam parallel to the axis of the paraboloid (a principle technologically applied in the construction of automobile headlights, searchlights, and other devices).

The explanations just considered may be conceived, then, as deductive arguments whose conclusion is the explanandum sentence, E, and whose premise-set, the explanans, consists of general laws, $L_1, L_2, \ldots, L_r$ and of other statements, $C_1, C_2, \ldots, C_k$, which make assertions about particular facts. The form of such arguments, which thus constitute one type of scientific explanation, can be represented by the following schema:

$$\text{D-N]} \quad \left. \begin{array}{c} L_1, L_2, \ldots, L_r \\[1em] \underline{C_1, C_2, \ldots, C_k} \\ E \end{array} \right\} \begin{array}{l} \text{Explanans} \\ \text{sentences} \\[1em] \text{Explanandum} \\ \text{sentence} \end{array}$$

Explanatory accounts of this kind will be called explanations by deductive subsumption under general laws, or *deductive-nomological explanations.* (The root of the term "nomological" is the Greek word 'nomos,' for law.) The laws invoked in a scientific explanation will also be called *covering laws* for the explanandum phenomenon, and the explanatory argument will be said to subsume the explanandum under those laws.

The explanandum phenomenon in a deductive-nomological explanation may be an event occurring at a particular place and time, such as the outcome of Périer's experiment. Or it may be some regularity found in nature, such as certain characteristics generally displayed by rainbows; or a uniformity expressed by an empirical law such as Galileo's or Kepler's laws. Deductive explanations of such uniformities will then invoke laws of broader scope, such as the laws of reflection and refraction, or Newton's laws of motion and of gravitation. As this use of Newton's laws illustrates, empirical laws are often explained by means of theoretical principles that refer to structures and processes underlying the uniformities in question. . . .

Deductive-nomological explanations satisfy the requirement of explanatory relevance in the strongest possible sense: the explanatory information they provide implies the explanandum sentence deductively and thus offers logically conclusive grounds why the explanandum phenomenon is to be expected. (We will soon encounter other scientific explanations, which fulfill the requirement only in a weaker, inductive, sense.) And the testability requirement is met as well, since the explanans implies among other things that under the specified conditions, the explanandum phenomenon occurs.

Some scientific explanations conform to the pattern (D-N) quite closely. This is so, particularly, when certain quantitative features of a phenomenon are explained by mathematical derivation from covering general laws, as in the case of reflection in spherical and paraboloidal mirrors. Or take the celebrated explanation, propounded by Leverrier (and independently by Adams), of peculiar irregularities in the motion of the planet Uranus, which on the current Newtonian theory could not be accounted for by the gravitational attraction of the other planets then known. Leverrier conjectured that they resulted from the gravitational pull of an as yet undetected outer planet, and he computed the position, mass, and other characteristics which that planet would have to possess to account in quantitative detail for the observed irregularities. His explanation was strikingly confirmed by the discovery, at the predicted location, of a new planet, Neptune, which had the quantitative characteristics attributed to it by Leverrier. Here again, the explanation has the character of a deductive argument whose premises include general laws—specifically, Newton's laws of gravitation and of motion—as well as statements specifying various quantitative particulars about the disturbing planet.

Not infrequently, however, deductive-nomological explanations are stated in an elliptical form: they omit mention of certain assumptions that are presupposed by the explanation but are simply taken for granted in the given context. Such explanations are sometimes expressed in the form 'E because C,' where E is the event to be explained and C is some antecedent or concomitant event or state of affairs. Take, for example, the statement: 'The slush on the sidewalk remained liquid during the frost because it had been sprinkled with salt'. This explanation does not explicitly mention any laws, but it tacitly presupposes at least one: that the freezing point of water is lowered whenever salt is dissolved in it. Indeed, it is precisely by virtue of this law that the sprinkling of salt acquires the explanatory, and specifically causative, role that the elliptical because-statement ascribes to it. That statement, incidentally, is elliptical also in other respects; for example, it tacitly takes for granted, and leaves unmentioned, certain assumptions about the prevailing physical conditions, such as the temperature's not dropping to a very low point. And if nomic and other assumptions thus omitted are added to the statement that salt had been sprinkled on the slush, we obtain the premises for a deductive-nomological explanation of the fact that the slush remained liquid.

Similar comments apply to Semmelweis's explanation that childbed fever was caused by decomposed animal matter introduced into the bloodstream through open wound surfaces. Thus formulated, the explanation makes no mention of general laws; but it presupposes that such contamination of the bloodstream generally leads to blood poisoning attended by the characteristic symptoms of childbed fever, for this is implied by the assertion that the contamination causes puerperal fever. The generalization was no doubt taken for granted by Semmelweis, to whom the cause of Kolletschka's fatal illness presented no etiological problem: given that infectious matter was introduced into the bloodstream, blood poisoning would result. (Kolletschka was by no means the first one to die of blood poisoning resulting from a cut with an infected scalpel. And by a tragic irony, Semmelweis himself was to suffer the same fate.) But once the tacit premise is made explicit, the explanation is seen to involve reference to general laws.

As the preceding examples illustrate, corresponding general laws are always presupposed by an explanatory statement to the effect that a particular event of a certain kind G (e.g., expansion of a gas under constant pressure; flow of a current in a wire loop) was caused by an event of another kind, F (e.g., heating of the gas; motion of the loop across a magnetic field). To see this, we need not enter into the complex ramifications of the notion of cause; it suffices to note that the general maxim "Same cause, same effect", when applied to such explanatory statements, yields the implied claim that whenever an event of kind F occurs, it is accompanied by an event of kind G.

To say that an explanation rests on general laws is not to say that its discovery required the discovery of the laws. The crucial new insight achieved by an explanation will sometimes lie in the discovery of some particular fact (e.g., the presence of an undetected outer planet; infectious matter adhering to the hands of examining physicians) which, by virtue of antecedently accepted general laws, accounts for the explanandum phenomenon. In other cases, such as that of the lines in the hydrogen spectrum, the explanatory achievement does lie in the discovery of a covering law (Balmer's) and eventually of an explanatory theory (such as Bohr's); in yet other cases, the major accomplishment of an explanation may lie in showing that, and exactly how, the explanandum phenomenon can be accounted for by reference to laws and data about particular facts that are already available: this is illustrated by the explanatory derivation of the reflection laws for spherical and paraboloidal mirrors from the basic law of geometrical optics in conjunction with statements about the geometrical characteristics of the mirrors.

An explanatory problem does not by itself determine what kind of discovery is required for its solution. Thus, Leverrier discovered deviations from the theoretically expected course also in the motion of the planet Mercury; and as in the case of Uranus, he tried to explain these as resulting from the gravitational pull of an as yet undetected planet, Vulcan, which would have to be a very dense and very small object between the sun and Mercury. But no such planet was found, and a satisfactory explanation was provided only much later by the general theory of relativity, which accounted for the irregularities not by reference to some disturbing particular factor, but by means of a new system of laws.

As we have seen, laws play an essential role in deductive-nomological explanations. They provide the link by reason of which particular circumstances (described by $C_1, C_2, \ldots, C_k$) can serve to explain the occurrence of a given event. And when the explanandum is not a particular event, but a uniformity such as those represented by characteristics mentioned earlier of spherical and paraboloidal mirrors, the explanatory laws exhibit a system of more comprehensive uniformities, of which the given one is but a special case.

The laws required for deductive-nomological explanations share a basic characteristic: they are, as we shall say, statements of universal form. Broadly speaking, a statement of this kind asserts a uniform connection between different empirical phenomena or between different aspects of an empirical phenomenon. It is a statement to the effect that whenever and wherever conditions of a specified kind F occur, then so will, always and without exception, certain conditions of another kind, G. (Not all scientific laws are of this type. In the sections that follow, we will encounter laws of probabilistic form, and explanations based on them.)

Here are some examples of statements of universal form: whenever the temperature of a gas increases while its pressure remains constant, its volume increases; whenever a solid is dissolved in a liquid, the boiling point of the liquid is raised; whenever a ray of light is reflected at a plane surface, the angle of reflection equals the angle of incidence; whenever a magnetic iron rod is broken in two, the pieces are magnets again; whenever a body falls freely from rest in a vacuum near the surface of the earth, the distance it covers in t seconds is $16t^2$ feet. Most of the laws of the natural sciences are quantitative: they assert specific mathematical connections between different quantitative characteristics of physical systems (e.g., between volume, temperature, and pressure of a gas) or of processes (e.g., between time and distance in free fall in Galileo's law; between the period of revolution of a planet and its mean distance from the sun, in Kepler's third law; between the angles of incidence and refraction in Snell's law).

Strictly speaking, a statement asserting some uniform connection will be considered a law only if there are reasons to assume it is true: we would not normally speak of false laws of nature. But if this requirement were rigidly observed, then the statements commonly referred to as Galileo's and Kepler's laws would not qualify as laws; for according to current physical knowledge, they hold only approximately; and as we shall see later, physical theory explains why this is so. Analogous remarks apply to the laws of geometrical optics. For example, even in a homogeneous medium, light does not move strictly in straight lines: it can bend around corners. We shall therefore use the word 'law' somewhat liberally, applying the term also to certain statements of the kind here referred to, which, on theoretical grounds, are known to hold only approximately and with certain qualifications.

We saw that the laws invoked in deductive-nomological explanations have the basic form: 'In all cases when conditions of kind F are realized, conditions of kind G are realized as well'. But, interestingly, not all statements of this universal form, even if true, can qualify as laws of nature. For example, the sentence 'All rocks in this box contain iron' is of universal form (F is the condition of being a rock in the box, G that of containing iron); yet even if true, it would not be regarded as a law, but as an assertion of something that "happens to be the case", as an "accidental generalization". Or consider the statement: 'All bodies consisting of pure gold have a mass of less than 100,000 kilograms'. No doubt all bodies of gold ever examined by man conform to it; thus, there is considerable confirmatory evidence for it and no disconfirming instances are known. Indeed, it is quite possible that never in the history of the universe has there been or will there be a body of pure gold with a mass of 100,000 kilograms or more. In this case, the proposed generalization would not only be well confirmed, but true. And yet, we would presumably regard its truth as accidental, on the ground that nothing in the basic laws of nature as conceived in contemporary science precludes the possibility of there being—or even the possibility of our producing—a solid gold object with a mass exceeding 100,000 kilograms.

Thus, a scientific law cannot be adequately defined as a true statement of universal form: this characterization expresses a necessary, but not a sufficient, condition for laws of the kind here under discussion.

What distinguishes genuine laws from accidental generalizations? This intriguing problem has been intensively discussed in recent years. Let us look briefly at some of the principal ideas that have emerged from the debate, which is still continuing.

One telling and suggestive difference, noted by Nelson Goodman,[3] is this: a law can, whereas an accidental generalization cannot, serve to support counterfactual conditionals, i.e., statements of the form 'If A were (had been) the case, then B would be (would have been) the case', where in fact A is not (has not been) the case. Thus, the assertion 'If this paraffin candle had been put into a kettle of boiling water, it would have melted' could be supported by adducing the law that paraffin is liquid above 60 degrees centigrade (and the fact that the boiling point of water is 100 degrees centigrade). But the statement 'All rocks in this box contain iron' could not be used similarly to support the counterfactual statement 'If this pebble had been put into the box, it would contain iron'. Similarly, a law, in contrast to an accidentally true generalization, can support subjunctive conditionals, i.e., sentences of the type 'If A should come to pass, then so

would B', where it is left open whether or not A will in fact come to pass. The statement 'If this paraffin candle should be put into boiling water then it would melt' is an example.

Closely related to this difference is another one, which is of special interest to us: a law can, whereas an accidental generalization cannot, serve as a basis for an explanation. Thus, the melting of a particular paraffin candle that was put into boiling water can be explained, in conformity with the schema (D-N), by reference to the particular facts just mentioned and to the law that paraffin melts when its temperature is raised above 60 degrees centigrade. But the fact that a particular rock in the box contains iron cannot be analogously explained by reference to the general statement that all rocks in the box contain iron.

It might seem plausible to say, by way of a further distinction, that the latter statement simply serves as a conveniently brief formulation of a finite conjunction of this kind: 'Rock r_1 contains iron, and rock r_2 contains iron, . . . , and rock r_{63} contains iron'; whereas the generalization about paraffin refers to a potentially infinite set of particular cases and therefore cannot be paraphrased by a finite conjunction of statements describing individual instances. This distinction is suggestive, but it is overstated. For to begin with, the generalization 'All rocks in this box contain iron' does not in fact tell us how many rocks there are in the box, nor does it name any particular rocks r_1, r_2, etc. Hence, the general sentence is not logically equivalent to a finite conjunction of the kind just mentioned. To formulate a suitable conjunction, we need additional information, which might be obtained by counting and labeling the rocks in the box. Besides, our generalization 'All bodies of pure gold have a mass of less than 100,000 kilograms' would not count as a law even if there were infinitely many bodies of gold in the world. Thus, the criterion we have under consideration fails on several grounds.

Finally, let us note that a statement of universal form may qualify as a law even if it actually has no instances whatever. As an example, consider the sentence: 'On any celestial body that has the same radius as the earth but twice its mass, free fall from rest conforms to the formula $s = 32t^2$'. There might well be no celestial object in the entire universe that has the specified size and mass, and yet the statement has the character of a law. For it (or rather, a close approximation of it, as in the case of Galileo's law) follows from the Newtonian theory of gravitation and of motion in conjunction with the statement that the acceleration of free fall on the earth is 32 feet per second per second; thus, it has strong theoretical support, just like our earlier law for free fall on the moon.

A law, we noted, can support subjunctive and counterfactual conditional statements about potential instances, i.e., about particular cases that might occur, or that might have occurred but did not. In similar fashion, Newton's theory supports our general statement in a subjunctive version that suggests its lawlike status, namely: 'On any celestial body that there may be which has the same size as the earth but twice its mass, free fall would conform to the formula $s = 32t^2$'. By contrast, the generalization about the rocks cannot be paraphrased as asserting that any rock that might be in this box would contain iron, nor of course would this latter claim have any theoretical support.

Similarly, we would not use our generalization about the mass of gold bodies—let us call it H—to support statements such as this: 'Two bodies of pure gold whose individual masses add up to more than 100,000 kilograms cannot be fused to form one body; or if fusion should be possible, then the mass of the resulting body will be less than 100,000 kg', for the basic physical and chemical theories of matter that are currently accepted do not preclude the kind of fusion here considered, and they do not imply that there would be a mass loss of the sort here referred to. Hence, even if the generalization H should be true, i.e., if no exceptions to it should ever occur, this would constitute a mere accident or coincidence as judged by current theory, which permits the occurrence of exceptions to H.

Thus, whether a statement of universal form counts as a law will depend in part upon the scientific theories accepted at the time. This is not to say that "empirical generalizations"—statements of universal form that are empirically well confirmed but have no basis in theory—never qualify as laws: Galileo's, Kepler's, and Boyle's laws, for example, were accepted as such before they received theoretical grounding. The relevance of theory is rather this: a statement of universal form, whether empirically confirmed or as yet untested, will qualify as a law if it is implied by an accepted theory (statements of this kind are often referred to as theoretical laws); but even if it is empirically well confirmed and presumably true in fact, it will not qualify as a law if it rules out certain hypothetical occurrences (such as the fusion of two gold bodies with a resulting mass of more than 100,000 kilograms, in the case of our generalization H) which an accepted theory qualifies as possible.[4]

Not all scientific explanations are based on laws of strictly universal form. Thus, little Jim's getting the measles might be explained by saying that he caught the disease from his brother, who had a bad case of the measles some days earlier. This account again links the explanandum event to an earlier occurrence, Jim's exposure to the measles; the latter is said to provide an explanation because there is a connection between exposure to the measles and contracting the disease. That connection cannot be expressed by a law of universal form, however; for not every case of exposure to the measles produces contagion. What can be claimed is only that persons exposed to the measles will contract the disease with high probability, i.e., in a high percentage of all cases. General statements of this type . . . will be called *laws of probabilistic form* or *probabilistic laws,* for short.

In our illustration, then, the explanans consists of the probabilistic law just mentioned and the statement that Jim was exposed to the measles. In contrast to the case of deductive-nomological explanation, these explanans statements do not deductively imply the explanandum statement that Jim got the measles; for in deductive inferences from true premises, the conclusion is invariably true, whereas in our example, it is clearly possible that the explanans statements might be true and yet the explanandum statement false. We will say, for short, that the explanans implies the explanandum, not with "deductive certainty", but only with near-certainty or with high probability.

The resulting explanatory argument may be schematized as follows:

The probability for persons exposed to the measles to catch the disease is high.

Jim was exposed to the measles.

$$\overline{}\quad \text{[makes highly probable]}$$

Jim caught the measles.

In the customary presentation of a deductive argument, which was used, for example, in the schema (D-N) above, the conclusion is separated from the premises by a single line, which serves to indicate that the premises logically imply the conclusion. The double line used in our latest schema is meant to indicate analogously that the "premises" (the explanans) make the "conclusion" (the explanandum sentence) more or less probable; the degree of probability is suggested by the notation in brackets.

Arguments of this kind will be called *probabilistic explanations*. As our discussion shows, a probabilistic explanation of a particular event shares certain basic characteristics with the corresponding deductive-nomological type of explanation. In both cases, the given event is explained by reference to others, with which the explanandum event is connected by laws. But in one case, the laws are of universal form; in the other, of probabilistic form. And while a deductive explanation shows that, on the information contained in the explanans, the explanandum was to be expected with "deductive certainty", an inductive explanation shows only that, on the information contained in the explanans, the explanandum was to be expected with

high probability, and perhaps with "practical certainty"; it is in this manner that the latter argument meets the requirement of explanatory relevance.

NOTES

1. From Holton and Roller, *Foundations of Modern Physical Science,* p. 160.
2. The derivation of the laws of reflection for the curved surfaces referred to in this example and in the next one is simply and lucidly set forth in Chap. 17 of Morris Kline, *Mathematics and the Physical World* (New York: Thomas Y. Crowell Company, 1959).
3. In his essay, "The Problem of Counterfactual Conditionals," reprinted as the first chapter of his book, *Fact, Fiction, and Forecast,* 2nd ed. (Indianapolis: The Bobbs-Merrill Co., Inc., 1965). This work raises fascinating basic problems concerning laws, counterfactual statements, and inductive reasoning, and examines them from an advanced analytic point of view.
4. For a fuller analysis of the concept of law, and for further bibliographic references, see E. Nagel, *The Structure of Science* (New York: Harcourt, Brace & World, Inc., 1961), Chap. 4.

13

WESLEY SALMON

Why Ask, "Why?"?

Concerning the first order question "Why?", I have raised the second order question "Why ask, 'Why?'?" to which you might naturally respond with the third order question "Why ask, 'Why ask,

Proceedings and Addresses of the American Philosophical Association 51 (August 1978), pp. 683–705. © Wesley C. Salmon. Reprinted by permission. All rights reserved.

"Why?'"?" But this way lies madness, to say nothing of an infinite regress. While an infinite sequence of nested intervals may converge upon a point, the series of nested questions just initiated has no point to it, and so we had better cut it off without delay. The answer to the very natural third order question is this: the question "Why ask, 'Why?'?" expresses a deep philosophical perplexity which I believe to be both significant in its own right and highly relevant to certain current philosophical discussions. I want to share it with you.

The problems I shall be discussing pertain mainly to scientific explanation, but before turning to them, I should remark that I am fully aware that many—perhaps most—why-questions are requests for some sort of *justification* (Why did one employee receive a larger raise than another? Because she had been paid less than a male colleague for doing the same kind of job.) or *consolation* (Why, asked Job, was I singled out for such extraordinary misfortune and suffering?). Since I have neither the time nor the talent to deal with questions of this sort, I shall not pursue them further, except to remark that the seeds of endless philosophical confusion can be sown by failing carefully to distinguish them from requests for scientific explanation.

Let me put the question I do want to discuss to you this way. Suppose you had achieved the epistemic status of Laplace's demon—the hypothetical super-intelligence who knows all of nature's regularities, and the precise state of the universe in full detail at some particular moment (say now, according to some suitable simultaneity slice of the universe). Possessing the requisite logical and mathematical skill, you would be able to predict any future occurrence, and you would be able to retrodict any past event. Given this sort of apparent omniscience, would your scientific knowledge be complete, or would it still leave something to be desired? Laplace asked no more of his demon; should we place further demands upon ourselves? And if so, what should be the nature of the additional demands?

If we look at most contemporary philosophy of science texts, we find an immediate *affirmative* answer to this question. Science, the majority say, has at least two principal aims—prediction (construed broadly enough to include inference from the observed to the unobserved, regardless of temporal relations) and explanation. The first of these provides knowledge of *what* happens; the second is supposed to furnish knowledge of *why* things happen as they do. This is not a new idea. In the *Posterior Analytics,* Aristotle distinguishes syllogisms which provide scientific understanding from those which do not.[1] In the Port Royal Logic, Arnauld distinguishes demonstrations which merely convince the mind from those which also enlighten the mind.[2]

This view has not been universally adopted. It was not long ago that we often heard statements to the effect that the business of science is to predict, not to explain. Scientific knowledge is descriptive—it tells us *what* and *how*. If we seek explanations—if we want to know why—we must go outside of science, perhaps to metaphysics or theology. In his preface to the third edition (1911) of *The Grammar of Science,* Karl Pearson wrote, "Nobody believes now that science *explains* anything; we all look upon it as a shorthand description, as an economy of thought."[3] This doctrine is not very popular nowadays. It is now fashionable to say that science aims not merely at describing the world—it also provides *understanding, comprehension,* and *enlightenment.* Science presumably accomplishes such high-sounding goals by supplying scientific explanations.

The current attitude leaves us with a deep and perplexing question, namely, if explanation does involve something over and above mere description, just what sort of thing is it? The use of such honorific near-synonyms as "understanding," "comprehension," and "enlightenment" makes it sound important and desirable, but does not help at all in the philosophical analysis of explanation—scientific or other. What, over and above its complete descriptive knowledge of the world, would Laplace's demon require in order to achieve understanding? I hope you can see that this is a real problem, especially for those who hold what I shall call "the inferential view" of scientific explanation, for Laplace's demon can infer every fact about the universe, past, present, and future. If the problem does not seem acute, I would quote a remark made by Russell about Zeno's paradox of the flying arrow—"The more the difficulty is meditated, the more real it becomes."[4]

It is not my intention this evening to discuss the details of the various formal models of scientific explanation which have been advanced in the last three decades.[5] Instead, I want to consider the general conceptions which lie beneath the most influential theories of scientific explanation. Two powerful intuitions seem to have guided much of the discussion. Although they have given rise to disparate basic conceptions and considerable controversy, both are, in my opinion, quite sound. Moreover, it seems to me, both can be incorporated into a single overall theory of scientific explanation.

1. The first of these intuitions is the notion that the explanation of a phenomenon essentially involves *locating and identifying its cause or causes*. This intuition seems to arise rather directly from common sense, and from various contexts in which scientific knowledge is applied to concrete situations. It is strongly supported by a number of paradigms, the most convincing of which are explanations of particular occurrences. To explain a given airplane crash, for example, we seek "the cause"—a mechanical failure, perhaps, or pilot error. To explain a person's death again we seek the cause—strangulation or drowning, for instance. I shall call the general view of scientific explanation which comes more or less directly from this fundamental intuition *the causal conception;* Michael Scriven has been one of its chief advocates.[6]

2. The second of these basic intuitions is the notion that all scientific explanation involves *subsumption under laws*. This intuition seems to arise from consideration of developments in theoretical science. It has led to the general "covering law" conception of explanation, as well as to several formal "models" of explanation. According to this view, a fact is subsumed under one or more general laws if the assertion of its occurrence follows, either deductively or inductively, from statements of the laws (in conjunction, in some cases, with other premises). Since this view takes explanations to be arguments, I shall call it *the inferential conception;* Carl G. Hempel has been one of its ablest champions.[7]

Although the proponents of this inferential conception have often chosen to illustrate it with explanations of particular occurrences—e.g., why did the bunsen flame turn yellow on this particular occasion?—the paradigms which give it strongest support are explanations of general regularities. When we look to the history of science for the most outstanding cases of scientific explanations, such examples as Newton's explanation of Kepler's laws of planetary motion or Maxwell's electromagnetic explanation of optical phenomena come immediately to mind.

It is easy to guess how Laplace might have reacted to my question about his demon, and to the two basic intuitions I have just mentioned. The super-intelligence would have everything needed to provide scientific explanations. When, to mention one of Laplace's favorite examples, a seemingly haphazard phenomenon, such as the appearance of a comet, occurs, it can be explained by showing that it actually conforms to natural laws.[8] On Laplace's assumption of determinism, the demon possesses explanations of all happenings in the entire history of the world—past, present, and future. Explanation, for Laplace, seemed to consist in showing how events conform to the laws of nature, and these very laws provide the causal connections among the various states of the world. The Laplacian version of explanation thus seems to conform both to the causal conception and to the inferential conception.

Why, you might well ask, is not the Laplacian view of scientific explanation basically sound? Why do twentieth century philosophers find it necessary to engage in lengthy disputes over this matter? There are, I think, three fundamental reasons: (1) the causal conception faces the difficulty that no adequate treatment of causation has yet been offered; (2) the inferential conception suffers from the fact that it seriously misconstrues the nature of subsumption under laws; and (3) both conceptions have overlooked a central explanatory principle.

The inferential view, as elaborated in detail by Hempel and others, has been the dominant theory of scientific explanation in recent years—indeed, it has become virtually "the received view." From that standpoint, anyone who had attained the epistemic status of Laplace's demon could use the known laws and initial conditions to predict a future event, and when the event comes to pass, the argument which enabled us to predict it would ipso facto constitute an explanation of it. If, as Laplace believed, determinism is true, then every *future* event would thus be amenable to deductive-nomological explanation.

When, however, we consider the explanation of past events—events which occurred earlier than our initial conditions—we find a strange disparity. Although, by applying known laws, we can reliably *retrodict* any past occurrence on the basis of facts subsequent to the event, our intuitions rebel at the idea that we can *explain* events in terms of subsequent conditions. Thus, although our inferences to future events qualify as explanations according to the inferential conception, our inferences to the past do not. Laplace's demon can, of course, construct explanations of past events by inferring the existence of still earlier conditions and, with the aid of the known laws, deducing the occurrence of the events to be explained from these conditions which held in the more remote past. But if, as the inferential conception maintains, explanations are essentially inferences, such an approach to explanation of past events seems strangely roundabout. Explanations demand an asymmetry not present in inferences.

When we drop the fiction of Laplace's demon, and relinquish the assumption of determinism, the asymmetry becomes even more striking. The demon can predict the future and retrodict the past with complete precision and reliability. We cannot. When we consider the comparative difficulty of prediction vs. retrodiction, it turns out that retrodiction enjoys a tremendous advantage. We have records of the past—tree rings, diaries, fossils—but none of the future. As a result, we can have extensive and detailed knowledge of the past which has no counterpart in knowledge about the future. From a newspaper account of an accident, we can retrodict all sorts of details which could not have been predicted an hour before the collision. But the newspaper story—even though it may *report* the explanation of the accident—surely does not *constitute* the explanation. We see that *inference* has a preferred temporal direction, and that *explanation* also has a preferred temporal direction. The fact that these two are opposite to each other is one thing which makes me seriously doubt that explanations are essentially arguments.[9] As we shall see, however, denying that explanations are arguments does not mean that we must give up the *covering law* conception. Subsumption under laws can take a different form.

Having considered a number of preliminaries, I should now like to turn my attention to an attempt to outline a general theory of causal explanation. I shall not be trying to articulate a formal model; I shall be focusing upon general conceptions and fundamental principles rather than technical details. I am not suggesting, of course, that the technical details are dispensable—merely that this is not the time or place to try to go into them. Let me say at the outset that I shall be relying very heavily upon works by Russell (especially, *The Analysis of Matter* and *Human Knowledge, Its Scope and Limits*) and Reichenbach (especially, *The Direction of Time*). Although, to the best of my knowledge, neither of these authors ever published an article, or a book, or

a chapter of a book devoted explicitly to scientific explanation, nevertheless, it seems to me that a rather appealing theory of causal explanation can be constructed by putting together the insights expressed in the aforementioned works.

Developments in twentieth-century science should prepare us for the eventuality that some of our scientific explanations will have to be statistical—not merely because our knowledge is incomplete (as Laplace would have maintained), but rather, because nature itself is inherently statistical. Some of the laws used in explaining particular events will be statistical, and some of the regularities we wish to explain will also be statistical. I have been urging that causal considerations play a crucial role in explanation; indeed, I have just said that regularities—and this certainly includes statistical regularities—require causal explanation. I do not believe there is any conflict here. It seems to me that, by employing a statistical conception of causation along the lines developed by Patrick Suppes and Hans Reichenbach,[10] it is possible to fit together harmoniously the causal and statistical factors in explanatory contexts. Let me attempt to illustrate this point by discussing a concrete example.

A good deal of attention has recently been given in the press to cases of leukemia in military personnel who witnessed an atomic bomb test (code name "Smokey") at close range in 1957.[11] Statistical studies of the survivors of the bombings of Hiroshima and Nagasaki have established the fact that exposure to high levels of radiation, such as occur in an atomic blast, is statistically relevant to the occurrence of leukemia—indeed, that the probability of leukemia is closely correlated with the distance from the explosion.[12] A clear pattern of statistical relevance relations is exhibited here. If a particular person contracts leukemia, this fact may be explained by citing the fact that he was, say, 2 kilometers from the hypocenter at the time of the explosion. This relationship is further explained by the fact that individuals located at specific distances from atomic blasts of specified magnitude receive certain high doses of radiation.

This tragic example has several features to which I should like to call special attention:

1. The location of the individual at the time of the blast is statistically relevant to the occurrence of

leukemia; the probability of leukemia for a person located 2 kilometers from the hypocenter of an atomic blast is radically different from the probability of the disease in the population at large. Notice that the probability of such an individual contracting leukemia is not high; it is much smaller than one-half—indeed, in the case of Smokey it is much less than 1/100. But it is markedly higher than for a random member of the entire human population. It is the *statistical relevance* of exposure to an atomic blast, not a high probability, which has explanatory force.[13] Such examples defy explanation according to an inferential view which requires high inductive probability for statistical explanation.[14] The case of leukemia is subsumed under a statistical regularity, but it does not "follow inductively" from the explanatory facts.

2. There is a *causal process* which connects the occurrence of the bomb blast with the physiological harm done to people at some distance from the explosion. High energy radiation, released in the nuclear reactions, traverses the space between the blast and the individual. Although some of the details may not yet be known, it is a well-established fact that such radiation does interact with cells in a way which makes them susceptible to leukemia at some later time.

3. At each end of the causal process—i.e., the transmission of radiation from the bomb to the person—there is a *causal interaction*. The radiation is emitted as a result of a nuclear interaction when the bomb explodes, and it is absorbed by cells in the body of the victim. Each of these interactions may well be irreducibly statistical and indeterministic, but that is no reason to deny that they are causal.

4. The causal processes begin at a central place, and they travel outward at a finite velocity. A rather complex set of statistical relevance relations is explained by the propagation of a process, or set of processes, from a common central event.

In undertaking a general characterization of causal explanation, we must begin by carefully distinguishing between causal processes and causal interactions. The transmission of light from one place to another, and the motion of a material particle, are obvious examples of causal processes. The collision of two billiard balls, and the emission or absorption of a photon, are standard examples of causal inter-

actions. Interactions are the sorts of things we are inclined to identify as events. Relative to a particular context, an event is comparatively small in its spatial and temporal dimensions; processes typically have much larger durations, and they may be more extended in space as well. A light ray, traveling to earth from a distant star, is a process which covers a large distance and lasts for a long time. What I am calling a "causal process" is similar to what Russell called a "causal line."[15]

When we attempt to identify causal processes, it is of crucial importance to distinguish them from such pseudo-processes as a shadow moving across the landscape. This can best be done, I believe, by invoking Reichenbach's *mark criterion*.[16] Causal processes are capable of propagating marks or modifications imposed upon them; pseudo-processes are not. An automobile traveling along a road is an example of a causal process. If a fender is scraped as a result of a collision with a stone wall, the mark of that collision will be carried on by the car long after the interaction with the wall occurred. The shadow of a car moving along the shoulder is a pseudo-process. If it is deformed as it encounters a stone wall, it will immediately resume its former shape as soon as it passes by the wall. It will not transmit a mark or modification. For this reason, we say that a causal process can transmit information or causal influence; a pseudo-process cannot.[17]

When I say that a causal process has the capability of transmitting a causal influence, it might be supposed that I am introducing precisely the sort of mysterious power Hume warned us against. It seems to me that this danger can be circumvented by employing an adaptation of the "at-at" theory of motion, which Russell used so effectively in dealing with Zeno's paradox of the flying arrow.[18] The flying arrow—which is, by the way, a causal process—gets from one place to another by being *at* the appropriate intermediate points of space *at* the appropriate instants of time. Nothing more is involved in getting *from* one point *to* another. A mark, analogously, can be said to be propagated from the point of interaction at which it is imposed to later stages in the process if it appears *at* the appropriate intermediate stages in the process *at* the appropriate times without additional interactions which regenerate the mark. The precise formulation of this condition is a bit tricky, but I believe the basic idea is simple, and that the details can be worked out.[19]

If this analysis of causal processes is satisfactory, we have an answer to the question, raised by Hume, concerning the connection between cause and effect. If we think of a cause as one event, and of an effect as a distinct event, then the connection between them is simply a spatio-temporally continuous causal process. This sort of answer did not occur to Hume because he did not distinguish between causal processes and causal interactions. When he tried to analyze the connections between distinct events, he treated them as if they were chains of events with discrete links, rather than processes analogous to continuous filaments. I am inclined to attribute considerable philosophical significance to the fact that each link in a chain has adjacent links, while the points in a continuum do not have next-door neighbors. This consideration played an important role in Russell's discussion of Zeno's paradoxes.[20]

After distinguishing between causal interactions and causal processes, and after introducing a criterion by means of which to discriminate the pseudo-processes from the genuine causal processes, we must consider certain configurations of processes which have special explanatory import. Russell noted that we often find similar structures grouped symmetrically about a center—for example, concentric waves moving across an otherwise smooth surface of a pond, or sound waves moving out from a central region, or perceptions of many people viewing a stage from different seats in a theatre. In such cases, Russell postulates the existence of a central event—a pebble dropped into the pond, a starter's gun going off at a race-track, or a play being performed upon the stage—from which the complex array emanates.[21] It is noteworthy that Russell never suggests that the central event is to be explained on the basis of convergence of influences from remote regions upon that locale.

Reichenbach articulated a closely related idea in his *principle of the common cause.* If two or more events of certain types occur at different places, but occur at the same time more frequently than is to be expected if they occurred independently, then this apparent coincidence is to be explained in terms of a common causal antecedent.[22] If, for example, all of the electric lights in a particular area go out simultaneously, we do not believe that they just happened by chance to burn out at the same time. We attribute the coincidence to a common cause such as a blown fuse, a downed transmission line, or trouble at the generating station. If all of the students in a dormitory fall ill on the same night, it is attributed to spoiled food in the meal which all of them ate. Russell's similar structures arranged symmetrically about a center obviously qualify as the sorts of coincidences which require common causes for their explanations.[23] . . .

Now, consider another phenomenon which appears to be of an altogether different sort. If an electric current is passed through an electrolytic solution—for example, one containing a silver salt—a certain amount of metallic silver is deposited on the cathode. The amount deposited is proportional to the amount of electric charge which passes through the solution. In constructing a causal explanation of this phenomenon (known as electrolysis), we postulate that charged ions travel through the solution, and that the amount of charge required to deposit a singly charged ion is equal to the charge on the electron. The magnitude of the electron charge was empirically determined through the work of J. J. Thomson and Robert Millikan. The amount of electric charge required to deposit one mole of a monovalent metal is known as the Faraday, and by experimental determination, it is equal to 96,487 coulombs. When this number is divided by the charge on the electron (-1.602×10^{-19} coulombs), the result is Avogadro's number. Indeed, the Faraday is simply Avogadro's number of electron charges.

The fundamental fact to which I wish to call attention is that the value of Avogadro's number ascertained from the analysis of Brownian motion agrees, within the limits of experimental error, with the value obtained by electrolytic measurement. Without a common causal antecedent, such agreement would constitute a remarkable coincidence. The point may be put in this way. From the molecular kinetic theory of gases we can derive the statement form, "The number of molecules in a mole of gas is _____." From the electrochemical theory of electrolysis, we can derive the statement form, "The number of electron charges in a Faraday is _____." The astonishing fact is that the same number fills both blanks. In my opinion, the instrumentalist cannot, with impunity, ignore what must be an amazing correspondence between what happens when one scientist is watching smoke particles dancing in a container of gas while another scientist in a different laboratory is observing the electroplating of silver. Without an underlying causal mechanism—of the sort involved in the postulation of atoms, molecules, and ions—the coincidence would be as miraculous as if the number of grapes harvested in California in any given year were equal, up to the limits of observational error, to the number of coffee beans produced in Brazil in the same year. Avogadro's number, I must add, can be ascertained in a variety of other ways as well—e.g., X-ray diffraction from crystals—which also appear to be entirely different unless we postulate the existence of atoms, molecules, and ions. The principle of the common cause thus seems to apply directly to the explanation of observable regularities by appeal to unobservable entities. In this instance, to be sure, the common cause is not some sort of event; it is rather a common constant underlying structure which manifests itself in a variety of different situations.

Let me now summarize the picture of scientific explanation I have tried to outline. If we wish to explain a particular event, such as death by leukemia of GI Joe, we begin by assembling the factors statistically relevant to that occurrence—for example, his distance from the atomic explosion, the magnitude of the blast, and the type of shelter he was in. There will be many others, no doubt, but these will do for purposes of illustration. We must also obtain the probability values associated with the relevancy relations. The statistical relevance relations are statistical regularities, and we proceed to explain them. Although this differs substantially from things I have said previously, I no longer believe that the assemblage of relevant factors provides a complete explanation—or much of anything in the way of an explanation.[24] We do, I believe, have a bona fide explanation of an event if we have a complete set of statistically relevant factors, the pertinent probability values, *and* causal explanations of the relevance relations. Subsumption of a particular occurrence under statistical regularities—which, we recall, does not imply anything about the construction of deductive or inductive arguments—is a necessary part of any adequate explanation of its occurrence, but it is not the whole story. The causal explanation of the regularity is also needed. This claim, it should be noted, is in direct conflict with the received view, according to which the mere subsumption—deductive or inductive—of an event under a lawful regularity constitutes a complete explanation. One can, according to the received view, go on to ask for an explanation of any law used to explain a given event, but that is a different explanation. I am suggesting, on the contrary, that if the regularity invoked is not a causal regularity, then a causal explanation of that very regularity must be made part of the explanation of the event. . . .

On the basis of the foregoing characterization of scientific explanation, how should we answer the question posed at the outset? What does Laplace's demon lack, if anything, with respect to the explanatory aim of science? Several items may be mentioned. The demon *may* lack an adequate recognition of the distinction between causal laws and non-causal regularities; it *may* lack adequate knowledge of causal processes and of their ability to propagate causal influence; and it *may* lack adequate appreciation of the role of causal interactions in *producing* changes and regularities in the world. None of these capabilities was explicitly demanded by Laplace, for his analysis of causal relations in general was rather superficial.

What does scientific explanation offer, over and above the inferential capacity of prediction and retrodiction, at which the Laplacian demon excelled? It provides knowledge of the mechanisms of *production* and *propagation* of structure in the world. That goes some distance beyond mere recognition of regularities, and of the possibility of subsuming particular phenomena thereunder. It is my view that knowledge of the mechanisms of production and propagation of structure in the world yields scientific understanding, and that this is what we seek when we pose explanation-seeking why-questions. The answers are well worth having. That is why we ask, not only "What?" but "Why?"

NOTES

1. Book I.2, 71b, 17–24.
2. Antoine Arnauld, *The Art of Thinking* (Indianapolis: Bobbs-Merrill, 1964), p. 330. "Such demonstrations may convince the mind, but they do not enlighten it; and enlightenment ought to be the

principal fruit of true knowledge. Our minds are unsatisfied unless they know not only *that* a thing is but *why* it is."

3. Karl Pearson, *The Grammar of Science*, 3rd ed. (New York: Meridian Books, 1957), p. xi. The first edition appeared in 1892, the second in 1899, and the third was first published in 1911. In the preface to the third edition, Pearson remarked, just before the statement quoted in the text, "Reading the book again after many years, it was surprising to find how the heterodoxy of the 'eighties had become the commonplace and accepted doctrine of today." Since the "commonplace and accepted doctrine" of 1911 has again become heterodox, one wonders to what extent such changes in philosophic doctrine are mere matters of changing fashion.

4. Bertrand Russell, *Our Knowledge of the External World* (London: George Allen & Unwin Ltd., 1922), p. 179.

5. The classic paper by Carl G. Hempel and Paul Oppenheim, "Studies in the Logic of Explanation," which has served as the point of departure for almost all subsequent discussion, was first published just thirty years ago in 1948 in *Philosophy of Science*, Vol. 15, pp. 135–175.

6. See, for example, his recent paper, "Causation as Explanation," *Nous*, Vol. 9 (1975), pp. 3–16.

7. Hempel's conceptions have been most thoroughly elaborated in his monographic essay, "Aspects of Scientific Explanation," in *Aspects of Scientific Explanation and Other Essays in the Philosophy of Science* (New York: Free Press, 1965), pp. 331–496.

8. P. S. Laplace, *A Philosophical Essay on Probabilities* (New York: Dover Publications, 1951), pp. 3–6.

9. In "A Third Dogma of Empiricism" in Robert Butts and Jaakko Hintikka, eds., *Basic Problems in Methodology and Linguistics* (Dordrecht: D. Reidel Publishing Co., 1977), pp. 149–166, I have given an extended systematic critique of the thesis (dogma?) that scientific explanations are arguments.

10. Patrick Suppes, *A Probabilistic Theory of Causation* (Amsterdam: North-Holland Publishing Co., 1970); Hans Reichenbach, *The Direction of Time* (Berkeley & Los Angeles: University of California Press, 1956), Chap. IV.

11. See *Nature*, Vol. 271 (2 Feb. 1978), p. 399.

12. Irving Copi, *Introduction to Logic*, 4th ed. (New York: Macmillan Publishing Co., 1972), pp. 396–397, cites this example from *No More War* by Linus Pauling.

13. According to the article in *Nature* (note 11), "the eight reported cases of leukaemia among 2235 [soldiers] was 'out of the normal range'." Dr. Karl Z. Morgan "had 'no doubt whatever' that [the] radiation had caused the leukaemia now found in those who had taken part in the manoeuvers."

14. Hempel's inductive-statistical model, as formulated in "Aspects of Scientific Explanation" (1965), embodied such a high probability requirement, but in "Nachwort 1976" inserted into a German translation of this article (*Aspekte wissenschaftlicher Erklärung*, Walter de Gruyter, 1977) this requirement is retracted.

15. Bertrand Russell, *Human Knowledge, Its Scope and Limits* (New York: Simon and Schuster, 1948), p. 459.

16. Hans Reichenbach, *The Philosophy of Space and Time* (New York: Dover Publications, 1958), Sec. 21.

17. See my "Theoretical Explanation" Sec. 3, pp. 129–134, in Stephan Körner, ed., *Explanation* (Oxford: Basil Blackwell, 1975), for a more detailed discussion of this distinction. It is an unfortunate lacuna in Russell's discussion of causal lines—though one which can easily be repaired—that he does not notice the distinction between causal processes and pseudo-processes.

18. See Wesley C. Salmon, ed., *Zeno's Paradoxes* (Indianapolis: Bobbs-Merrill, 1970), p. 23, for a description of this "theory."

19. I have made an attempt to elaborate this idea in "An 'At-At' Theory of Causal Influence," *Philosophy of Science*, Vol. 44, No. 2 (June 1977), pp. 215–224. Because of a criticism due to Nancy Cartwright, I now realize that the formulation given in this article is not entirely satisfactory, but I think the difficulty can be repaired.

20. Russell, *Our Knowledge of the External World*, Lecture VI, "The Problem of Infinity Considered Historically." The relevant portions are reprinted in my anthology, *Zeno's Paradoxes*.

21. Russell, *Human Knowledge*, pp. 460–475.

22. Reichenbach, *The Direction of Time*, Sec. 19.

23. In "Theoretical Explanation" I discuss the explanatory import of the common cause principle in greater detail.

24. Compare Wesley C. Salmon, et al., *Statistical Explanation and Statistical Relevance* (Pittsburgh: University of Pittsburgh Press, 1971), p. 78. There I ask, "What more could one ask of an explanation?" The present paper attempts to present at least part of the answer.

14

BAS VAN FRAASSEN

The Pragmatics of Explanation

Traditionally, theories are said to bear two sorts of relation to the observable phenomena: *description* and *explanation*. Description can be more or less accurate, more or less informative; as a minimum, the facts must 'be allowed by' the theory (fit some of its models), as a maximum the theory actually implies the facts in question. But in addition to a (more or less informative) description, the theory may provide an explanation. This is something 'over and above' description; for example, Boyle's law describes the relationship between the pressure, temperature, and volume of a contained gas, but does not explain it—kinetic theory explains it. The conclusion was drawn, correctly I think, that even if two theories are strictly empirically equivalent they may differ in that one can be used to answer a given request for explanation while the other cannot.

Many attempts were made to account for such 'explanatory power' purely in terms of those features and resources of a theory that make it informative (that is, allow it to give better descriptions). On Hempel's view, Boyle's law does explain these empirical facts about gases, but minimally. The kinetic theory is perhaps better *qua* explanation simply because it gives so much more information about the behavior of gases, relates the three quantities in question to other observable quantities, has a beautiful simplicity, unifies our over-all picture of the world, and so on. The use of more sophisticated statistical relationships by Wesley Salmon and James Greeno (as well as by I. J. Good, whose theory of

such concepts as weight of evidence, corroboration, explanatory power, and so on deserves more attention from philosophers), are all efforts along this line. If they had succeeded, an empiricist could rest easy with the subject of explanation.

But these attempts ran into seemingly insuperable difficulties. The conviction grew that explanatory power is something quite irreducible, a special feature differing in kind from empirical adequacy and strength. An inspection of examples defeats any attempt to identify the ability to explain with any complex of those more familiar and down-to-earth virtues that are used to evaluate the theory *qua* description. Simultaneously it was argued that what science is really after is understanding, that this consists in being in a position to explain, hence what science is really after goes well beyond empirical adequacy and strength. Finally, since the theory's ability to explain provides a clear reason for accepting it, it was argued that explanatory power is evidence for the *truth* of the theory, special evidence that goes beyond any evidence we may have for the theory's empirical adequacy.

Around the turn of the century, Pierre Duhem had already tried to debunk this view of science by arguing that explanation is not an aim of science. In retrospect, he fostered that explanation–mysticism which he attacked. For he was at pains to grant that explanatory power does not consist in resources for description. He argued that only metaphysical theories explain, and that metaphysics is an enterprise foreign to science. But fifty years later, Quine having argued that there is no demarcation between science and philosophy, and the difficulties of the ametaphysical stance of the positivist-oriented philosophies having made a return to metaphysics tempting, one noticed that scientific activity does involve explanation, and Duhem's argument was deftly reversed.

Once you decide that explanation is something irreducible and special, the door is opened to elaboration by means of further concepts pertaining thereto, all equally irreducible and special. The premises of an explanation have to include lawlike statements; a statement is lawlike exactly if it implies some non-trivial counterfactual conditional statement; but it can do so only by asserting relationships of necessity in nature. Not all classes correspond to genuine properties; properties and propensities figure in explanation. Not everyone has joined this return to essentialism or neo-Aristotelian realism, but some eminent realists have publicly explored or advocated it.

Even more moderate elaborations of the concept of explanation make mysterious distinctions. Not every explanation is a scientific explanation. Well then, that irreducible explanation-relationship comes in several distinct types, one of them being scientific. A scientific explanation has a special form, and adduces only special sorts of information to explain—information about causal connections and causal processes. Of course, a causal relationship is just what 'because' must denote; and since the *summum bonum* of science is explanation, science must be attempting even to describe something beyond the observable phenomena, namely causal relationships and processes.

These last two paragraphs describe the flights of fancy that become appropriate if explanation is a relationship *sui generis* between theory and fact. But there is no direct evidence for them at all, because if you ask a scientist to explain something to you, the information he gives you is not different in kind (and does not sound or look different) from the information he gives you when you ask for a description. Similarly in 'ordinary' explanations: the information I adduce to explain the rise in oil prices, is information I would have given you to a battery of requests for description of oil supplies, oil producers, and oil consumption. To call an explanation scientific, is to say nothing about its form or the sort of information adduced, but only that the explanation draws on science to get this information (at least to some extent) and, more importantly, that the criteria of evaluation of how good an explanation it is, are being applied using a scientific theory. . . .

The discussion of explanation went wrong at the very beginning when explanation was conceived of as a relationship like description: a relation between theory and fact. Really it is a three-term relation, between theory, fact, and context. No wonder that no single relation between theory and fact ever managed to fit more than a few examples! Being an explanation is essentially relative, for an explanation is an *answer*. (In just that sense, being a daughter is something relative: every woman is a daughter, and every daughter is a woman, yet being a daughter is not the same as being a woman.) Since an explanation is an answer, it is evaluated *vis-à-vis* a question, which is a request for information. But exactly what is requested, by means of the question 'Why is it the case that *P*?', differs from context to context. In addition, the background theory plus data relative to which the question is evaluated, as arising or not arising, depends on the context. And even what part of that background information is to be used to evaluate how good the answer is, *qua* answer to that question, is a contextually determined factor. So to say that a given theory can be used to explain a certain fact, is always elliptic for: there is a proposition which is a telling answer, relative to this theory, to the request for information about certain facts (those counted as relevant for *this* question) that bear on a comparison between this fact which is the case, and certain (contextually specified) alternatives which are not the case.

So scientific explanation is not (pure) science but an application of science. It is a use of science to satisfy certain of our desires; and these desires are quite specific in a specific context, but they are always desires for descriptive information. (Recall: every daughter is a woman.) The exact content of the desire, and the evaluation of how well it is satisfied, varies from context to context. It is not a single desire, the same in all cases, for a very special sort of thing, but rather, in each case, a different desire for something of a quite familiar sort.

Hence there can be no question at all of explanatory power as such (just as it would be silly to speak of the 'control power' of a theory, although of course we rely on theories to gain control over nature and circumstances). Nor can there be any question of explanatory success as providing evidence for the truth of a theory that goes beyond any evidence we have for its providing an adequate description of the phenomena. For in each case, a success of explanation is a success of adequate and informative description. And while it is true that we seek for explanation, the value of this search for science is that the search for explanation is *ipso facto* a search for empirically adequate, empirically strong theories.

15

PHILIP KITCHER

Explanatory Unification

1. THE DECLINE AND FALL OF THE COVERING LAW MODEL

One of the great apparent triumphs of logical empiricism was its official theory of explanation. In a series of lucid studies (Hempel 1965, Chapters 9, 10, 12; Hempel 1962; Hempel 1966), C. G. Hempel showed how to articulate precisely an idea which had received a hazy formulation from traditional empiricists such as Hume and Mill. The picture of explanation which Hempel presented, the *covering law model*, begins with the idea that explanation is derivation. When a scientist explains a phenomenon, he derives (deductively or inductively) a sentence describing that phenomenon (the *explanandum* sentence) from a set of sentences (the *explanans*) which must contain at least one general law.

Today the model has fallen on hard times. Yet it was never the empiricists' whole story about explanation. Behind the official model stood an unofficial model, a view of explanation which was not treated precisely, but which sometimes emerged in discussions of theoretical explanation. In contrast-

ing scientific explanation with the idea of reducing unfamiliar phenomena to familiar phenomena, Hempel suggests this unofficial view: "What scientific explanation, especially theoretical explanation, aims at is not [an] intuitive and highly subjective kind of understanding, but an objective kind of insight that is achieved by a systematic unification, by exhibiting the phenomena as manifestations of common, underlying structures and processes that conform to specific, testable, basic principles" (Hempel 1966, p. 83; see also Hempel 1965, pp. 345, 444). Herbert Feigl makes a similar point: "The aim of scientific explanation throughout the ages has been *unification*, i.e., the comprehending of a maximum of facts and regularities in terms of a minimum of theoretical concepts and assumptions" (Feigl 1970, p. 12).

This unofficial view, which regards explanation as unification, is, I think, more promising than the official view. My aim in this paper is to develop the view and to present its virtues. Since the picture of explanation which results is rather complex, my exposition will be programmatic, but I shall try to show that the unofficial view can avoid some prominent shortcomings of the covering law model.

Philosophy of Science 48 (1981), pp. 507–531. Reprinted with permission from the University of Chicago Press.

Why should we want an account of scientific explanation? Two reasons present themselves. Firstly, we would like to understand and to evaluate the popular claim that the natural sciences do not merely pile up unrelated items of knowledge of more or less practical significance, but that they increase our understanding of the world. A theory of explanation should show us *how* scientific explanation advances our understanding. (Michael Friedman cogently presents this demand in his (1974).) Secondly, an account of explanation ought to enable us to comprehend and to arbitrate disputes in past and present science. Embryonic theories are often defended by appeal to their explanatory power. A theory of explanation should enable us to judge the adequacy of the defense.

The covering law model satisfies neither of these *desiderata*. Its difficulties stem from the fact that, when it is viewed as providing a set of necessary *and sufficient* conditions for explanation, it is far too liberal. Many derivations which are intuitively nonexplanatory meet the conditions of the model. Unable to make relatively gross distinctions, the model is quite powerless to adjudicate the more subtle considerations about explanatory adequacy which are the focus of scientific debate. Moreover, our ability to derive a description of a phenomenon from a set of premises *containing a law* seems quite tangential to our understanding of the phenomenon. Why should it be that exactly those derivations which employ laws advance our understanding?

The unofficial theory appears to do better. As Friedman points out, we can easily connect the notion of unification with that of understanding. (However, as I have argued in my (1976), Friedman's analysis of unification is faulty; the account of unification offered below is indirectly defended by my diagnosis of the problems for his approach.) Furthermore, as we shall see below, the acceptance of some major programs of scientific research—such as the Newtonian program of eighteenth century physics and chemistry, and the Darwinian program of nineteenth century biology—depended on recognizing promises for unifying, and thereby explaining, the phenomena. Reasonable skepticism may protest at this point that the attractions of the unofficial view stem from its unclarity. Let us see.

2. EXPLANATION: SOME PRAGMATIC ISSUES

Our first task is to formulate the problem of scientific explanation clearly, filtering out a host of issues which need not concern us here. The most obvious way in which to categorize explanation is to view it as an activity. In this activity we answer the actual or anticipated questions of an actual or anticipated audience. We do so by presenting reasons. We draw on the beliefs we hold, frequently using or adapting arguments furnished to us by the sciences.

Recognizing the connection between explanations and arguments, proponents of the covering law model (and other writers on explanation) have identified explanations as special types of arguments. But although I shall follow the covering law model in employing the notion of argument to characterize that of explanation, I shall not adopt the ontological thesis that explanations are arguments. Following Peter Achinstein's thorough discussion of ontological issues concerning explanation in his (1977), I shall suppose that an explanation is an ordered pair consisting of a proposition and an act type.[1] The relevance of arguments to explanation resides in the fact that what makes an ordered pair (p, explaining q) an explanation is that a sentence expressing p bears an appropriate relation to a particular argument. (Achinstein shows how the central idea of the covering law model can be viewed in this way.) So I am supposing that there are acts of explanation which draw on arguments supplied by science, reformulating the traditional problem of explanation as the question: What features should a scientific argument have if it is to serve as the basis for an act of explanation?[2]

The complex relation between scientific explanation and scientific argument may be illuminated by a simple example. Imagine a mythical Galileo confronted by a mythical fusilier who wants to know why his gun attains maximum range when it is mounted on a flat plain, if the barrel is elevated at 45° to the horizontal. Galileo reformulates this question as the question of why an ideal projectile, projected with fixed velocity from a perfectly smooth horizontal plane and subject only to gravitational acceleration, attains maximum range when the angle of elevation of the projection is 45°. He defends this reformulation by arguing that the effects of air resistance in the case of the actual projectile, the cannonball, are insignificant, and that the curvature of the earth and the unevenness of the ground can be neglected. He then selects a kinematical argument which shows that, for fixed velocity, an ideal projectile attains maximum range when the angle of elevation is 45°. He adapts this argument by explaining to the fusilier some unfamiliar terms ('uniform acceleration', let us say), motivating some problematic principles (such as the law of composition of velocities), and by omitting some obvious computational steps. Both Galileo and the fusilier depart satisfied.

The most general problem of scientific explanation is to determine the conditions which must be met if science is to be used in answering an explanation-seeking question Q. I shall restrict my attention to explanation-seeking why-questions, and I shall attempt to determine the conditions under which an argument whose conclusion is S can be used to answer the question "Why is it the case that S?". More colloquially, my project will be that of deciding when an argument explains why its conclusion is true.[3]

We leave on one side a number of interesting, and difficult issues. So, for example, I shall not discuss the general relation between explanation-seeking questions and the arguments which can be used to answer them, nor the pragmatic conditions governing the idealization of questions and the adaptation of scientific arguments to the needs of the audience. (For illuminating discussions of some of these issues, see Bromberger 1962.) Given that so much is dismissed, does anything remain?

In a provocative article (van Fraassen 1977), Bas van Fraassen denies, in effect, that there are any issues about scientific explanation other than the pragmatic questions I have just banished. After a survey of attempts to provide a theory of explanation he appears to conclude that the idea that explanatory power is a special virtue of theories is a myth. We accept scientific theories on the basis of their empirical adequacy and simplicity, and, having done so, we use the arguments with which they supply us to give explanations. This activity of applying scientific arguments in explanation accords with extra-scientific, "pragmatic", conditions. Moreover, our views about these extra-scientific factors are re-

vised in the light of our acceptance of new theories: ". . . science schools our imagination so as to revise just those prior judgments of what satisfies and eliminates wonder" (van Fraassen 1977, p. 150). Thus there are no context-independent conditions, beyond those of simplicity and empirical adequacy which distinguish arguments for use in explanation.

van Fraassen's approach does not fit well with some examples from the history of science—such as the acceptance of Newtonian theory of matter and Darwin's theory of evolution—examples in which the explanatory promise of a theory was appreciated in advance of the articulation of a theory with predictive power. Moreover, the account I shall offer provides an answer to skepticism that no "global constraints" (van Fraassen 1977, p. 146) on explanation can avoid the familiar problems of asymmetry and irrelevance, problems which bedevil the covering law model. I shall try to respond to van Fraassen's challenge by showing that there are certain context-independent features of arguments which distinguish them for application in response to explanation-seeking why-questions, and that we can assess theories (including embryonic theories) by their ability to provide us with such arguments. Hence I think that it is possible to defend the thesis that historical appeals to the explanatory power of theories involve recognition of a virtue over and beyond considerations of simplicity and predictive power.

Resuming our main theme, we can use the example of Galileo and the fusilier to achieve a further refinement of our problem. Galileo selects and adapts an argument from his new kinematics—that is, he draws an argument from a set of arguments available for explanatory purposes, a set which I shall call the *explanatory store*. We may think of the sciences not as providing us with many unrelated individual arguments which can be used in individual acts of explanation, but as offering a reserve of explanatory arguments, which we may tap as need arises. Approaching the issue in this way, we shall be led to present our problem as that of specifying the conditions which must be met by the explanatory store.

The set of arguments which science supplies for adaptation in acts of explanation will change with our changing beliefs. Therefore the appropriate *analysandum* is the notion of the store of arguments relative to a set of accepted sentences. Suppose that, at the point in the history of inquiry which interests us, the set of accepted sentences is K. (I shall assume, for simplicity's sake, that K is consistent. Should our beliefs be inconsistent then it is more appropriate to regard K as some tidied version of our beliefs.) The general problem I have set is that of specifying $E(K)$, the *explanatory store over K*, which is the set of arguments acceptable as the basis for acts of explanation by those whose beliefs are exactly the members of K. (For the purposes of this paper I shall assume that, for each K there is exactly one $E(K)$.)

The unofficial view answers the problem: for each K, $E(K)$ is the set of arguments which best unifies K. My task is to articulate the answer. I begin by looking at two historical episodes in which the desire for unification played a crucial role. In both cases, we find three important features: (i) prior to the articulation of a theory with high predictive power, certain proposals for theory construction are favored on grounds of their explanatory promise; (ii) the explanatory power of embryonic theories is explicitly tied to the notion of unification; (iii) particular features of the theories are taken to support their claims to unification. Recognition of (i) and (ii) will illustrate points that have already been made, while (iii) will point towards an analysis of the concept of unification.

3. A NEWTONIAN PROGRAM

Newton's achievements in dynamics, astronomy and optics inspired some of his successors to undertake an ambitious program which I shall call "dynamic corpuscularianism."[4] *Principia* had shown how to obtain the motions of bodies from a knowledge of the forces acting on them, and had also demonstrated the possibility of dealing with gravitational systems in a unified way. The next step would be to isolate a few basic force laws, akin to the law of universal gravitation, so that, applying the basic laws to specifications of the dispositions of the ultimate parts of bodies, all of the phenomena of nature could be derived. Chemical reactions, for example, might be understood in terms of the rearrangement of ultimate parts under the action of cohesive and repulsive forces. The phenomena of reflection, refraction and diffraction of light might be viewed as resulting from a special force of attraction between light corpuscles and ordinary matter. These speculations encouraged eighteenth century Newtonians to construct very general hypotheses about interatomic forces—even in the absence of any confirming evidence for the existence of such forces.

In the preface to *Principia*, Newton had already indicated that he took dynamic corpuscularianism to be a program deserving the attention of the scientific community:

> I wish we could derive the rest of the phenomena of Nature by the same kind of reasoning from mechanical principles, for I am induced by many reasons to suspect that they may all depend upon certain forces by which the particles of bodies, by some causes hitherto unknown, are either mutually impelled towards one another, and cohere in regular figures, or are repelled and recede from one another (Newton 1962, p. xviii. See also Newton 1952, pp. 401–402).

This, and other influential passages, inspired Newton's successors to try to complete the unification of science by finding further force laws analogous to the law of universal gravitation. Dynamic corpuscularianism remained popular so long as there was promise of significant unification. Its appeal began to fade only when repeated attempts to specify force laws were found to invoke so many different (apparently incompatible) attractive and repulsive forces that the goal of unification appeared unlikely. Yet that goal could still motivate renewed efforts to implement the program. In the second half of the eighteenth century Boscovich revived dynamic corpuscularian hopes by claiming that the whole of natural philosophy can be reduced to "one law of forces existing in nature".[5]

The passage I have quoted from Newton suggests the nature of the unification that was being sought. *Principia* had exhibited how one style of argument, one "kind of reasoning from mechanical principles", could be used in the derivation of descriptions of many, diverse, phenomena. The unifying power of Newton's work consisted in its demonstration that one *pattern* of argument could be used again and again in the derivation of a wide range of accepted sentences. In searching for force laws analogous to the law of universal gravitation,

Newton's successors were trying to generalize the pattern of argument presented in *Principia,* so that one "kind of reasoning" would suffice to derive all phenomena of motion. If, furthermore, the facts studied by chemistry, optics, physiology and so forth, could be related to facts about particle motion, then one general pattern of argument would be used in the derivation of all phenomena. I suggest that this is the ideal of unification at which Newton's immediate successors aimed, which came to seem less likely to be attained as the eighteenth century wore on, and which Boscovich's work endeavored, with some success, to reinstate.

4. THE RECEPTION OF DARWIN'S EVOLUTIONARY THEORY

The picture of unification which emerges from the last section may be summarized quite simply: a theory unifies our beliefs when it provides one (or more generally, a few) pattern(s) of argument which can be used in the derivation of a large number of sentences which we accept. I shall try to develop this idea more precisely in later sections. But first I want to show how a different example suggests the same view of unification.

In several places, Darwin claims that his conclusion that species evolve through natural selection should be accepted because of its explanatory power, that " . . . the doctrine must sink or swim according as it groups and explains phenomena" (F. Darwin 1887; Vol. 2, p. 155, quoted in Hull 1974, p. 292). Yet, as he often laments, he is unable to provide any complete derivation of any biological phenomenon—our ignorance of the appropriate facts and regularities is "profound". How, then, can he contend that the primary virtue of the new theory is its explanatory power?

The answer lies in the fact that Darwin's evolutionary theory promises to unify a host of biological phenomena (C. Darwin 1964, pp. 243–244). The eventual unification would consist in derivations of descriptions of these phenomena which would instantiate a common pattern. When Darwin expounds his doctrine what he offers us is the pattern. Instead of detailed explanations of the presence of some particular trait in some particular species, Dar-

win presents two "imaginary examples" (C. Darwin 1964; pp. 90–96) and a diagram, which shows, in a general way, the evolution of species *represented by schematic letters* (1964, pp. 116–126). In doing so, he exhibits a pattern of argument, which, he maintains, can be instantiated, *in principle,* by a complete and rigorous derivation of descriptions of the characteristics of any current species. The derivation would employ the principle of natural selection—as well as premises describing ancestral forms and the nature of their environment and the (unknown) laws of variation and inheritance. In place of detailed evolutionary stories, Darwin offers *explanation-sketches.* By showing how a particular characteristic would be advantageous to a particular species, he indicates an explanation of the emergence of that characteristic in the species, suggesting the outline of an argument instantiating the general pattern.

From this perspective, much of Darwin's argumentation in the *Origin* (and in other works) becomes readily comprehensible. Darwin attempts to show how his pattern can be applied to a host of biological phenomena. He claims that, by using arguments which instantiate the pattern, we can account for analogous variations in kindred species, for the greater variability of specific (as opposed to generic) characteristics, for the facts about geographical distribution and so forth. But he is also required to resist challenges that the pattern cannot be applied in some cases, that premises for arguments instantiating the pattern will not be forthcoming. So, for example, Darwin must show how evolutionary stories, fashioned after his pattern, can be told to account for the emergence of complex organs. In both aspects of his argument, whether he is responding to those who would limit the application of his pattern or whether he is campaigning for its use within a realm of biological phenomena, Darwin has the same goal. He aims to show that his theory should be accepted because it unifies and explains.

5. ARGUMENT PATTERNS

Our two historical examples[6] have led us to the conclusion that the notion of an argument pattern is central to that of explanatory unification. Quite different considerations could easily have pointed us in the same direction. If someone were to distinguish between the explanatory worth of two arguments instantiating a common pattern, then we would regard that person as an explanatory deviant. To grasp the concept of explanation is to see that if one accepts an argument as explanatory, one is thereby committed to accepting as explanatory other arguments which instantiate the same pattern. . . .

6. EXPLANATION AS UNIFICATION

As I have posed it, the problem of explanation is to specify which set of arguments we ought to accept for explanatory purposes given that we hold certain sentences to be true. Obviously this formulation can encourage confusion: we must not think of a scientific community as *first* deciding what sentences it will accept and *then* adopting the appropriate set of arguments. The Newtonian and Darwinian examples should convince us that the promise of explanatory power enters into the modification of our beliefs. . . .

7. ASYMMETRY, IRRELEVANCE AND ACCIDENTAL GENERALIZATION

Some familiar difficulties beset the covering law model. The *asymmetry problem* arises because some scientific laws have the logical form of equivalences. Such laws can be used "in either direction." Thus a law asserting that the satisfaction of a condition C_1 is equivalent to the satisfaction of a condition C_2 can be used in two different kinds of argument. From a premise asserting that an object meets C_1, we can use the law to infer that it meets C_2; conversely, from a premise asserting that an object meets C_2, we can use the law to infer that it meets C_1. The asymmetry problem is generated by noting that in many such cases one of these derivations can be used in giving explanations while the other cannot.

Consider a hoary example. (For further examples, see Bromberger 1966.) We can explain why a simple pendulum has the period it does by deriving a specification of the period from a specification of the length and the law which relates length and period. But we cannot explain the length of the pendulum by deriving a specification of the length from a specification of the period and the same law. What accounts for our different assessment of these two arguments? Why does it seem that one is explanatory while the other "gets things backwards"? The covering law model fails to distinguish the two, and thus fails to provide answers.

The *irrelevance problem* is equally vexing. The problem arises because we can sometimes find a lawlike connection between an accidental and irrelevant occurrence and an event or state which would have come about independently of that occurrence. Imagine that Milo the magician waves his hands over a sample of table salt, thereby "hexing" it. It is true (and I shall suppose, lawlike) that all hexed samples of table salt dissolve when placed in water. Hence we can construct a derivation of the dissolving of Milo's hexed sample of salt by citing the circumstances of the hexing. Although this derivation fits the covering law model, it is, by our ordinary lights, nonexplanatory. (This example is given by Wesley Salmon in his (1970); Salmon attributes it to Henry Kyburg. For more examples, see Achinstein 1971.)

The covering law model explicitly debars a further type of derivation which any account of explanation ought to exclude. Arguments whose premises contain no laws, but which make essential use of accidental generalizations are intuitively nonexplanatory. Thus, if we derive the conclusion that Horace is bald from premises stating that Horace is a member of the Greenbury School Board and that all members of the Greenbury School Board are bald we do not thereby explain why Horace is bald. (See Hempel 1965, p. 339.) We shall have to show that our account does not admit as explanatory derivations of this kind.

I want to show that the account of explanation I have sketched contains sufficient resources to solve these problems.[7] In each case we shall pursue a common strategy. Faced with an argument we want to exclude from the explanatory store we endeavor to show that any set of arguments containing the unwanted argument could not provide the best unification of our beliefs. Specifically, we shall try to show either that any such set of arguments will be more limited than some other set with an equally satisfactory basis, or that the basis of the set must fare worse according to the criterion of using the smallest number of most stringent patterns. . . . In actual practice, this strategy for exclusion is less complicated than one might fear, and, as we shall see, its applications to the examples just discussed brings out what is intuitively wrong with the derivations we reject.

Consider first the irrelevance problem. Suppose that we were to accept as explanatory the argument which derives a description of the dissolving of the salt from a description of Milo's act of hexing. What will be our policy for explaining the dissolving of samples of salt which have not been hexed? If we offer the usual chemical arguments in these latter cases then we shall commit ourselves to an inflated basis for the set of arguments we accept as explanatory. For, unlike the person who explains *all* cases of dissolving of samples of salt by using the standard chemical pattern of argument, we shall be committed to the use of two different patterns of argument in covering such cases. Nor is the use of the extra pattern of argument offset by its applicability in explaining other phenomena. Our policy employs one extra pattern of argument without extending the range of things we can derive from our favored set of arguments. Conversely, if we eschew the standard chemical pattern of argument (just using the pattern which appeals to the hexing) we shall find ourselves unable to apply our favored pattern to cases in which the sample of salt dissolved has not been hexed. Moreover, the pattern we use will not fall under the more general patterns we employ to explain chemical phenomena such as solution, precipitation and so forth. Hence the unifying power of the basis for our preferred set of arguments will be less than that of the basis for the set of arguments we normally accept as explanatory.[8]

If we explain the dissolving of the sample of salt which Milo has hexed by appealing to the hexing then we are faced with the problems of explaining the dissolving of unhexed samples of salt. . . . We have two options: (a) to adopt two patterns of argument corresponding to the two kinds of case; (b) to adopt one pattern of argument whose instantiations apply just to the cases of hexed salt. The general moral is that appeals to hexing fasten on a local and accidental feature of the cases of solution. By contrast our standard arguments instantiate a pattern which can be generally applied.[9]

A similar strategy succeeds with the asymmetry problem. We have general ways of explaining why bodies have the dimensions they do. Our practice is to describe the circumstances leading to the formation of the object in question and then to show how it has since been modified. Let us call explanations of this kind "origin and development derivations." (In some cases, the details of the original formation of the object are more important; with other objects, features of its subsequent modification are crucial.) Suppose now that we admit as explanatory a derivation of the length of a simple pendulum from a specification of the period. Then we shall either have to explain the lengths of *non*swinging bodies by employing quite a different style of explanation (an origin and development derivation) or we shall have to forgo explaining the lengths of such bodies. The situation is exactly parallel to that of the irrelevance problem. Admitting the argument which is intuitively nonexplanatory saddles us with a set of arguments which is less good at unifying our beliefs than the set we normally choose for explanatory purposes.

Our approach also solves a more refined version of the pendulum problem (given by Paul Teller in his (1974)). Many bodies which are not currently executing pendulum motion *could* be making small oscillations, and, were they to do so, the period of their motion would be functionally related to their dimensions. For such bodies we can specify the *dispositional period* as the period which the body would have if it were to execute small oscillations. Someone may now suggest that we can construct derivations of the dimensions of bodies from specifications of their dispositional periods, thereby generating an argument pattern which can be applied as generally as that instantiated in origin and development explanations. This suggestion is mistaken. There are some objects—such as the Earth and the Crab Nebula—which *could not* be pendulums, and for which the notion of a dispositional period makes no sense. Hence, the argument pattern proposed cannot entirely supplant our origin and development derivations, and, in consequence, acceptance of it would fail to achieve the best unification of our beliefs.

The problem posed by accidental generalizations can be handled in parallel fashion. We have a general pattern of argument, using principles of physiology, which we apply to explain cases of baldness. This pattern is generally applicable, whereas that which derives ascriptions of baldness using the principle that all members of the Greenbury School Board are bald is not. . . .

Of course, this does not show that an account of explanation along the lines I have suggested would sanction only derivations which satisfy the conditions imposed by the covering law model. For I have not argued that an explanatory derivation need contain *any* sentence of universal form. What *does* seem to follow from the account of explanation as unification is that explanatory arguments must not use accidental generalization, and, in this respect, the new account appears to underscore and generalize an important insight of the covering law model. Moreover, our success with the problems of asymmetry and irrelevance indicates that, even in the absence of a detailed account of the notion of stringency and of the way in which generality of the consequence set is weighed against paucity and stringency of the patterns in the basis, the view of explanation as unification has the resources to solve some traditional difficulties for theories of explanation. . . .

8. CONCLUSIONS

I have sketched an account of explanation as unification, attempting to show that such an account has the resources to provide insight into episodes in the history of science and to overcome some traditional problems for the covering law model. In conclusion, let me indicate very briefly how my view of explanation as unification suggests how scientific explanation yields understanding. By using a few patterns of argument in the derivation of many beliefs we minimize the number of *types* of premises we must take as underived. That is, we reduce, in so far as possible, the number of types of facts we must accept as brute. Hence we can endorse something close to Friedman's view of the merits of explanatory unification (Friedman 1974, pp. 18–19).

Quite evidently, I have only *sketched* an account of explanation. To provide precise analyses of the notions I have introduced, the basic approach to explanation offered here must be refined against concrete exam-

ples of scientific practice. What needs to be done is to look closely at the argument patterns favored by scientists and attempt to understand what characteristics they share. If I am right, the scientific search for explanation is governed by a maxim, once formulated succinctly by E. M. Forster. Only connect.

NOTES

1. Strictly speaking, this is one of two views which emerge from Achinstein's discussion and which he regards as equally satisfactory. As Achinstein goes on to point out, either of these ontological theses can be developed to capture the central idea of the covering law model.

2. To pose the problem in this way we may still invite the charge that *arguments* should not be viewed as the bases for acts of explanation. Many of the criticisms leveled against the covering law model by Wesley Salmon in his seminal paper on statistical explanation (Salmon 1970) can be reformulated to support this charge. My discussion in section 7 will show how some of the difficulties raised by Salmon for the covering law model do not bedevil my account. However, I shall not respond directly to the points about statistical explanation and statistical inference advanced by Salmon and by Richard Jeffrey in his (1970). I believe that Peter Railton has shown how these specific difficulties concerning statistical explanation can be accommodated by an approach which takes explanations to be (or be based on) arguments (see Railton 1978), and that the account offered in section 4 of his paper can be adapted to complement my own.

3. Of course, in restricting my attention to why-questions I am following the tradition of philosophical discussion of scientific explanation: as Bromberger notes in section IV of his (1966) not all explanations are directed at why-questions, but attempts to characterize explanatory responses to why-questions have a special interest for the philosophy of science because of the connection to a range of methodological issues. I believe that the account of explanation offered in the present paper could be extended to cover explanatory answers to some other kinds of questions (such as how-questions). But I do want to disavow the claim that unification is relevant to all types of explanation. If one believes that explanations are sometimes offered in response to what-questions (for example), so that it is correct to talk of someone explaining what a gene is, then one should

allow that some types of explanation can be characterized independently of the notions of unification or of argument. I ignore these kinds of explanation in part because they lack the methodological significance of explanations directed at why-questions and in part because the problem of characterizing explanatory answers to what-questions seems so much less recalcitrant than that of characterizing explanatory answers to why-questions (for a similar assessment, see Belnap and Steel 1976, pp. 86–87). Thus I would regard a full account of explanation as a heterogeneous affair, because the conditions required of adequate answers to different types of questions are rather different, and I intend the present essay to make a proposal about how *part* of this account (the most interesting part) should be developed.

4. For illuminating accounts of Newton's influence on eighteenth century research see Cohen (1956) and Schofield (1969). I have simplified the discussion by considering only *one* of the programs which eighteenth century scientists derived from Newton's work. A more extended treatment would reveal the existence of several different approaches aimed at unifying science, and I believe that the theory of explanation proposed in this paper may help in the historical task of understanding the diverse aspirations of different Newtonians. (For the problems involved in this enterprise, see Heimann and McGuire 1971.)

5. See Boscovich (1966) Part III, especially p. 134. For an introduction to Boscovich's work, see the essays by L. L. Whyte and Z. Markovic in Whyte (1961). For the influence of Boscovich on British science, see the essays of Pearce Williams and Schofield in the same volume, and Schofield (1969).

6. The examples could easily be multiplied. I think it is possible to understand the structure and explanatory power of such theories as modern evolutionary theory, transmission genetics, plate tectonics and sociobiology in the terms I develop here.

7. More exactly, I shall try to show that my account can solve some of the principal versions of these difficulties which have been used to discredit the covering law model. I believe that it can also overcome more refined versions of the problems than I consider here, but to demonstrate that would require a more lengthy exposition.

8. There is an objection to this line of reasoning. Can't we view the arguments $<(x)((Sx \& Hx) \to Dx), Sa \& Ha, Da>$, $<(x)((Sx \& {\sim}Hx) \to Dx), Sb \& {\sim}Hb, Db>$ as instantiating a common pattern? I reply that, insofar as we can view these arguments as instantiating a common pattern, the standard pair of comparable (low-level) derivations—$<(x)(Sx{\to}Dx), Sa, Da>$, $<(x)(Sx{\to}Dx), Sb, Db>$—share a more stringent common pattern. Hence, incorporating the deviant derivations in the explanatory store would give us an inferior basis. We can justify the claim that the pattern instantiated by the standard pair of derivations is more stringent than that shared by the deviant derivations, by noting that representation of the deviant pattern would compel us to broaden our conception of schematic sentence, and, even were we to do so, the deviant pattern would contain a "degree of freedom" which the standard pattern lacks. For a representation of the deviant "pattern" would take the form $<(x)((Sx \& \alpha Hx) \to Dx), Sa \& \alpha Ha, Da>$, where '$\alpha$' is to be replaced uniformly either with the null symbol or with '$\sim$'. Even if we waive my requirement that, in schematic sentences, we substitute for *non-logical* vocabulary, it is evident that this "pattern" is more accommodating than the standard pattern.

9. However, the strategy I have recommended will not avail with a different type of case. Suppose that a deviant wants to explain the dissolving of the salt by appealing to some property which holds universally. That is, the "explanatory" arguments are to begin from some premise such as "$(x)((x$ is a sample of salt $\& x$ does not violate conservation of energy$) \to x$ dissolves in water)" or "$(x)((x$ is a sample of salt $\& x = x) \to x$ dissolves in water)." I would handle these cases somewhat differently. If the deviant's explanatory store were to be as unified as our own, then it would contain arguments corresponding to ours in which a redundant conjunct systematically occurred, and I think it would be plausible to invoke a criterion of simplicity to advocate dropping that conjunct.

REFERENCES

Achinstein, P. (1971), *Law and Explanation.* Oxford: Oxford University Press.

Achinstein, P. (1977), "What Is an Explanation?," *American Philosophical Quarterly* 14: pp. 1–15.

Belnap, N., and Steel, T. B. (1976), T*he Logic of Questions and Answers.* New Haven: Yale University Press.

Boscovich, R. J. (1966), *A Theory of Natural Philosophy* (trans. J. M. Child). Cambridge: M.I.T. Press.

Bromberger, S. (1962), "An Approach to Explanation," in R. J. Butler (ed.), *Analytical Philosophy* (First Series). Oxford: Blackwell.

Bromberger, S. (1966), "Why-Questions," in R. Colodny (ed.), *Mind and Cosmos.* Pittsburgh: University of Pittsburgh Press.

Cohen, I. B. (1956), *Franklin and Newton*. Philadelphia: American Philosophical Society.

Darwin, C. (1964), *On the Origin of Species*, Facsimile of the First Edition, edited by E. Mayr. Cambridge: Harvard University Press.

Darwin, F. (1887), *The Life and Letters of Charles Darwin*. London: John Murray.

Eberle, R., Kaplan, D., and Montague, R. (1961), "Hempel and Oppenheim on Explanation," *Philosophy of Science* 28: pp. 418–428.

Feigl, H. (1970), "The 'Orthodox' View of Theories: Remarks in Defense as Well as Critique," in M. Radner and S. Winokur (eds.), *Minnesota Studies in the Philosophy of Science,* Volume IV. Minneapolis: University of Minnesota Press.

Friedman, M. (1974), "Explanation and Scientific Understanding," *Journal of Philosophy* LXXI, pp. 5–19.

Heimann, P., and McGuire, J. E. (1971), "Newtonian Forces and Lockean Powers," *Historical Studies in the Physical Sciences* 3, pp. 233–306.

Hempel, C. G. (1965), *Aspects of Scientific Explanation*. New York: The Free Press.

Hempel, C. G. (1962), "Deductive-Nonlogical vs. Statistical Explanation," in H. Feigl and G. Maxwell (eds.), *Minnesota Studies in the Philosophy of Science,* Volume III. Minneapolis: University of Minnesota Press.

Hempel, C. G. (1966), *Philosophy of Natural Science*. Englewood Cliffs: Prentice-Hall.

Hull, D. (ed.) (1974), *Darwin and His Critics*. Cambridge: Harvard University Press.

Jeffrey, R. (1970), "Statistical Explanation vs. Statistical Inference," in N. Rescher (ed.), *Essays in Honor of Carl G. Hempel*. Dordrecht: D. Reidel.

Kitcher, P. S. (1976), "Explanation, Conjunction and Unification," *Journal of Philosophy,* LXXIII: pp. 207–212.

Lavoisier, A. (1862), *Oeuvres*. Paris.

Newton, I. (1962), *The Mathematical Principles of Natural Philosophy* (trans. A. Motte and F. Cajori). Berkeley: University of California Press.

Newton, I. (1952), *Opticks*. New York: Dover.

Railton, P. (1978), "A Deductive-Nomological Model of Probabilistic Explanation," *Philosophy of Science* 45: pp. 206–226.

Salmon, W. (1970), "Statistical Explanation," in R. Colodny (ed.), *The Nature and Function of Scientific Theories*. Pittsburgh: University of Pittsburgh Press.

Schofield, R. E. (1969), *Mechanism and Materialism*. Princeton: Princeton University Press.

Teller, P. (1974), "On Why-Questions," *Noûs* VIII: pp. 371–380.

van Fraassen, B. (1977), "The Pragmatics of Explanation," *American Philosophical Quarterly* 14: pp. 143–150.

Whyte, L. L. (ed.) (1961), *Roger Joseph Boscovich*. London: Allen and Unwin.

16

MERILEE H. SALMON

Explanation in the Social Sciences

1. INTRODUCTION

Disagreements about explanation in the social sciences are closely bound up with views about whether or not the so-called social sciences really are sciences.

The dispute has a long history. J. S. Mill, following Hume and the philosophers of the French Enlightenment, maintains that a science of human nature is impossible (1874, 586). He believes that the thoughts and feelings of humans are the causes of their actions. On this basis, he argues that we can investigate the causal connections between thoughts and actions by employing the same canons of inference (Mill's Methods) that we use to discover and justify causal regularities in the physical world.

Mill recognizes that the complexity of human behavior impedes the development of causal explanations. Nevertheless, he believes that at least an *inexact* science of human behavior is possible. Whether a science is "exact" or "inexact" depends on how accurate the predictions of the science are. Mill doubts that the science of human behavior will ever become as exact as the physical science of astronomy, for example, because human actions are subject to so many unknown, and possibly unknowable, circumstances. In addition, even when the circumstances surrounding behavior are known, we are sometimes unable to describe or measure them accurately.

Mill points out, however, that the accumulation and interaction of many minor causal forces simi-

larly hinder accurate prediction in some physical sciences. Thus, the main laws that govern the movement of tides are known, but, because of irregularities in shorelines and ocean floors, as well as changes in direction of winds, precise predictions of tidal movements are not possible. Still, Mill says, no one doubts that tidology is a science, and hence the failure of prediction is no barrier in principle to the development of an *inexact* human science. With this limitation in mind, Mill urges scientists to investigate human behavior with the aim of uncovering general laws.

Mill says that to discover patterns of connection between human thought and human action we must first study history to discern some regular connections. In Mill's version of what later came to be called a *covering-law model of explanation,* he suggests that the regularities revealed by historical studies ("the lowest kind of empirical laws") are themselves to be explained by showing that they are derived from laws of character development. Laws of character development, in turn, are to be explained by showing how they result from the operation of general laws of the mind (1874, 589).

Mill believes that fundamental causes of human behavior are probably mental rather than physical, but he does not think that this makes explanation of human behavior significantly different from explanation in the physical sciences. For Mill, subsumption under causal generalizations is at the heart of explanation. When we explain an event (either a physical occurrence or a human action) we must subsume it under an appropriate causal generalization; when we explain a generalization we must subsume it under a more general law or set of laws. *Explanation* thus is possible in inexact as well as

Scientific Explanation, Minnesota Studies in the Philosophy of Science, Vol. XIII (Minneapolis: University of Minnesota Press, 1989), pp. 384–409. Philip Kitcher and Wesley C. Salmon, eds. Reprinted by permission of the publisher.

exact sciences, though in the latter more precise *predictions* are possible.

C. G. Hempel's work on explanation in the social sciences lies squarely in the tradition of Mill. Hempel argues that insofar as explanations in history and other social sciences are complete, rather than elliptical or partial, these explanations require relevant universal or statistical generalizations. Hempel recognizes that many of the generalizations invoked in explanations in the social sciences are vague, common-sensical claims, unlike the more precise and well-confirmed generalizations in the physical sciences. He points out, though, that when the generalizations are not well founded, the explanations that they underlie are accordingly weakened (1962, 15–18). Both Hempel and Mill insist that explanatory laws have empirical content, but Hempel, unlike Mill, countenances *noncausal* explanatory laws.

Hempel agrees with Mill that human action can be explained by reference to mental causes, such as motives, beliefs, desires, and reasons. Similarly, he agrees with Mill that future investigations might show that mental concepts are "reducible" in accord with some materialist program. However, both Hempel and Mill recognize that even if successful materialist reductions were forthcoming, explanations in terms of mental causation would not thereby be rendered obsolete. Just as in the physical sciences, an explanation at one level (such as, for example, explanation of the behavior of gases in terms of pressure and temperature) can be correct and informative even when the phenomena can also be explained at a "deeper" level (as when the behavior of gases is explained in terms of molecular motion).

Although both Mill and Hempel embrace nonmaterialist causal explanations of behavior, their shared belief that explanations of human behavior are fundamentally similar to explanations of physical phenomena is challenged by critics who argue that the ability of humans to exercise free choice sets them apart from the rest of nature. Since humans are able to make decisions and carry out plans in accord with their own reasons rather than some external constraint, the critics say, voluntary human behavior, unlike physical phenomena, cannot be subsumed under laws. Thus, the critics conclude that covering-law explanations cannot account for human behavior.

Mill was aware of this objection, but thought that it was based on a misunderstanding about the possibility of *accurate prediction* in science. He believed he had solved the problem by showing that even in the physical sciences precise predictions are not always possible. He argued that since the absence of precise predictions is not an impediment to the construction of inexact physical sciences, it cannot prevent the construction of inexact human sciences.

Hempel responds differently to the criticism that some distinctive form of explanation is required to account for human actions. He argues that explanations in terms of mental causes, such as motivating reasons, have the same *logical* structure as covering-law explanations in the physical sciences. A human action, like a physical event, he says, is explained when it is shown to follow from explanatory facts that include at least one law statement. Hempel presents the following model for explaining behavior that is a result of rational deliberation:

Agent A was in a situation of type C. [Initial Condition]

A was disposed to act rationally. [Initial Condition]

Any person who is disposed to act rationally will, when in situations of type C, invariably (with high probability) do X. [General law]

A does X. [Event to be explained] (1962, 27)

In Hempel's model, agents are considered rational if they are disposed to take appropriate means to achieve their chosen ends. In contrast to Dray's (1957) account, Hempel does not assume that agents always act rationally, but instead he regards the attribution of rationality to an agent as an explicit initial condition in the explanation. Hempel adds that covering-law explanations with the same logical form as explanations of rational behavior can be framed for behavior that is not a result of "rationality and more or less explicit deliberation, but . . . other dispositional features, such as character and emotional make-up" (1962, 27).

Hempel's models display clearly what he means when he says that explanations in physical and social sciences are similar to one another: Each has the logical structure of an *argument* (inductive or deductive) in which the event to be explained is the conclusion and some initial conditions and law or laws constitute the explanatory premises. Hempel's models so aptly clarify and refine respectable popular intuitions about the nature of explanations in the physical sciences that they have become a focus for discussion of alternate views about explanation in social sciences. In what follows, I will look at some objections to his approach and some alternative accounts of explanation in the social sciences.

The objections to Hempel's account that I will discuss fall roughly into three categories. The first position, *interpretativism,* regards explanations of human purposive action as having an entirely different structure from causal explanations or any other explanations that appeal to *laws.* Interpretativists deny that there are empirical laws connecting reasons with actions; they say that to suppose there could be such laws involves committing a *logical* error.

The second type of objection can be called "nomological skepticism." It is dominated by the worry that there are no *laws* in social science available for constructing covering-law explanations. Some nomological skeptics admit the eventual possibility of discovering laws, while others are less sanguine. The skeptics' doubts about the possibility of finding appropriate laws are pragmatic in contrast to the logical concerns of the interpretativists. Skeptics believe that the great variability and complexity of human behavior pose practical barriers to framing generalizations that are at once informative and true. Although both Mill and Hempel have addressed these practical concerns, the skeptics remain unconvinced.

The third type of objection, proposed by *critical theorists,* sees lawful explanation as a threat to human autonomy. Critical theorists worry about the ethical implications of trying to explain human behavior in the same manner that we explain the actions (or movements) of nonconscious physical objects.

Aside from the three types of objections just described, other fundamentally important criticisms of Hempel's models raise questions about whether laws are *parts* of explanations or whether instead laws justify or underlie explanations (Scriven 1959; Humphreys 1989), whether explanations are arguments (Jeffrey 1969), whether a statistical explana-

tion must show that the event to be explained is highly probable (Salmon 1965), and whether the pragmatic features of explanation have been adequately addressed (van Fraassen 1980). As the present volume attests, the models originally proposed by Hempel more than twenty years ago cannot survive these criticisms intact. Not surprisingly, an adequate account of scientific explanation must go beyond that early work.

Nevertheless, we can look at criticism of explanation in the social sciences without exploring these refinements. For insofar as newer models are causal models of explanation or covering-law models of explanation (see, for example, Glymour et al. 1987) the objections raised to them by interpretativists, nomological skeptics, and critical theorists will not be very different from their objections to Hempel's original models. Since this paper focuses on lawful explanation in the social sciences, I will limit the scope of this discussion to the criticisms raised by interpretativists, nomological skeptics, and critical theorists.

2. INTERPRETATIVISM

Interpretativists reject as logically confused the claim that explanations in the social sciences are fundamentally similar to those in the physical sciences. Their position goes back at least to Dilthey, whose main work begins just after Mill's death. A forceful exposition of interpretativism is found in R. G. Collingwood (1946).

Collingwood uses the terms "history" and "historical thought" to refer to all studies of human affairs, including not only history, but also (parts of) anthropology, sociology, political science, psychology, and economics. The contrast class, called here "physical science," includes physics, chemistry, and even the so-called historical sciences, such as evolutionary biology and geology.

The events studied by history have, according to Collingwood, an "inside" as well as an "outside." The outsides of events consist of everything belonging to them that can be described in terms of bodies or their movements. The insides of events can be described only in terms of the thoughts of the agent (or agents) who are responsible for the events. Human behavior requires for its complete description not only the account of bodily motions involved but also an account of the beliefs and desires of the agent. For example, the physical description of one person cutting another with a knife does not distinguish an act of surgery from an act of assault, a ritual act, or an accidental cutting. A more detailed physical description of the "outside" may provide clues that will help to ascertain the intention of the cutter, but until that intention is uncovered, Collingwood would say, we simply do not know what action took place.

Collingwood argues on the basis of the difference between events with only an outside and those with both an inside and an outside that there is a fundamental difference between the search for so-called causes of human behavior (the *reasons* for the behavior) and causes of physical events. In the physical sciences, the event is perceived, and its cause (a separate event) is sought. This investigation takes the scientist beyond the original event in order to relate it to other separate events, thus bringing it under a general law of nature.

In contrast, when studying human actions, the historian must look inside the event to discover the thought that is expressed *in* the event. The event studied is not really separate from the thought, but is the mere expression of the thought. The thought is the event's "inside," that which makes the event what it is. From his claim that the reasons for actions are not external to the actions, Collingwood concludes the relation between the two cannot be governed by laws of cause and effect. Thus, he says, causal laws play no essential role in the explanation of human behavior.

When an action is described as the sort of action it is, the description itself includes the reason for the action. In the case of purposive human behavior, the acts of describing and explaining are therefore one and the same. The reason for the act is not viewed as some cause that is separate from it, but rather is *logically* or *meaningfully* related to the act. The reason gives meaning to the act, and makes it the sort of act it is. Accordingly, Collingwood would say that the model of searching for regular connections between actions and reasons, as proposed by Mill and, later, Hempel, simply makes no sense.

Among contemporary philosophers, Peter Winch (1958) is a leading advocate of the interpretativist position. Like Collingwood, he denies the role of anything analogous to a law of nature in explaining human behavior, and he regards the term "social science" as misleading for this reason. Winch's view is that "social relations really exist only in and through the ideas that are current in a society; [and] . . . that social relations fall into the same logical category as relations between ideas" (1958, 133). Causality is thus no more an appropriate category for understanding social relationships than it is for understanding mathematical relationships. Whereas Collingwood pays special attention to individual beliefs and desires, Winch, following Wittgenstein, emphasizes the social character of human action and thus he focuses on the importance of *rules* or norms of behavior.

Winch's broad notion of a rule covers not only formal regulations, such as traffic rules and tax deadlines, but also unstated cultural norms or conventions, such as those governing the appropriate distance between speakers engaged in a face-to-face conversation. In addition, "rule" embraces practices and institutions such as religion, democratic government, and money. Some rules are regulative, such as the rule to stop for red lights, while others are constitutive of the practices they embrace.

Consider, for example, the act of offering a sacrifice. This act is possible only within a certain type of institution in which particular kinds of behavior, such as killing animals in a prescribed way and under certain conditions, count as offering sacrifices. Without such an institution or social practice, it makes no sense to call the killing of the animal in that way a sacrifice. The point of these constitutive rules is that whether an action is performed and what kind of action it is depends not only on individual intentions, but also on the social set-up. Social relations, however, according to Winch, do not *cause* the act; they rather constitute it by giving it the meaning it has.

Obviously, we cannot understand what is going on in a sacrifice if we do not have the concept of a sacrifice. To say that the concept of sacrifice depends on the social set-up, however, goes far beyond the claim that, as individuals, we can acquire the concept only through socialization. For, not only how we come to learn the concept, but also the very meaning of the concept depends on the possibility of social relations of a certain type.

To understand human behavior, Winch says, we require more than just abstract knowledge of the rules of a society. We also need to understand what counts as following a rule in a particular case. To grasp this, he says, we must somehow come to share the viewpoints, attitudes, and feelings of the actors. For example, consider a society which has a norm that requires showing respect to elders. Knowing this rule and having a physical description of some bit of behavior are not enough to figure out whether or not this behavior in the presence of an elder counts as showing respect. We also require some knowledge of the beliefs and attitudes of the person engaging in the behavior. Was this person aware an elder was present? Did the person know that form of address was considered disrespectful? Was the person merely careless in his manner? Was the form of address considered disrespectful by the speaker or the elder?

Even in one's own society, working out such matters can require complicated negotiations. Understanding human behavior in exotic societies is doubly problematic, and those who attempt it must be on guard against ethnocentrism. Winch's sensitivity to the importance of rules and how they are applied, as Papineau notes (1978, 96–97), accords well with the concerns of contemporary cognitive anthropologists. Ward Goodenough (1957), for example, says that culture consists not of "things, people, behavior or emotions but the forms or organizations of these things in the minds of people" (quoted in Frake 1969, 38). Cognitive anthropologists like Goodenough see their task as uncovering these forms (i.e., rules) primarily on the basis of what people say about how they categorize and organize the furniture of their worlds in applying the rules.

Winch does not deny that we can predict human behavior. He admits that with a knowledge of the rules of a society and an understanding of how the rules are applied, reliable prediction is often possible. He insists, however, that successful prediction of behavior is unlike predictive success in the physical sciences. The difference is not just that physical science yields more accurate predictions, for in some cases, as in tidology, it does not. The point is that in the physical sciences, *causal regularities* are the basis of the predictions, whereas insofar as behavior is rule-governed, it is not subject to causal laws. Rules, according to interpretativists, function as standards or norms of behavior; they give meaning to behavior, but do not cause it. Although we can predict behavior from a knowledge both of the rules of a society and of how those rules are translated into actual behavior, the prediction is not based on a causal relation between the rule and the behavior. The concepts of cause and effect, Winch insists, simply do not apply to the relation between a reason for action and the action, any more than the concepts of cause and effect apply to the relation between being a Euclidean triangle and having internal angles with a sum of 180 degrees.

Interpretativists acknowledge that discerning the rules that underlie human behavior is an important goal. They insist, however, that these rules have a logically different character from causal laws—or other empirical laws—discovered in the physical sciences. Because interpretativists see the relation between reasons and behavior as a logical relation rather than a causal relation, they reject the covering-law models of explanation.

Drawing support from interpretative philosophers, some social scientists have also rejected covering-law models of explanation. Clifford Geertz, for example, claims to do so in *The Interpretation of Cultures* (1975).

Anthropology, for Geertz (in contrast to cognitive anthropologists like Goodenough), is ethnography. Ethnography, he says, consists in interpreting the flow of social discourse. This interpretation itself consists of inscribing or recording the flow, "fixing" it so that it can be shared and reexamined long after the actual events take place. A major problem for the ethnographer is that of finding the appropriate general concepts for describing or classifying the observed behavior. For example, an ethnographer might observe that meetings are taking place and that political issues are being discussed. Further features noted might include low voices and concern with arrangements to secure secrecy. Is the appropriate *thick description* (Geertz borrows Ryle's expression) or interpretation of such activities that of "fomenting a rebellion" or something less inflammatory?

Geertz pointedly calls the ethnographer's activity of fixing the flow "interpretation" rather than explanation. Following Dray's (1957) advice to historians, Geertz believes that anthropologists should primarily be concerned with assigning observed behavior to appropriate concepts. Like Dray, he denies that covering laws play any role in this activity. He adds that any true statements that are general enough to serve in covering-law explanations in anthropology are either hopelessly vague or trivial. In such passages, Geertz's rejection of the possibility of lawful explanation in anthropology echoes Scriven's criticism of Hempel's models in "Truisms as the Grounds for Historical Explanation" (1959), and also reflects the position I have characterized as nomological skepticism.

In the same work, however, Geertz also says that anthropologists are interested not only in particular ethnographic studies (which he calls microscopic work) but also in going beyond ethnographies to compare, to contrast, and to generalize. Indeed, he acknowledges that some generalization occurs even within microscopic studies. As Geertz says, "We begin with our own interpretations of what our informants are up to, or what they think they are up to, and then systematize these." In another place, where he is talking about doing ethnography, he says:

> Looking at the ordinary in places where it takes unaccustomed forms brings out not, as has so often been claimed, the arbitrariness of human behavior . . . but the degree to which its meaning varies according to the pattern of life by which it is informed. Understanding a people's culture exposes their normalness without reducing their particularity. . . . It renders them accessible; setting them in the frame of their own banalities, it dissolves their opacity. (1975, 14)

However, Geertz's remarks are ambiguous, for they lend themselves either to an interpretativist or to a nomothetic (covering-law) construction. The nomothetically inclined reader can say: "What can it mean to expose the normality of a people if we do not know what normality is and have no concern with the normal?" Some account of what is normal provides the framework for our recognition and understanding of traits in another culture. But to have

an account of what is normal is to be in possession of at least some statistical laws of human nature, to know that humans usually believe, say, or do thus and so under such and such circumstances. Moreover, at least part of what it is to interpret or to understand something is to place it within an intelligible framework or to see how it fits into a pattern. But intelligible frameworks and patterns are the sorts of regularities that, according to Hempel, are expressed in statistical or universal laws.

Winch could object to such a nomothetic reading of Geertz's position on the grounds that the reading involves a serious misunderstanding of the type of regularity or normality present. The appropriate regularities, according to the interpretativist, are not causal regularities, but rules or norms of behavior. A grasp of these is crucial for understanding behavior, but norms and rules are not explanatory laws, and so cannot play that role in explanation.

Geertz, however, in contrast to Winch, does not want to deny that anthropology is a science, albeit a different and "softer" kind of science than physics. In discussing how anthropological theories are constructed, Geertz says that whereas in covering-law models individual cases are subsumed under *laws*, the generalizations that constitute theory in anthropology are generalizations *within* cases rather than generalizations *across* cases. Amplifying this point, Geertz goes on to say that the type of generalization he refers to is called, in medical science, "clinical inference." He says that anthropologists should look for "intelligible patterns" into which otherwise unrelated bits of information can be fitted rather than "universal laws" that can be combined with initial conditions to *predict* cultural patterns.

Geertz's rejection of prediction as a goal for anthropology provides an important clue to his dismissal of the covering-law models. He apparently believes that covering-law explanation commits one to an objectionable view about the close relationship between explanation and prediction.

Hempel, it is true, argues for a kind of symmetry between scientific explanation and scientific prediction. He regards scientific explanations of individual events as arguments to the effect that the event to be explained was to be expected on the grounds of laws and initial conditions, and scientific predictions as arguments to the effect that the event predicted is to be expected on the grounds of laws and initial conditions. Given Hempel's analysis, explanation and prediction do not differ in their logical structure, but only in the temporal order of the events to be explained or predicted and in the starting point of our knowledge of the situation. In explanation, we are aware that the event has occurred, we search for the relevant laws, and we try to reconstruct the initial conditions from which the event can be derived. In prediction, we are aware of the initial conditions, and with knowledge of the relevant laws, we derive the occurrence of the predicted future event.

It is important, however, not to misread Hempel's thesis of the symmetry of explanation and prediction as asserting that whenever an event can be explained it could have been predicted. Hempel, in response to criticisms of his symmetry thesis by Scriven and others, defends "the conditional thesis that an adequate explanatory argument must be such that it could have served to predict the explanandum event *if the* information included in the explanans had been known and taken into account before the occurrence of that event" (1965, 371). At the same time, he insists that his symmetry thesis does *not* require that we always can know independently of the occurrence of the explanandum event that all the conditions required for an explanation are realized. There are clearly situations in which it is only "after the fact" that we are aware that some of the initial conditions were present. For example, before a murder occurs there may have been no way to recognize that the rage of the murderer was sufficient to motivate him to commit the act, or to know that he was capable of murder.

Hempel's version of the symmetry thesis does *not* claim that because the act can be explained in terms of laws and initial conditions, that these laws and initial conditions could have been discovered before the act occurred. To hold that any act which can be explained after the fact could have been predicted before the fact (without further qualification) requires a much stronger symmetry thesis. This stronger version of symmetry is indefensible, and should not be attributed to Hempel. (Hempel's symmetry thesis is closely linked with his understanding of explanations as *arguments*. For a weaker version of symmetry, compatible with the view that explanations are not arguments, see Salmon 1970.)

Geertz apparently regards the indefensible strong symmetry thesis as a feature of "the covering-law model of explanation." He specifically rejects attempts to establish covering laws in anthropology to *predict* behavior because he sees the goal of social science as understanding rather than prediction. He then goes on to reject explanation as a goal because, guided by the strong symmetry thesis, he erroneously identifies lawful explanation with the ability to make predictions.

Ironically, when Geertz says that clinical inference provides a model for understanding anthropology, he seriously undercuts his interpretativist stance. Clinical inference involves diagnosis of symptoms, such as the diagnosis of measles on the basis of its telltale signs. Diagnosis, as Geertz correctly notes, is not concerned with predicting. It is directed toward analyzing, interpreting, or explicating the complex under consideration. Geertz sees this activity as the essential concern of anthropologists and therefore argues that anthropology is a diagnostic science rather than a nomothetic one.

Hempel shows, however, that cases of diagnosis or clinical inference fall easily within the scope of covering-law explanation (1965, 454–455). Moreover, in several places where Hempel discusses "explanation by concept" and other forms of explanation in the social sciences, he offers an account of explanation that deemphasizes the relation between explanation and prediction, and focuses instead on an explanation's power to increase our understanding. Near the end of "Aspects of Scientific Explanation," for example, he says that explanation "seeks to provide a systematic understanding of empirical phenomena by showing how they fit into a nomic nexus" (1965, 488). The same emphasis on understanding rather than prediction is apparent in "Explanation in Science and in History" (1962, 9).

Hempel's arguments strongly support the position that scientific "understanding" is just as dependent on laws as scientific explanation. If Hempel is correct, Geertz's form of interpretativism is very different from Winch's, for it is just as dependent on laws as nomological explanation. Geertz, however, as noted earlier, has also questioned whether there can be any nontrivial laws in anthropology. This nomothetic skepticism is a sticking point for many social scientists. In the next sections, we will consider some attempts to find and characterize appropriate laws for explaining (and understanding) human behavior.

3. RATIONALITY AND EXPLANATIONS OF BEHAVIOR

When Hempel wrote "The Function of General Laws in History" (1965, 251–252), he believed that it would be possible to discover causal laws of human behavior though he did not specify any particular form that these laws would take. At that time there was widespread hope that behaviorism would soon deliver what it promised. Hempel reflects this optimism when he suggests that we may someday find behavioristic stimulus-response laws or laws of learned behavior that connect circumstances and actions. If statements describing such regularities—which I take to be causal—were available, they would be suitable covering laws for constructing explanations of human actions. Similarly, though Hempel does not say so here, if causal laws connecting brain states and human *actions* (not mere bodily movements) were available, these could also be used in covering-law explanations of behavior. However, neither of these materialist programs has so far succeeded in supplying the sorts of laws which the program's early adherents hoped would be forthcoming.

Hempel clearly regards instances of his proposed law schema—"Any person who is disposed rationally will, when in circumstances of type C, invariably (with high probability) do X" (1962, 28–29)—as having empirical content, whether or not the law is causal. In keeping with his understanding of rationality and other character traits as *dispositional* properties, Hempel does not regard them as completely definable in terms of manifest behavior. He says instead that they may be partially definable by means of reduction sentences. Following Carnap's account of reduction sentences, Hempel says that the connections between dispositional properties (e.g., rationality, fearlessness) and the actions that are symptomatic of these dispositions are stated in claims that express either necessary or sufficient conditions for the presence of the given disposition in terms of the manifest behavior.

Reduction statements seem to be analytic, for they offer at least partial definitions of dispositional properties. Yet not all these statements can qualify as analytic, Hempel says, for in conjunction they imply nonanalytic statements of connection between various manifest characteristics. For example, if a specific form of behavior A is a sufficient condition for the presence of rationality, and another specific form of behavior B is a necessary condition of rationality, then the claim "Whenever A is present B will be also," which is a consequence of the conjunction of the two reduction sentences, and which asserts that two types of behavior are always found together, "will normally turn out to be synthetic" (1962, 28–29). On this basis, Hempel claims that the generalizations which state that in situations of a certain sort, rational agents will act in specified ways are empirical.

The success of Hempel's defense of the empirical character of the laws that connect dispositions with behavior is obviously tied to the possibility of finding suitable reduction sentences for characterizing dispositional properties such as rationality, fearlessness, and the like. This is, I believe, so closely linked with behaviorism that it inherits the problems associated with that program. Moreover, Hempel admits in "A Logical Appraisal of Operationism" (1965, 133) that the ability to derive synthetic statements from pairs of reduction sentences casts some doubt on the "advisability or even the possibility" of preserving the distinction between analytic and synthetic sentences in a logical reconstruction of science. Despite his concession to the blurring of this traditional empiricist distinction, he maintains his original standard requiring empirical content in explanatory laws.

It is fair to say that behaviorism no longer enjoys the sway it once held. In a more contemporary attempt to understand the nature of rationality, many philosophers have rejected the behaviorist approach and turned to a decision-theoretical analysis of rationality. However, if we try to understand "rationality" in decision-theoretic terms rather than behavioristic terms, the empirical content in the law (or law schema)—"Any person who is disposed rationally will, when in circumstances of type C, invariably (with high probability) do X"—remains elusive.

The decision-theoretic approach to rationality is intended to be applicable under the assumption that an agent is acting independently and with only probabilistic knowledge of the outcomes of various actions. Under such circumstances, a person is said to behave rationally just in case the person acts so as to maximize expected utility. If we adopt this criterion, then to say that if an agent is disposed to act rationally in circumstances C, the agent will do X, is equivalent to saying that in circumstances C, action X maximizes expected utility (or, in Papineau's (1978) phrase, "expected desirability").

The expected desirability of an action is the sum of the products of the probabilities and the values for each of the possible outcomes of the action. Maximizing expected desirability just means choosing the action with a sum that is not lower than the sum of any other action. The statement that an agent is in circumstances C (i.e., the agent holds various beliefs to which probabilities are assigned and also has desires with values attached to them) is clearly empirical. However, whether or not actions maximize expected desirability under such specified circumstances is not an empirical matter, but rather a judgment based on a calculation which is determined by the criterion of rationality.

The decision-theoretic account of rationality does not actually require agents to perform calculations. However, to be rational, actions must accord with the results of such calculations, had the calculations been performed. This means that *rational* agents must choose actions that maximize expected desirability, for they could not do otherwise and still be rational agents.

Thus, if we use the decision theorist's analysis of rationality, instances of Hempel's schema for a law ("Any person who is disposed to act rationally will when in circumstances C invariably (with high probability) do X") fail to meet his criterion of empirical content. But if the laws have no empirical content, they cannot ground genuine scientific explanation, even though they may be the basis for another form of explanation, similar to explanation in mathematics, where the laws are not empirical.

The criterion of rationality offered by decision theory is minimal in the sense that agents need only take some available means to achieve whatever ends they desire, or at least to avoid frustrating those ends. The decision-theoretical account of rationality does not assume that agents make good use of available evidence when forming beliefs. Even if agents' beliefs are based on prejudice or ignorance, or if their desires are peculiar or hard to comprehend, their behavior can be rational. Moreover, as mentioned before, to be rational in this sense, agents need not assign explicit probabilities to beliefs or quantify values, or even make rough or precise calculations of expected desirability. It is enough for agents to act *as if* they were maximizing desirability, given their beliefs and desires.

If we strengthen the decision theorist's standard of rationality to require that rational agents use an objective physical basis instead of a purely personal consideration to assign probabilities to various possible outcomes of an action, we invoke some probabilistic laws. However, these are not covering laws; the probabilistic laws are used only to assess the truth of the initial condition which states that the agent is disposed to act rationally. As such they are not even part of the covering-law explanation itself.

David Papineau adopts the decision-theoretical approach to rationality in his (1978) attempt to reconcile the interpretativist position with nomothetic explanation of human action. Papineau presents a covering-law model of explanation of human action that differs from Hempel's in several respects. According to Hempel, an agent's disposition to act rationally is an initial condition that must be established empirically. For Papineau, agents always act rationally. According to him, the lawlike generalization that grounds explanations of behavior is: "Agents always perform those actions with greatest expected desirability" (1978, 81). This law, he believes, is implicit in ordinary explanations of individual human behavior. Accordingly, for Papineau, acting out of a character trait, such as fearlessness, would not contrast with his acting rationally, as it might for Hempel. (For Hempel, humans may be rational at some times, and not at others; for Papineau, it is a contingent, but universal truth that humans behave rationally. We must remember though that the two differ about the meaning of rationality.)

Papineau agrees with Winch that rules do not *cause* behavior, for otherwise, he says, humans would be mere puppets of their cultural milieu, unable to violate social norms. Obviously people can disregard norms, whereas causal laws cannot be violated. Papineau acknowledges that the existence of rules plays an important role in forming agents' beliefs and desires, and that the interpretativist has something important to say about this (1978, ch. 4; also see Braybrooke, 1987, 112–116). However, Papineau says that the information which guides our attributions of beliefs and desires to agents is not actually part of the explanation of human behavior. All that is required for explanation, in addition to the generalization about maximizing expected desirability, is an account of what beliefs and desires an individual has (these are the initial conditions), not an account of how the agent came to have them. Papineau does not deny the importance and relevance of the interpretativists' concerns, but he sees them as supplementing, rather than conflicting with, lawful explanation.

Papineau is aware of the apparent lack of empirical content in his proposed law: "Agents always perform those actions with greatest expected desirability," and he tries to defend his model of explanation against this criticism. Although, he says, it is true that we use this general principle to infer the nature of agents' beliefs and desires, it is legitimate to do so as long as we do not include the action we are trying to account for in establishing those attitudes. We infer beliefs and desires from agents' past actions, aided by knowledge of the norms of their society, and then—attributing to them those attitudes—explain their present actions in the light of that knowledge.

Nevertheless, Papineau recognizes that sometimes people do not act in accord with the beliefs and desires that have been correctly attributed to them in the past. And when this happens, he admits that we do use the anomalous action to infer the beliefs and desires that underlie it. His argument to legitimize this move appeals to Lakatos's understanding of a theoretical "core statement" that is maintained in the face of presumptive counterexamples by revising various auxiliary assumptions (see, for example, Lakatos 1970). It is a commonplace that people's beliefs and desires do change over time, so we must be prepared to take account of such revisions. However, he says, "this preparedness does not condemn the overall theory as unscientific—none of the most revered theories in the history of science would ever have survived if their proponents had not been similarly prepared to defend their central tenets from the phenomena by revising auxiliary hypotheses" (1978, 88).

Papineau does not mean to countenance ad hoc revisions to save his generalization that agents act always so as to maximize expected desirability. He says that if we attribute to an agent extraordinary desires or beliefs that we would not expect to be available to the agent, we should be prepared to give some account of the circumstances that will lead to new and independently testable propositions. We are bound to do this, just as any physical scientist is required to do so in a "progressive" research program (1978, 88).

Ultimately, however, Papineau urges us to accept this theory of human action because we do not have a better theory available. Behaviorism is impoverished, he says; physiological accounts are woefully undeveloped, and the rule-governed account of behavior must be integrated into the decision-theoretical model if humans are not to be understood as cultural puppets who rigidly conform to norms, rules, and conventions.

Even if interpretativists were willing to accept Lakatos's account of the nature of theories in physical science, I doubt that Papineau's arguments would undermine their insistence that human behavior is not subject to causal laws, for he does not address the claim that reasons for behavior are *logically* different from causes. At the same time, those who urge nomological explanations of human behavior will be disappointed in Papineau's failure to put forth an explanatory generalization with empirical content.

One can accept a decision-theoretical account of the nature of rationality, while nevertheless recognizing, as Donald Davidson (1980) does, that no *criterion* of rationality, however satisfactory it is, can function as a law in covering-law explanations of human behavior.

In a series of papers (1980, especially essays 1, 7, 12, 14), Davidson refutes the interpretativist claim that it is a logical error to suppose that reasons can cause actions. Interpretativists have typically argued that if things are logically related (or connected by relations of meaning), then they cannot be causally related. Davidson shows that whether or not two events are related causally depends on what the world is like, whereas various linguistic accounts of the relationship may be classified as "analytic" or "synthetic." For example, suppose that a rusting understructure causes a bridge to collapse. The sentence "The rusting of the iron understructure caused the bridge to collapse" would normally be regarded as synthetic, whereas "The cause of the bridge's collapse caused the bridge to collapse," would normally be regarded as analytic. Yet the causal relationship between rusting and collapsing is a feature of the world, not of either sentence.

However, Davidson does not believe that reason explanations can be understood as covering-law explanations that invoke an implicit law connecting *types of reasons* with *types of behavior*. Thus, suppose that Jason's running in the marathon can be (truly) explained by his desire to prove his self-worth. Davidson does not deny that there is a causal law connecting Jason's reason (his desire to prove his self-worth) for running the marathon with his running, for he agrees with Hempel that "if *A* causes *B*, there must be descriptions of *A* and *B* which show that *A* and *B* fall under a law" (1980, 262), and he also agrees that explanations in terms of reasons are—if correct—causal explanations. However, he does say that we hardly ever, if ever, know what that empirical law is, and also argues that the (unknown) law does not have the form of a regular connection between a psychological cause and an action, such as "Whenever anyone wants to prove his worth, he runs a marathon." He argues further that no matter how carefully the circumstances surrounding, for example, someone's wanting to prove his worth and the opportunities to do so may be qualified, the resulting expression is not a psychophysical law.

Obviously it is not possible here to develop or to criticize Davidson's arguments with the care they deserve. It is worth pointing out, though, that whereas Davidson refutes a major premise of the interpretativists (i.e., that events cannot be causally related if descriptions of them are logically related), he seems to agree with Winch in saying that Mill was barking up the wrong tree in his search for explanatory *laws* that have the form of connecting mental causes with behavioral effects.

Davidson argues that explanations that cite reasons are informative because they tell us a lot about the individuals whose acts they are invoked to explain, rather than a lot about general connections between reasons and actions. For example, if the explanation of why Jason ran the marathon is correct, then we have achieved some understanding of what motivates him and how he expresses this. Furthermore, with this knowledge, we can make reliable conditional predictions of how he would behave in other circumstances.

Although Davidson says that his reflections reinforce Hempel's view "that reason explanations do not differ in their general logical character from explanation in physics or elsewhere" (1980, 274), his arguments against the possibility of covering laws that connect descriptions of psychological states and behavior in such explanations represent a significant departure from the tradition of Mill and Hempel. Although Davidson delineates the special character of reason explanations differently from the interpretativists, they can perhaps take some comfort in his recognition of the special or anomalous nature of explanations in terms of reasons.

4. THE EXISTENCE OF APPROPRIATE LAWS

Those who deny that there is a strong similarity between *reason* explanations of human action and explanations in physical science should remember that not all explanations in the social sciences appeal to laws connecting reasons (or dispositional properties such as fear) with actions. Consider, for example, the tentative laws concerning social structure proposed by G. P. Murdock (1949). These statements do not refer to beliefs or desires of any individuals. The generalizations relate various systems of kinship in different societies to differing forms of marriage, to patterns of postmarital residence, to rules of descent, and to forms of family. These generalizations are not causal, for they do not attempt to assign temporal priority, nor do they cite any mechanisms for the regularities they describe. In many cases, Murdock's generalizations state the coexistence of some rules with other rules, but they do not seem themselves to express norms or rules of any society. By analogy with some physical *structural* laws, such as "All copper conducts electricity," Murdock's proposed laws can also be classified as "structural laws." Other candidates for structural laws in the social sciences are the "law of evolutionary potential" (the more specialized the system, the less likely it is that evolution to the next stage will occur) proposed by Sahlins and Service (1960) and the "law of cultural diffusion" (the greater the distance between two groups in time and space, the more unlikely it is that diffusion will take place between them) (Sanders and Price 1968).

Murdock's proposed laws are generally regarded by anthropologists as problematic. He tried to verify his generalizations by using information that had been recorded by scores of anthropologists who differed widely in methods and theoretical presuppositions. As a result, serious questions can be raised about whether terms used to characterize the data on which Murdock based his generalizations were employed consistently. While anthropologists have justifiably criticized the design of Murdock's studies, there seems to be no reason in principle why structural generalizations in the social sciences cannot be framed and tested.

Not all social regularities that are well established are good candidates for *explanatory laws*. Some regularities raise more questions than they answer; they require rather than provide explanations. This may be true of Murdock's generalizations. Current examples of nonexplanatory generalizations are those puzzling but well-supported statements that describe patterned connections between birth order, sibling intelligence, and achievement. However, when correlations like these are strongly supported across many cultures, social scientists are stimulated to search for deeper regularities to explain them (Converse 1986). The deeper regularities, if discovered, might be causal laws or other structural laws. It is not clear in advance of their discovery whether these deeper generalizations will refer to any individual human reasons for acting.

In opposition to Geertz's pessimism about the ability of the generalizations framed by social scientists to "travel well," that is, to apply to situations other than those which gave rise to their formulation, Converse is optimistic. He believes that social scientists will eventually discover useful high-level generalizations. He says, on the basis of his own (admittedly limited) experience, that "we shall discover patterns of strong regularity, for which we are theoretically quite unprepared, yet which reproduce themselves in surprising degree from world to world and hence urgently demand explanation" (1986, 58).

Anthropologist Melford Spiro also says that it is too early to abandon hope of finding any panhuman generalizations that are not either "false—because ethnocentric—or trivial and vacuous." He says:

> That any or all of the generalizations and theories of the social sciences (including anthropology) may be culture-bound is the rock upon which anthropology, conceived as a theoretical discipline, was founded. But the proper scholarly response to this healthy skepticism is not, surely, their a priori rejection, but rather the development of a research program for their empirical assessment (1986, 269).

Despite the optimism of Spiro and Converse about the future of the social sciences, neither they nor anyone else can deny the superior status of the laws now available to contemporary physical sciences. Two standard responses to the paucity of interesting, well-supported generalizations in the social sciences are that the data are far more intractable than in the physical sciences, and that we have not been trying long enough. Converse, for example, says that given the complexity of the data, "it would not surprise me if social science took five-hundred years to match the accomplishment of the first fifty years of physics" (1986, 48). However, Alasdair MacIntyre (1984), analyzes the situation rather differently.

First of all, MacIntyre claims that the salient fact about the social sciences is "the absence of the discovery of any law-like generalizations whatsoever" (1984, 88). MacIntyre thinks that this reflects a systematic misrepresentation of the aim and character of generalizations in the social sciences rather than a failure of social scientists. He says that although some "highly interesting" generalizations have been offered that are well supported by confirming instances, they all share features that distinguish them from *law-like* generalizations:

1. They coexist in their disciplines with recognized counterexamples;

2. They lack both universal quantifiers and scope modifiers (i.e., they contain unspecified *ceteris paribus* clauses);

3. They do not entail any well-defined set of counterfactual conditionals (1984, 90–91).

MacIntyre presents four "typical" examples of generalization in the social sciences, including Oscar Newman's generalization that "the crime rate rises in high-rise buildings with the height of a building up to a height of thirteen floors, but at more than thirteen floors levels off."

In response to a rather obvious objection to (1)—that most laws of social science are probabilistic rather than universal—MacIntyre replies that this misses the point. He says that probabilistic generalizations of natural science express universal quantification over sets, not over individuals, so they are subject to refutation in "precisely the same way and to the same degree" as nonprobabilistic laws. He concludes from this that

> [W]e throw no light on the status of the characteristic generalizations of the social sciences by calling them probabilistic; for they are as different from the generalizations of statistical mechanics as they are from the generalizations of Newtonian mechanics or of the gas law equations. (1984, 91)

MacIntyre's position represents a serious misunderstanding of the nature of statistical laws. These laws do not state some universal generalization about sets; they state the *probability* (where this is greater than 0% and less than 100%) that different types of events will occur together. Of course, the probabilistic connections in the physical sciences are usually stated numerically, and in this they differ from many generalizations in the social sciences. But statistical generalizations need not be stated numerically; the probabilistic connection can be conveyed with such expressions as "usually" and "for the most part."

Moreover, statistical generalizations in physics, or in any other field, are not confirmed or disconfirmed in exactly the same way as universal generalizations. A universal generalization can be overthrown by a single genuine counterexample that cannot be accommodated by a suitable revision of auxiliary hypotheses. In contrast, *any* distribution in a given sample is compatible with a statistical generalization.

MacIntyre is correct in saying that statistical generalizations in social science are different from those in statistical mechanics. But this is because the probabilistic laws in statistical mechanics are based on a theory that is well tested and supported by far more elaborate and conclusive evidence than is presently available for any statistical generalizations in the social sciences. The retarded development of theories in the social sciences may be at least partly attributed to a scarcity of resources for investigation along with greater complexity of data.

MacIntyre's point about "scope modifiers" is one that has received much attention from those concerned to point out the differences between the social and physical sciences (e.g., Scriven 1959). In the physical sciences, the exact conditions under which the law is supposed to apply are presumably explicit, whereas in the social sciences, vague clauses specifying "under normal conditions" or some such equivalent, are substituted. This difference can be interpreted in several ways. MacIntyre regards the *ceteris paribus* clauses as required because of the ineliminability of *Fortuna*, or basic unpredictability, in human life. Hempel points outs, however, the widespread use in physical science of *provisoes*, which is his term for assumptions "which are essential, but generally unstated presuppositions of theoretical inferences" (1988). Hempel supports his point with an example from the theory of magnetism:

> The theory clearly allows for the possibility that two bar magnets, suspended by fine threads close to each other at the same level, will not arrange themselves in a straight line; for example if a strong magnetic field of suitable direction should be present in addition, then the bars would orient themselves so as to be parallel to each other; similarly a strong air current would foil the prediction, and so forth.

Hempel says that the laws of magnetism neither state precisely how such conditions would interfere with the results, nor do they guarantee that such conditions will not occur. Yet such *ceteris paribus* clauses are surely implicit.

Ceteris paribus clauses sometimes result from inadequate information (the complexity of the data again) about the precise boundary conditions under which a given lawlike statement is applicable. *Fortuna* plays no role here. With *ceteris paribus* clauses, proposed laws can be stated tentatively, while research proceeds to attempt to sharpen and refine the spheres of application. As Converse points out

(1986, 50), such tidying up occurs in the physical sciences as well, as shown by work that had to be done by astrophysicists as a result of information brought back from recent space explorations of distant planets. Whether or not *ceteris paribus* clauses are *stated*, then, rather than whether they are required, seems to distinguish the social sciences from the physical sciences.

MacIntyre's third point—the failure of generalizations in the social sciences to support counterfactuals—raises a complicated issue. One common way of attempting to distinguish genuine laws from "coincidental" generalizations (a thorny, and as yet unresolved problem) is by appealing to the former's ability to support counterfactuals. So, on this view, saying that a generalization cannot support counterfactuals is just another way of saying it is not a law. However, since we do not have any widely accepted account of what it is to support counterfactuals that is independent of our understanding of causal laws, it is not clear what MacIntyre's remarks about the inability of generalizations in the social sciences to support counterfactuals adds to his claim that these generalizations are not laws.

In any case, MacIntyre admits that the probabilistic laws of quantum mechanics do support counterfactuals. It is reasonable to claim that these laws also contain elements of "essential unpredictability," for the laws cannot predict the behavior of individual atoms or even sets of atoms.

Furthermore, if we do not equate the possibility of explanation with that of accurate prediction (and MacIntyre agrees that the strong symmetry thesis is indefensible) then essential unpredictability poses no barrier to lawful explanation in the social sciences.

Nothing that is said here supports the view that Oscar Newman's generalization about crime rates in high-rise dwellings, quoted above, is a genuine law or that it could play a role in a covering-law explanation. (However, it may have considerable practical predictive value for those contemplating designs of housing projects.) As it stands, Newman's generalization states an interesting correlation. We want to know how well it stands up in new situations. More than that, even if it does not apply beyond the observed instances—if the generalization is no more than a summary—we want to know *why* the correlation exists for those instances. This is the kind of generalization that can lead us to form interesting and testable causal hypotheses about connections between criminal behavior and features of living situations. These can stimulate the acquisition of new data and further refinements of the hypotheses, or the formulation of additional hypotheses. Ultimately, this process could lead to the establishment of laws that are very different in form (not merely refined in terms of scope) from the generalization that initiated the inquiry.

5. ETHICAL ISSUES

Central to MacIntyre's discussion of the character of generalizations in social science is an attack on systems of bureaucratic managerial expertise. He says that those who aspire to this expertise misrepresent the character of generalizations in social science by presenting them as "laws" similar to laws of physical science. Social scientists do this, MacIntyre claims, to acquire and hang on to the power that goes along with knowledge of reliable predictive generalizations.

The ethical problems that trouble MacIntyre are of paramount importance to the critique of social science put forth by a group of philosophers, known as critical theorists, who are associated with the Frankfurt School. While the views of this group—which includes Horkheimer, Adorno, Marcuse, Habermas, Apel, and others—are not monolithic, certain themes are pervasive. These writers regard any attempt to model social science on the pattern of the physical sciences as both erroneous and immoral. Like the interpretativists, they believe that covering-law explanations in social science involve a fundamental confusion between natural (causal) laws and normative rules. In addition, they complain that explanations in the physical sciences are divorced from historical concerns, and that this cannot be so in the social sciences. Most critical theorists would agree, for example, with Gadamer's characterization of physical science: "It is the aim of science to so objectify experience that it no longer contains any historical element. The scientific experiment does this by its methodological procedure" (1975, 311, quoted in Grünbaum 1984, 16).

Grünbaum refutes this characterization, using examples from classical electrodynamics and other fields (1984, 17–19) to show that laws of physical sciences do embody historical and contextual features. Grünbaum also criticizes the attempts of critical theorists to argue that the historical elements in physical theories are not historical in the relevant sense (1984, 19–20), and points out that Habermas bases this so-called lack of symmetry between physics and psychoanalysis on the platitude that Freudian narratives are *psychological* (1984, 21).

Karl-Otto Apel's defense of the asymmetry between history and physical science departs somewhat from the statements of Habermas that Grünbaum criticizes. It brings out, perhaps, more clearly the worry about loss of human autonomy that is the real concern of critical theorists:

> It is true, I think, that physics has to deal with irreversibility in the sense of the second principle of thermodynamics. . . . But, in this very sense of irreversibility, physics may suppose nature's being definitely determined concerning its future and thus having no history in a sense that would resist nomological objectification.
>
> Contrary to this, social science . . . must not only suppose irreversibility—in the sense of a statistically determined process—*but irreversibility, in the sense of the advance of human knowledge influencing the process of history in an irreversible manner.* (Apel 1979, 20, emphasis mine)

Apel then goes on to talk about the problem presented to social science (but not to physical science) by Merton's theorem concerning self-fulfilling and self-destroying prophecy.

Apel's remarks suggest that critical theorists' talk about the special "historic" character of social science, here and elsewhere, really amounts to the recognition that humans are often able to use their knowledge of what has happened to redirect the course of events, and to change what would have been otherwise had they not been aware of what was going on and had they not formed goals of their own. Since humans are agents with purposes, they enter into the molding of their own histories in a way not possible by any nonthinking part of nature.

Grünbaum does include examples of "feedback" systems in his account of how past states count in the determination of present behavior (1984, 19), but it is at least arguable that the concept of "purposive behavior" is not entirely captured in the descriptions of mechanical feedback systems (Taylor 1966). Apparently, the critical theorists use the term "history" in a special way to refer to accounts of autonomous human behavior. In this, they are similar to Collingwood, who does not apply the term to any processes of nature—even geological and evolutionary processes—that do not involve human intentions:

> The processes of nature can therefore be properly described as sequences of mere events, but those of history cannot. They are not processes of mere events but processes of actions, which have an inner side, consisting of processes of thought; and what the historian is looking for is these processes of thought. All history is the history of thought. (Collingwood 1946, 215)

In addition to using "history" in this special way, critical theorists are concerned that the "regularities" observed in our (corrupt) social system—that are the result of unfortunate historical circumstances and that can be changed—are in danger of being presented by a nomothetic social science as exactly analogous to unchangeable laws of nature. The reaction of critical theorists to the attempt to discover laws and to construct nomological explanations in the *physical* sciences ranges from Horkheimer's acceptance of the goal to Marcuse's outright condemnation (see Lesnoff 1979, 98). Critical theorists, however, agree in rejecting nomological explanation in the social sciences.

Deductive-nomological explanations (covering-law explanations in which the laws are universal generalizations) are supposed to show that the event to be explained *resulted from* the particular circumstances, in accord with the relevant laws cited in an explanation of it (Hempel 1962, 10). Since the description of the event in a successful deductive-nomological explanation follows logically from the explanatory statements, it is plausible, using a *modal* conception of explanation, to say that, *given the circumstances and the operative laws, the event had to occur.* Leaving aside the point that in the social sciences explanations are much more likely to be probabilistic than deductive-nomological, the critical theorists, I believe, mistakenly read this feature of "necessity" in deductive-nomological explanations as an attempt to take what is the case (i.e., the event to be explained), and show that it *must be* the case, in the sense that the event was inevitable and could not have been otherwise, even if circumstances had been different. The mistake here is similar to the incorrect belief that any conclusion of a correct deductive argument is *necessary* just because the conclusion follows *necessarily* from the premises. If deductive-nomological explanation is misinterpreted in this way, it seems to present a challenge to humans' abilities to intervene and change circumstances.

However, such an understanding of scientific explanation is mistaken. A social science that is committed to providing scientific explanations is not thereby committed to serve the ends of regimes that want to maintain their dominance by making any existing social arrangements seem *necessary.*

In the same vein, critical theorists protest that scientific explanations are merely descriptions of the status quo, since scientific explanations fail to present a range of possible alternatives to what is in fact the case. However, it is not at all clear that explanations should tell us what could be or might have been; the goal seems rather to say *why* things are as they are. Understanding why things are as they are is, after all, often a prerequisite for changing the way things are.

In part, the complaint that nomological science is oriented only toward description rather than understanding or explanation is based on incorrectly identifying science with technology, and mistaking the goals of technology—prediction and control of the environment—for the goals of science. In the grips of this mistake some version of the following argument is adopted by critical theorists:

> Physical science aims only at prediction and control of the physical environment. Therefore, a social science that is similar to physical science in its methods and aims has as its goal the prediction and control of the behavior of other humans.

Such a science would be an inherently manipulative—and thus ethically unsavory—enterprise (Habermas 1984, 389).

The picture of science as mere technique—prediction and control—is obviously inadequate as well as somewhat at odds with the critical theorists' own view that science is committed to the status quo and

insensitive to what might be. Manipulation is, after all, often directed toward other ends than the maintenance of the status quo.

We have already discussed the differences between explanation and prediction, and have rejected the strong symmetry thesis. We can sometimes reliably predict outcomes on the basis of regularities that give no understanding of the situation. We can also have significant understanding and be unable to use this knowledge for reliable prediction. It is difficult to make the case that all of science is directed toward prediction and control of the environment. Certainly those scientists who are engaged in pure research cannot always spell out immediate practical applications of that research when asked to do so.

If, as I believe, the critical theorists' assessment of the nature of physical science is grossly inaccurate, their ethical worries may nevertheless be well founded. For certainly predictive knowledge and control are highly valued byproducts of scientific knowledge. Furthermore, if scientists pretend to have the power of prediction and control when they do not, or if they capitalize on the laymen's respect for science to claim that scientific expertise grants them moral expertise as well, then they behave unethically.

It would be naïve to suppose that an increase in understanding is the only aim of scientists or even the chief aim of most scientists. Fame and money motivate scientists as they do all humans. It has been argued (Bourdieu 1975; Horton 1982) that the struggle for status rather than a pure concern for truth is dramatically more pronounced in the social sciences than in the physical sciences. Concern for status is often shown in attacks on credentials and other forms of name-calling. Horton recognizes these features in the conduct of social scientists and blames this behavior on the comparative lack of agreement about what constitutes normal social science, and consequently what counts as outstanding achievement. However, this cannot be the whole story, for among physical scientists as well, attempts to increase one's status by denigrating the credentials of others is all too common. *The Double Helix* (Watson 1968) shocked many nonscientists with this revelation, but it came as no surprise to those working in the field. The heated dispute about whether or not the impact of a comet caused the extinction of the dinosaurs provides a current public example of name-calling among physical scientists that can hardly be overlooked by anyone who reads newspapers.

Social scientists are painfully aware that it would be a serious deception to put forth the present findings of their disciplines in the same light as well-founded physical theories. Unfortunately, overconfidence in and misuse of the predictive power of science is a feature of bureaucracy, as MacIntyre notes. However, bureaucratic overconfidence is not confined to the pronouncements of social scientists, as the investigation into the tragic failure of the space shuttle in 1986 attests. MacIntyre and the critical theorists do raise our awareness that such abuses occur, and that is helpful. But the occasional occurrence of abuses does not prove that a search for scientific laws and scientific explanations in physical science or in social science is unethical.

6. CONCLUSION

Hempel's account of covering-law explanation in the social sciences, which is similar to his account of explanation in the physical sciences, was chosen because of its clarity and importance as a point of departure for discussion of contemporary views of explanation in the social sciences. Responses to Hempel's models of explanation by interpretativists, nomological skeptics, and critical theorists were presented and criticized.

From the array of accounts of scientific explanation presented in this volume, it should be apparent that no consensus about these matters exists or is likely to be reached any time soon. In the absence of a completely acceptable account of scientific explanation, we have only approximations. Yet, despite protests of the critics of causal and nomological explanation in the social sciences, the best approximations to a satisfactory philosophical theory of explanation seem to embrace successful explanations in the social sciences as well as successful explanations in the physical sciences. None of the critics, I believe, has demonstrated that the admitted differences between our social environment and our physical environment compel us to seek entirely different methods of understanding each.

REFERENCES

Apel, K.-O. 1979. Types of Social Science in the Light of Human Cognitive Interests. In Brown, pp. 3–50.

Braybrooke, D. 1987. *Philosophy of Social Science*. Englewood Cliffs, NJ: Prentice-Hall.

Bourdieu, P. 1975. The Specificity of the Scientific Field and the Social Conditions of the Progress of Reason. *Social Science Information* 14:19–47.

Brown, S. C., ed. 1979. *Philosophical Disputes in the Social Sciences*. Sussex: Harvester Press; Atlantic Highlands, NJ: Humanities Press.

Canfield, J., ed. 1966. *Purpose in Nature*. Englewood Cliffs, NJ: Prentice-Hall.

Collingwood, R. G. 1946. *The Idea of History*. Oxford: The University Press.

Colodny, R., ed. 1962. *Frontiers of Science and Philosophy*. Pittsburgh: University of Pittsburgh Press.

Converse, P. 1986. Generalization and the Social Psychology of "Other Worlds." In Fiske and Shweder, pp. 42–60.

Davidson, D. 1980. *Essays on Actions and Events*. Oxford: Clarendon Press.

Dray, W. 1957. *Laws and Explanation in History*. Oxford: The University Press.

Fiske, D. W., and Shweder, R. A. 1986. *Metatheory in Social Science*. Chicago and London: University of Chicago Press.

Frake, C. O. 1969. The Ethnographic Study of Cognitive Systems. In Tyler, pp. 28–41.

Gadamer, H. G. 1975. *Truth and Method*. New York: Seabury Press.

Gardiner, P., ed. 1959. *Theories of History*. New York: The Free Press.

Geertz, C. 1975. *The Interpretation of Cultures*. London: Hutchinson.

Glymour, C., et al. 1987. *Discovering Causal Structure*. San Diego: Academic Press.

Goodenough, W. 1975. Cultural Anthropology and Linguistics. In *Georgetown University Monograph Series on Language and Linguistics* No. 9, pp. 167–73.

Grünbaum, A. 1984. *The Foundation of Psychoanalysis*. Berkeley and Los Angeles: University of California Press.

Habermas, J. 1981. *The Theory of Communicative Action*, Vol. 1. Trans. by T. McCarthy. Boston: Beacon Press.

Hempel, C. G. 1962. Explanation in Science and in History. In Colodny, pp. 1–33.

———. 1965. *Aspects of Scientific Explanation.* New York: The Free Press.

———. 1988. Provisoes: A Problem Concerning the Inferential Function of Scientific Theories. *Erkenntnis* 28:147–64. (Also to appear in *The Limitations of Deductivism.* A. Grünbaum and W. Salmon, eds., Berkeley and Los Angeles: University of California Press.)

Hollis, M., and Lukes, S., eds. 1982. *Rationality and Relativism.* Cambridge, MA: The MIT Press.

Horton, R. 1982. Tradition and Modernity Revisited. In Hollis and Lukes, pp. 201–60.

Humphreys, P. 1989. Scientific Explanation: The Causes, Some of the Causes, and Nothing but the Causes. In *Scientific Explanation,* Minnesota Studies in the Philosophy of Science, Vol. XIII (Minneapolis: University of Minnesota Press, 1989), chap. 4.

Jeffrey, R. 1969. Statistical Explanation vs. Statistical Inference. In Rescher, ed., pp. 104–13.

Lakatos, I. 1970. Falsification and the Methodology of Scientific Research Programmes. In Lakatos and Musgrave, pp. 91–195.

———, and Musgrave, A., eds. 1970. *Criticism and the Growth of Knowledge.* Cambridge: the University Press.

Lesnoff, M. 1979. Technique, Critique, and Social Science. In Brown, pp. 89–116.

MacIntyre, A. 1984. *After Virtue.* Notre Dame: University of Notre Dame Press.

Mill, J. S. 1874. *Logic,* 8th ed. New York: Harper Bros.

Murdock, G. P. 1949. *Social Structure.* New York: Macmillan.

Papineau, D. 1978. *For Science in the Social Sciences.* New York: St. Martin's Press.

Rescher, N., ed. 1969. *Essays in Honor of Carl G. Hempel.* Dordrecht: D. Reidel.

Sahlins, M., and Service, E. 1960. *Evolution and Culture.* Ann Arbor: University of Michigan Press.

Salmon, W. C. 1965. The Status of Prior Probabilities in Statistical Explanation. *Philosophy of Science* 32:137–46.

———. 1970. Statistical Explanation. In Colodny, ed., pp. 173–231.

Sanders, W. T., and Price, B. 1968. *Mesoamerica: The Evolution of a Civilization.* New York: Random House.

Scriven, M. 1959. Truisms as the Grounds for Historical Explanation. In Gardiner, pp. 443–71.

Spiro, M. 1986. Cultural Relativism and the Future of Anthropology. *Cultural Anthropology* I, 3:259–86.

Taylor, R. 1966. Comments on a Mechanistic Conception of Purposefulness. In Canfield, pp. 17–26.

Tyler, S., ed. 1969. *Cognitive Anthropology.* New York: Holt, Rinehart, Winston.

van Fraassen, B. 1980. *The Scientific Image.* Oxford: the University Press.

Watson, J. 1968. *The Double Helix.* Boston: Atheneum Press.

Winch, P. 1958. *The Idea of a Social Science and Its Relation to Philosophy.* London: Routledge & Kegan Paul.

PART 3 SUGGESTIONS FOR FURTHER READING

Achinstein, Peter. "Can There Be a Model of Explanation?" *Theory and Decision* 13 (1981): 201–227.

Armstrong, David. *What Is a Law of Nature?* Cambridge: Cambridge University Press, 1983.

Hemple, Carl. *Aspects of Scientific Explanation.* New York: Free Press, 1965.

Kitcher, P., and W. Salmon, eds. *Scientific Explanation.* Minneapolis: University of Minnesota Press, 1989.

McIntyre, Lee. *Laws and Explanation in the Social Sciences: Defending a Science of Human Behavior.* Boulder, CO: Westview Press, 1996.

Pitt, J., ed. *Theories of Explanation.* New York: Oxford University Press, 1988.

Ruben, David-Hillel. *Explaining Explanation.* Oxford: Oxford University Press, 1994.

———, ed. *Explanation.* Oxford: Oxford University Press, 1993.

van Fraassen, Bas. *The Scientific Image.* Oxford: Clarendon Press, 1980.

———. *Laws and Symmetry.* Oxford: Clarendon Press, 1989.

PART 4

THE UNITY OF SCIENCE
Are All Sciences Reducible to Physics?

The more unified our knowledge, the greater our understanding. A complete understanding of the world, then, would require a completely unified science. The positivists thought that all of science could be unified because all sciences were, in principle, reducible to physics.

According to the positivists, one theory is reducible to another if the laws of the former are derivable from those of the latter. For example, chemistry is reducible to physics because the laws of chemistry are derivable from the laws of physics. Because the laws of chemistry and physics mention different objects, however, the laws of chemistry are not derivable from the laws of physics alone. Certain additional statements are needed. These include "bridge laws" that relate chemical terms to physical terms as well as boundary conditions that identify the circumstances under which physical events will produce chemical events. Schematically, then, theoretical reduction has the following form:

Lower-level laws

Bridge laws

Boundary conditions

Therefore, higher-level laws

The notion that higher-level sciences are reducible to lower-level ones is based on a conception of nature as consisting of a hierarchy of increasingly complex entities. In this view, the higher-level entities are composed of lower-level ones. For example, organisms are composed of cells, cells are composed of molecules, and molecules are composed of atoms. Difference sciences study entities at different levels. Biology, for example, studies organisms and cells, chemistry studies molecules, and physics studies atoms. Because the behavior of each type of entity is supposedly determined by the behavior of the entities that make it up, the higher-level sciences should be reducible to the lower-level ones.

Paul Oppenheim and Hilary Putnam present a classic statement of what's involved in reducing one theory to another. They claim that reducing all sciences to physics not only represents an ideal state, but also is a trend of current research. So the belief that all sciences can one day be reduced to physics need not be taken as an article of faith. Because it is the simplest hypothesis, and because there is both direct and indirect evidence to support it, they claim we are justified in believing that such a reduction will occur.

Jerry Fodor is not convinced by Oppenheim and Putnam's arguments. He claims that not all sciences are reducible to physics because the behavior of higher-level entities is not always determined by the behavior of lower-level ones. Consider money, for example. Economists formulate laws regarding money, such as Gresham's law, which says that bad money will drive good money out of circulation. This law holds no matter what the money is made of, be it gold, silver, paper, sea shells, beads, or other objects. As a result, it's doubtful that the laws of economics are reducible to or explainable in terms of the laws of physics.

More generally, the reason for this irreducibility is that the kinds, classes, or groupings of objects that are significant at one level are not necessarily important for a science at a different level. Rocks and dust are importantly different for geologists, but they're pretty much the same for physicists—they are made up of similar atoms. So physics doesn't recognize the difference between geologically very different categories.

Darden and Maull argue that even if a unified science cannot be achieved by reducing higher-level sciences to lower-level ones, it may still be achieved through the development of what they call "interfield theories." An interfield theory makes explicit the relations between fields. A field, according to Darden and Maull, is an area of science consisting of a central problem, a common domain of facts, and a set of techniques for explaining those facts. By their lights, then, cytology (the study of cells) and genetics (the study of inherited characteristics) constitute fields. The chromosome theory of heredity united these fields by showing that the factors governing inherited characteristics lay on the chromosomes. The unification thus achieved brought about a better understanding of the mechanism of heredity, although neither cytology nor genetics was reduced to the other.

John Dupré does not deny that interfield theories produce a valuable form of unification. He maintains, however, that even if every field were related by an interfield theory to some other, we still might not have a unified science because the related fields may not all be scientific. There could be an interfield theory relating linguistics to literary theory, for example. But literary theory is not normally considered a part of science. The price of a unified knowledge, therefore, may be a disunified science.

The fact that nonscientific theories can be united to scientific ones poses a problem for those, such as Kitcher, who claim that what makes a theory pseudoscientific is that it can't be made part of a unified science. Since any field can potentially be unified with any other, via interfield theories, Dupré suggests that what distinguishes science from pseudoscience is not the content of its theories but the character of its practitioners. If they do not possess the appropriate intellectual virtues—if they are not sensitive to empirical facts, if they do not work from plausible background assumptions, if they are not sensitive to criticism—they are not doing science.

George Reisch finds this characterization of science too permissive. Without a hard-and-fast criterion for distinguishing science from nonscience, we will be hard-pressed to avoid giving creationism equal billing in our textbooks. He argues that creationism is not a science because its basic beliefs are inconsistent with those that connect and unify the existing sciences. Unfortunately, we do not have a general theory specifying what sorts of connections are required to make something a science. Nevertheless, Reisch argues, we do not need such a theory to reject creationism as unscientific. The inability of scientists to integrate creationism into current theories is insufficient.

17

PAUL OPPENHEIM AND HILARY PUTNAM

Unity of Science as a Working Hypothesis

1. INTRODUCTION

1.1. The expression "Unity of Science" is often encountered, but its precise content is difficult to specify in a satisfactory manner. It is the *aim of this paper* to formulate a precise concept of Unity of Science; and to examine to what extent that unity can be attained.

A concern with Unity of Science hardly needs justification. We are guided especially by the conviction that Science of Science, i.e., the meta-scientific study of major aspects of science, is the natural means for counterbalancing specialization by promoting the integration of scientific knowledge. The desirability of this goal is widely recognized; for example, many universities have programs with this end in view; but it is often pursued by means different from the one just mentioned, and the conception of the Unity of Science might be especially suited as an organizing principle for an enterprise of this kind.

1.2. As a preliminary, we will distinguish, in order of increasing strength, three broad concepts of Unity of Science:

First, Unity of Science in the weakest sense is attained to the extent to which all the terms of science[1] are reduced to the terms of some one discipline (e.g., physics, or psychology). This concept of *Unity of Language* (12) may be replaced by a number of sub-concepts depending on the manner in which one specifies the notion of "reduction" involved. Certain authors, for example, construe reduction as the *definition* of the terms of science by means of those in the selected basic discipline (reduction by means of biconditionals (47)); and some of these require the definitions in question to be analytic, or "true in virtue of the meanings of the terms involved" (epistemological reduction); others impose no such restriction upon the biconditionals effecting reduction. The notion of reduction we shall employ is a wider one, and is designed to include reduction by means of biconditionals as a special case.

Second, Unity of Science in a stronger sense (because it implies Unity of Language, whereas the reverse is not the case) is represented by *Unity of Laws* (12). It is attained to the extent to which the laws of science become reduced to the laws of some one discipline. If the ideal of such an all-comprehensive explanatory system were realized, one could call it *Unitary Science* (18, 19, 20, 80). The exact meaning of 'Unity of Laws' depends, again, on the concept of "reduction" employed.

Third, Unity of Science in the strongest sense is realized if the laws of science are not only reduced to the laws of some one discipline, but the laws of that discipline are in some intuitive sense "unified" or "connected." It is difficult to see how this last requirement can be made precise; and it will not be imposed here. Nevertheless, trivial realizations of "Unity of Science" will be excluded, for example, the simple conjunction of several branches of science does not *reduce* the particular branches in the sense we shall specify.

1.3. In the present paper, the term 'Unity of Science' will be used in two senses, to refer, first, to an ideal *state* of science, and, second, to a pervasive *trend* within science, seeking the attainment of that ideal.

Concepts, Theories, and the Mind-Body Problem, Minnesota Studies in the Philosophy of Science, Vol. II, H. Feigl, M. Scriven, and G. Maxwell, eds. (Minneapolis: University of Minnesota Press, 1958), pp. 3–36. Reprinted by permission of the publisher.

In the first sense, 'Unity of Science' means the state of unitary science. It involves the two constituents mentioned above: unity of vocabulary, or "Unity of Language"; and unity of explanatory principles, or "Unity of Laws." That Unity of Science, in this sense, can be fully realized constitutes an over-arching meta-scientific hypothesis which enables one to see a unity in scientific activities that might otherwise appear disconnected or unrelated, and which encourages the construction of a unified body of knowledge.

In the second sense, 'Unity of Science' exists as a trend within scientific inquiry, whether or not unitary science is ever attained, and notwithstanding the simultaneous existence (and, of course, legitimacy) of other, even *incompatible,* trends.

1.4. The expression 'Unity of Science' is employed in various other senses, of which two will be briefly mentioned in order to distinguish them from the sense with which we are concerned. In the first place, what is sometimes referred to is something that we may call the *Unity of Method* in science. This might be represented by the thesis that all the empirical sciences employ the same standards of explanation, of significance, of evidence, etc.

In the second place, a radical reductionist thesis (of an alleged "logical," not an empirical kind) is sometimes referred to as the thesis of the Unity of Science. Sometimes the "reduction" asserted is the definability of all the terms of science in terms of *sensationalistic predicates* (10); sometimes the notion of "reduction" is wider (11) and predicates referring to *observable qualities of physical things* are taken as basic (12). These theses are epistemological ones, and ones which today appear doubtful. The epistemological uses of the terms 'reduction,' 'physicalism,' 'Unity of Science,' etc., should be carefully distinguished from the use of these terms in the present paper.

2. UNITY OF SCIENCE AND MICRO–REDUCTION

2.1. In this paper we shall employ a concept of reduction introduced by Kemeny and Oppenheim in their paper on the subject (47), to which the reader is referred for a more detailed exposition. The principal requirements may be summarized as follows: given two theories T_1 and T_2, T_2 is said to be *reduced* to T_1 if and only if:

1. The vocabulary of T_2 contains terms not in the vocabulary of T_1.

2. Any observational data explainable by T_2 are explainable by T_1.

3. T_1 is at least as well systematized as T_2. (T_1 is normally more complicated than T_2; but this is allowable, because the reducing theory normally explains more than the reduced theory. However, the "ratio," so to speak, of simplicity to explanatory power should be at least as great in the case of the reducing theory as in the case of the reduced theory.)[2]

Kemeny and Oppenheim also define the reduction of a branch of science B_2 by another branch B_1 (e.g., the reduction of chemistry to physics). Their procedure is as follows: take the accepted theories of B_2 at a given time t as T_2. Then *B_2 is reduced to B_1 at time t* if and only if there is some theory T_1 in B_1 at t such that T_1 reduces T_2 (47). Analogously, if *some* of the theories of B_2 are reduced by some T_1 belonging to branch B_1 at t, we shall speak of a *partial reduction* of B_2 to B_1 at t. This approach presupposes (1) the familiar assumption that some division of the total vocabulary of both branches into theoretical and observational terms is given, and (2) that the two branches have the same observational vocabulary.

2.2. The essential feature of a *micro*-reduction is that the branch B_1 deals with the parts of the objects dealt with by B_2. We must suppose that corresponding to each branch we have a specific universe of discourse U_{Bi};[3] and that we have a part-whole relation, Pt (75; 76, especially p. 91). Under the following conditions we shall say that the reduction of B_2 to B_1[4] is a *micro-reduction*: B_2 is reduced to B_1; and the objects in the universe of discourse of B_2 are wholes which possess a decomposition (75; 76, especially p. 91) into proper parts all of which belong to the universe of discourse of B_1. For example, let us suppose B_2 is a branch of science which has multicellular living things as its universe of discourse. Let B_1 be a branch with cells as its universe of discourse. Then the things in the universe of discourse of B_2 can be decomposed into proper parts belonging to the universe of discourse of B_1. If, in addition, it is the case that B_1 reduces B_2 at the time t, we shall say that B_1 *micro-reduces* B_2 *at time t.*

We shall also say that a branch B_1 is a *potential micro-reducer* of a branch B_2 if the objects in the universe of discourse of B_2 are wholes which possess a decomposition into proper parts all of which belong to the universe of discourse of B_1. The definition is the same as the definition of 'micro-reduces' except for the omission of the clause 'B_2 is reduced to B_1.'

Any micro-reduction constitutes a step in the direction of *Unity of Language* in science. For, if B_1 reduces B_2, it explains everything that B_2 does (and normally, more besides). Then, even if we cannot define in B_1 analogues for some of the theoretical terms of B_2, we can *use B_1 in place of B_2.* Thus any reduction, in the sense explained, permits "reduction" of the total vocabulary of science by making it possible to dispense with some terms.[5] Not every reduction moves in the direction of Unity of Science; for instance reductions *within* a branch lead to a simplification of the vocabulary of science, but they do not necessarily lead in the direction of Unity of Science as we have characterized it (although they may at times fit into that trend). However, *micro*-reductions, and even partial micro-reductions, insofar as they permit us to replace some of the terms of one branch of science by terms of another, *do* move in this direction.

Likewise, the micro-reduction of B_2 to B_1 moves in the direction of *Unity of Laws*; for it "reduces" the total number of scientific laws by making it possible, in principle, to dispense with the laws of B_2 and explain the relevant observations by using B_1.

The relations 'micro-reduces' and 'potential micro-reducer' have very simple properties: (1) they are transitive (this follows from the transitivity of the relations 'reduces' and 'Pt'); (2) they are irreflexive (no branch can micro-reduce itself); (3) they are asymmetric (if B_1 micro-reduces B_2, B_2 never micro-reduces B_1). The two latter properties are not purely formal; however, they require for their derivation only the (certainly true) empirical assumption that there does not exist in infinite descending chain of proper parts, i.e., a series of things x_1, x_2, x_3, . . . such that x_2 is a proper part of x_1, x_3 is a proper part of x_2, etc.

The just-mentioned *formal* property of the relation 'micro-reduces'—its transitivity—is of great importance for the program of Unity of Science. It means that micro-reductions have a *cumulative* character. That is, if a branch B_3 is micro-reduced to B_2, and B_2 is in turn micro-reduced to B_1, then B_3 is automatically micro-reduced to B_1. This simple fact is sometimes overlooked in objections[6] to the theoretical possibility of attaining unitary science by means of micro-reduction. Thus it has been contended that one manifestly cannot explain human behavior by reference to the laws of atomic physics. It would indeed be fantastic to suppose that the simplest regularity in the field of psychology could be explained directly—i.e., "skipping" intervening branches of science—by employing subatomic theories. But one may believe in the attainability of unitary science without thereby committing oneself to this absurdity. It is not absurd to suppose that psychological laws may eventually be explained in terms of the behavior of individual neurons in the brain; that the behavior of individual cells—including neurons—may eventually be explained in terms of their biochemical constitution; and that the behavior of molecules—including the macro-molecules that make up living cells—may eventually be explained in terms of atomic physics. If this is achieved, then psychological laws will have in *principle,* been reduced to laws of atomic physics, although it would nevertheless be hopelessly impractical to try to derive the behavior of a single human being directly from his constitution in terms of elementary particles.

2.3. *Unitary* science certainly does not exist today. But will it ever be attained? It is useful to divide this question into two subquestions: (1) If unitary science can be attained at all, *how* can it be attained? (2) *Can* it be attained at all?

First of all, there are various abstractly possible ways in which unitary science might be attained. However, it seems very doubtful, to say the least, that a branch B_2 could be reduced to a branch B_1, if the things in the universe of discourse of B_2 are not themselves in the universe of discourse of B_1 and also do not possess a decomposition into parts in the universe of discourse of B_1. ("They don't speak about the same things.")

It does not follow that B_1 must be a potential *micro*-reducer of B_2, i.e., that all reductions are micro-reductions.

There are many cases in which the reducing theory and the reduced theory belong to the same branch, or to branches with the same universe of discourse. When we come, however, to branches with different universes—say, physics and psychology—it seems clear that the possibility of reduction depends on the existence of a structural connection between the universes *via* the 'Pt' relation. Thus one cannot plausibly suppose—for the present at least—that the behavior of inorganic matter is explainable by reference to psychological laws; for inorganic materials do not consist of living parts. One supposes that psychology may be reducible to physics, but not that physics may be reducible to psychology!

Thus, the only method of attaining unitary science that appears to be seriously available at present is micro-reduction.

To turn now to our second question, *can* unitary science be attained? We certainly do not wish to maintain that it has been *established* that this is the case. But it does not follow, as some philosophers seem to think, that a tentative acceptance of the hypothesis that unitary science can be attained is therefore a mere "act of faith." We believe that this hypothesis is *credible;*[7] and we shall attempt to support this in the latter part of this paper, by providing empirical, methodological and pragmatic reasons in its support. We therefore think the assumption that unitary science can be attained through cumulative micro-reduction recommends itself *as a working hypothesis.*[8] That is, we believe that it is in accord with the standards of reasonable scientific judgment to tentatively accept this hypothesis and to work on the assumption that further progress can be made in this direction, without claiming that its truth has been established, or denying that success may finally elude us.

3. REDUCTIVE LEVELS

3.1. As a basis for our further discussion, we wish to consider now the possibility of ordering branches in such a way as to indicate the major potential micro-reductions standing between the present situation and the state of unitary science. The most natural way to do this is by their universes of discourse.

We offer, therefore, a system of *reductive levels* so chosen that a branch with the things of a given level as its universe of discourse will always be a potential micro-reducer of any branch with things of the next higher level (if there is one) as its universe of discourse.

Certain conditions of adequacy follow immediately from our aim. Thus:

1. There must be several levels.

2. The number of levels must be finite.

3. There must be a unique lowest level (i.e., a unique "beginner" under the relation 'potential micro-reducer'); this means that success at transforming all the *potential* micro-reductions connecting these branches into *actual* micro-reductions must, *ipso facto,* mean reduction to a single branch.

4. Any thing of any level except the lowest must possess a decomposition into things belonging to the next lower level. In this sense each level will be as it were a "common denominator" for the level immediately above it.

5. Nothing on any level should have a part on any higher level.

6. The levels must be selected in a way which is "natural"[9] and justifiable from the standpoint of present-day empirical science. In particular, the step from any one of our reductive levels to the next lower level must correspond to what is, scientifically speaking, a crucial step in the trend toward over-all physicalistic reduction.

The accompanying list gives the levels we shall employ;[10] the reader may verify that the six conditions we have listed are all satisfied.

6 . . . . Social groups
5 . . . . (Multicellular)
            living things
4 . . . . Cells
3 . . . . Molecules
2 . . . . Atoms
1 . . . . Elementary particles

Any whole which possesses a decomposition into parts all of which are on a given level, will be counted as also belonging to that level. Thus each level includes all higher levels. However, the highest level to which a thing belongs will be considered the "proper" level of that thing.

This inclusion relation among our levels reflects the fact that scientific laws which apply to the things of a given level and to all combinations of those things also apply to all things of higher level. Thus a physicist, when he speaks about "all physical objects," is also speaking about living things—but not qua living things.

We maintain that each of our levels is *necessary* in the sense that it would be utopian to suppose that one might reduce all of the major theories or a whole branch concerned with any one of our six levels to a theory concerned with a lower level, *skipping* entirely the *immediately* lower level; and we maintain that our levels are *sufficient* in the sense that it would not be utopian to suppose that a major theory on any one of our levels *might* be directly reduced to the next lower level. (Although this is *not* to deny that it may be convenient, in special cases, to introduce intervening steps.)

However, this contention is significant only if we suppose some set of *predicates* to be associated with each of these levels. Otherwise, as has been pointed out,[11] *trivial* micro-reductions would be possible; e.g., we might introduce the property "Tran" (namely, the property of being an atom of a transparent substance) and then "explain the transparency of water in terms of properties on the atomic level," namely, by the hypothesis that all atoms of water have the property Tran. More explicitly, the explanation would consist of the statements

a. (x) (x is transparent $\equiv$ (y) (y is an atom of x $\supset$ Tran(y))

b. (x) (x is water $\supset$ (y) (y is an atom of x $\supset$ Tran(y))

To exclude such trivial "micro-reductions," we shall suppose that with each level there is associated a list of the theoretical predicates normally employed to characterize things on that level at present (e.g., with level 1, there would be associated the predicates used to specify spatio-temporal coordinates, mass-energy, and electric charge). And when we speak of a theory concerning a given level, we will mean not only a theory whose universe of discourse is that level, but one whose predicates belong to the appropriate list. Unless the hypothesis that theories concerning level n + 1 can be reduced by a theory concerning level n is restricted in this way, it lacks any clear empirical significance.

3.2. If the "part-whole" ('Pt') relation is understood in the wide sense, that x Pt y holds if x is spatially or temporally contained in y, then everything, continuous or discontinuous, belongs to one or another reductive level; in particular to level 1 (at least), since it is a whole consisting of elementary particles. However, one may wish to understand 'whole' in a narrower sense (as "structured organization of elements"[12]). Such a specialization involves two essential steps: (1) the construction of a calculus with such a narrower notion as its primitive concept, and (2) the definition of a particular 'Pt' relation satisfying the axioms of the calculus.

Then the problem will arise that some things do not belong to *any* level. Hence a theory dealing with such things might not be micro-reduced even if all the micro-reductions indicated by our system of levels were accomplished; and for this reason, unitary science might not be attained.

For a trivial example, "a man in a phone booth" is an aggregate of things on different levels which we would not regard as a whole in such a narrower sense. Thus, such an "object" does not belong to any reductive level; although the "phone booth" belongs to level 3 and the man belongs to level 5.

The problem posed by such aggregates is not serious, however. We may safely make the assumption that the behavior of "man in phone booths" (to be carefully distinguished from "men in phone booths") could be completely explained given (a) a complete physicochemical theory (i.e., a theory of levels up to 3, including "phone booths"), and (b) a complete individual psychology (or more generally, a theory of levels up to 5). With this assumption in force, we are able to say: If we can construct a theory that explains the behavior of all the objects in our system of levels, then it will also handle the aggregates of such objects. . . .

4. THE CREDIBILITY OF OUR WORKING HYPOTHESIS

4.1. John Stuart Mill asserts (55, Book VI, Chapter 7) that since (in our wording) human social groups are wholes whose parts are individual persons, the "laws of the phenomena of society" are "derived from and may be resolved into the laws of the nature of individual man." In our terminology, this is to suggest that it is a logical truth that theories concerning social groups (level 6) can be *micro-reduced* by theories concerning individual living things (level 5); and, *mutatis mutandis,* it would have to be a logical truth that theories concerning any other level can be micro-reduced by theories concerning the next lower level. As a consequence, what we have called the "working hypothesis" that unitary science can be attained would likewise be a logical truth.

Mill's contention is, however, not so much *wrong* as it is vague. What is one to count as "the nature of individual man"? As pointed out above (section 3.1) the question whether theories concerning a given reductive level can be reduced by a theory concerning the next lower level has empirical content only if the theoretical vocabularies are specified; that is, only if one associates with each level, as we have supposed to be done, a particular set of theoretical concepts. Given, e.g., a sociological theory T_2, the question whether there exists a true psychological theory T_1 *in a particular vocabulary* which reduces T_2 is an empirical question. Thus our "working hypothesis" is one that can only be justified on empirical grounds.

Among the factors on which the degree of credibility of *any* empirical hypothesis depends are (45, p. 307) the *simplicity* of the hypothesis, the *variety* of the evidence, its *reliability,* and last but not least, the *factual support* afforded by the evidence. We proceed to discuss each of these factors.

4.2 As for the *simplicity*[13] of the hypothesis that unitary science can be attained, it suffices to consider the traditional alternatives mentioned by those who oppose it. "Hypotheses" such as Psychism and Neo-Vitalism assert that the various objects studied by contemporary science have special parts or attributes, unknown to present-day science, in addition to those indicated in our system of reductive levels. For example, men are said to have not only

cells as parts; there is also an immaterial "psyche"; living things are animated by "entelechies" or "vital forces"; social groups are moved by "group minds." But, in none of these cases are we provided *at present* with postulates or coordinating definitions which would permit the derivation of testable predictions. Hence, the claims made for the hypothetical entities just mentioned lack any clear scientific meaning; and as a consequence, the question of supporting evidence cannot even be raised.

On the other hand, if the effort at micro-reduction should seem to fail, we cannot preclude the introduction of theories postulating presently unknown relevant parts or presently unknown relevant attributes for some or all of the objects studied by science. Such theories are perfectly admissible, provided they have genuine explanatory value. For example, Dalton's chemical theory of molecules might not be reducible to the best available theory of atoms at a given time if the latter theory ignores the existence of the electrical properties of atoms. Thus the hypothesis of micro-reducibility,[14] as the meaning is specified at a particular time, may be false because of the insufficiency of the theoretical apparatus of the reducing branch.

Of course, a new working hypothesis of micro-reducibility, obtained by enlarging the list of attributes associated with the lowest level, might then be correct. However, if there are presently unknown attributes of a more radical kind (e.g., attributes which are relevant for explaining the behavior of living, but not of non-living things), then no such simple "repair" would seem possible. In this sense, Unity of Science is an alternative to the view that it will eventually be necessary to *bifurcate* the conceptual system of science, by the postulation of new entities or new attributes unrelated to those needed for the study of inanimate phenomena.

4.3. The requirement that there be *variety* of evidence assumes a simple form in our present case. If all the past successes referred to a single pair of levels, then this would be poor evidence indeed that theories concerning each level can be reduced by theories concerning a lower level. For example, if all the past successes were on the atomic level, we should hardly regard as justified the inference that laws concerning social groups can be explained by reference to the "individual psychology" of the members of those groups. Thus, the first requirement is that one should be able to provide examples of successful micro-reductions between several pairs of levels, preferably between all pairs.

Second, within a given level what is required is, preferably, examples of different kinds, rather than a repetition of essentially the same example many times. In short, one wants good evidence that *all* the phenomena of the given level can be micro-reduced.

We shall present below a survey of the past successes in each level. This survey is, of course, only a sketch; the successful micro-reductions and projected micro-reductions in biochemistry alone would fill a large book. But even from this sketch it will be apparent, we believe, how great the variety of these successful micro-reductions is in both the respects discussed.

4.4. Moreover, we shall, of course, present only evidence from authorities regarded as *reliable* in the particular area from which the theory or experiment involved is drawn.

4.5. The important factor *factual support* is discussed only briefly now, because we shall devote to it many of the following pages and would otherwise interrupt our presentation.

The first question raised in connection with any hypothesis is, of course, what *factual support* it possesses, that is, what confirmatory or disconfirmatory evidence is available. The evidence supporting a hypothesis is conveniently subdivided into that providing *direct* and that providing *indirect* factual support. By the direct factual support for a hypothesis we mean, roughly,[15] the proportion of confirmatory as opposed to disconfirmatory instances. By the indirect factual support, we mean the inductive support obtained from other well-confirmed hypotheses that lend credibility to the given hypothesis. While intuitively adequate quantitative measures of direct factual support have been worked out by Kemeny and Oppenheim,[16] no such measures exist for indirect factual support. The present paper will rely only on intuitive judgements of these magnitudes, and will not assume that quantitative explicata will be worked out.

As our hypothesis is that theories of each reductive level can be micro-reduced by theories of the next lower level, a "confirming instance" is simply any successful micro-reduction between any two of our levels. The *direct* factual support for our hypothesis is thus provided by the *past successes* at reducing laws about the things on each level by means of laws referring to the parts on lower (usually, the next lower) levels. In the sequel, we shall survey the past successes with respect to each pair of levels.

As *indirect* factual support, we shall cite evidence supporting the hypothesis that each reductive level is, in evolution and ontogenesis (in a wide sense presently to be specified) prior to the one above it. The hypothesis of *evolution* means here that (for n = 1, . . . , 5) there was a time when there were things of level n, but no things of any higher level. This hypothesis is highly speculative on levels 1 and 2; fortunately the micro-reducibility of the molecular to the atomic level and of the atomic level to the elementary particle level is relatively well established on other grounds.

Similarly, the hypothesis of ontogenesis is that, in certain cases, for any *particular* object on level n, there was a time when it did not exist, but when some of its parts on the next lower level existed; and that it developed or was causally produced out of these parts.[17]

The reason for our regarding evolution and ontogenesis as providing indirect factual support for the Unity of Science hypothesis may be formulated as follows:

Let us, as is customary in science, assume causal determination as a guiding principle; i.e., let us assume that things that appear later in time can be accounted for in terms of things and processes at earlier times. Then, if we find that there was a time when a certain whole did not exist, and that things on a lower level came together to form that whole, it is very natural to suppose that the characteristics can be micro-reduced by a theory involving only characteristics of the parts.

For the same reason, we may cite as further indirect factual support for the hypothesis of empirical Unity of Science the various successes at *synthesizing* things of each level out of things on the next lower level. Synthesis strongly increases the evidence that the characteristics of the whole in question are causally determined by the characteristics, including spatio-temporal arrangement, of its parts by showing that the object is produced, under controlled laboratory conditions, whenever parts with those characteristics are arranged in that way.

The consideration just outlined seems to us to constitute an argument against the view that, as objects of a given level combine to form wholes belonging to a higher level, there appear certain new phenomena which are "emergent" (35, p. 151; 76, p. 93) in the sense of being forever irreducible to laws governing the phenomena on the level of the parts. What our argument opposes is not, of course, the obviously true statement that there are many phenomena which are not reducible by currently available theories pertaining to lower levels; our working hypothesis rejects merely the claim of absolute irreducibility, unless such a claim is supported by a theory which has a sufficiently high degree of credibility; thus far we are not aware of any such theory. It is not sufficient, for example, simply to advance the claim that certain phenomena considered to be specifically human, such as the use of verbal language, in an abstract and generalized way, can never be explained on the basis of neurophysiological theories, or to make the claim that this conceptual capacity distinguishes man in principle and not only in degree from non-human animals.

4.6. Let us mention in passing certain *pragmatic* and *methodological* points of view which speak in favor of our working hypothesis:

1. It is of *practical* value, because it provides a good synopsis of scientific activity and of the relations among the several scientific disciplines.

2. It is, as has often been remarked, *fruitful* in the sense of stimulating many different kinds of scientific research. By way of contrast, belief in the *irreducibility* of various phenomena has yet to yield a single accepted scientific theory.

3. It corresponds *methodologically* to what might be called the "Democritean tendency" in science; that is, the pervasive methodological tendency[18] to try, insofar as is possible, to explain apparently dissimilar phenomena in terms of qualitatively identical parts and their spatio-temporal relations. . . .

5. CONCLUDING REMARKS

The possibility that all science may one day be reduced to microphysics (in the sense in which chemistry seems today to be reduced to it), and the presence of a unifying trend toward micro-reduction running through much of scientific activity, have often been noticed both by specialists in the various sciences and by meta-scientists. But these opinions have, in general, been expressed in a more or less vague manner and without very deep-going justification. It has been our aim, first, to provide precise definitions for the crucial concepts involved, and, second, to reply to the frequently made accusations that belief in the attainability of unitary science is "a mere act of faith." We hope to have shown that, on the contrary, a tentative acceptance of this belief, an acceptance of it as a working hypothesis, is *justified*, and that the hypothesis is *credible*, partly on methodological grounds (e.g., the simplicity of the hypothesis, as opposed to the bifurcation that rival suppositions create in the conceptual system of science), and partly because there is really a large mass of direct and indirect evidence in its favor.

NOTES

1. Science, in the wider sense, may be understood as including the formal disciplines, mathematics, and logic, as well as the empirical ones. In this paper, we shall be concerned with science only in the sense of empirical disciplines, including the sociohumanistic ones.

2. By a "theory" (in the widest sense) we mean any hypothesis, generalization, or law (whether deterministic or statistical), or any conjunction of these; likewise by "phenomena" (in the widest sense) we shall mean either particular occurrences or theoretically formulated general patterns. Throughout this paper, "explanation" ("explainable" etc.) is used as defined in Hempel and Oppenheim (35). As to "explanatory power," there is a definite connection with "systematic power." See Kemeny and Oppenheim (46, 47).

3. If we are willing to adopt a "Taxonomic System" for classifying all the things dealt with by science, then the various classes and subclasses in such a system could represent the possible "universes of discourse." In this case, the U_{Bi} of any branch

would be associated with the extension of a taxo-nomic term in the sense of Oppenheim (62).

4. Henceforth, we shall as a rule omit the clause 'at time t'.

5. Oppenheim (62, section 3) has a method for measuring such a reduction.

6. Of course, in some cases, such "skipping" does occur in the process of micro-reduction, as shall be illustrated later on.

7. As to degree of *credibility*, see Kemeny and Oppenheim (45, especially p. 307).

8. The "acceptance, as an overall fundamental working hypothesis, of the reduction theory, with physical science as most general, to which all others are reducible; with biological science less general; and with social science least general of all," has been emphasized by Hockett (37, especially p. 571).

9. As to *natural*, see Hempel (33, p. 52), and Hempel and Oppenheim (34, pp. 107, 110).

10. Many well-known hierarchical orders of the same kind (including some compatible with ours) can be found in modern writings. It suffices to give the following quotation from an article by L. von Bertalanffy (95, p. 164): "Reality, in the modern conception, appears as a tremendous hierarchical order of organized entities, leading, in a superposition of many levels, from physical and chemical to biological and sociological systems. Unity of Science is granted, not by an utopian reduction of all sciences to physics and chemistry, but by the structural uniformities of the different levels of reality." As to the last sentence, we refer in the last paragraph of section 2.2 to the problem noted. Von Bertalanffy has done pioneer work in developing a General System Theory which, in spite of some differences of emphasis, is an interesting contribution to our problem.

11. The following example is a slight modification of the one given in Hempel and Oppenheim (35, p. 148). See also Rescher and Oppenheim (76, pp. 93, 94).

12. See Rescher and Oppenheim (76, p. 100), and Rescher (75). Of course, nothing is intrinsically a "true" whole; the characterization of certain things as "wholes" is always a function of the point of view, i.e., of the particular 'Pt' relation selected. For instance, if a taxonomic system is given, it is very natural to define 'Pt' so that the "wholes" will correspond to the things of the system. Similarly for *aggregate* see Rescher and Oppenheim (76, p. 90, n. 1).

13. See Kemeny and Oppenheim (47, n. 6). A suggestive characterization of *simplicity* in terms of the "entropy" of a theory has been put forward by Rothstein (78). Using Rothstein's terms, we may say that any micro-reduction moves in the direction of lower entropy (greater organization).

14. The statement that B_2 is *micro-reducible* to B_1 means (according to the analysis we adopt here) that some *true* theory belonging to B_1—i.e., some true theory with the appropriate vocabulary and universe of discourse, whether accepted or not, and whether it is ever even written down or not—micro-reduces every true theory of B_2. This seems to be what people have in mind when they assert that a given B_2 may not be reduced to a given B_1 at a certain time, but may nonetheless be reducible (micro-reducible) to it.

15. See Kemeny and Oppenheim (45, p. 307); also for "related concepts," like Carnap's "degree of confirmation" see Carnap (13).

16. As to degree of credibility see Kemeny and Oppenheim (45, especially p. 307).

17. Using a term introduced by Kurt Lewin (48), we can also say in such a case: any particular object on level n is *genidentical* with these parts.

18. Though we cannot accept Sir Arthur Eddington's idealistic implications, we quote from his *Philosophy of Physical Science* (17, p. 125): "I conclude therefore that our engrained form of thought is such that we shall not rest satisfied until we are able to represent all physical phenomena as an interplay of a vast number of structural units intrinsically alike. All the diversity of phenomena will be then seen to correspond to different forms of relatedness of these units or, as we should usually say, different configurations."

REFERENCES

1. Blum, H. F. *Time's Arrow and Evolution*. Princeton: Princeton Univ. Press, 1951.

2. Blum, H. F. "Perspectives in Evolution," *American Scientist*, 43:595–610 (1955).

3. Bonner, J. T. *Morphogenesis*. Princeton: Princeton Univ. Press, 1952.

4. Born, M. "The Interpretation of Quantum Mechanics," *British Journal for the Philosophy of Science*, 3:95–106 (1953).

5. Brazier, M. A. B. *The Electric Activity of the Nervous System*. London: Sir Isaac Pitman & Sons, Ltd., 1951.

6. Broad, C. D. The Mind and Its Place in Nature. New York: Harcourt, Brace, 1925.

7. Burkholder, P. R. "Cooperation and Conflict among Primitive Organisms," *American Scientist,* 40:601–631 (1952).

8. Burnet, M. "How Antibodies Are Made," *Scientific American,* 191:74–78 (November 1954).

9. Calvin, M. "Chemical Evolution and the Origin of Life," *American Scientist,* 44:248–263 (1956).

10. Carnap, R. *Der logische Aufbau der Welt.* Berlin-Schlachtensee: Im Weltkreis-Verlag, 1928. Summary in N. Goodman, *The Structure of Appearances,* pp. 114–146. Cambridge: Harvard Univ. Press, 1951.

11. Carnap, R. "Testability and Meaning," *Philosophy of Science,* 3:419–471 (1936), and 4:2–40 (1937). Reprinted by Graduate Philosophy Club, Yale University, New Haven, 1950.

12. Carnap, R. *Logical Foundations of the Unity of Science, International Encyclopedia of Unified Science,* Vol. I, pp. 42–62. Chicago: Univ. of Chicago Press, 1938.

13. Carnap, R. *Logical Foundations of Probability.* Chicago: Univ. of Chicago Press, 1950.

14. Comte, Auguste. *Cours de Philosophie Positive.* 6 Vols. Paris: Bachelier, 1830–42.

15. Crick, F. H. C. "The Structure of Hereditary Material," *Scientific American,* 191:54–61 (October 1954).

16. Dodd, S. C. "A Mass-Time Triangle," *Philosophy of Science,* 11:233–244 (1944).

17. Eddington, Sir Arthur. *The Philosophy of Physical Science.* Cambridge: Cambridge University Press, 1949.

18. Feigl, H. "Logical Empiricism," in D. D. Runes (ed.), *Twentieth Century Philosophy,* pp. 371–416. New York: Philosophical Library, 1943. Reprinted in H. Feigl and W. Sellars (eds.), *Readings in Philosophical Analysis.* New York: Appleton-Century-Crofts, 1949.

19. Feigl, H. "Unity of Science and Unitary Science," in H. Feigl and M. Brodbeck (eds.), *Readings in the Philosophy of Science,* pp. 382–384. New York: Appleton-Century-Crofts, 1953.

20. Feigl, H. "Functionalism, Psychological Theory and the Uniting Sciences: Some Discussion Remarks," *Psychological Review,* 62:232–235 (1955).

21. Flint, R. *Philosophy as Scientia Scientiarum and the History of the Sciences.* New York: Scribner, 1904.

22. Fox, S. W. "The Evolution of Protein Molecules and Thermal Synthesis of Biochemical Substances," *American Scientist,* 44:347–359 (1956).

23. Fraenkel-Conrat, H. "Rebuilding a Virus," *Scientific American,* 194:42–47 (June 1956).

24. Gamow, G. "The Origin and Evolution of the Universe," *American Scientist,* 39:393–406 (1951).

25. Gamow, G. *The Creation of the Universe.* New York: Viking Press, 1952.

26. Gamow, G. "The Evolutionary Universe," *Scientific American,* 195:136–154 (September 1956).

27. Goldschmidt, R. B. "Evolution, as Viewed by One Geneticist," *American Scientist,* 40:84–98 (1952).

28. Goldschmidt, R. B. *Theoretical Genetics.* Berkeley and Los Angeles: Univ. of California Press, 1955.

29. Guhl, A. M. "The Social Order of Chickens," *Scientific American,* 194:42–46 (February 1956).

30. Haskins, C. P. *Of Societies and Man.* New York: Norton & Co., 1951.

31. Hayek, F. A. *Individualism and the Economic Order.* Chicago: Univ. of Chicago Press, 1948.

32. Hebb, D. O. *The Organization of Behavior.* New York: Wiley, 1949.

33. Hempel, C. G. *Fundamentals of Concept Formation in the Empirical Sciences,* Vol. II, No. 7 of *International Encyclopedia of Unified Science.* Chicago: Univ. of Chicago Press, 1952.

34. Hempel, C. G., and P. Oppenheim. *Der Typusbegriff im Lichte der neuen Logik; wissenschaftstheoretische Untersuchungen zur Konstitutionsforschung und Psychologie.* Leiden: A. W. Sythoff, 1936.

35. Hempel, C. G., and P. Oppenheim. "Studies in the Logic of Explanation," *Philosophy of Science,* 15:135–175 (1948).

36. Hoagland, H. "The Weber-Fechner Law and the All-or-None Theory," *Journal of General Psychology,* 3:351–373 (1930).

37. Hockett, C. H. "Biophysics, Linguistics, and the Unity of Science," *American Scientist,* 36:558–572 (1948).

38. Horowitz, N. H. "The Gene," *Scientific American,* 195:78–90 (October 1956).

39. Hull, C. L. *Principles of Animal Behavior.* New York: D. Appleton-Century, Inc., 1943.

40. Jacobson, H. "Information, Reproduction, and the Origin of Life," *American Scientist,* 43:119–127 (1955).

41. Jeffress, L. A. *Cerebral Mechanisms in Behavior; the Hixon Symposium.* New York: Wiley, 1951.

42. Johnson, M. "The Meaning of Time and Space in Philosophies of Science," *American Scientist,* 39:412–431 (1951).

43. Kartman, L. "Metaphorical Appeals in Biological Thought," *American Scientist,* 44:296–301 (1956).

44. Kemeny, J. G. "Man Viewed as a Machine," *Scientific American,* 192:58–66 (April 1955).

45. Kemeny, J. G., and P. Oppenheim. "Degree of Factual Support," *Philosophy of Science,* 19:307–324 (1952).

46. Kemeny, J. G., and P. Oppenheim. "Systematic Power," *Philosophy of Science*, 22:27–33 (1955).

47. Kemeny, J. G., and P. Oppenheim. "On Reduction," *Philosophical Studies*, 7:6–19 (1956).

48. Lewin, Kurt. *Der Begriff der Genese*. Berlin: Verlag von Julius Springer, 1922.

49. Linderstrom-Lang, K. U. "How Is a Protein Made?" *American Scientist*, 41:100–106 (1953).

50. Lindsey, A. W. *Organic Evolution*. St. Louis: C. V. Mosby Company, 1952.

51. Lippitt, R. "Field Theory and Experiment in Social Psychology," *American Journal of Sociology*, 45:26–79 (1939).

52. MacCorquodale, K., and P. E. Meehl. "On a Distinction Between Hypothetical Constructs and Intervening Variables," *Psychological Review*, 55:95–105 (1948).

53. McCulloch, W. S., and W. Pitts. "A Logical Calculus of the Ideas Immanent in Nervous Activity," *Bulletin of Mathematical Biophysics*, 5:115–133 (1943).

54. Mannheim, K. *Ideology and Utopia*. New York: Harcourt, Brace, 1936.

55. Mill, John Stuart. *System of Logic*. New York: Harper, 1848 (1st ed. London, 1843).

56. Miller, S. L. "A Production of Amino Acids Under Possible Primitive Earth Conditions," *Science*, 117:528–529 (1953).

57. Miller, S. L. "Production of Some Organic Compounds Under Possible Primitive Earth Conditions," *Journal of the American Chemical Society*, 77:2351–2361 (1955).

58. Moscana, A. "Development of Heterotypic Combinations of Dissociated Embryonic Chick Cells," *Proceedings of the Society for Experimental Biology and Medicine*, 92:410–416 (1956).

59. Needham, J. *Time*. New York: Macmillan, 1943.

60. Nogushi, J., and T. Hayakawa. Letter to the Editor, *Journal of the American Chemical Society*, 76:2846–2848 (1954).

61. Oparin, A. I. *The Origin of Life*. New York: Macmillan, 1938 (Dover Publications, Inc. edition, 1953).

62. Oppenheim, P. "Dimensions of Knowledge," *Revue Internationale de Philosophie*, Fascicule 40, Section 7 (1957).

63. Pauling, L. "Chemical Achievement and Hope for the Future," *American Scientist*, 36:51–58 (1948).

64. Pauling, L. "Quantum Theory and Chemistry," *Science*, 113:92–94 (1951).

65. Pauling, L., D. H. Campbell, and D. Pressmann. "The Nature of Forces between Antigen and Antibody and of the Precipitation Reaction," *Physical Review*, 63:203–219 (1943).

66. Pauling, L., and R. B. Corey. "Two Hydrogen-Bonded Spiral Configurations of the Polypeptide Chain," *Journal of the American Chemical Society*, 72:5349 (1950).

67. Pauling, L., and R. B. Corey. "Atomic Coordination and Structure Factors for Two Helical Configurations," *Proceedings of the National Academy of Science* (U.S.), 37:235 (1951).

68. Pauling, L., and M. Delbrück. "The Nature of Intermolecular Forces Operative in Biological Processes," *Science*, 92:585–586 (1940).

69. Penfield, W. "The Cerebral Cortex and the Mind of Man," in P. Laslett (ed.), *The Physical Basis of Mind*, pp. 56–64. Oxford: Blackwell, 1950.

70. Piaget, J. *The Moral Judgment of the Child*. London: Kegan Paul, Trench, Trubner and Company, Ltd., 1932.

71. Piaget, J. *The Language and Thought of the Child*. London: Kegan Paul, Trench, Trubner and Company, New York: Harcourt, Brace, 1926.

72. Platt, J. R. "Amplification Aspects of Biological Response and Mental Activity," *American Scientist*, 44:180–197 (1956).

73. Probability Approach in Psychology (Symposium), *Psychological Review*, 62:193–242 (1955).

74. Rashevsky, N. Papers in general of Rashevsky, published in the *Bulletin of Mathematical Biophysics*, 5 (1943).

75. Rescher, N., "Axioms of the Part Relation," *Philosophical Studies*, 6:8–11 (1955).

76. Rescher, N. and P. Oppenheim. "Logical Analysis of Gestalt Concepts," *British Journal for the Philosophy of Science*, 6:89–106 (1955).

77. Rosenblueth, A. *The Transmission of Nerve Impulses at Neuroeffector Junctions and Peripheral Synapses*. New York: Technological Press of MIT and Wiley, 1950.

78. Rothstein, J. *Communication, Organization, and Science*. Indian Hills, Colorado: Falcon's Wing Press, 1957.

79. Scriven, M. "A Possible Distinction Between Traditional Scientific Disciplines and the Study of Human Behavior," in H. Feigl and M. Scriven (eds.), Vol. I, *Minnesota Studies in the Philosophy of Science*, pp. 330–339. Minneapolis: Univ. of Minnesota Press, 1956.

80. Sellars, W. "A Semantic Solution of the Mind-Body Problem," *Methodos*, 5:45–84 (1953).

81. Sellars, W. "Empiricism and the Philosophy of Mind," in H. Feigl and M. Scriven (eds.), *Minnesota Studies in the Philosophy of Science*, Vol. I, pp. 253–329. Minneapolis: Univ. of Minnesota Press, 1956.

82. Shannon, C. E., and J. McCarthy (eds.), *Automata Studies*. Princeton: Princeton Univ. Press, 1956.

83. Sherif, M. "Experiments in Group Conflict," *Scientific American*, 195:54–58 (November 1956).

84. Sherif, M., and C. W. Sherif. *An Outline of Social Psychology*. New York: Harper, 1956.

85. Sherrington, Charles. *The Integrative Action of the Nervous System*. New Haven: Yale Univ. Press, 1948.

86. Simmel, G. *Sociologie*. Leipzig: Juncker und Humblot, 1908.

87. Simpson, G. G., C. S. Pittendrigh, and C. H. Tiffany. *Life*. New York: Harcourt, Brace, 1957.

88. Stanley, W. M. "The Structure of Viruses," reprinted from publication No. 14 of the American Association for the Advancement of Science, *The Cell and Protoplasm*, pp. 120–135 (reprint consulted) (1940).

89. Timoféeff-Ressovsky, N. W. *Experimentelle Mutationsforschung in der Vererbungslehre*. Dresden und Leipzig: Verlag von Theodor Steinkopff, 1937.

90. Tolman, E. C. *Purposive Behavior in Animals and Men*. New York: The Century Company, 1932.

91. Turing, A. M. "On Computable Numbers, With an Application to the Entscheidungsproblem," *Proceedings of the London Mathematical Society*, Ser. 2, 42:230–265 (1936).

92. Turing, A. M. "A Correction," *Proceedings of the London Mathematical Society*, Ser. 2, 43:544–546 (1937).

93. Vannerus, A. *Vetenskapssystematik*. Stockholm: Aktiebolaget Ljus, 1907.

94. Veblen, T. *The Theory of the Leisure Class*. London: Macmillan, 1899.

95. Von Bertalanffy, L. "An Outline of General System Theory," *British Journal for the Philosophy of Science*, 1:134–165 (1950).

96. Von Neumann, John. "The General and Logical Theory of Automata," in L. A. Jeffress (ed.), *Cerebral Mechanisms in Behavior; The Hixon Symposium*, pp. 20–41. New York: John Wiley and Sons, Inc., 1951.

18

JERRY FODOR

Special Sciences

A typical thesis of positivistic philosophy of science is that all true theories in the special sciences should reduce to physical theories in the long run. This is intended to be an empirical thesis, and part of the evidence which supports it is provided by such scientific successes as the molecular theory of heat and the physical explanation of the chemical bond. But the philosophical popularity of the reductivist program cannot be explained by reference to these achievements alone. The development of science has witnessed the proliferation of specialized disciplines at least as often as it has witnessed their reduction to physics, so the widespread enthusiasm for reduction can hardly be a mere induction over its past successes.

I think that many philosophers who accept reductivism do so primarily because they wish to endorse

Synthese 28 (1974), pp. 97–115. With kind permission from Kluwer Academic Publishers.

the generality of physics *vis à vis* the special sciences: roughly, the view that all events which fall under the laws of any science are physical events and hence fall under the laws of physics.[1] For such philosophers, saying that physics is basic science and saying that theories in the special sciences must reduce to physical theories have seemed to be two ways of saying the same thing, so that the latter doctrine has come to be a standard construal of the former.

In what follows, I shall argue that this is a considerable confusion. What has traditionally been called 'the unity of science' is a much stronger, and much less plausible, thesis than the generality of physics. If this is true it is important. Though reductionism is an empirical doctrine, it is intended to play a regulative role in scientific practice. Reducibility to physics is taken to be a *constraint* upon the acceptability of theories in the special sciences, with the curious consequence that the more the special sciences succeed, the more they ought to disappear. Methodological problems about psychology, in particular, arise in just this way: the assumption that the subject-matter of psychology is part of the subject-matter of physics is taken to imply that psychological theories must reduce to physical theories, and it is this latter principle that makes the trouble. I want to avoid the trouble by challenging the inference.

I

Reductivism is the view that all the special sciences reduce to physics. The sense of "reduce to" is, however, proprietary. It can be characterized as follows.[2]

Let

(1) $S_1x \rightarrow S_2x$

be a law of the special science S. ((1) is intended to be read as something like 'all S_1 situations bring about S_2 situations'. I assume that a science is individuated largely by reference to its typical predicates, hence that if S is a special science 'S_1' and 'S_2' are not predicates of basic physics. I also assume that the 'all' which quantifies laws of the special sciences needs to be taken with a grain of salt; such laws are typically *not* exceptionless. This is a point to which I shall return at length.) A necessary and sufficient condition of the reduction of (1) to a law

of physics is that the formulae (2) and (3) be laws, and a necessary and sufficient condition of the reduction of S to physics is that all its laws be so reducible.[3]

(2a) $S_1x \leftrightarrows P_1x$
(2b) $S_2x \leftrightarrows P_2x$
(3) $P_1x \rightarrow P_2x$.

'P_1' and 'P_2' are supposed to be predicates of physics, and (3) is supposed to be a physical law. Formulae like (2) are often called 'bridge' laws. Their characteristic feature is that they contain predicates of both the reduced and the reducing science. Bridge laws like (2) are thus contrasted with 'proper' laws like (1) and (3). The upshot of the remarks so far is that the reduction of a science requires that any formula which appears as the antecedent or consequent of one of its proper laws must appear as the reduced formula in some bridge law or other.[4]

Several points about the connective '$\rightarrow$' are in order. First, whatever other properties that connective may have, it is universally agreed that it must be transitive. This is important because it is usually assumed that the reduction of some of the special sciences proceeds via bridge laws which connect their predicates with those of intermediate reducing theories. Thus, psychology is presumed to reduce to physics via, say, neurology, biochemistry, and other local stops. The present point is that this makes no difference to the logic of the situation so long as the transitivity of '$\rightarrow$' is assumed. Bridge laws which connect the predicates of S to those of $S^\star$ will satisfy the constraints upon the reduction of S to physics so long as there are other bridge laws which, directly or indirectly, connect the predicates of $S^\star$ to physical predicates.

There are, however, quite serious open questions about the interpretations of '$\rightarrow$' in bridge laws. What turns on these questions is the respect in which reductivism is taken to be a physicalist thesis.

To begin with, if we read '$\rightarrow$' as 'brings about' or 'causes' in proper laws, we will have to have some other connective for bridge laws, since bringing about and causing are presumably *a*symmetric, while bridge laws express symmetric relations. Moreover, if '$\rightarrow$' in bridge laws is interpreted as any relation other than identity, the truth of reductivism will only guarantee the truth of a weak version of physicalism, and this would fail to express the underlying ontological bias of the reductivist program.

If bridge laws are not identity statements, then formulae like (2) claim at most that, by law, x's satisfaction of a P predicate and x's satisfaction of an S predicate are causally correlated. It follows from this that it is nomologically necessary that S and P predicates apply to the same things (i.e., that S predicates apply to a subset of the things that P predicates apply to). But, of course, this is compatible with a non-physicalist ontology since it is compatible with the possibility that x's satisfying S should not itself *be* a physical event. On this interpretation, the truth of reductivism does *not* guarantee the generality of physics *vis à vis* the special sciences since there are some events (satisfactions of S predicates) which fall in the domains of a special science (S) but not in the domain of physics. (One could imagine, for example, a doctrine according to which physical and psychological predicates are both held to apply to organisms, but where it is denied that the event which consists of an organism's satisfying a psychological predicate is, in any sense, a physical event. The up-shot would be a kind of psychophysical dualism of a non-Cartesian variety; a dualism of events and/or properties rather than substances.)

Given these sorts of considerations, many philosophers have held that bridge laws like (2) ought to be taken to express contingent event identities, so that one would read (2a) in some such fashion as 'every event which consists of x's satisfying S_1 is identical to some event which consists of x's satisfying P_1 and vice versa'. On this reading, the truth of reductivism would entail that every event that falls under any scientific law is a physical event, thereby simultaneously expressing the ontological bias of reductivism and guaranteeing the generality of physics *vis a vis* the special sciences.

If the bridge laws express event identities, and if every event that falls under the proper laws of a special science falls under a bridge law, we get the truth of a doctrine that I shall call 'token physicalism'. Token physicalism is simply the claim that all the events that the sciences talk about are physical events. There are three things to notice about token physicalism.

First, it is weaker than what is usually called 'materialism'. Materialism claims *both* that token physicalism is true *and* that every event falls under the laws of some science or other. One could therefore be a token physicalist without being a materialist, though I don't see why anyone would bother.

Second, token physicalism is weaker than what might be called 'type physicalism,' the doctrine, roughly, that every *property* mentioned in the laws of any science is a physical property. Token physicalism does not entail type physicalism because the contingent identity of a pair of events presumably does not guarantee the identity of the properties whose instantiation constitutes the events; not even where the event identity is nomologically necessary. On the other hand, if every event is the instantiation of a property, then type physicalism does entail token physicalism: two events will be identical when they consist of the instantiation of the same property by the same individual at the same time.

Third, token physicalism is weaker than reductivism. Since this point is, in a certain sense, the burden of the argument to follow, I shan't labor it here. But, as a first approximation, reductivism is the conjunction of token physicalism with the assumption that there are natural kind predicates in an ideally completed physics which correspond to each natural kind predicate in any ideally completed special science. It will be one of my morals that the truth of reductivism cannot be inferred from the assumption that token physicalism is true. Reductivism is a sufficient, but not a necessary, condition for token physicalism.

In what follows, I shall assume a reading of reductivism which entails token physicalism. Bridge laws thus state nomologically necessary contingent event identities, and a reduction of psychology to neurology would entail that any event which consists of the instantiation of a psychological property is identical with some event which consists of the instantiation of some neurological property.

Where we have got to is this: reductivism entails the generality of physics in at least the sense that any event which falls within the universe of discourse of a special science will also fall within the universe of discourse of physics. Moreover, any prediction which follows from the laws of a special science and a statement of initial conditions will also follow from a theory which consists of physics and the bridge laws, together with the statement of initial conditions. Finally, since 'reduces to' is supposed to be an asymmetric relation, it will also turn out that physics is *the* basic science; that is, if reductivism is true, physics is the only science that is general in the sense just specified. I now want to argue that reductivism is too strong a constraint upon the unity of science, but that the relevantly weaker doctrine will preserve the desired consequences of reductivism: token physicalism, the generality of physics, and its basic position among the sciences.

II

Every science implies a taxonomy of the events in its universe of discourse. In particular, every science employs a descriptive vocabulary of theoretical and observation predicates such that events fall under the laws of the science by virtue of satisfying those predicates. Patently, not every true description of an event is a description in such a vocabulary. For example, there are a large number of events which consist of things having been transported to a distance of less than three miles from the Eiffel Tower. I take it, however, that there is no science which contains 'is transported to a distance of less than three miles from the Eiffel Tower' as part of its descriptive vocabulary. Equivalently, I take it that there is no natural law which applies to events in virtue of their being instantiations of the property *is transported to a distance of less than three miles from the Eiffel Tower* (though I suppose it is conceivable that there is some law that applies to events in virtue of their being instantiations of some distinct but co-extensive property). By way of abbreviating these facts, I shall say that the property *is transported* . . . does not determine a *natural kind*, and that predicates which express that property are not natural kind predicates.

If I knew what a law is, and if I believed that scientific theories consist just of bodies of laws, then I could say that P is a natural kind predicate relative to S iff S contains proper laws of the form $P_x \rightarrow \alpha_x$ or $\alpha_x \rightarrow P_x$; roughly, the natural kind predicates of a science are the ones whose terms are the bound variables in its proper laws. I am inclined to say this even in my present state of ignorance, accepting the consequence that it makes the murky notion of a natural kind viciously dependent on the equally murky notions *law* and *theory*. There is no firm footing here. If we disagree about what is a natural kind, we will probably also disagree about what is a law, and for the same reasons. I don't know how to break out of this circle, but I think that there are interesting things to say about which circle we are in.

For example, we can now characterize the respect in which reductivism is too strong a construal of the doctrine of the unity of science. If reductivism is true, then *every* natural kind is, or is co-extensive with, a physical natural kind. (Every natural kind *is* a physical natural kind if bridge laws express property identities, and every natural kind is co-extensive with a physical natural kind if bridge laws express event identities.) This follows immediately from the reductivist premise that every predicate which appears as the antecedent or consequent of a law of the special sciences must appear as one of the reduced predicates in some bridge, together with the assumption that the natural kind predicates are the ones whose terms are the bound variables in proper laws. If, in short, some physical law is related to each law of a special science in the way that (3) is related to (1), then every natural kind predicate of a special science is related to a natural kind predicate of physics in the way that (2) relates 'S_1' and 'S_2' to 'P_1' and 'P_2'.

I now want to suggest some reasons for believing that this consequence of reductivism is intolerable. These are not supposed to be knock-down reasons; they couldn't be, given that the question whether reductivism is too strong is finally an *empirical* question. (The world could turn out to be such that every natural kind corresponds to a physical natural kind, just as it could turn out to be such that the property *is transported to a distance of less than three miles from the Eiffel Tower* determines a natural kind in, say, hydrodynamics. It's just that, as things stand, it seems very unlikely that the world *will* turn out to be either of these ways.)

The reason it is unlikely that every natural kind corresponds to a physical natural kind is just that (a) interesting generalizations (e.g., counter-factual supporting generalizations) can often be made about events whose physical descriptions have nothing in common, (b) it is often the case that *whether* the physical descriptions of the events subsumed by these generalizations have anything in common is, in an obvious sense, entirely irrelevant to the truth of the generalizations, or to their interestingness, or to their degree of confirmation or, indeed, to any of their epistemologically important properties, and (c) the special sciences are very much in the business of making generalizations of this kind.

I take it that these remarks are obvious to the point of self-certification; they leap to the eye as soon as one makes the (apparently radical) move of taking the special sciences at all seriously. Suppose, for example, that Gresham's 'law' really is true. (If one doesn't like Gresham's law, then any true generalization of any conceivable future economics will probably do as well.) Gresham's law says something about what will happen in monetary exchanges under certain conditions. I am willing to believe that physics is general *in the sense that it implies that any event which consists of a monetary exchange* (hence any event which falls under Gresham's law) *has a true description in the vocabulary of physics and in virtue of which it falls under the laws of physics.* But banal considerations suggest that a description which covers all such events must be wildly disjunctive. Some monetary exchanges involve strings of wampum. Some involve dollar bills. And some involve signing one's name to a check. What are the chances that a disjunction of physical predicates which covers all these events (i.e., a disjunctive predicate which can form the right hand side of a bridge law of the form '*x* is a monetary exchange ⇆ . . .') expresses a physical natural kind? In particular, what are the chances that such a predicate forms the antecedent or consequent of some proper law of physics? The point is that monetary exchanges have interesting things in common; Gresham's law, if true, says what one of these interesting things is. But what is interesting about monetary exchanges is surely not their commonalities under *physical* description. A natural kind like a monetary exchange *could* turn out to be co-extensive with a physical natural kind; but if it did, that would be an accident on a cosmic scale.

In fact, the situation for reductivism is still worse than the discussion thus far suggests. For, reductivism claims not only that all natural kinds are co-extensive with physical natural kinds, but that the co-extensions are nomologically necessary; bridge laws are *laws*. So, if Gresham's law is true, it follows that there is a (bridge) law of nature such that '*x* is a monetary exchange ⇆ *x* is *P*', where *P* is a term for a physical natural kind. But, surely, there is no such law. If there were, then *P* would have to cover not only all the systems of monetary exchange that there *are*, but also all the systems of monetary exchange that there *could be*; a law must succeed with the counterfactuals. What physical predicate is a candidate for '*P*' in '*x* is a nomologically possible monetary exchange iff P_x'?

To summarize: an immortal econophysicist might, when the whole show is over, find a predicate in physics that was, in brute fact, co-extensive with 'is a monetary exchange'. If physics is general—if the ontological biases of reductivism are true—then there must *be* such a predicate. But (a) to paraphrase a remark Donald Davidson made in a slightly different context, nothing but brute enumeration could convince us of this brute co-extensivity, and (b) there would seem to be no chance at all that the physical predicate employed in stating the co-extensivity is a natural kind term, and (c) there is still less chance that the co-extension would be lawful (i.e., that it would hold not only for the nomologically possible world that turned out to be real, but for any nomologically possible world at all).

I take it that the preceding discussion strongly suggests that economics is not reducible to physics in the proprietary sense of reduction involved in claims for the unity of science. There is, I suspect, nothing special about economics in this respect; the reasons why economics is unlikely to reduce to physics are paralleled by those which suggest that psychology is unlikely to reduce to neurology.

If psychology is reducible to neurology, then for every psychological natural kind predicate there is a co-extensive neurological natural kind predicate, and the generalization which states this co-extension is a law. Clearly, many psychologists believe something of the sort. There are departments of 'psychobiology' or 'psychology and brain science' in universities throughout the world whose very existence is an institutionalized gamble that such lawful co-extensions can be found. Yet, as has been frequently remarked in recent discussions of materialism, there are good grounds for hedging these bets. There are no firm data for any but the grossest correspondence between types of psychological states and types of neurological states, and it is entirely possible that the nervous system of higher organisms characteristically achieves a given psychological end by a wide variety of neurological means. If so, then the attempt to pair neurological structures with psychological functions is foredoomed. Physiological psychologists of the stature of Karl Lashley have held precisely this view.

The present point is that the reductivist program in psychology is, in any event, *not* to be defended on ontological grounds. Even if (token) psychological events are (token) neurological events, it does not follow that the natural kind predicates of psychology are co-extensive with the natural kind predicates of any other discipline (including physics). That is, the assumption that every psychological event is a physical event does not guarantee that physics (or, *a fortiori,* any other discipline more general than psychology) can provide an appropriate vocabulary for psychological theories. I emphasize this point because I am convinced that the make-or-break commitment of many physiological psychologists to the reductivist program stems precisely from having confused that program with (token) physicalism.

What I have been doubting is that there are neurological natural kinds co-extensive with psychological natural kinds. What seems increasingly clear is that, even if there is such a co-extension, it cannot be lawlike. For, it seems increasingly likely that there are nomologically possible systems other than organisms (namely, automata) which satisfy natural kind predicates in psychology, and which satisfy no neurological predicates at all. Now, as Putnam has emphasized, if there are any such systems, then there are probably vast numbers, since equivalent automata can be made out of practically anything. If this observation is correct, then there can be no serious hope that the class of automata whose psychology is effectively identical to that of some organism can be described by *physical* natural kind predicates (though, of course, if token physicalism is true, that class can be picked out by some physical predicate or other). The upshot is that the classical formulation of the unity of science is at the mercy of progress in the field of computer simula-

tion. This is, of course, simply to say that that formulation was too strong. The unity of science was intended to be an empirical hypothesis, defeasible by possible scientific findings. But no one had it in mind that it should be defeated by Newell, Shaw and Simon.

I have thus far argued that psychological reductivism (the doctrine that every psychological natural kind is, or is co-extensive with, a neurological natural kind) is not equivalent to, and cannot be inferred from, token physicalism (the doctrine that every psychological event is a neurological event). It may, however, be argued that one might as well take the doctrines to be equivalent since the only possible *evidence* one could have for token physicalism would also be evidence for reductivism: namely, the discovery of type-to-type psychophysical correlations.

A moment's consideration shows, however, that this argument is not well taken. If type-to-type psychophysical correlations would be evidence for token physicalism, so would correlations of other specifiable kinds.

We have type-to-type correlations where, for every *n*-tuple of events that are of the same psychological kind, there is a correlated *n*-tuple of events that are of the same neurological kind. Imagine a world in which such correlations are *not* forthcoming. What is found, instead, is that for every *n*-tuple of type identical psychological events, there is a spatiotemporally correlated *n*-tuple of type *distinct* neurological events. That is, every psychological event is paired with some neurological event or other, but psychological events of the same kind may be paired with neurological events of different kinds. My present point is that such pairings would provide as much support for token physicalism as type-to-type pairings do *so long as we are able to show that the type distinct neurological events paired with a given kind of psychological event are identical in respect of whatever properties are relevant to type-identification in psychology.* Suppose, for purposes of explication, that psychological events are type identified by reference to their behavioral consequences.[5] Then what is required of all the neurological events paired with a class of type homogeneous psychological events is only that they be identical in respect of their behavioral consequences. To put it briefly, type identical events do not, of course, have *all* their properties in common, and type distinct events must nevertheless be identical in *some* of their properties. The empirical confirmation of token physicalism does not depend on showing that the neurological counterparts of type identical psychological events are themselves type identical. What needs to be shown is only that they are identical in respect of those properties which determine which kind of *psychological* event a given event is.

Could we have evidence that an otherwise heterogeneous set of neurological events have these kinds of properties in common? Of course we could. The neurological theory might itself explain why an *n*-tuple of neurologically type distinct events are identical in their behavioral consequences, or, indeed, in respect of any of indefinitely many other such relational properties. And, if the neurological theory failed to do so, some science more basic than neurology might succeed.

My point in all this is, once again, not that correlations between type homogeneous psychological states and type heterogeneous neurological states would prove that token physicalism is true. It is only that such correlations might give us as much reason to be token physicalists as type-to-type correlations would. If this is correct, then the epistemological arguments from token physicalism to reductivism must be wrong.

It seems to me (to put the point quite generally) that the classical construal of the unity of science has really misconstrued the *goal* of scientific reduction. The point of reduction is *not* primarily to find some natural kind predicate of physics co-extensive with each natural kind predicate of a reduced science. It is, rather, to explicate the physical mechanisms whereby events conform to the laws of the special sciences. I have been arguing that there is no logical or epistemological reason why success in the second of these projects should require success in the first, and that the two are likely to come apart *in fact* wherever the physical mechanisms whereby events conform to a law of the special sciences are heterogeneous.

NOTES

1. I shall usually assume that sciences are about events, in at least the sense that it is the occurrence of events that makes the laws of a science true. But I shall be pretty free with the relation between events, states, things and properties. I shall even permit myself some latitude in construing the relation between properties and predicates. I realize that all these relations are problems, but they aren't my problem in this paper. Explanation has to *start* somewhere, too.

2. The version of reductionism I shall be concerned with is a stronger one than many philosophers of science hold; a point worth emphasizing since my argument will be precisely that it is too strong to get away with. Still, I think that what I shall be attacking is what many people have in mind when they refer to the unity of science, and I suspect (though I shan't try to prove it) that many of the liberalized versions suffer from the same basic defect as what I take to be the classical form of the doctrine.

3. There is an implicit assumption that a science simply *is* a formulation of a set of laws. I think this assumption is implausible, but it is usually made when the unity of science is discussed, and it is neutral so far as the main argument of this paper is concerned.

4. I shall sometimes refer to 'the predicate which constitutes the antecedent or consequent of a law'. This is shorthand for 'the predicate such that the antecedent or consequent of a law consists of that predicate, together with its bound variables and the quantifiers which bind them'. (Truth functions of elementary predicates are, of course, themselves predicates in this usage.)

5. I don't think there is any chance at all that this is true. What is more likely is that type-identification for psychological states can be carried out in terms of the 'total states' of an abstract automaton which models the organism. For discussion, see Block and Fodor (1972).

BIBLIOGRAPHY

Block, N., and Fodor, J., 'What Psychological States Are Not,' *Philosophical Review* **81** (1972) 159–181.

19

L. DARDEN AND N. L. MAULL

Interfield Theories

INTRODUCTION

Interactions between different areas or branches or fields of science have often been obscured by current emphasis on the relations between different scientific theories. Although some philosophers have indicated that different branches may be related, the actual focus has been on the relations between theories within the branches. For example, Ernest Nagel has discussed the reduction of one branch of science to another ([27], ch. 11). But the relation that Nagel describes is really nothing more than the derivational reduction of the *theory* or *experimental law* of one branch of science to the theory of another branch.

We, in contrast to Nagel, are interested in the interrelations between the areas of science that we call *fields*. For example, cytology, genetics, and biochemistry are more naturally called fields than theories. Fields may have theories within them, such as the classical theory of the gene in genetics; such theories we call *intra*field theories. In addition, and more important for our purposes here, interrelations between fields may be established via *interfield theories*. For example, the fields of genetics and cytology are related via the chromosome theory of Mendelian heredity. The existence of such interfield theories has been obscured by analyses such as Nagel's that erroneously conflate theories and fields and see interrelations as derivational reductions.

The purpose of this paper is, first, to draw the distinction between field and *intra*field theory, and,

then, more importantly, to discuss the generation of heretofore unrecognized *inter*field theories and their functions in relating two fields.[1] Finally we wish to mention the implications of this analysis for reduction accounts and for unity and progress in science.

By analysis of a number of examples we will show that a field is an area of science consisting of the following elements: a central problem, a domain consisting of items taken to be facts related to that problem,[2] general explanatory factors and goals providing expectations as to how the problem is to be solved,[3] techniques and methods, and, sometimes, but not always, concepts, laws, and theories which are related to the problem and which attempt to realize the explanatory goals. A special vocabulary is often associated with the characteristic elements of a field.[4] Of course, we could attempt to associate institutional and sociological factors with the elements of a field, but such an attempt would fail to serve the purpose of our discussion. We are interested in conceptual, not sociological, or institutional, change. Thus, the elements of a field are conceptual, not sociological, of primary interest to the philosopher, not the sociologist.

The elements are also historical. Fields emerge in science, evolve, sometimes even cease to be. (We have not yet explored the latter phenomenon of decline.) Although any or all of the elements of the field may have existed separately in science, they must be brought together in a fruitful way for the field to emerge. Such an emergence is marked by the recognition of a promising way to solve an important problem and the initiation of a line of research in that direction. For instance, what comes to be the central problem of a field may have been a long-unsolved puzzle and the techniques may have been used elsewhere, but the field emerges when

Philosophy of Science 44 (1977), pp. 43-64. Reprinted with permission from the University of Chicago Press.

someone sees that those techniques yield information relevant to the problem. Or, perhaps, a new concept is proposed, giving new insight into an old puzzling problem and generating a line of research.

Because the convergence of the elements of a field can be identified historically, the emergence of a field can often be dated. Then, scientists who were part of the new field can be identified: they used the techniques of the field to solve its central problem. Others who had worked on the central problem in other ways or who had used the techniques for other purposes were not members of the field. The lone precursor who worked on the problem with the techniques but did not found an ongoing line of research, may, with hindsight, be called a geneticist, a biochemist, or whatever, even though the field (and perhaps even the term designating the field) did not exist at the time.

Of the terms current in philosophy of science which refer to categories broader than theory, the one which has most similarities to 'field' is Stephen Toulmin's 'discipline'.[5] What Toulmin classes as a discipline, we would probably also class as a field. Included in his examples of disciplines are genetics along with physics, atomic physics, chemistry, biochemistry, biology, and evolutionary biology ([38], pp. 141, 145, 146, 180). From these examples we see that Toulmin encounters a difficulty which is also a problem for an analysis in terms of fields: criteria seem to be needed to distinguish between disciplines, subdisciplines, and supradisciplines. For instance, is atomic physics a subdiscipline of physics or is atomic physics the discipline and physics the supradiscipline? Toulmin gives no way of distinguishing disciplines from smaller or larger units; as a result, his examples are somewhat confusing.

In this paper we discuss fields that are within the broader scientific areas of biology and chemistry, but we do not give criteria for distinguishing more inclusive from less inclusive categories. We suspect that the level at which an analysis is carried out may depend on the questions being asked and the historical period being examined. For example, historical examinations of science in the nineteenth century might well ask the question—when did biology emerge as a field in science? But twentieth century historians of biology are more likely to treat fields within what has become the broader area of biology. Thus, the formulation of time-independent criteria for the delineating of fields will be difficult and might even serve to obscure important aspects of the historical development of science.

Toulmin's lists of the components of a discipline are numerous: "body of concepts, methods, and fundamental aims" ([38], p. 139); "a communal tradition of procedures and techniques for dealing with theoretical or practical problems" ([38], p. 142); "(i) the current explanatory goals of the science, (ii) its current repertory of concepts and explanatory procedures, and (iii) the accumulated experience of the scientists working in this particular discipline" ([38], p. 175). Taken collectively, these components are similar to the elements of a field.[6]

Although similarities exist between field and discipline, there are several reasons why we have not adopted Toulmin's term. First, even though the components of disciplines are similar to those of fields, they are not identical. The central problem, domain, and techniques will play important roles in our analysis; they are not found in Toulmin's lists. Furthermore, we find it difficult to use or to analyze such components as "accumulated experience of scientists." But most important, Toulmin's notion of a discipline is embedded in an epistemology: knowledge is the result of a selection process much like the selection processes proposed by evolutionary theory for biological organisms. Although Toulmin's analysis is provocative, we would rather not commit our analysis to his "evolutionary epistemology." The legitimate use of an evolutionary analogy, we believe, can only be discovered by a detailed investigation of science, for example by examination of interactions among the elements of fields.

Examples of fields will now be examined in more detail. Cytology in its early days had the central problem—what are the basic units of organisms? This problem was solved by the postulation of the cell theory and its subsequent elaboration and confirmation in the nineteenth century. Afterwards, the problem for cytologists (or cell biologists as they have come to be called) became the characterization of different types of cells, of organelles within cells, and of their various functions. The problem is tackled primarily with the technique of microscopic analysis.

The field of genetics, on the other hand, has as its central problem the explanation of patterns of inheritance of characteristics. The characteristics may be either gross phenotypic differences, such as eye color in the fly *Drosophila*, as investigated in classical genetics, or molecular differences, such as loss of enzyme activity, as investigated in modern transmission genetics. The patterns of inheritance are investigated with the technique of artificial breeding. The laws of segregation and independent assortment (Mendel's Laws), once their scope was known and they were well-confirmed, became part of the domain to be explained. For many of the early geneticists, though not all, the goal was to solve the central problem by the formulation of a theory involving material units of heredity (genes) as explanatory factors. In attempting to realize the goal, T. H. Morgan and his associates formulated the theory of the gene of classical genetics. Extension of the theory and techniques from *Drosophila*, Morgan's model organism, to microorganisms marked the modern phases of the field of genetics, a phase which may be called modern transmission genetics.

The central problem of biochemistry is the determination of a network of interactions between the molecules of cellular systems and their molecular environment; these molecules and their interrelations are the items of the domain. As was the case with genetics in which laws became part of the domain, here too, the solution to a problem may contribute new domain items. For example, the Krebs cycle was part of the solution to the problem of determining the interactions between molecules and became, in turn, part of the domain of biochemistry; its relation to other complex pathways then posed a new problem. Many techniques of biochemistry are aimed at the reproduction of *in vivo* systems *in vitro*, that is, the "test tube" simulation of the chemical reactions that occur in living things.

The determination of the structure and three-dimensional configuration of molecules has become the concern of physical chemistry.[7] Thus, the central problem of physical chemistry is the determination of the interactions of all parts of a molecule relative to one another, under varying conditions. The domain of physical chemistry is the parts of molecules and their interactions. Physical chemistry has evolved complex techniques for the determination

of the structure and conformation of molecules: x-ray diffraction, mass spectrometry, electron microscopy, and the measurement of optical rotation.

With these examples of fields in mind we may contrast fields and *intra*field theories. A field at one point in time may not contain a theory, or may consist of several competing theories, or may have one rather successful theory. Well-confirmed laws and theories may become part of the domain and a more encompassing theory be sought to explain them. Although theories within a field may compete with one another, in general, fields do not compete, nor do theories in different fields compete. Furthermore, one field does not reduce another field; reduction in the sense of derivation would be impossible between such elements of a field as techniques and explanatory goals.[8]

Even though fields do not bear the relations formerly thought to exist between theories, fields may be related to one another. Indeed, our main concern here is with the relations between fields which serve to generate a different type of theory, the *interfield theory*, which sets out and explains the relations between fields. Our task now is to discuss the conditions which lead to the generation of interfield theories. The discussion of general features of generation will be followed by examples of interfield theories: the chromosome theory of Mendelian heredity bridging the fields of cytology and genetics; the operon theory relating the fields of genetics and biochemistry; and the theory of allosteric regulation connecting the fields of biochemistry and physical chemistry. The examples will then serve as a basis for characterizing the general functions of interfield theories.

THE GENERATION OF INTERFIELD THEORIES

An interfield theory functions to make explicit and explain relations between fields. Relations between fields may be of several types; among them are the following:

1. A field may provide a *specification of the physical location* of an entity or process postulated in another field. For example, in its

earliest formulation, the chromosome theory of Mendelian heredity postulated that the Mendelian genes were *in* or *on* the chromosomes; cytology provided the physical location of the genes. With more specific knowledge, the theory explained the relation in more detail: the genes are part of (in) the chromosomes. Thus, the relation became more specific, a *part-whole* relation.

2. A field may provide the *physical nature* of an entity or process postulated in another field. Thus, for example, biochemistry provided the physical nature of the repressor, an entity postulated in the operon theory.

3. A field may investigate the *structure* of entities or processes, the *function* of which is investigated in another field. Physical chemistry provides the structure of molecules whose function is described biochemically.

4. Fields may be linked *causally*, the entities postulated in one field providing the causes of effects investigated in the other. For example, the theory of allosteric regulation provides a causal explanation of the interaction between the physicochemical structure of certain enzymes and a characteristic biochemical pattern of their activity.

These types of relations are not necessarily mutually exclusive; as the examples indicate, structure-function relations may also be causal.

Several different types of reasons may exist for generating an interfield theory to make explicit such relations between fields. First, relationships between two fields may already be known to exist prior to the formulation of the interfield theory. We shall refer to such pre-established relationships as *background knowledge*. For example, prior to the proposal of the operon theory, the fields of genetics and biochemistry were known to be related; to cite one of many instances, the physical nature of the gene was specified biochemically as DNA. Thus, further relations could be expected between the fields and might lead to the generation of an interfield theory.

Secondly, a stronger reason for proposing an interfield theory exists when two fields *share an interest in explaining different aspects of the same phenomenon.*[9] For example, genetics and cytology

shared an interest in explaining the phenomenon of heredity, but genetics did so by breeding organisms and explaining the patterns of inheritance of characters with postulated genes. Cytology, on the other hand, investigated the location of the heredity material within the cell using microscopic techniques. Since they were both working on the problem of explaining the phenomenon of heredity, a relation between them was expected to exist.

Furthermore, *questions arise in each field which are not answerable using the concepts and techniques of that field.* These questions direct the search for an interfield theory. For example, in genetics the question arose: where are the genes located? But no means of solving that question within genetics were present since the field did not have the techniques or concepts for determining physical location. Cytology did have such means.

In brief, an interfield theory is likely to be generated when background knowledge indicates that relations already exist between the fields, when the fields share an interest in explaining different aspects of the same phenomenon, and when questions arise about that phenomenon within a field which cannot be answered with the techniques and concepts of that field.

Questions about the relations between fields pose an *interfield theoretical problem:* how are the relations between the fields to be explained? The solution to an interfield theoretical problem is an interfield theory. Dudley Shapere, in discussing theoretical problems, says: "Theoretical problems call for answers in terms of ideas different from those used in characterizing the domain items. . . . These new ideas, moreover, are expected to "account for" the domain . . ." ([34] p. 22). Shapere's analysis is for an *intra*field theory, in out terminology, but we may extend it to our case by saying that new ideas are introduced to account for the relationships between the two different domains of the different fields. The new idea which the theory introduces gives the nature of the relations between two fields, such as the types of relations discussed in (1), (2), (3) and (4) above.[10]

Suppose a relation between fields is suspected to exist because the fields share an interest in explaining aspects of the same phenomenon. Familiar types of relations between fields can then be considered, for example, causal, part-whole, or structure-function. The most likely relation (as indicated by considerations which will not be examined in this paper) can be chosen and particularized for the case in point. Thus a new idea is introduced specifying the nature of the relations between fields, that is, an interfield theory is formulated.

We will now turn to the examination of detailed examples of interfield theories in order to illustrate the general features of their generation just discussed and to analyze their functions in more detail.

THE CHROMOSOME THEORY OF MENDELIAN HEREDITY

Cytology emerged as a field in the 1820s and 30s with improvements in the microscope and the proposal of the cell theory. By the late 1800s, as a result of their investigations of the structures within cells, cytologists asked the following question: where within the germ cells is the hereditary material located? A widely accepted answer by 1900 proposed the chromosomes (darkly staining bodies within the nuclei of cells) as the likely location. (For further discussion see [40], [6], and [15].)

On the other hand, theories of heredity had been proposed in the late nineteenth century but none had the necessary ties to experimental data to give rise to a field of heredity until the discovery of (what have come to be called) Mendel's laws in 1900. Although Mendel had worked with garden peas, noted their hereditary characteristics, crossed them artificially, and proposed a law—he formulated only one—characterizing the patterns of inheritance, he did not found a field. Genetics emerged between 1900 and 1905 with the independent discovery of Mendel's law by Hugo de Vries and Carl Correns and with the promulgation of Mendel's experimental approach by William Bateson. Although Bateson did not (for reasons too complex to examine here), other geneticists postulated (what have come to be called) genes as the causes of hereditary characteristics. (For further discussion see [7], [8], and [10].)

Thus, by 1903 cytology and genetics had both investigated hereditary phenomena but asked different questions about it. At least some geneticists postulated Mendelian units to account for the patterns of inheritance of observed characteristics. Cytologists, on the other hand, proposed the chromosomes as the location of the hereditary material in the germ cells. Genes were, thus, hypothetical entities with known functions; chromosomes were entities visible with the light microscope with a postulated function.

But questions arose in each field which were not answerable with the techniques and concepts of the field itself. Genetics was unable to answer the question: where are the genes located? Its techniques were those of artificial breeding which provided no way of determining physical location. Cytology was known to provide a way of investigating the cells and their contents and thus was the natural field to turn to in search of an answer to the question about the location of genes. On the other hand, cytologists had no way of investigating the functioning of chromosomes in producing *individual* hereditary characteristics. Theodor Boveri had, however, investigated the loss of one (or more) *entire* chromosome(s) and the changes in *many* characteristics in the developing embryo that such loss produced ([2]).

In addition to there being questions in each field which could not be answered within that field, more important in the historical generation of the chromosome theory was the fact that properties of the chromosomes and genes showed striking similarities. At least three properties of chromosomes and genes had been found independently in the two fields (see items 1, 2 and 3 of Table 1). Both Walter Sutton ([37]) and Theodor Boveri ([3]), pp. 117–118) were struck by the remarkable similarities and were independently led in 1903 and 1904 to postulate the chromosome theory of Mendelian heredity as a result. The theory (using the modern term 'gene') is the following: the genes are in or on the chromosomes. The theory solves the theoretical problem as to the nature of the relations between genes and chromosomes by introducing the *new idea* that the chromosomes are the physical location of the Mendelian genes. Although August Weismann ([39]) had, in 1892, postulated that the chromosomes were composed of hierarchies of hereditary units, which he called "biophores, determinants, and ids," his was not a theory of *Mendelian* units. That the units which obeyed Mendel's law were located in

TABLE 1.   Relations between Chromosomes and Genes

Chromosomes	*Genes*
1. Pure individuals (remain distinct, do not join)	1. Pure individuals (remain distinct, no hybrids)
2. Found in pairs (in diploid organisms prior to gametogenesis and after fertilization)	2. Found in pairs (in diploid organisms prior to segregation and after fertilization)
3. The reducing division results in one-half to gametes	3. Segregation results in one-half to gametes
4. *Prediction:* Random distribution of maternal and paternal chromosomes in formation of gametes	4. Characters from maternal and paternal lines found mixed in one individual offspring; independent assortment (often) of genes
5. Chromosome number smaller than gene number	5. *Prediction:* Some genes do not assort independently in inheritance; instead are linked on the same chromosome
6. Some chromosomes form chiasmata, areas of intertwining *Prediction:* An exchange of parts of chromosomes at chiasmata	6. More combinations of linked genes than number of chromosomes; "crossing-over" occurs

or on the chromosomes was a new idea proposed by Boveri and Sutton.

The ambiguity as to whether the genes were "in" or "on" the chromosomes was resolved in favor of the "in" with further development of the theory in the hands of T. H. Morgan and his associates in the 1910s and 20s ([24], [25], [26]). Thus, the relationship between genes and chromosomes postulated by the interfield theory became that of part to whole, and the theory explained the correlated properties because parts would be expected to share at least some properties of their wholes.

But the theory did more than explain properties of genes and chromosomes already known. It also functioned to predict new items for the domains of each field on the basis of knowledge of the other. For example, item 4 of Table 1 is a prediction Sutton made about the behavior of chromosomes on the basis of the behavior of genes. This prediction corrected a misconception of cytologists. Mistakenly, Sutton said ([37], p. 29), cytologists prior to the formation of the chromosome theory of Mendelian heredity had thought that the sets of chromosomes from the mother and father remained intact in their offspring and separated as units in the formation of gametes (sexual cells; in animals, eggs and sperm) in offspring. However, the independent assortment of hereditary characteristics, and therefore the genes which cause them, led to a reexamination of the behavior of the chromosomes, with the

subsequent finding that the maternal and paternal chromosomes are distributed randomly in the formation of gametes ([37], [5]). The prediction for cytology of random segregation of chromosomes as a result of independent assortment of genes was thus substantiated.

Predictions went both ways. The knowledge from cytology of the small number of chromosomes compared to larger numbers of genes led both Boveri and Sutton to the prediction that some genes would be linked in inheritance, in other words, that exceptions to independent assortment would occur. The finding of linked genes substantiated this prediction. (See item 5 of Table 1.) The finding of predictions made on the basis of the theory served to provide support for the theory. As a result, both genetic and cytological evidence provided confirmation.

Not only did the theory predict new domain items for each field, it also served to focus attention on previously known but neglected items. For example, item 6 shows the correlation between the new finding in genetics that some genes "cross-over" or become unlinked and the previously known chiasmata, or areas of intertwining between chromosomes. Chiasmata had been seen by cytologists prior to their correlation with crossing-over, but no function for them was known so they had not been considered important. With the correlation to a property of genes by Morgan ([24]), chiasmata took on a new significance and subsequent investigation showed that they were indeed areas of exchange between parts of chromosomes as predicted by the genetic evidence. This is an example of the change in relative importance of a type of domain item; an item previously considered peripheral became a center of investigation.[11]

After the formulation of the theory and its confirmation, new findings about genes raised parallel questions about chromosomes and vice versa. New types of experiments were designed using the techniques from both fields. Calvin Bridges, a coworker of T. H. Morgan, was one of the most successful practitioners of the new method of research ([4]). Thus, the theory generated a new line of research coordinating the techniques and findings of both fields.

In summary, the chromosome theory of Mendelian heredity is an interfield theory bridging the fields of genetics and cytology. It was generated to unify the knowledge of heredity found in both fields and thereby to explain the similar properties of chromosomes and genes. It functioned to focus attention on previously neglected items of the domains and to predict new items for the domains of each field. It further served to generate a new line of research coordinating the fields of cytology and genetics. Success in finding the predictions of the theory and in developing the common line of research resulted in the confirmation of the theory and the fruitful bridging of two fields of science. . . .

THE FUNCTION OF INTERFIELD THEORIES

In summary, an interfield theory functions in some or all of the following ways:

a. To solve (perhaps "correctly") the theoretical problem which led to its generation, that is, to introduce a new idea as to the nature of the relations between fields;

b. To answer questions which, although they arise within a field, cannot be answered using the concepts and techniques of that field alone;

c. To focus attention on previously neglected items of the domains of one or both fields;

d. To predict new items for the domains of one or both fields;

e. To generate new lines of research which may, in turn, lead to another interfield theory.[12]

CONCLUSION

Because our examples represent significant developments in science, because they have important similarities in generation and function, and because other examples of theories sharing these characteristics may be found in other cases, we have attempted to set forth the characteristics common to this type of theory, the interfield theory.

NOTES

1. We are not using 'theory' in the sense of 'deductive system'. In spite of the difficulties of providing a general analysis of 'theory', we have retained the term because the developments with which we are concerned are called 'theories' by their originators. Furthermore, the interfield developments are solutions to theoretical problems, as we shall see.

2. 'Domain' is here used in the sense analyzed by Dudley Shapere ([33]).

3. Stephen Toulmin emphasizes the importance of explanatory problems and goals in [38], chs. 2, 3.

4. A special vocabulary is not a formal language, but a specialized part of the natural language. Nor is a special vocabulary a theoretical vocabulary, since the terms in it may be associated with the domain or techniques of the field as well as its problem solutions. A term may become part of the special vocabulary by specialization within the field (e.g., 'mutation' became specialized in genetics) or may be introduced as a new term (e.g., 'epistasis' was introduced in genetics). Further examples of such terms are the following: in genetics—'test cross', 'cis', 'trans', 'locus'; in cytology—'meiosis', 'mitosis', 'karyotype', 'chromosome'; in biochemistry—'respiratory quotient', 'ligase', 'citric acid cycle'; in physical chemistry—'bond angle', 'secondary structure', and 'optical rotation'. For further discussion of special vocabularies see [20] and [21].

5. Other current broader categories include Imre Lakatos's "research program", in [19] and Thomas Kuhn's "paradigm" or "disciplinary matrix" in [18]. These have fewer similarities to fields and are fraught with more difficulties than Toulmin's analysis. For further discussion see [8].

Two further comments about fields are in order here. This analysis is not intended as a demarcation between science and nonscience. There may well be fields of nonscience with some of the same elements indicated here for scientific fields. Criteria other than those we have provided would be necessary to distinguish nonscientific fields from scientific ones. Secondly, this analysis does not presuppose that all of science can be neatly divided into mutually exclusive fields. Such division would not be expected of things which evolve. Further investigation is necessary to determine the limits of applicability of the term 'field' in other cases.

6. In comparing fields and disciplines. Toulmin said that the field is what the discipline is concerned with. "A discipline is an activity." (Private conversation with LD, 12 March 1974).

7. Determination of the structure and three-dimensional configuration of molecules was not, at the turn of the century, the concern of physical chemists, like Wilhelm Ostwald, who were interested only in energy relations in biological systems. However, organic chemists (for example, Emil Fischer) were interested in the structural analysis of molecules. But by 1910 even Ostwald, who had been influenced by Ernst Mach's positivistic view of science, admitted that molecules exist, thus removing skepticism about the application of the kinetic techniques of physical chemistry to problems about the structure of molecules. For an excellent account of the interaction between physical chemistry and organic chemistry, see [12].

8. We are not taking a stand as to whether, in some possible cases, a theory in one field may be derived (in the sense of reduction) from a theory in another field. However, our examples here do not indicate that any such reduction has occurred; on the contrary, a main point of this paper is that an analysis in terms of interfield theories, not reductions, is the appropriate analysis in these important cases in biology.

9. To say that two different fields share an interest in "the same phenomenon" is only to say that scientists believed and had good reasons to believe that they were dealing with the same phenomenon.

10. Shapere's analysis differs from ours in another respect. For Shapere, items that are related to one another and demand explanation make up *one* domain. In fact, "related items about which there is a problem demanding a theory as an answer" is one of the alternative definitions of a domain supplied by Shapere. Hence, once relations are seen between two domains of two different fields, Shapere would probably regard the situation as the formulation of a new single domain encompassing the other two. However, we have introduced discussion of a field and domains that are characteristic of particular fields, distinctions not explicitly discussed by Shapere. And our primary concern is with such relationships between fields. Hence, although it is possible to say that a new domain is formed as a result of the discovery of relations between different domains, we do not wish to so characterize the situation. We prefer to regard the domains of different fields as separate but related and the interfield theory as providing an explanation of the relations.

11. Shifts in importance of domain items were discussed by Shapere ([33], pp. 532–533).

12. In the above discussion we have not viewed the interfield theory as functioning to establish a new, third field with relations to the previously separate fields. We believe the situation is made unnecessarily complicated by doing so. However, just as Shapere might regard the two domains as joined (see above, note 10, the interfield theory could be seen as forming a new field. For example, 'cytogenetics' and 'molecular genetics' seem to be used to refer to the areas between genetics and cytology related by the chromosome theory and between modern transmission genetics and biochemistry related, in part, by the operon theory. Some scientists who used techniques from both original fields, such as Bridges did, might be considered members of the "interfield theory field." This cumbersome analysis is a possible interpretation but not the best, we think.

REFERENCES

[1] Bourgeois, S., Cohen, M., and Orgel, L. "Suppression of and Complementation among Mutants of the Regulatory Gene of the Lactose Operon of *Escherichia coli*." *Journal of Molecular Biology* 14 (1965): 300–302.

[2] Boveri, T. "On Multipolar Mitosis as a Means of Analysis of the Cell Nucleus." In *Foundations of Ex-*

perimental Embryology. Edited by B. H. Willier and J. Oppenheimer. Englewood Cliffs, N.J.: Prentice-Hall, 1964, pages 75–97.

[3] Boveri, T. *Ergebnisse über die Konstitution der chromatischen Substanz des Zellkerns*. Jena: G. Fischer, 1904.

[4] Bridges, C. B. "Non-disjunction as Proof of the Chromosome Theory of Heredity." *Genetics* 1 (1916): 1–52, 107–163.

[5] Carothers, E. E. "The Mendelian Ratio in Relation to Certain Orthopteran Chromosomes," *Journal of Morphology* 24 (1913): 487–509.

[6] Coleman, W. "Cell, Nucleus, and Inheritance: A Historical Study." *Proceedings of the American Philosophical Society* 109 (1965): 125–158.

[7] Coleman, W. "William Bateson and Conservative Thought in Science." *Centaurus* 15 (1970): 228–314.

[8] Darden, L. "Reasoning in Scientific Change: The Field of Genetics at Its Beginnings." Unpublished Ph.D. Dissertation. University of Chicago, 1974.

[9] Dienert, F. "Sur la Fermentation du Galactose et sur l'Accountamance des levures à ce Sucre." *Annales de l'Institute Pasteur* 14 (1900): 138–189.

[10] Dunn, L. C. *A Short History of Genetics*. New York: McGraw-Hill, 1965.

[11] Fischer, E. "Einfluss der Konfiguration auf die Wirkung der Enzyme." *Berichte der deutschen chemische Gesellschaft* 27 (1894): 2985–2993.

[12] Fruton, J. S. *Molecules and Life: Historical Essays on the Interplay of Chemistry and Biology*. New York: Wiley-Interscience, 1972.

[13] Gilbert, W., and Müller-Hill, B. "Isolation of the Lac Repressor." *Proceedings of the National Academy of Sciences, U.S.A.* 56 (1966): 1891–1898.

[14] Gilbert, W., and Müller-Hill, B. "The Lac Operator Is DNA." *Proceedings of the National Academy of Sciences, U.S.A.* 58 (1967): 2415–2421.

[15] Hughes, Arthur. *A History of Cytology*. New York: Abelard-Schuman, 1959.

[16] Jacob, F., and Monod, J. "Genetic Regulatory Mechanisms in the Synthesis of Proteins." *Journal of Molecular Biology* 3 (1961): 318–356.

[17] Koshland, D. E. "Protein Shape and Biological Control." *Scientific American* 229 (1973): 52–64.

[18] Kuhn, T. *The Structure of Scientific Revolutions* (2nd ed.). Chicago: University of Chicago Press, 1970.

[19] Lakatos, I. "Falsification and the Methodology of Scientific Research Programs." In *Criticism and the Growth of Knowledge*. Edited by I. Lakatos and Alan

Musgrave. Cambridge, England: University Press, 1970, pages 91–195.

[20] Maull Roth, N. "Progress in Modern Biology: An Alternative to Reduction." Unpublished Ph.D. Dissertation. University of Chicago, 1974.

[21] Maull, N. "Unifying Science without Reduction." Forthcoming in *Studies in the History and Philosophy of Science*.

[22] Monod, J., Changeux, J.-P., and Jacob, F. "Allosteric Proteins and Cellular Control Systems." *Journal of Molecular Biology* 6 (1963): 306–329.

[23] Monod, J., Wyman, J., and Changeux, J.-P., "On the Nature of Allosteric Transitions: A Plausible Model." *Journal of Molecular Biology* 12 (1965): 88–118.

[24] Morgan, T. H. "An Attempt to Analyze the Constitution of the Chromosomes on the Basis of Sex-Limited Inheritance in *Drosophila*." *Journal of Experimental Zoology* 11 (1911): 365–413.

[25] Morgan, T. H. *The Theory of the Gene*. New Haven: Yale University Press, 1926.

[26] Morgan, T. H., Sturtevant, A. H., Muller, H. J., and Bridges, C. B. *The Mechanism of Mendelian Heredity*. New York: Henry Holt and Co., 1915.

[27] Nagel, E. *The Structure of Science*. New York: Harcourt, Brace and World, Inc., 1961.

[28] Oppenheim, P., and Putnam, H. "Unity of Science as a Working Hypothesis." In *Concepts, Theories and the Mind-Body Problem. Minnesota Studies in the Philosophy of Science*. Vol. II, edited by H. Feigl, M. Scriven, and G. Maxwell. Minneapolis: University of Minnesota Press, 1958, pages 3–36.

[29] Pardee, A., Jacob, F., and Monod, J. "The Genetic Control and Cytoplasmic Expression of 'Inducibility' in the Synthesis of β galactosidase by *E. coli*." *Journal of Molecular Biology* 1 (1959): 165–178.

[30] Schaffner, K. "Logic of Discovery and Justification in Regulatory Genetics." *Studies in the History and Philosophy of Science* 4 (1974): 349–385.

[31] Schaffner, K. "The Peripherality of Reduction in the Development of Molecular Biology." *Journal of the History of Biology* 7 (1974): 111–139.

[32] Schaffner, K. "The Unity of Science and Theory Construction in Molecular Biology." In *AAAS 1969: Boston Studies in the Philosophy of Science*. Vol. XI, edited by R. S. Cohen and R. J. Seeger. Dordrecht: D. Reidel Publishing, 1974.

[33] Shapere, D. "Scientific Theories and Their Domains." In *The Structure of Scientific Theories*, edited by F. Suppe. Urbana: University of Illinois Press, 1974, pages 518–565.

[34] Shapere, D. Unpublished MS. Presented at IUHPS-LMPS Conference on Relations Between History and Philosophy of Science. Jyväskylä, Finland, 1973.

[35] Stent, G. "The Operon: On Its Third Anniversary." *Science* 144 (1964): 816–820.

[36] Suppe, F. *The Structure of Scientific Theories.* Urbana: University of Illinois Press, 1974.

[37] Sutton, W. "The Chromosomes in Heredity." *Biological Bulletin* 4 (1903): 231–251. Reprinted in

Classic Papers in Genetics, edited by J. A. Peters. Englewood Cliffs, N.J.: Prentice-Hall, 1959, pages 27–41.

[38] Toulmin, S. *Human Understanding.* Vol. I. Princeton: Princeton University Press, 1972.

[39] Weismann, A. *The Germ-Plasm, A Theory of Heredity.* Translated by W. N. Parker and H. Röonfeldt. New York: Charles Scribner's Sons, 1892.

[40] Wilson, E. *The Cell in Development and Inheritance.* (2nd ed.). New York: Macmillan, 1900.

20

JOHN DUPRÉ

The Disunity of Science

A number of philosophers have discussed processes of scientific *unification,* processes that form links of various kinds between parts of science or extend the range of application of scientific theories or techniques. However, the identification of such processes need have little bearing on the questions about overall scientific unity that are my present concern. To clarify what kind of scientific unity would be required to create a presumption of credibility for arbitrary parts of the sociological whole, it will be helpful to look at some recent philosophical appeals to scientific unification that cannot be expected to fill such a role.

One account of scientific unification, explicitly opposed to any kind of traditional reductionism, has been presented by Lindley Darden and Nancy Maull (Darden and Maull, 1977; Maull, 1977). Darden and Maull distance themselves from the re-

ductionist tradition by criticizing the assumption that areas of scientific investigation should be identified with particular *theories.* Instead, they propose a more complex and generally more local classification of scientific domains, into scientific *fields.* A scientific field is to be identified by a number of elements: a central problem, a domain of facts related to that problem, explanatory goals and expectations, appropriate techniques and methods, and sometimes concepts, laws, and theories thought relevant to the realization of the explanatory goals (Darden and Maull, 1977, p. 44). Examples they discuss include cytology, in which the central problem is the characterization of different kinds of cells and their functions; and genetics, for which the central problem is the explanation of patterns of inheritance.

Various relations may exist between fields. One field may aim to explain the nature or structure of entities postulated in another field, or there may be causal relations between entities postulated in different fields (Darden and Maull, 1977, p. 49). Such

The Disorder of Things (Cambridge: Harvard University Press, 1993), pp. 221–243. © 1993 by the President and Fellows of Harvard College. Reprinted by permission of the publisher.

relations may pose problems the solution of which calls for what they call *interfield* theories. An example is the chromosome theory of Mendelian heredity, as a theory lying between the fields of cytology and genetics. These interconnections, finally, suggest a possible reconception of the unity of science. When we move from the classical reductionist's hierarchy of scientific theories to Darden and Maull's more realistic classification of science in terms of fields, we note that there are no sharp boundaries between these parts. Indeed many important theories lie precisely in the interstices between such fields. Thus, Darden and Maull hypothesize, "unity in science is a complex network of relationships between fields effected by interfield theories" (p. 61).

I agree with most of this. Certainly some much more fine-grained classification of scientific domains, more closely related to real scientific practice, strikes me as a great improvement on the broad categories assumed by classical reductionism. And I have no objection to the idea that there may be close connections of the kind Darden and Maull describe between such fields. The only point I want to insist on, possibly though not certainly in disagreement with Darden and Maull, is that the consequences of these proposals do nothing to encourage any kind of global thesis of the unity of science.

First, as Darden and Maull explicitly state (1977, p. 45, n. 5), their conception does nothing to address the demarcation problem, to distinguish science from nonscience or pseudoscience. So it does not say what fields might, in their sense, be unified. Their ideas do, perhaps, suggest a bolder thesis. Might the continued development of links between theories eventually provide a densely connected network such that all and only the fields that truly belonged to science were connected to this network? Exploration of this possibility will begin to reveal some of the difficulties with even such a weak conception of scientific unity.

To begin with, it seems entirely possible that the extrapolation of this process of connection of fields by interfield theories would produce not one connected whole, but a number of such wholes. Clearly, the possibility of such a bridge between cytology and genetics or even—to take the more traditionally hierarchical example described by Maull (1977) as generating inter*level* theories—between genetics and

biochemistry does not show that there might be a theory linking, say, electronics and cultural anthropology; or even that there might be a sequence of theories leading from one member of this pair to the other.

But even supposing that there were, in the ideal future of science, enough interfield theories to provide a sequence of links connecting any two scientific fields, this would not be sufficient to establish, in any interesting sense, the unity of science. For it is hard to imagine that the conception of an interfield theory capable of realizing this possibility could be sufficiently restrictive to exclude connections reaching out to areas that we would not normally think of as part of science. Thus, for example, there is a natural area of theoretical investigation lying between linguistics and literary theory, the latter of which is not generally considered to be part of science. Or, for that matter, there are many theories that lie between fields of philosophy and of science. The Ghiselin/Hull theory of species as individuals, for example, might reasonably be said to lie between metaphysics and systematics. Many philosophers would no doubt welcome the discovery that philosophy was just an area of science. But the conclusion of this whole line of thought now seems to be just that no form of knowledge production can be entirely isolated from all others. This seems far too banal an observation to glorify with the title "unity of science"; and certainly it does nothing to legitimate the parts of such a unified science.

The problem might be put as follows. Darden and Maull successfully identify a kind of connection that can be forged between different parts of science. But their characterization of these connections presupposes that the two fields being connected are, indeed, part of science. In the absence of a solution to the demarcation problem, this does nothing to illuminate the question how many such connections would be needed to form a unified science. Only a characterization of interfield links with the consequence that all and only parts of science are connected can unify science. And this, finally, will be possible only if science is antecedently a unified whole, which is, of course, the question at issue.

A quite different appeal to the unification of science, again explicitly opposed to reductionism, is to be found in the work of Philip Kitcher (1981).

Kitcher's discussion starts with the question, deriving from well-known difficulties with positivist accounts of explanation, What makes an argument to a certain sentence expressing some fact an explanation of that fact? The answer he develops is, very roughly, that arguments are accepted when they instantiate argument patterns that, in turn, are accepted because of their capacity to generate large numbers of conclusions that we take to be true. Kitcher sees scientific progress, at least in the sense of the provision of increasingly better explanations, as consisting in the production of theories with ever-wider ranges of explanatory potential. If progress involves movement toward theories of ever-broader range of application, it would thereby seem to involve a move in the direction of scientific unification. And indeed Kitcher does claim explicitly that unification should be taken to be a criterion of scientific explanation.

It is not relevant to this discussion to address the important issues about explanation that are the main targets of Kitcher's account. For however convincing is the case for the thesis that, at a *local* level, extension of the range of application of a kind of explanation is crucial in achieving acceptance for scientific innovations, it implies nothing about the probability that science will eventually undergo *global* unification. Progress (if such there be) in, say, biology, could go on indefinitely instantiated by a sequence of theories able to explain ever more *biological* facts, but would do so by achieving ever-more-powerful understandings of the same domain of phenomena. Kitcher's argument implies that *if* a theory were introduced that explained in approximately the same way, for example, a large range of both chemical and biological phenomena, it would be accepted over a theory that explained only the same biological phenomena. But there is no reason to believe that such theories are generally available, and much reason to doubt it.

The situation is parallel to that for another desideratum often proposed for scientific theories, simplicity. If this is indeed a desideratum, then if two theories have equal merit in terms of every other relevant characteristic, but one is simpler than the other, we should accept the simpler theory. But this principle tells us nothing about how simple or complex the best theory may ultimately turn out to be. If the world is very complex, then presumably simple

theories will fail dismally in other respects, such as empirical adequacy. Thus, similarly, I have no objection to unification as a desideratum of scientific theorizing. But how much unification is consistent with otherwise adequate theorizing is another matter.

A general moral can be drawn from the discussion of the (quite different) ideas of Kitcher and of Darden and Maull. Unification may be significant for science in many ways. But an account of the significance of unification tells us nothing about the existence of unity. Treaties may exist, and be a wonderful thing. But these facts do not imply that there is a world government.

TOWARD A PLURALISTIC EPISTEMOLOGY

Science, construed simply as the set of knowledge-claiming practices that are accorded that title, is a mixed bag. The role of theory, evidence, and institutional norms will vary greatly from one area of science to the next. My suggestion that science should be seen as a family resemblance concept seems to imply not merely that no strong version of scientific unity of the kind advocated by classical reductionists can be sustained, but that there can be no possible answer to the demarcation problem. But as I have also indicated, I find Popper's motivation for attending to this question to be entirely convincing. Although I do not share Popper's particular hostilities to Marxism and psychoanalysis, which provide the negative test cases for his account, I do think it is vitally important, especially given the social and political power that presently accrues to successful claimants to the title of scientist, that we develop critical principles for assessing the validity of such claims. It would strike me, for example, as a fatal flaw in my position if it led to the conclusion that nothing could be said in explanation of the epistemic superiority of the theory of evolution over the apparently competing claims of creationists.

While it is clear that my Wittgensteinian thesis does indeed exclude a simple criterion of demarcation, I offer instead, and once more, the consolations of pluralism. I suggest that we try to replace the kind of epistemology that unites pure descriptivism and scientistic apologetics with something more like a virtue epistemology. There are many possible and actual such virtues: sensitivity to empirical fact, plausible background assumptions, coherence with other things we know, exposure to criticism from the widest variety of sources, and no doubt others. Some of the things we call "science" have many such virtues, others very few. To take the example mentioned above, it will hardly be difficult to demonstrate in such terms the greater credibility earned by the subtle argumentation and herculean marshaling of empirical facts of a Darwin, followed by a century and more of further empirical research and theoretical criticism, than that due to the attempt to ground historical matters of fact on the oracular interpretations of an ancient book of often unknown or dubious provenance. Such an approach would at the very least have the capacity to capture the rich variety of projects of inquiry, without conceding that anything goes. And . . . it could provide an epistemological standard for science that would be overtly and unashamedly normative.

One further consequence of my position should be mentioned. Earlier . . . I distinguished two possible conceptions of the demarcation problem: the differentiation of good from bad science, and the distinction of science from nonscience. The schematic suggestions in the previous paragraph gesture toward a solution to the former problem. But the approach I advocate also implies that a solution to the second problem is unlikely to be forthcoming. Many plausible epistemic virtues will be exemplified as much by practices not traditionally included within science as by paradigmatic scientific disciplines. Many works of philosophy or literary criticism, even, will be more closely connected to empirical fact, coherent with other things we know, and exposed to criticism from different sources than large parts of, say, macroeconomics or theoretical ecology. In general, I can imagine no reason why a ranking of projects of inquiry in terms of a plausible set of epistemic virtues (let alone epistemic and social virtues) would end up with most of the traditional sciences gathered at the top. No sharp distinction between science and lesser forms of knowledge production can survive this reconception of epistemic merit. It might fairly be said, if paradoxically, that with the disunity of science comes a kind of unity of knowledge.

REFERENCES

Darden, Lindley, and Nancy Maull. [1977]. "Interfield Theories." *Philosophy of Science* 43: 44–64.

Kitcher, Philip. [1981]. "Explanatory Unification." *Philosophy of Science* 48: 507–531.

Maull, Nancy. [1977]. "Unifying Science without Reduction." *Studies in the History and Philosophy of Science* 8: 143–162.

21

GEORGE A. REISCH

Pluralism, Logical Empiricism, and the Problem of Pseudoscience

1. INTRODUCTION

In this paper I will make three claims. First, I will argue that recent and growing interest in "pluralism" in philosophy of science effectively capitulates to demands by supporters of creation-science (hereafter: creationists) that creation-science be accepted as a legitimate branch of science. To make this point, I will examine writings on pluralism in the philosophy of biology by John Dupré and Philip Kitcher. Second, I will argue that long-overlooked aspects of logical empiricism, in particular the views of Otto Neurath, provide resources to effectively demarcate creation-science from genuine science. The third claim is implied throughout. Once I have resuscitated a logical empiricist approach to the demarcation problem, it should become clear that the widespread rejection (by pluralists and postmodernists) of the logical empiricist ideal of unity of science has been hasty. Since the problem of creation-science is a live problem, and its outcome potentially very dangerous for the future of science education, this rejection has also been potentially dangerous for science.

2. DEMARCATION AND UNITY

In recent years, creation-science has been kept out of public schools in the United States by legal decisions holding that it is not scientific.[1] These decisions hold instead that it is a set of religious beliefs and that teaching them in public schools violates the Establishment Clause. Because local legislatures continue to promote or pass legislation in support of creation-science, it is likely that it will be kept in check only if courts can continue to demarcate it from genuine scientific theory.[2] I will argue here that even though Kitcher and Dupré are foes of creation-science, their pluralist views do not support this line of defense against it.

In order to situate my argument, I will outline two ways to specify the unity of science and establish a boundary demarcating science from non-science. The first I call "simple unification" (or "simple demarcation") as imagined by Popper or early logical empiricism. One identifies a property or properties shared by all scientific disciplines or theories—such as falsifiability, or reducibility to basic observation reports—and lacked by those which are unscientific. In this way, one specifies a boundary around the sciences which demarcates them from non-science, and within which they can become more strongly unified and interconnected.

The second way to specify the unity of the sciences begins with identifying these interconnections among them. Call this "network unification" (or "network demarcation"). These connections can be methodological, theoretical, based on similarity of models used in different sciences, on relations among different scientific practices or laboratory cultures, or even on the frequency of journal-citations between and among different disciplines. As a result, different kinds of unity of science may result. For example, the use of radiocarbon dating methodologically connects physics to paleontology and physical anthropology. Quantum mechanics theoretically connects parts of physics and chemistry.

Philosophy of Science, 65 (June 1998), pp. 333–348. Reprinted by permission of the University of Chicago Press.

In any case, network unification does not involve drawing a fixed boundary around the (genuine) sciences, for what counts as scientific can change according to these connections. If, for example, an astronomer convincingly demonstrated that zodiacal signs or observed conjunctions of planets were useful for pursuing substantive astronomical questions, the network unificationist could expand the boundary to include these parts of astrology that were previously considered unscientific.

3. PLURALISM, SPECIES–CONCEPTS, AND DEMARCATION

Like most contemporary philosophers of science, Kitcher and Dupré reject simple demarcation. Kitcher writes,

> One of the great morals of the demise of logical positivism was the difficulty—or, to put it bluntly, apparent impossibility—of articulating a criterion for distinguishing genuine science. (Kitcher 1993, 195)

Dupré also notes that attempts to demarcate science from non-science have failed. He remains skeptical as to whether this distinction is even coherent and recommends, instead, a family resemblance conception of the sciences (Dupré 1993, 222):

> Science, to borrow an important idea from the later Wittgenstein, is best seen as a family resemblance concept. That is, there will be a number, perhaps an indefinite number, of features characteristic of parts of science, and every part of science will have some of these features, but very probably none will have all. (Dupré 1993, 242)

This family resemblance conception must be distinguished from the network unification conception described above, for the family resemblance view does not support demarcation. Since there are no disciplines or kinds of inquiry which altogether lack those "features characteristic of parts of science," there are no disciplines which are unambiguously unscientific. Network demarcation, on the other hand, has more bite. By requiring that one specify the interconnections that variously connect the (genuine) sciences, one can rule out those trivial or vague characteristics

(the goal of attaining knowledge, for example) that might seem to render creation-science, pyramid-power, and particle physics epistemological sisters.

If simple demarcation is impossible, and if the family resemblance view also cannot support demarcation, then network demarcation remains the only way to demarcate creation-science from real science. This was the route Kitcher took in his own sustained critique of creation-science:

> One important theme I shall emphasize is that, although the creationist campaign is advertised as an assault on evolutionary theory, it really constitutes an attack on the whole of science. Evolutionary biology is intertwined with other sciences, ranging from nuclear physics and astronomy to molecular biology and geology. If evolutionary biology is to be dismissed, then the fundamental principles of other sciences will have to be excised. All other major fields of science will have to be trimmed—or, more exactly, mutilated—to fit the creationists' bill. Moreover, in attacking the methods of evolutionary biology, creationists are actually criticizing methods that are used throughout science. As I shall argue extensively, there is no basis for separating the procedures and practices of evolutionary biology from those that are fundamental to all sciences. (1982, 4–5)

His argument rests on the implicit claims: (1) that science's theories and methods are interconnected; (2) that theories and methods of creation-science are largely inconsistent with those of evolutionary theory and others sciences; and (3) that science as a whole *ought to remain* as unified as possible. Thus, creation-science can be rejected because its acceptance would effectively ruin science by forcing ill-advised revisions throughout the rest of the network. This third claim is crucial: only because the sciences should be as unified and as internally consistent as possible would "all other major fields of science . . . *have to be trimmed*" (emphasis added) if creation-science were accepted.

This is the right approach to take. Yet, as I will show below, Kitcher's and Dupré's pluralisms block this approach by denying this crucial, third claim. In general, pluralists believe neither that the sciences are unified, nor that they ought to be unified. In the cases of Kitcher and Dupré, this can be seen in their contributions to ongoing debate about the problem of species-concepts.

For decades, biologists, evolutionists, ecologists, geneticists, and others hae been arguing over how the term "species" is best defined. Are species populations that have descended from common ancestors? Are they populations that can interbreed (or *do* interbreed)? Are they those populations carrying (in some sense) the same genes, or fulfilling certain ecological roles in a given environment?[3] There are many different ways to define "species" and there has been no widespread agreement about which is best.

Dupré offers an explanation for this circumstance: there simply *are* many, metaphysically different kinds of species. Consider, for example, his reply to David Hull, who worries that pluralism about species-concepts may be just a facile retreat from challenging theoretical problems (Hull 1987, 181). Only those blinded by the ideal of unity of science, Dupré says, will be moved by such worries. They are compelling

> only if one is already committed to the view that science requires, in the end, a unified biology with a wholly univocal concept of the species. But . . . biological science encompasses projects with various goals and addresses very disparate phenomena. Thus my motivation for pluralism is not methodological ("let a thousand flowers bloom, even the weediest ones"), but ontological. It is that the complexity and variety of the biological world is such that only a pluralistic approach is likely to prove adequate for its investigation. (1993, 53).

Since the biological world is ontologically plural, our biological sciences should be pluralistic (and disunified) as well.

Kitcher's view is similar. On one occasion, he indicates "some puzzles" about dominant concepts of species and of populations. Because different formulations of these concepts can emphasize different aspects of a population's history or behaviors, they can lead to incompatible classifications of lineages. On one view, for instance, a lineage may count as a distinct species; on another it may not. Or, whether or not a lineage counts as a species may depend on seemingly insignificant and irrelevant properties of

other lineages (1989, 202–203). Puzzles like this lead Kitcher to endorse pluralism:

> I believe that there is no single, objectively right, way to segment the entire lineage into species. Various ways of proceeding offer partial solutions, emphasizing some biological features of the situation and downplaying others. I propose (once again) that we take a pluralistic view of species, allowing that there are equally legitimate alternative ways of segmenting lineages— and indeed legitimate ways of dividing organisms into species that do not treat species as historical entities at all. (1989, 204)

Note that this formulation of pluralism is designed to solve a methodological problem—it "offers a way out of an apparent difficulty in segmenting lineages" (1989, 204–205).

Kitcher's and Dupré's pluralisms are different in scope and in motivation. Kitcher's is directed more narrowly to the species-concept problem in evolutionary theory, and it is explicitly methodologically supported. Dupré's pluralism is broader and embraces concept-pluralism and theory-pluralism throughout science. And, his pluralism is explicitly motivated by metaphysical considerations.

Despite these differences, however, Kitcher and Dupré both downplay the ideal of unity that drives Kitcher's earlier argument against creation-science. Importantly, their pluralistic views about the species-problem are not just descriptive. They do not merely register the fact that very different conceptions of species are available. Any interpretation of the issue, including pluralism's opposite, "monism" or "unificationism," must accept this fact about the current state of evolutionary theory. At issue, instead, is the attitude or orientation one recommends for its future: Will progress lay in efforts to unify and harmonize these different meanings of "species"? Or will it lay in accepting this plurality as an epistemically healthy situation? Dupré and Kitcher take the second position.

As a result, Kitcher's argument no longer goes through: if we follow Dupré and Kitcher, the acceptance of creation-science as a legitimate science would *not* mutilate evolutionary theory. If it includes an array of different, incompatible species-concepts, acceptance of creation-science would mean merely that we add more concepts to this array. One could no longer reject them on the ground that, because they are incompatible with the rest of evolutionary theory, their acceptance would make future unification more difficult or impossible. Unification, for pluralists, is not a paramount scientific ideal. Nor, under a generalized pluralism like Dupré's, would all of the sciences interconnected with evolutionary theory have to be mutilated. In his 1993 essay, Dupré rejects reductionistic links among scientific laws (Part II) and he rejects the idea that the sciences are unified by cannons of methodology or by identifiable sociological properties (Chapter 10). Contrary to Kitcher, Dupré does not believe the sciences are intertwined with each other.

What would a creationist species-concept be like? Kitcher notes that most creationists are

> perfectly willing to allow for descent with modification, even to suppose that a single species can split into two descendant species. What they deny is that one *kind* of organism can evolve from another *kind*. (1982, 143)

A creationist conception of species would treat them as independent metaphysical kinds, and would perhaps dovetail with creationist theories specifying the limits within which species can split or evolve. The important point is that pluralism supplies no grounds to reject such a concept as unscientific. One might like to say that this conception of species is false of real species; that it is incompatible with others entertained by evolutionary theory. Under Kitcher's and Dupré's pluralisms, however, this strategy is ruled out. They hold that inconsistencies and incompatibilities among species-concepts are proper and routine.

4. PLURALISM ARGUABLY SUPPORTS CREATION-SCIENCE

Besides blocking the demarcationist argument that it is unscientific, pluralism offers creation-science positive support. First, creationists hold that creation-science and evolutionary theory are similar in kind, that both are "models" or "scientific models" (or, according to some, that both are "religious"). Thus, they argue, creation-science should have "equal time" with evolutionary theory in classrooms.[4] By blocking demarcation, pluralism agrees that creation-science is similar in kind to genuine science.

Second, metaphysical pluralism supports creation-science by suggesting that its concepts correspond to real, ontologically robust objects in the world. Consider Dupré's (1993) arguments for the disunity of science. Science is disunified and disordered because the world is ontologically plural and disunified. Why is the world so metaphysically diverse? At least in part, Dupré suggests, we make it so ourselves:

> If . . . human action is causally efficacious, so that the structure of things is partly determined by our own choices, then there will be different patterns of things to explore corresponding to the different projects we have for making the world the ways we want it to be. (1993, 259)

Given the variety of human interests and practices, the world's radical metaphysical plurality should not surprise us. Accordingly, there are "many equally legitimate ways of dividing the world into kinds" (1993, 6); there are "countless kinds of things" (1993, 1); there is "equal reality and causal efficacy of objects both large and small" (1993, 7).

This is good news for creationists. If their interests, choices, and goals of inquiry help determine the way the world is, then creation-science should enjoy as robust an ontological foundation as any other domain of inquiry. All that is required is for creation-scientists to be allowed to pursue their epistemic goals, teach their theories, recruit new talent in schools, and so on. In this way, creation-science can create its own niche within our metaphysically diverse world. To Dupré's claim that "the ideal hare that the physiologist might construct of ideal cells is just not the same as the ideal hare that is hunted by the ecologist's ideal lynx[11] (1993, 118), creationists could add that all of these ideal kinds are yet again distinct from God's own, designed and created some six thousand years ago.

It has been argued that Kitcher's pluralism is also intertwined with similar metaphysical consequences. Consider how his pluralism toward species-concepts works in light of his well-known explanatory unificationism:

> Consider science as a sequence of practices that attempt to incorporate true statements (insofar as is possible) and to articulate the best unification of them (insofar as is possible). As this sequence proceeds, certain features of the organization may

stabilize: predicates of particular types may be used in explanatory schemata and employed in inductive generalization; particular schemata may endure (possibly embedded in more powerful schemata). The "joints of nature" and the "objective dependencies" are the reflections of these stable elements. The natural kinds would be the extensions of the predicates that figured in our explanatory schemata and were counted as the projectible limit, as our practices developed to embrace more and more phenomena. (1993, 172)

Each part of this project essentially involves our own efforts. To speak of the scientifically known world, therefore, we must speak of the ways we have come to know it: "The causal structure of the world, the division of things into kinds, the objective dependencies among phenomena are all generated from our efforts at organization" (1993, 172). P. Kyle Stanford noted recently that, given Kitcher's pluralism about species-concepts, this interpretation of science implies that species are both humanly constructed and metaphysically real. That is,

the legitimate interests of biologists *constitute* those divisions of organisms recognized as species. Thus, as the course of biological inquiry proceeds, we do not decide that we were previously *mistaken* about which groups of organisms were species; rather, as our explana-tory and practical interests change, which divisions of organisms *actually are* species changes as well. (Stanford 1995, 85, 83)

Like Dupré's metaphysical pluralism, and for the same reason, these implications of Kitcher's pluralism are friendly to creation-science. Unless Kitcher and Dupré can show that the epistemic interests and efforts of creationists to structure the world are somehow not legitimate or genuine—unless, that is, they offer some means of demarcation—their pluralist programs appear to support prospects for creation-science.

5. HOW DO KITCHER AND DUPRÉ CLAIM TO REJECT CREATION-SCIENCE?

Dupré and Kitcher have no love for creation-science. Dupré rejects it not on the grounds that it is unsci-

entific, but on the grounds that it constitutes very bad science. He upholds a pluralist "virtue epistemology" in which the sciences exhibit an array of epistemic virtues, none of which is manifest in all of the sciences. Since creation-science has very few of these virtues, it should be rejected:

It would strike me . . . as a fatal flaw in my position if it led to the conclusion that nothing could be said in explanation of the epistemic superiority of the theory of evolution over the apparently competing claims of creationists. (Dupré 1993, 242)

Even though demarcation is impossible—"no sharp distinction between science and lesser forms of knowledge production can survive this reconception of epistemic merit" (1993, 243)—creation-science should still have few devotees in a pluralist intellectual climate.

Unfortunately, this argument is not joined to the legal and constitutional realities surrounding the problem. Dupré's pluralist virtue epistemology does not support the prevailing argument that creation-science is unscientific, much less that it is a set of religious beliefs. Since, as Michael Ruse notes, "the U.S. constitution does not bar the teaching of weak science,"[5] there would be no legal obstacles keeping creation-science from biology classrooms were the debate framed within Dupré's pluralism. If Richard Lewontin is right that the creationist crusade aims less at achieving quality science education and more at wresting curricular control away from governmental authorities and other elites (Lewontin 1997, 28), creationists would be happy to have creation-science appear in schools as a scientific theory that happens to enjoy less epistemic prestige than others. But they would be unwilling to lower epistemic standards so far as to teach creation-science explicitly as non-science. They have framed their crusade within the old-fashioned demarcation problem and they hope to gain for their beliefs the prestige and authority of genuine science. That is why a demarcationist approach to the problem will be more effective than Dupré's.

Kitcher has also argued that creation-science is extremely poor science (1982). In addition, he has more recently suggested a different analysis of the problem. Creationists, he observes, typically ignore the many defects in their own arguments for the alleged frailty or incoherence of evolutionary theory. When confronted with these problems, creationists typically do not try to modify and improve their arguments—they just repeat them. This leads Kitcher to claim that the issue is psychological, not epistemological. The "apsychologistic character of twentieth-century philosophy of science" has distracted us from the fact that "the primary division is a psychological one between *scientists* and *pseudo-scientists*." This division does not involve the "logical characteristics" (1993, 195) of creationist views but rather a division between the psychological styles, propensities and (perhaps) aptitudes of genuine scientists, on the one hand, and supporters of creation-science, on the other. Still, this approach does not get us around Ruse's point: the constitution does not bar pseudoscientists from teaching bad science, either.

6. PHILOSOPHY AND POLITICS

Before I suggest how parts of logical empiricism can be resuscitated to meet this problem, I would like to reply to an obvious criticism. One could claim simply that whether or not creation-science is taught in schools is an educational and political question that is independent from scientific and epistemological questions. On this view, the problem of creation-science that I pose for Kitcher and Dupré is not a philosophical problem to which their views need respond.

Several considerations tell against the value of this distinction between politics and philosophy. First, in this case, relying on this dichotomy may lead to a practical absurdity. The growth of creation-science could lead to diminished political, social, and financial support for scientific research and education. In the extreme, science as we know it may cease to exist in an inhospitable cultural and religious climate. If so, the diminished growth and respect for science would be an absurdly high price to pay for keeping scientific and epistemological matters unsullied by question of politics and policy.

Second, this notion that philosophy of science should remain aloof from all things political is surprisingly recent. How and to what extent philosophy

of science should engage culture and politics were real questions for the founders of professional philosophy of science, namely European logical empiricists and their early converts in America. It was only in the late 1950s that philosophy of science—the field and this journal—adopted an explicitly politically neutral posture (Howard 1996). Before this time, however, thinkers as different as Otto Neurath and the neo-pragmatist Charles Morris hoped that their efforts as philosophers and as editors of logical empiricism's flagship, the *International Encyclopedia of Unified Science*, would have real political social effects (Cartwright et al. 1996; Reisch 1994, 1995, 1997). It would be mistaken, therefore, to maintain that Kitcher's and Dupré's views must be evaluated independently of their political and practical implications on the grounds that professional philosophy of science *has never* aimed to engage political and social problems.

Finally, some philosophers have already begun to re-engage philosophy and politics. Dupré himself connects his metaphysics and epistemology to the social and political effects they might have. He believes the classical ideal of unified science is socially mischievous and, partly on this basis, recommends a pluralistic, disunified view of the sciences.[6] Ironically, Dupré's liberal social values are similar to those of the early logical empiricists who championed the unity of science. Since his metaphysical pluralism is sympathetic to creation-science, however, it threatens to render creation-science a Trojan Horse. Admitted through the gates on the basis of pluralism and the disunity of science, creationists will likely reject Dupré's social and cultural pluralism.[7]

7. HOW TO SALVAGE THE LOGICAL EMPIRICIST APPROACH TO DEMARCATION

I have claimed that Kitcher's 1982 essay specified the correct, network demarcationist approach: creation-science is not scientific because its basic principles and beliefs are incompatible with those that connect and unify the extant sciences. Under pluralism, however, unity is no longer a paramount scientific goal and this demarcationist strategy withers. I will argue here that unity should be upheld as an ideal for science and, in addition, that the demarcation problem should be reframed as a scientific problem.

My argument draws upon the work of Otto Neurath, who was also a self-avowed pluralist. But there are some important differences between Neurath's pluralism, which I will sketch and recommend here, and Kitcher's and Dupré's pluralisms. Neurath's views are closely connected to his anti-philosophical sensibilities. Although he stands as a founder of professional philosophy of science, and an ardent champion of the unity of science, Neurath did not believe that philosophy of science would play any special role in unifying the sciences. For example, he saw the *International Encyclopedia of Unified Science* as forum for scientists and scientifically inclined philosophers to discuss problems encountered while bridging and connecting different sciences. As I have argued elsewhere, he especially rejected the idea that the *Encyclopedia* would operate as a blueprint for the future of science, a blueprint drawn by philosophers posing as far-seeing architects (Reisch 1997a). This conception of the project dangerously threatened to obscure the many, plural ways (some of them not foreseeable today) in which the sciences may come to be reconfigured and unified in the future.

Another aspect of Neurath's pluralism is his critique of "pseudorationalism." He often ridiculed the ideal of epistemic certainty and the notion that philosophy has its own, non-scientific domain of objects and concepts to investigate. (He even suggested that the word "philosophy" be placed on his infamous *Index verborum prohibitorum* (see Reisch 1997b), believing that the discipline would lose its autonomy and dissolve—though not without metaphysical residue—into various scientific disciplines). Pseudorationalists failed to see that all human knowledge—whether it belongs to philosophy, the human sciences, or the natural sciences; whether it be theoretical or applied—resides in the same, human plane, subject to the vagaries of historical, cultural, and economic change: "All of science and its parts are always under discussion. *Alles fliesst*" (1937, 85).

From Neurath's point of view, therefore, the Unity of Science Movement would remain embedded *within* historical processes—the very ones the Movement aimed to shape and guide—that cannot be viewed or understood from any external, or higher, epistemological platform.[8] As a result—and this is the important point for my argument below—the Movement would have to operate without the benefit of such a vantage point and without the theories of science that it might make available. No theory of science would guide the Unity of Science Movement.

What would Neurath have said about creation-science? He would have derided it as "metaphysics." For Neurath, "metaphysical" ideas or methods were those that are unscientific and "isolated" from the existing network of (more or less) unified sciences.[9] He described the development of science in a sway that is suggestive of network demarcation and of Kitcher's image of that network mutilated to accommodate creation-science:

> If a statement is made, it is to be confronted with the totality of existing statements. If it agrees with them, it is joined to them; if it does not agree, it is called 'untrue' and rejected; or the existing complex of statements of science is modified so that the new statement can be incorporated (Neurath 1931, 53).

The remainder is metaphysics—those statements that are either senseless or whose compatibility with the rest of science cannot in principle be decided (Neurath 1937, 83). "Metaphysical" language is thus "isolated" from science.[10] In the case of creation-science, statements about immaterial agencies and the creation of things through supernatural processes, or conceptions of species that do not "agree" with those already accepted, would render it "isolated" from the existing network of sciences.

At this point, however, we encounter familiar problems. Neurath's comments are vague. Precisely what sorts of connections and agreements count? Don't some of the more speculative branches of genuine science (say, e.g., string theory in physics) appear in some ways as isolated from other sciences as does creation-science? Ideally, ambiguities like these could be settled with a widely accepted, general theory of science. Yet, none is available. It seems, therefore, that Neurath's approach leads us to Kitcher's and Dupré's conclusion: demarcation is simply an impossible philosophical task.

But Neurath provides us with another approach. We can still demarcate creation-science from science by reconfiguring that task according to Neurath's unificationist project: if his Unity of Science Movement was to proceed *without* relying on a theory of science, perhaps it is a mistake to suppose in the first place that the demarcation problem requires a general theory of science for its solution. If Neurath was right that scientists, and not philosophers, are best equipped to develop and unify the sciences, then perhaps they are also best equipped to determine what is scientific and what is not. Network demarcation should be a scientific problem.

By speaking of "scientists" and "philosophers," I am using shorthand. The relevant distinction is not between the professions or their members but rather between two ways of framing the demarcation question. The issue is whether or not we attempt demarcation *on the basis of a philosophical theory of science*. When framed as a scientific, and not a philosophical, problem, the demarcation question becomes much like Kitcher's (non-pluralistic) formulation of the issue: Can the central concepts, methods, and theories of creation-science be incorporated into the network of existing sciences? Most scientists—and anyone scientifically literate—would answer "no." Creation-science, therefore, should be considered non-scientific.

This approach escapes the liabilities of pluralism that I described earlier. Since it does not accept that demarcation is impossible, this strategy does not capitulate to claims that creation-science is scientific; nor does it rest on metaphysical theories which might seem to secure an ontological foundation for creation-science. This approach also overcomes the objections that will likely be raised against it. One may object, for instance, that this distinction between "philosophical" and "scientific" ways of framing the question is strained, that any demarcation must appeal to certain philosophical models or theories or intuitions about what is and what is not scientific. This is not true. Scientists routinely make decisions affecting the very content of the sciences—thereby determining, on this view, what is and what is not scientific—and they do so without having (or desiring) general, powerful theories of science of the sort that philosophers traditionally seek to develop.

Another objection is that this proposal relinquishes normative authority. If scientists were to become interested in creation-science then, according to this proposal, it would count as genuine science. This inference is correct, but it does not count as an objection. Most, if not all, scientists are disinterested in creation-science (except, perhaps, as an anti-scientific social phenomenon). That is the practical and legal strength of this approach. Even if scientists of the future were to convert to creation-science, it presents a problem *now,* and this approach provides us tools to frame and manage that problem, now.

This last point, finally, underscores one of the more subtle but distinctive features of Neurathian philosophy (or anti-philosophy) of science: Why, Neurath would ask the objector, would one even hope that philosophy of science could assuage this worry about scientists of the future turning creationist? The very idea exposes a rather medieval conception of philosophy—that all branches of knowledge are somehow controlled by, or justified by, this "queen" of the sciences. Instead, Neurath anticipated today's naturalism (see, e.g., Uebel 1991, 1992) and deflationism (Fine 1986) which sees science as a practice that has a life independent of philosophy. It need not answer to philosophers' intuitions that some form of inquiry is or is not "scientific" (or "rational" or "progressive"). That philosophy can or should have such a strong purchase on science is an idea Neurath would urge us to reject as a metaphysical dream, "isolated" from a properly empirical, scientific view of the world.

NOTES

1. The "balanced treatment" acts of Arkansas and Louisiana, both of which supported the teaching of creation-science in biology classes, were ruled unconstitutional by a federal court (in 1982), and the U.S. Supreme Court (in 1986), respectively. Other relevant cases include Webster v. New Lenox School District (1990) in which the Seventh Circuit Court of Appeals found that school districts may prohibit the teaching of creation-science (in order that the Establishment Clause not be violated) and Peloza v. Capistrano School District (1994) in which the Ninth Circuit Court of Appeals found that a teacher's right to free exercise of religion is not violated by requiring the teacher to present evolution ("Six Significant Court Decisions regarding Evolution/Creation Issues", publication of the National Center for Science Education, Feb. 1996).

Recently, Tennessee, Alabama, and Hall County, Georgia, either voted on or passed bills sympathetic to creation-science and critical of the presentation of evolutionary theory as established 'fact' ("Dumping on Darwin" *Time*, 147, no. 12, March 18, 1996).

2. Others have also recently noted this circumstance. See, e.g., Holton 1993, 181–183; Sorell 1991, 177.

3. For an overview of this controversy, and papers by the major players, see *Biology and Philosophy*, v. 2, 1987.

4. For evolutionary theory and creation-science as "models," see Act 590 of 1981, 73rd General Assembly, State of Arkansas, in La Follette 1983, 17; for both as religious beliefs, see Judge Overton's opinion in the Arkansas case, in La Follette 1983, 45–73, esp. 51, 62–63.

5. Ruse 1996, 357. Ruse's comment is directed against Larry Laudan. Like Dupré, Laudan wants judgments about the quality of science to determine what gets taught in the schools (Laudan 1996, 354).

6. Advocates of unity of science, Dupré says, perpetuate bad politics as well as bad metaphysics; reductionism and unity of science incline us to think of social minorities as homogenous classes, or as monolithic natural kinds, with essential features reducible to sociological, psychological, biological, chemical, and ultimately physical differentia (Dupré 1993, 253). As a result, allegiance to the ideal of unity makes us more likely to stereotype and discriminate against these groups. Without it, on the other hand, we would incline to "skepticism about the extent to which homogeneity among the members of any kind should be assumed" (1993, 252), and we would be more likely to respect variety, individuality, and civil liberty.

7. This is because, unlike Dupré, creationists tend to accept the reality of monolithic, natural kinds. Scientifically, they construe species as immutable, historically stable kinds. Sociopolitically, they see feminists, homosexuals, communists, pagans, and even evolutionists as (heretical) homogenous kinds standing in the way of their goals. (That moral and "scientific" issues are strongly connected in creationist ideology is indicated, for example, in the Arkansas act which notes that the teaching of evolutionary theory interferes with "moral training" and "religious training" of students (Act 590, in La Follette 1983, 18). On the classification of evolutionists as "satanically inspired," see Holton 1993, 182.)

8. This image is suggested by Neurath's Protagorean proclamation in the Vienna Circle's manifesto, *Wissenschaftliche Weltauffassung*: "All experience forms a complex network, which cannot always be surveyed and can often be grasped only in parts. Everything is accessible to man; and man is the measure of all things" (Neurath, Carnap, and Hahn 1929, 306). This same embeddedness informs Neurath's critique of the concept of truth: in legal proceedings, he wrote, this term has a useful meaning: a judge or jury conventionally serves as a final, external standard to test claims. However, "in the republic of the sciences other customs are valid" (1944, 13) because there is no judge. Different scientists routinely argue about what is and is not "true," so this concept cannot be appealed to as an unproblematic arbiter. Neurath complained similarly about Bertrand Russell's casual use of "true" and "false," in his *Inquiry Concerning Meaning and Truth* (in which Neurath came in for some stinging criticism (Russell 1940, 143–149)): "he speaks of 'error' and 'knowledge,'" Neurath wrote to Carnap. "Who is in the [judge's] chair? Russell personally?" "There is no point outside the 'world' from where we may judge on TRUE and FALSE." (Neurath to Carnap, July 17, 1942, University of Pittsburgh, Archives for Scientific Philosophy, ASP 102-56-04. Quoted by permission of the University of Pittsburgh. All rights reserved.)

9. No mention of Neurath and his contempt for metaphysics can exclude this often cited anecdote: "I remember in the WIENER KREIS as we read Wittgenstein, I again and again said: that is Metaphysics. It became dull and [Hans] Hahn suggested I should speak of M. only to shorten the sounds[.] And since I too often said "M" he suggested I should only remark when I am satisfied by saying 'NM'—such have been our jokes." (Neurath to Carnap & Morris, Nov. 18, 1944, Charles Morris Papers, Peirce Edition Project, Indiana University Purdue University Indianapolis; published with permission of the Peirce Edition Project.)

10. See Neurath 1944, 9–10. Neurath notes that the statement "we call 'isolated' ones" are "formerly metaphysical ones." (Neurath to Herbert Feigl, 26 Feb. 1943, Charles Morris Papers, Peirce Edition Project (IUPUI); published with permission of the Peirce Edition Project.)

REFERENCES

Cartwright, Nancy, Jordi Cat, Lola Fleck, and Thomas Uebel (1996), *Otto Neurath: Philosophy between Science and Politics*. Cambridge: Cambridge University Press.

Dupré, John (1993), *The Disorder of Things: Metaphysical Foundations of the Disunity of Science*. Cambridge, MA: Harvard University Press.

Fine, Arthur (1986), *The Shaky Game*. Chicago: University of Chicago Press.

Holton, Gerald (1993), *Science and Anti-Science*. Cambridge, MA: Harvard University Press.

Howard, Don (1996), "Philosophy of Science and Social Responsibility: Some Historical Reflections", ms. read at Philosophy of Science Association annual meeting, Cleveland, Ohio, Nov. 1996.

Hull, David (1987), "Genealogical Actors in Ecological Roles", *Biology and Philosophy* 2:168–184.

Kitcher, Philip (1982), *Abusing Science: The Case against Creationism*. Cambridge, MA: MIT Press.

———. (1989), "Some Puzzles about Species", in Michael Ruse (ed.), *What the Philosophy of Biology Is*. Dordrecht: Kluwer, pp. 183–208.

———. (1993), *The Advancement of Science*. New York: Oxford University Press.

Laudan, Larry (1996), "Science at the Bar—Causes for Concern", in Ruse (ed.) (1996), 351–355.

La Follette, Marcel C. (ed.) (1983), *Creationism, Science, and the Law: The Arkansas Case*. Cambridge, MA: MIT Press.

Lewontin, Richard (1997), "Billions and Billions of Demons", *New York Review of Books*. January 9, 1997:28–32.

Neurath, Otto, Rudolf Carnap, and Hans Hahn (1929), "The Scientific Conception of the World: The Vienna Circle", in Neurath 1973, 299–319.

Neurath, Otto (1931), "Physicalism," in Neurath 1983, 52–57.

———. (1937), *"Prognosen und Terminologie in Physik, Biologie, Soziologie," Trauvauz du IX Congres International de Philosophie, IV. L'Unite de la Science: la methode et les methodes. Actualities Scientifiques et Industrielles*, No 533, Paris: Hermann & Cᵢᵉ, pp. 77–85.

———. (1944), "Foundations of the Social Sciences", *International Encyclopedia of Unified Science* v. 2, no. 1. Chicago: University of Chicago Press, 1–51.

———. (1973), *Empiricism and Sociology*. Marie Neurath and Robert S. Cohen (eds.). Translated by M. Neurath and Paul Foulkes. Boston: Reidel.

———. (1983), *Philosophical Papers: 1913–1946*. R. S. Cohen and M. Neurath (eds. and trans.). Boston: Reidel.

Reisch, George (1994), "Planning Science: Otto Neurath and the *International Encyclopedia of Unified Science*", *British Journal for the History of Science* 27:153–175.

———. (1995), *"A History of the International Encyclopedia of Unified Science"*, unpublished Ph.D. dissertation. University of Chicago.

———. (1997a), "How Postmodern Was Neurath's Idea of Unity of Science?" *Studies in History and Philosophy of Science* 28:439–451.

———. (1997b), "Economics, Epistemologist . . . and Censor?: On Otto Neurath's *Index Verborum Prohibitorum*", *Perspectives on Science*.

Ruse, Michael (1996), "Pro Judice", in M. Ruse (ed.) (1966), 356–362.

———. (ed.) (1996), *But Is It Science? The Philosophical Question in the Creation/Evolution Controversy*. Amherst, NY: Prometheus Books.

Russell, Bertrand (1940), *An Inquiry into Meaning and Truth*. London: George Allen and Unwin, Ltd.

Sorell, Tom (1991), *Scientism: Philosophy and the Infatuation with Science*. New York: Routledge.

Stanford, P. Kyle (1995), "For Pluralism and against Realism about Species", *Philosophy of Science* 62:70–91.

Uebel, Thomas (1991), "Neurath's Programme for Naturalistic Epistemology", *Studies in History and Philosophy of Science* 22:623–646.

———. (1992), *Overcoming Logical Positivism from Within: The Emergence of Neurath's Naturalism in the Vienna Circle's Protocol Sentence Debate*. Atlanta: Rodopi.

PART 4 SUGGESTIONS FOR FURTHER READING

Hull, David. "Reduction in Genetics—Biology or Philosophy?" *Philosophy of Science* 39 (1972): 491–499

———. "The Reduction of Mendelian to Molecular Genetics." In *Philosophy of Biological Science*. Englewood Cliffs, NJ: Prentice-Hall, 1974.

Kincaid, Harold. *Individualism and the Unity of Science*. New York: Rowman & Littlefield, 1997.

Rasenberg, Alexander. *Instrumental Biology: or the Disunity of Science*. Chicago: University of Chicago Press, 1994.

Sklar, Lawrence. "Types of Intertheoretic Reduction." *British Journal for the Philosophy of Science* 18 (1967): 109–124.

Spector, Marshall. *Concepts of Reduction in Physical Science*. Philadelphia: Temple University Press, 1978.

Waters, C. Kenneth. "Genes Made Molecular." *Philosophy of Science* 61 (1994): 163–185.

Wilson, E. O. *Consilience: The Unity of Knowledge*. New York: Vintage, 1998.

PART 5

THEORY AND OBSERVATION
Is Seeing Believing?

Logical positivism grew out of discussions among a group of thinkers in Vienna during the first three decades of the twentieth century. Known as the Vienna Circle, this group included mathematician Hans Hahn; economist Otto Neurath; physicist Philipp Frank; and philosophers Moritz Schlick, Rudolf Carnap, and Herbert Feigl, among others. What brought these men together was a common desire to develop a comprehensive and coherent view of the nature and scope of scientific knowledge.

The term "positivism" derives from the philosophy of Auguste Comte (1798–1857), who maintained that all knowledge is scientific knowledge and that the function of philosophy should be to articulate the general principles common to all the sciences. The logical positivists were sympathetic to these views, but they owe a greater intellectual debt to physicist Ernst Mach (1838–1916), who held that sense experience is the only source of knowledge and that the division of sciences into various branches is arbitrary and artificial. The view that all knowledge is based on sense experience is known as empiricism. Although there were empiricists in ancient Greece, empiricism received its classic formulation at the hands of the British philosophers John Locke (1632–1704), George Berkeley (1685–1753), and David Hume (1711–1766.) The main problem facing empiricists is that of explicating how a claim must be related to experience in order to count as knowledge. The logical positivists attempted to use the new symbolic logic developed by Bertrand Russell (1872–1969) and Alfred North Whitehead (1861–1947) to solve this problem (hence the term "logical"). A description of classical empiricism helps put logical positivism in perspective.

The central claim of classical empiricism is that all of our ideas are derived from sense experience. As Locke put it, the mind is a blank slate (a *tabula rasa*) that contains nothing more than what the senses have put there. Classical empiricism recognizes two basic types of ideas: simple and complex. Simple ideas—those that do not contain other ideas as constituents—correspond to physical sensations, such as hot, cold, light, dark, and so on. Complex ideas are built out of simple ideas. The idea of an apple, for example, is composed of the ideas of a certain shape, size, color, taste, texture, and other qualities. Thus all complex ideas are analyzable into simple ideas that represent sensations.

The British empiricists drew two important corollaries from this view of the nature of thought: (1) An idea is real only if it is derived from or reducible to sense impressions, and (2) a term is meaningful only if it stands for a real idea. Hume used these principles to demonstrate that terms such as "liberty," "causality," and "self" are not meaningful, for they do not stand for real ideas. As he says in his *Enquiries Concerning the Human Understanding,*

When we entertain, therefore, any suspicion that a philosophical term is employed without any meaning or idea (as is but too frequent), we need but enquire, *from what impression is that supposed idea derived?* And if it be impossible to assign any, this will serve to confirm our suspicion.[1]

Such terms, according to Hume, should be purged from our language. He writes,

When we run over libraries, persuaded of these principles, what havoc must we make? If we take in our hand any volume; of divinity or school metaphysics, for instance; let us ask, *Does it contain any abstract reasoning concerning quantity or number?* No. *Does it contain any experimental reasoning concerning matter of fact and existence?* No. Commit it then to the flames: for it can contain nothing but sophistry and illusion.[2]

According to Hume, people who use terms that are not derived from sensations don't know what they are talking about, because they aren't talking about anything. Just as the British empiricists thought that all meaningful terms stood for ideas that are derivable from sense experience, the logical positivists thought that all meaningful sentences represent states of affairs that are verifiable through sense experience. If there is no way to tell whether a sentence is true, it doesn't tell us anything about the world. And if it doesn't tell us anything about the world, it is cognitively meaningless. A nonverifiable sentence, such as "The nothing nothings itself," may arouse certain emotions in us and thus have emotive meaning. But it can have no cognitive meaning because there is no way to establish its truth. For the logical positivists, then, a sentence is (cognitively) meaningful only if it is verifiable, and the meaning of a sentence is the set of conditions that would verify it. These views became known as the *verifiability theory of meaning.*

The logical positivists believed that some sentences are directly verifiable. These are known as *observation sentences,* or *protocol sentences.* Rudolf Carnap thought that observation sentences report sensory experiences (for example, "There is a sensation of red, here, now"). Others, such as Otto Neurath, believed that observation sentences report states of affairs that can be directly perceived by means of our unaided senses (for example, "The cat is on the mat"). Despite these differences, both believed that observation sentences form the basis of our knowledge of the world.

Non-observation sentences—those that are not directly verifiable—are called *theoretical sentences.* These sentences usually contain theoretical terms such as "electron" that refer to things that cannot be directly perceived. According to the

---

[1] David Hume, *Enquiries Concerning the Human Understanding and Concerning the Principles of Morals,* ed. L. A. Selby-Bigge (Oxford: Clarendon Press, 1972), p. 22.
[2] *Ibid.,* p. 365.

verifiability theory of meaning, theoretical sentences are meaningful only if they imply the right sort of observation sentences and thus are verifiable. To establish the legitimacy of scientific theorizing, then, the logical positivists tried to show how the truth of theoretical sentences depends on the truth of observation sentences. In the first reading, Rudolf Carnap provides a brief description of this project. The original idea that theoretical sentences are logically equivalent to sets of observation sentences was soon found to be mistaken. Nevertheless, Carnap remained confident that the nature of the relationship can be specified.

Mary Hesse does not believe that the required connection between observation sentences and theoretical sentences can be established because she does not consider the distinction between observation sentences and theoretical sentences to be viable. Observation sentences supposedly differ from theoretical ones in that their truth is determined by experience alone. But experience alone can never establish the truth of a sentence—other beliefs always play a role. So even reports of sensation are shot through with theory.

Hesse points out that when we learn a language, we learn certain generalizations, or "laws," that relate one type of thing to another, such as "All mammals live on land." As we learn more about the world, however, these laws may change. And when they change, the meaning of the terms that figure into those laws may change. So the meaning of every term—including observation terms—is determined by our theories about the world. Whatever difference there is, then, between observation and theoretical terms cannot be a difference of kind, but one only of degree.

Just as the logical positivists maintained that the meaning of observation terms is independent of our theoretical beliefs, so they thought that the perception of physical objects is unaffected by our theoretical beliefs. For them, perception was a source of neutral data that could be used to adjudicate among competing theories. The theory that best explained the data would be the most plausible.

Norwood Russell Hanson, however, suggests that perception—like meaning—is theory laden. Consider, for example, the perception of an x-ray tube. Hanson contends that a physicist and an Eskimo baby would see very different things when looking at the tube. It's not that they would see the same thing and interpret it differently, because seeing is not a two-step process. We don't first see something and then interpret it. What we see is already interpreted. Our theories about the world structure what we perceive. Thus, all perception is fraught with theory.

Hanson argues that there are no theory-independent facts. Hesse argues that even if there were, there is no theory-independent language with which to express them. Thomas Kuhn uses these insights to construct a conception of science that is radically different from that of the positivists. According to Kuhn, science is essentially a puzzle-solving activity. This puzzle solving takes place in the context of what he calls a paradigm. Kuhn uses the word "paradigm" to refer to particular scientific theories within a discipline as well as to the disciplinary matrix itself, which includes the concepts, methods, and standards used to arrive at those theories. The notion of paradigms as specific accomplishments appears to be primary, however, because paradigmatic theories define by example how research in a particular discipline should be conducted. They serve as exemplars that young scientists are taught to emulate. Among the theories Kuhn cites as paradigms are

Newton's theory of gravity, Franklin's theory of electricity, and Copernicus's theory of the solar system.

Normal science, for Kuhn, consists not in trying to confirm or falsify theories, but in trying to solve the puzzles identified by a paradigm. Most theories make predictions that go beyond the data that gave rise to them. Scientists investigate those predictions to see whether they are borne out by the facts. If not, scientists have a puzzle on their hands that must be solved through some modification of the theory or the methodology. Most puzzles can be solved within the context of the paradigm. But sometimes a puzzle or anomaly develops that cannot be solved with the resources available in the paradigm. This leads to a crisis within the scientific community that can be resolved only by a scientific revolution that leads to the adoption of a new paradigm.

Adopting a new paradigm, however, is not a purely rational affair. Scientists cannot simply examine a number of competing paradigms and choose the one that best fits the data, for there are no neutral data available to all paradigms. Since all perception is theory laden, what counts as data will vary from paradigm to paradigm. Any choice among paradigms, therefore, cannot be made on rational grounds.

Even if there were a common set of facts available to all paradigms, there is no common language with which to express them and no common set of criteria by which to evaluate them. As a result, no paradigm can be considered any better than any other. To make an objective judgment about the relative merits of different paradigms, we need some neutral standard against which to judge them. But there is no such standard. So, Kuhn claims, different paradigms are incommensurable; they cannot be compared with one another.

Kuhn's incommensurability thesis undermines the notion of scientific progress. Because the positivists believed that there is a common set of data available to all theories and a common language with which to express the data, it made sense to speak of theories providing progressively better accounts of the data. If perception, meaning, and value are all theory laden, however, no such progress is possible. Science can no longer be viewed as providing increasingly better theories about the world.

Larry Laudan takes issue with Kuhn's incommensurability thesis. He agrees with Kuhn that we cannot judge the relative truthfulness of various theories or paradigms because we don't know what it means to say that one theory is more truthful than another. But he disagrees that this means that one theory or paradigm cannot be considered better than another. The relative worth of a theory should be judged not in terms of its truthfulness, but in terms of its problem-solving effectiveness. There are two types of problems a theory can solve: empirical problems, which have to do with how well the theory corresponds to the world, and conceptual problems, which have to do with how the theory coheres with our other beliefs. The best theory solves the largest number of important empirical problems and generates the smallest number of conceptual problems.

For Laudan, however, conceptual problems arise not only when a theory fails to cohere with other scientific beliefs. They can also arise when a theory fails to cohere with beliefs in other areas, such as philosophy, theology, or literary theory. So, for him, there is no fundamental difference between science and other forms of intellectual inquiry. Insofar as all of these disciplines attempt to explain some aspect of the world, they can be judged in terms of their problem-solving effectiveness.

22

RUDOLF CARNAP

The Methodological Character of Theoretical Concepts

OUR PROBLEMS

In discussions on the methodology of science, it is customary and useful to divide the language of science into two parts, the observation language and the theoretical language. The observation language uses terms designating observable properties and relations for the description of observable things or events. The theoretical language, on the other hand, contains terms which may refer to unobservable events, unobservable aspects or features of events, e.g., to micro-particles like electrons or atoms, to the electromagnetic field or the gravitational field in physics, to drives and potentials of various kinds in psychology, etc. In this article I shall try to clarify the nature of the theoretical language and its relation to the observation language. . . .

One of the main topics will be the problem of a criterion of significance for the theoretical language, i.e., exact conditions which terms and sentences of the theoretical language must fulfill in order to have a positive function for the explanation and prediction of observable events and thus to be acceptable as empirically meaningful. I shall leave aside the problem of a criterion of significance for the observation language, because there seem to be hardly any points of serious disagreement among philosophers today with respect to this problem, at least if the observation language is understood in the narrow sense indicated above. On

the other hand, the problem for the theoretical language is a very serious one. There are not only disagreements with respect to the exact location of the boundary line between the meaningful and the meaningless, but some philosophers are doubtful about the very possibility of drawing any boundary line. It is true that empiricists today generally agree that certain criteria previously proposed were too narrow; for example, the requirement that all theoretical terms should be definable on the basis of those of the observation language and that all theoretical sentences should be translatable into the observation language. We are aware at present that these requirements are too strong because the rules connecting the two languages (which we shall call "rules of correspondence") can give only a partial interpretation for the theoretical language. From this fact, some philosophers draw the conclusion that, once the earlier criteria are liberalized, we shall find a continuous line from terms which are closely connected with observations, e.g., 'mass' and 'temperature,' through more remote terms like 'electromagnetic field' and 'psi-function' in physics, to those terms which have no specifiable connection with observable events, e.g., terms in speculative metaphysics; therefore, meaningfulness seems to them merely a matter of degree. This skeptical position is maintained also by some empiricists; Hempel, for instance, has given clear and forceful arguments for this view. Although he still regards the basic idea of the empiricist meaning criterion as sound, he believes that deep-going modifications are necessary. First, the question of meaningfulness cannot, in his opinion, be raised for any single term or sentence but only for the whole system consisting of the theory, expressed in the theoretical language, and the correspondence

The Foundations of Science and the Concepts of Psychology and Psychoanalysis. Minnesota Studies in the Philosophy of Science, Vol. 1, Herbert Feigl and Michael Scriven, eds. (Minneapolis: University of Minnesota Press, 1956), pp. 38–43. Reprinted by permission of the publisher.

rules. And secondly, even for this system as a whole, he thinks that no sharp distinction between meaningful and meaningless can be drawn; we may, at best, say something about its confirmation on the basis of the available observational evidence, or about the degree of its explanatory or predictive power for observable events.

The skeptics do not, of course, deny that we can draw an exact boundary line if we want to. But they doubt whether any boundary line is an adequate explication of the distinction which empiricists had originally in mind. They believe that, if any boundary line

is drawn, it will be more or less arbitrary; and moreover, that it will turn out to be either too narrow or too wide. That it is too narrow means that some terms or sentences are excluded which are accepted by scientists as meaningful; that it is too wide means that some terms or sentences are included which scientifically thinking men would not accept as meaningful.

My attitude is more optimistic than that of the skeptics. I believe that, also in the theoretical language, it is possible to draw an adequate boundary line which separates the scientifically meaningful from the meaningless.

23

MARY HESSE

Is There an Independent Observation Language?

Of all the men of the century Faraday had the greatest power of drawing ideas out of his experiments and making his physical apparatus do his thinking, so that experimentation and inference were not two proceedings, but one.

-C. S. PEIRCE
Values in a Universe of Chance

I. OBSERVATION PREDICATES

Rapidity of progress, or at least change, in the analysis of scientific theory structure is indicated by the fact that only a few years ago the natural question to

The Nature and Function of Scientific Theories, Robert G. Colodny, ed. (Pittsburgh: University of Pittsburgh Press, 1970), pp. 35–43. © 1970 by University of Pittsburgh Press. Reprinted by permission of the publisher.

ask would have been, "Is there an independent theoretical language?" The assumption would have been that theoretical language in science is parasitic upon observation language, and probably ought to be eliminated from scientific discourse by disinterpretation and formalization, or by explicit definition in or reduction to observation language. Now, however, several radical and fashionable views place the onus on believers in an observation language to show that such a concept has any sense in the absence of a theory. It is time to pause and ask what motivated the distinction between a so-called theoretical language and an observation language in the first place, and whether its retention is not now more confusing than enlightening.

In the light of the importance of the distinction in the literature, it is surprisingly difficult to find any clear statement of what the two languages are supposed to consist of. In the classic works of twentieth-century philosophy of science, most accounts of the observation language were dependent on circular definitions of observability and its cognates, and the theoretical language was generally defined as consisting of those scientific terms which are not observational. We find quasi definitions of the following kind: "'Observation-statement' designates a statement which records an actual or possible observation"; "Experience, observation, and cognate terms will be used in the widest sense to cover observed facts about material objects or events in them as well as directly known facts about the contents or objects of immediate experience"; "The observation language uses terms designating observable properties and relations for the description of observable things or events"; *observables*, i.e., . . . things or events which are ascertainable by direct observation."[1] Even Nagel, who gives the most thorough account of the alleged distinction between theoretical and observation terms, seems to presuppose that there is nothing problematic about the "direct experimental evidence" for observation statements, or the "experimentally identifiable instances" of observation terms.[2]

In contrast with the allegedly clear and distinct character of the observation terms, the meanings of theoretical terms, such as "electron," "electromagnetic wave," and "wave function,"[3] were held to be obscure. Philosophers have dealt with theoretical terms by various methods, based on the assumption that they have to be explained by means of the observation terms as given. None of the suggested methods has, however, been shown to leave theoretical discourse uncrippled in some area of its use in science. What suggests itself, therefore, is that the presuppositions of all these methods themselves are false, namely,

a. that the meanings of the observation terms are unproblematic,

b. that the theoretical terms have to be understood by means of the observation terms, and

c. that there is, in any important sense, a distinction between two *languages* here, rather than different kinds of uses within the same language.

In other words, the fact that we somehow understand, learn, and use observation terms does not in the least imply that the way in which we understand, learn, and use them is either different from or irrelevant to the way we understand, learn, and use theoretical terms. Let us then subject the observation language to the same scrutiny which the theoretical language has received.

Rather than attacking directly the dual language view and its underlying empiricist assumptions, my strategy will be first to attempt to construct a different account of meaning and confirmation in the observation language. This project is not the ambitious one of a general theory of meaning, nor of the learning of language, but rather the modest one of finding conditions for understanding and use of terms in science—some specification, that is to say, in a limited area of discourse, of the "rules of usage" which distinguish meaningful discourse from mere vocal reflexes. In developing this alternative account I shall rely on ideas which have become familiar particularly in connection with Quine's discussions of language and meaning and the replies of his critics, whose significances for the logic of science seem not yet to have been exploited nor even fully understood.[4]

I shall consider, in particular, the predicate terms of the so-called observation language. But first something must be said to justify considering the problem as one of "words" and not of "sentences." It has often been argued that it is sentences that we learn, produce, understand, and respond to, rather than words, that is, that in theoretical discussion of language, sentences should be taken as units. There are, however, several reasons why this thesis, whether true or false, is irrelevant to the present problem, at least in its preliminary stages. The observation language of science is only a segment of the natural language in which it is expressed, and we may for the moment assume that rules of sentence formation and grammatical connectives are already given when we come to consider the use of observation predicates. Furthermore, since we are interested in alleged distinctions between the observation and theoretical languages, we are likely to find these distinctions in the characteristics of their respective predicates, not in the connectives which we may assume that they share. Finally, and most importantly, the present enterprise does not have the general positive aim of describing the entire structure of a language. It has rather the negative aim of showing that there are no terms in the observation language which are sufficiently accounted for by "direct observation," "experimentally identifiable instances," and the like. This can best be done by examining the hardest cases, that is, predicates which do appear to have direct empirical reference. No one would seriously put forward the direct-observation account of grammatical connectives; and if predicates are shown not to satisfy the account, it is likely that the same arguments will suffice to show that sentences do not satisfy it either.

So much for preliminaries. The thesis I am going to put forward can be briefly stated in two parts:

i. All descriptive predicates, including observation and theoretical predicates, must be introduced, learned, understood, and used, either by means of direct empirical associations in some physical situations, or by means of sentences containing other descriptive predicates which have already been so introduced, learned, understood, and used, or by means of both together. (Introduction, learning, understanding, and use of a word in a language will sometimes be summarized in what follows as the *function* of that word in the language.)

ii. No predicates, not even those of the observation language, can function by means of direct empirical associations alone.

The process of functioning in the language can be spelled out in more detail:

A. Some predicates are initially learned in empirical situations in which an association is established between some aspects of the situation and a certain word. Given that any word with extralinguistic reference is ever learned, this is a necessary statement and does not presuppose any particular theory about what an association is or how it is established. This question is one for psychology or linguistics rather than philosophy. Two necessary remarks can, however, be made about such learning:

1. Since every physical situation is indefinitely complex, the fact that the particular aspect to be associated with the word is identified out of a multiplicity of other aspects implies

that degrees of physical similarity and difference can be recognized between different situations.

2. Since every situation is in detail different from every other, the fact that the word can be correctly reused in a situation in which it was not learned has the same implication.

These remarks would seem to be necessarily implied in the premise that some words with reference are learned by empirical associations. They have not gone unchallenged, however, and it is possible to distinguish two sorts of objections to them. First, some writers, following Wittgenstein, have appeared to deny that physical similarity is necessary to the functioning of *any* word with extralinguistic reference. That similarity is not *sufficient,* I am about to argue, and I also agree that not all referring words need to be introduced in this way, but if *none* were, I am unable to conceive how an intersubjective descriptive language could ever get under way. The onus appears to rest upon those who reject similarity to show in what other way descriptive language is possible.[5] The other sort of objection is made by Popper, who argues that the notion of repetition of instances which is implied by (1) and (2) is essentially vacuous, because similarity is always similarity *in certain respects,* and "with a little ingenuity" we could always find similarities in *some* same respects between all members of any finite set of situations. That is to say, "anything can be said to be a repetition of anything else, if only we adopt the appropriate point of view."[6] But if this were true, it would make the learning process in empirical situations impossible. It would mean that however finitely large the number of presentations of a given situation-aspect, that aspect could never be identified as the desired one out of the indefinite number of other respects in which the presented situations are all similar. It would, of course, be possible to eliminate some other similarities by presenting further situations similar in the desired respect but not in others, but it would then be possible to find other respects in which all the situations, new and old, are similar— and so on without end.

However, Popper's admission that "a little ingenuity" may be required allows a less extreme interpretation of his argument, namely, that the physics and physiology of situations already give us some "point of view" with respect to which some pairs of situations are similar in more obvious respects than others, and one situation is more similar in some respect to another than it is in the same respect to a third. This is all that is required by the assertions (1) and (2). Popper has needlessly obscured the importance of these implications of the learning process by speaking as though, before any repetition can be recognized, we have to take thought, and *explicitly* adopt a point of view. If this were so, a regressive problem would arise about how we ever learn to apply the predicates in which we explicitly express that point of view. An immediate consequence of this is that there must be a stock of predicates in any descriptive language for which it is impossible to *specify* necessary and sufficient conditions of correct application. For if any such specification could be given for a particular predicate, it would introduce further predicates requiring to be learned in empirical situations for which there was no specification. Indeed such unspecified predicates would be expected to be in the majority, for those for which necessary and sufficient conditions can be given are dispensable except as a shorthand and, hence, essentially uninteresting. We must, therefore, conclude that the primary process of recognition of similarities and differences is necessarily *unverbalizable*. The emphasis here is of course on *primary*, because it may be perfectly possible to give empirical descriptions of the conditions, both psychological and physical, under which similarities are recognized, but such descriptions will themselves depend on further undescribable primary recognitions.

B. It may be thought that the primary process of classifying objects according to recognizable similarities and differences will provide us with exactly the independent observation predicates required by the traditional view. This, however, is to overlook a logical feature of relations of similarity and difference, namely, that they are not *transitive*. Two objects *a* and *b* may be judged to be similar to some degree in respect to predicate *P*, and may be placed in the class of objects to which *P* is applicable. But object *c* which is judged similar to *b* to the same degree may not be similar to *a* to the same or indeed to any degree. Think of judgments of similarity of three

shades of color. This leads to the conception of some objects as being more "central" to the *P*-class than others, and also implies that the process of classifying objects by recognition of similarities and differences is necessarily accompanied by some loss of (unverbalizable) information. For if *P* is a predicate whose conditions of applicability are dependent on the process just described, it is impossible to *specify* the degree to which an object satisfies *P* without introducing more predicates about which the same story would have to be told. Somewhere this potential regress must be stopped by some predicates whose application involves loss of information which is present to recognition but not verbalizable. However, as we shall see shortly, the primary recognition process, though necessary, is not sufficient for classification of objects as *P*, and the loss of information involved in classifying leaves room for changes in classification to take place under some circumstances. Hence primary recognitions do not provide a stable and independent list of primitive observation predicates.

C. It is likely that the examples that sprang to mind during the reading of the last section were such predicates as "red," "ball," and "teddy bear." But notice that nothing that has been said rules out the possibility of giving the same account of apparently much more complex words. "Chair," "dinner," and "mama" are early learned by this method, and it is not inconceivable that it could also be employed in first introducing "situation," "rule," "game," "stomachache," and even "heartache." This is not to say, of course, that complete fluency in using these words could be obtained by this method alone; indeed, I am now going to argue that complete fluency cannot be obtained in the use of *any* descriptive predicate by this method alone. It should only be noticed here that it is possible for any word in natural language having some extralinguistic reference to be introduced in suitable circumstances in some such way as described in section A.

D. As learning of the language proceeds, it is found that some of these predicates enter into general statements which are accepted as true and which we will call *laws:* "Balls are round"; "In summer leaves are green"; "Eating unripe apples leads to stomachache." It matters little whether some of these are what we would later come to call analytic statements; some, perhaps most, are synthetic. It is not necessary, either, that every such law should be *in fact* true, only that it is for the time being accepted as true by the language community. As we shall see later, any one of these laws may be *false* (although not all could be false at once). Making explicit these general laws is only a continuation and extension of the process already described as identifying and reidentifying proper occasions for the use of a predicate by means of physical similarity. For knowledge of the laws will now enable the language user to apply descriptions correctly in situations other than those in which he learned them, and even in situations where nobody could have learned them in the absence of laws, for example, "stomachache" of an absent individual known to have consumed a basketful of unripe apples, or even "composed of diatomic molecules" of the oxygen in the atmosphere. In other words, the laws enable generally correct inferences and predictions to be made about distant ("unobservable") states of affairs.

E. At this point the system of predicates and their relations in laws has become sufficiently complex to allow for the possibility of internal misfits and even contradictions. This possibility arises in various ways. It may happen that some of the applications of a word in situations turn out not to satisfy the laws which are true of other applications of the word. In such a case, since degrees of physical similarity are not transitive, a reclassification may take place in which a particular law is preserved in a subclass more closely related by similarity, at the expense of the full range of situations of application which are relatively less similar. An example of this would be the application of the word "element" to water, which becomes incorrect in order to preserve the truth of a system of laws regarding "element," namely, that elements cannot be chemically dissociated into parts which are themselves elements, that elements always enter as a whole into compounds, that every substance is constituted by one or more elements, and so on. On the other hand, the range of applications may be widened in conformity with a law, so that a previously incorrect application becomes correct. For example, "mammal" is correctly applied to whales, whereas it was previously thought that "Mammals live only on land" was a well-entrenched law pro-

viding criteria for correct use of "mammal." In such a case it is not adequate to counter with the suggestion that the correct use of "mammal" is *defined* in terms of animals which suckle their young, for it is conceivable that if other empirical facts had been different, the classification in terms of habitat would have been more useful and comprehensive than that in terms of milk production. And in regard to the first example, it cannot be maintained that it is the *defining* characteristics of "element" that are preserved at the expense of its application to water, because of the conditions mentioned it is not clear that any particular one of them is, or ever has been, taken as *the* defining characteristic; and since the various characteristics are logically independent, it is empirically possible that some might be satisfied and not others. *Which* is preserved will always depend on what system of laws is most convenient, most coherent, and most comprehensive. But the most telling objection to the suggestion that correct application is decided by definition is of course the general point made at the end of section A that there is always a large number of predicates for which *no* definition in terms of necessary and sufficient conditions of application can be given. For these predicates it is possible that the primary recognition of, for example, a whale as being sufficiently similar to some fish to justify its inclusion in the class of fish may be explicitly overridden in the interests of preserving a particular set of laws.

Properly understood, the point developed in the last paragraph should lead to a far-reaching reappraisal of orthodoxy regarding the theory-observation distinction. To summarize, it entails that no feature in the total landscape of functioning of a descriptive predicate is exempt from modification under pressure from its surroundings. That any empirical law may be abandoned in the face of counterexamples is trite, but it becomes less trite when the functioning of every predicate is found to depend essentially on some laws or other and when it is also the case that any "correct" situation of application—*even that in terms of which the term was originally introduced*—may become incorrect in order to preserve a system of laws and other applications. It is in this sense that I shall understand the "theory dependence" or "theory-ladenness" of all descriptive predicates.

One possible objection to this account is easily anticipated. It is not a *conventionalist* account, if by that we mean that any law can be assured of truth by sufficiently meddling with the meanings of its predicates. Such a view does not take seriously the systematic character of laws, for it contemplates preservation of the truth of a given law irrespective of its coherence with the rest of the system, that is, the preservation of simplicity and other desirable internal characteristics of the system. Nor does it take account of the fact that not all primary recognitions of empirical similarity can be overridden in the interest of preserving a given law, for it is upon the existence of some such recognitions that the whole possibility of language with empirical reference rests. The present account on the other hand demands both that laws shall remain connected in an economical and convenient system and that at least most of its predicates shall remain applicable, that is, that they shall continue to depend for applicability upon the primary recognitions of similarity and difference in terms of which they were learned. That it is possible to have such a system with a given set of laws and predicates is not a convention but a fact of the empirical world. And although this account allows that *any* of the situations of correct application may change, it cannot allow that *all* should change, at least not all at once. Perhaps it would even be true to say that only a small proportion of them can change at any one time, although it is conceivable that over long periods of time most or all of them might come to change piecemeal. It is likely that almost all the terms used by the alchemists that are still in use have now changed their situations of correct use quite radically, even though at any one time chemists were preserving most of them while modifying others.

NOTES

1. A. J. Ayer, *Language, Truth, and Logic*, 2d ed. (London: Gollancz, 1946), p. 11; R. B. Braithwaite, *Scientific Explanation* (New York: Cambridge University Press, 1953), p. 8; R. Carnap, "The Methodological Character of Theoretical Concepts," in *Minnesota Studies in the Philosophy of Science*, I, ed. H. Feigl and M. Scriven (Minneapolis: University of Minnesota Press, 1956), p. 38; C. G. Hempel, "The Theoreti-

cian's Dilemma," in *Minnesota Studies in the Philosophy of Science*, II, ed. H. Feigl, M. Scriven, and G. Maxwell (Minneapolis: University of Minnesota Press, 1958), p. 41.

2. E. Nagel, *The Structure of Science* (New York: Harcourt, Brace & World, 1961).

3. It would be possible to give examples from sciences other than physics: "adaptation," "function," "intention," "behavior," "unconscious mind"; but the question whether these are theoretical terms in the sense here distinguished from observation terms is controversial; so is the question whether, if they are, they are eliminable from their respective sciences. These questions would take us too far afield.

4. The account which follows is by no means original; in fact, versions of it have been lying about in literature for so long that it may even sound trite. It may be useful to bring together here references to those discussions which I have found particularly helpful, but no claim is made for exhaustiveness, especially in regard to work published after 1966. My own more recent developments of some of the ideas of the present essay are to be found in my chapter, "Duhem, Quine, and a New Empiricism," in *Knowledge and Necessity*, Royal Institute of Philosophy Lectures, Vol. 3, 1968–69 (London: Macmillan, 1970), p. 191, and in the reference in note 18 below.

   Among general works in analytic philosophy which contain clues for a revised analysis of the observation language, the following should be specially mentioned: L. Wittgenstein, *Philosophical Investigations*, trans. G. E. M. Anscombe (London and New York: Macmillan, 1953); F. Waismann, "Verifiability," in *Logic and Language*, 1st ser., ed. A. G. N. Flew (Oxford: Blackwell, 1952), p. 117; F. Waismann, *Principles of Linguistic Philosophy*, ed. R. Harré (London: Macmillan; New York: St. Martin's Press, 1965); Peter Geach, *Mental Acts* (London: Routledge & Kegan Paul, 1957); D. W. Hamlyn, *The Psychology of Perception* (London: Routledge & Kegan Paul, 1957); P. F. Strawson, *Individuals, an Essay in Descriptive Metaphysics* (London: Methuen, 1959); Renford Bambrough, "Universals and Family Resemblances," *Proceedings of the Aristotelian Society*, 62 (1961), pp. 207–22.

   Early discussions of observation and and experiment which distinguish the observational from the theoretical, without taking the meaning of observation terms for granted, are to be found in P. Duhem, *The Aim and Structure of Physical Theory*, 2d ed., trans. P. Wiener (Princeton: Princeton University Press, 1954) pt. II, chaps. 4–6; N. R. Campbell,

Physics, the Elements (Cambridge University Press, 1920; reprint ed., *Foundations of Science*, New York: Dover, 1957), chap. 2. In both these works experimental laws are analyzed in terms of the mutually supporting parts of a network of relationships among observation terms. The question of the reference of descriptive terms and the analogy of the network is pursued in W. V. O. Quine, "Two Dogmas of Empiricism," *Philosophical Review*, 60 (1951), p. 20, reprinted in *From a Logical Point of View* (Cambridge, Mass.: Harvard University Press, 1953), p. 10, and in his *Word and Object* (Cambridge, Mass.: MIT Press, 1960), chaps., 1–3. Not dissimilar theses with regard to the interrelations of meaning and inference are stated in W. Sellars, "Some Reflections on Language-Games," *Philosophy of Science*, 21 (1954), p. 204. Also see A. Kaplan, "Definition and Specification of Meaning," *Journal of Philosophy*, 43 (1946), p. 281, and A. Kaplan and H. F. Schott, "A Calculus for Empirical Classes," *Methodos*, 3 (1951), p. 165. (I owe the last two of these references to Dr. N. Jardine.)

A radical reinterpretation of the positivist understanding of the observation language is implied in Karl R. Popper, *The Logic of Scientific Discovery* (London: Hutchinson, 1959), where all observation terms are analyzed as "dispositional," i.e., lose their alleged "direct reference." In Popper's "The Aim of Science," *Ratio*, 1 (1957), p. 24, the mutual adjustments and corrections of theoretical and observational laws are illustrated. Both aspects of Popper's work are taken further by P. K. Feyerabend, "An Attempt at a Realistic Interpretation of Experience," *Proceedings of the Aristotelian Society*, 58 (1957–58), p. 143, and "Explanation, Reduction, and Empiricism," in *Minnesota Studies in the Philosophy of Science*, III, ed. H. Feigl and G. Maxwell (Minneapolis: University of Minnesota Press, 1962), p. 28, and subsequent writings where he attacks the alleged "stability" and theory independence of the observation language and the alleged deductive relation between theory and observation. The notion of theory independence of the observation language is also attacked from various points of view in N. R. Hanson, *Patterns of Discovery* (New York: Cambridge University Press, 1958); M. Scriven, "Definitions, Explanations, and Theories," in *Minnesota Studies in the Philosophy of Science*, II, ed. H. Feigl, M. Scriven, and G. Maxwell, p. 99; W. Sellars, "The Language of Theories," in *Current Issues in the Philosophy of Science*, ed. H. Feigl and G. Maxwell (New York: Holt, Rinehart and Winston, 1961), p. 57; and H.

Putnam, "What Theories Are Not," in *Logic, Methodology, and Philosophy of Science*, ed. E. Nagel, P. Suppes, and A. Tarski (Stanford, Calif.: Stanford University Press, 1962), p. 240, and his "The Analytic and the Synthetic," in *Minnesota Studies in the Philosophy of Science*, III, ed. H. Feigl and G. Maxwell, p. 358. There are further discussions of these issues in May Brodbeck, "Explanation, Prediction, and 'Imperfect' Knowledge," in *Minnesota Studies in the Philosophy of Science*, III, ed. H. Feigl and G. Maxwell, p. 231; Mary Hesse, "Theories, Dictionaries and Observation," *British Journal of the Philosophy of Science*, 9 (1958), pp. 12, 128, and her "Gilbert and the Historians," *British Journal of the Philosophy of Science*, 11 (1960), pp. 1, 130; P. Alexander, "Theory-Construction and Theory-Testing," *British Journal of the Philosophy of Science*, 9 (1958), p. 29, and his *Sensationalism and Scientific Explanation* (London: Routledge & Kegan Paul, 1963); Dudley Shapere, "Space, Time, and Language—An Examination of Some Problems and Methods of the Philosophy of Science," in *Philosophy of Science*, The Delaware Seminar, 2, ed. Bernard Baumrin (New York: Interscience, 1963), p. 139, and his "Meaning and Scientific Change," in *Mind and Cosmos*, ed. R. G. Colodny (Pittsburgh: University of Pittsburgh Press, 1966), p. 41.

Interpretations of the history of science in terms of successive "conceptual frameworks" or "paradigms" may be found in Stephen Toulmin, *Foresight and Understanding* (London: Hutchinson, 1961); T. S. Kuhn, *The Structure of Scientific Revolutions* (Chicago: University of Chicago, 1962); R. Harré, *Matter and Method* (London: Macmillan, 1964); Mary Hesse, *Forces and Fields* (London: Nelson, 1961; and Totowa, N. J.: Littlefield, 1965). Related to this view of the primacy of the theoretical models in the development of science are various general arguments on the role of physical models and metaphors in the structure of theories. See, for example, N. R. Campbell, *Physics, the Elements*, chap. 6; G. Buchdahl, "Theory Construction: The Work of N. R. Campbell," *Isis*, 55 (1964), p. 151; E. Hutten, "On Semantics and Physics," *Proceedings of the Aristotelian Society*, 49 (1948–49), p. 115, and his "The Role of Models in Physics," *British Journal of the Philosophy of Science*, 4 (1953), p. 284; Mary Hesse, "Models in Physics," *British Journal of the Philosophy of Science*, 4 (1953), p. 198, and her *Models and Analogies in Science* (London: Sheed and Ward, 1963; and Notre Dame, Ind.: University of Notre Dame Press,

1966); R. B. Braithwaite, *Scientific Explanation,* chap. 4, and his "Models in the Empirical Sciences," in *Logic, Methodology and Philosophy of Science,* ed. E. Nagel, P. Suppes, and A. Tarski, p. 224; M. Black, *Models and Metaphors* (Ithaca, N.Y.: Cornell University Press, 1962), chaps. 3, 13; E. Nagel, *The Structure of Science,* chap. 6; R. Harré, *Theories and Things* (London: Sheed and Ward, 1961); C. M. Turbayne, *The Myth of Metaphor* (New Haven and London: Yale University Press, 1962); D. Schon, *The Displacement of Concepts* (London: Tavistock Publications, 1963); P. Achinstein, "Theoretical Terms and Partial Interpretations," *British Journal of the Philosophy of Science,* 14 (1963), p. 89, and his "Theoretical Models," ibid., 16 (1965), p. 102; Marshall Spector, "Theory and Observation," ibid., 17 (1966), pp. 1, 89; E. McMullin, "What Do Physical Models Tell Us?" in *Logic, Methodology and Philosophy of Science,* ed. B. van Rootselaar and J. F. Stahl (Amsterdam: North Holland Publishing Co., 1968), p. 385.

   In *Concepts of Science* (Baltimore: Johns Hopkins Press, 1968), Peter Achinstein develops an analysis of meaning of the observation and theory languages, and a critique of the observation-theory distinction, from a point of view similar to that presented here.

5. See, for example, Alan Gauld, "Could a Machine Perceive?" *British Journal of the Philosophy of Science,* 17 (1964), p. 44, and especially p. 53.

   The a priori character of this account of descriptive language has been challenged by D. Davidson and by N. Chomsky. Davidson, in "The-ories of Meaning and Learnable Languages," *Logic, Methodology and Philosophy of Science,* ed. Y. Bar-Hillel (Amsterdam: North Holland Publishing Co., 1965), p. 383, claims that there is no need for a descriptive predicate to be learned in the presence of the object to which it is properly applied, since, for example, it might be learned in "a skillfully faked environment" (p. 386). This possibility does not, however, constitute an objection to the thesis that it must be learned in *some* empirical situation and that this situation must have some similarity with those situations in which the predicate is properly used. Chomsky, on the other hand ("Quine's Empirical Assumptions," *Synthese,* 19 [1968], p. 53) attacks what he regards as Quine's "Humean theory" of language acquisition by stimulus and conditioned response. But the necessity of the *similarity* condition for language learning does not depend on the particular empirical mechanism of learning. Learning by patterning the environment in terms of a set of "innate ideas" would depend equally upon subsequent application of the same pattern to similar features of the environment. Moreover, "similar" cannot just be *defined as* "properly ascribed the same descriptive predicate in the same language community," for, as is argued below, similarity is a matter of degree and is a nontransitive relation, whereas "properly ascribed the same descriptive predicate" is not. The two terms cannot, therefore, be synonymous. See also Quine's reply to Chomsky, ibid., p. 274.

6. Karl R. Popper, *The Logic of Scientific Discovery,* Appendix X, p. 422.

24

N. R. HANSON

Observation

Were the eye not attuned to the Sun,
The Sun could never be seen by it.

GOETHE[1]

Consider two microbiologists. They look at a prepared slide; when asked what they see, they may give different answers. One sees in the cell before him a cluster of foreign matter: it is an artifact, a coagulum resulting from inadequate staining techniques. This clot has no more to do with the cell, *in vivo*, than the scars left on it by the archaeologist's spade have to do with the original shape of some Grecian urn. The other biologist identifies the clot as a cell organ, a 'Golgi body'. As for techniques, he argues: 'The standard way of detecting a cell organ is by fixing and staining. Why single out this one technique as producing artifacts, while others disclose genuine organs?'

The controversy continues.[2] It involves the whole theory of microscopical technique; nor is it an obviously experimental issue. Yet it affects what scientists say they see. Perhaps there is a sense in which two such observers do not see the same thing, do not begin from the same data, though their eyesight is normal and they are visually aware of the same object.

Imagine these two observing a Protozoon—*Amoeba*. One sees a one-celled animal, the other a non-celled animal. The first sees *Amoeba* in all its analogies with different types of single cells: liver cells, nerve cells, epithelium cells. These have a wall, nucleus, cytoplasm, etc. Within this class *Amoeba* is distinguished only by its independence. The other, however, sees *Amoeba*'s homology not with single cells, but with whole animals. Like all animals *Amoeba* ingests its food, digests and assimilates it. It excretes, reproduces and is mobile—more like a complete animal than an individual tissue cell.

This is not an experimental issue, yet it can affect experiment. What either man regards as significant questions or relevant data can be determined by whether he stresses the first or the last term in 'unicellular animal'.[3]

Some philosophers have a formula ready for such situations: 'Of course they see the same thing. They make the same observation since they begin from the same visual data. But they interpret what they see differently. They construe the evidence in different ways.'[4] The task is then to show how these data are molded by different theories or interpretations or intellectual constructions.

Considerable philosophers have wrestled with this task. But in fact the formula they start from is too simple to allow a grasp of the nature of observation within physics. Perhaps the scientists cited above do not begin their inquiries from the same data, do not make the same observations, do not even see the same thing? Here many concepts run together. We must proceed carefully, for wherever it makes sense to say that two scientists looking at *x* do not see the same thing, there must always be a prior sense in which they do see the same thing. The issue is, then, 'Which of these senses is most illuminating for the understanding of observational physics?'

These biological examples are too complex. Let us consider Johannes Kepler: imagine him on a hill watching the dawn. With him is Tycho Brahe. Kepler

Patterns of Discovery (New York: Cambridge University Press, 1958), pp. 4–19. Reprinted with permission of the publisher.

regarded the sun as fixed: it was the earth that moved. But Tycho followed Ptolemy and Aristotle in this much at least: the earth was fixed and all other celestial bodies moved around it. *Do Kepler and Tycho see the same thing in the east at dawn?*

We might think this an experimental or observational question, unlike the questions 'Are there Golgi bodies?' and 'Are Protozoa one-celled or non-celled?'. Not so in the sixteenth and seventeenth centuries. Thus Galileo said to the Ptolemaist ' . . . neither Aristotle nor you can prove that the earth is *de facto* the center of the universe . . .'.[5] 'Do Kepler and Tycho see the same thing in the east at dawn?' is perhaps not a *de facto* question either, but rather the beginning of an examination of the concepts of seeing and observation.

The resultant discussion might run:

'Yes, they do.'

'No, they don't.'

'Yes, they do!'

'No, they don't!' . . .

That this is possible suggests that there may be reasons for both contentions.[6] Let us consider some points in support of the affirmative answer.

The physical processes involved when Kepler and Tycho watch the dawn are worth noting. Identical photons are emitted from the sun; these traverse solar space, and our atmosphere. The two astronomers have normal vision; hence these photons pass through the cornea, aqueous humor, iris, lens, and vitreous body of their eyes in the same way. Finally their retinas are affected. Similar electro-chemical changes occur in their selenium cells. The same configuration is etched on Kepler's retina as on Tycho's. So they see the same thing.

Locke sometimes spoke of seeing in this way: a man sees the sun if his is a normally formed retinal picture of the sun. Dr. Sir W. Russell Brain speaks of our retinal sensations as indicators and signals. Everything taking place behind the retina is, as he says, 'an intellectual operation based largely on non-visual experience . . .'.[7] What we *see* are the changes in the *tunica retina*. Dr. Ida Mann regards the macula of the eye as itself 'seeing details in bright light', and the rods as 'seeing approaching motor-cars'. Dr. Agnes Arber speaks of the eye as itself seeing.[8] Often, talk of seeing can direct attention to the retina. Normal people are distinguished from those

for whom no retinal pictures can form: we may say of the former that they can see whilst the latter cannot see. Reporting when a certain red dot can be seen may supply the occulist with direct information about the condition of one's retina.[9]

This need not be pursued, however. These writers speak carelessly: seeing the sun is not seeing retinal pictures of the sun. The retinal images which Kepler and Tycho have are four in number, inverted and quite tiny.[10] Astronomers cannot be referring to these when they say they see the sun. If they are hypnotized, drugged, drunk or distracted they may not see the sun, even though their retinas register its image in exactly the same way as usual.

Seeing is an experience. A retinal reaction is only a physical state—a photochemical excitation. Physiologists have not always appreciated the differences between experiences and physical states.[11] People, not their eyes, see. Cameras, and eyeballs, are blind. Attempts to locate within the organs of sight (or within the neurological reticulum behind the eyes) some nameable called 'seeing' may be dismissed. That Kepler and Tycho do, or do not, see the same thing cannot be supported by reference to the physical states of their retinas, optic nerves or visual cortices: there is more to seeing than meets the eyeball.

Naturally, Tycho and Kepler see the same physical object. They are both visually aware of the sun. If they are put into a dark room and asked to report when they see something—anything at all—they may both report the same object at the same time. Suppose that the only object to be seen is a certain lead cylinder. Both men see the same thing: namely this object—whatever it is. It is just here, however, that the difficulty arises, for while Tycho sees a mere pipe, Kepler will see a telescope, the instrument about which Galileo has written to him.

Unless both are visually aware of the same object there can be nothing of philosophical interest in the question whether or not they see the same thing. Unless they both see the sun in this prior sense our question cannot even strike a spark.

Nonetheless, both Tycho and Kepler have a common visual experience of some sort. This experience perhaps constitutes their seeing the same thing. Indeed, this may be a seeing logically more basic than anything expressed in the pronouncement 'I see the sun' (where each means something different by 'sun'). If what they meant by the word 'sun' were the only clue, then Tycho and Kepler could not be seeing the same thing, even though they were gazing at the same object.

If, however, we ask, not 'Do they see the same thing?' but rather 'What is it that they both see?', an unambiguous answer may be forthcoming. Tycho and Kepler are both aware of a brilliant yellow-white disc in a blue expanse over a green one. Such a 'sense-datum' picture is single and uninverted. To be unaware of it is not to have it. Either it dominates one's visual attention completely or it does not exist.

If Tycho and Kepler are aware of anything visual, it must be of some pattern of colors. What else could it be? We do not touch or hear with our eyes, we only take in light.[12] This private pattern is the same for both observers. Surely if asked to sketch the contents of their visual fields they would both draw a kind of semicircle on a horizon-line.[13] They say they see the sun. But they do not see every side of the sun at once; so what they really see is discoid to begin with. It is but a visual aspect of the sun. In any single observation the sun is a brilliantly luminescent disc, a penny painted with radium.

So something about their visual experiences at dawn is the same for both: a brilliant yellow-white disc centered between green and blue color patches. Sketches of what they both see could be identical—congruent. In this sense Tycho and Kepler see the same things at dawn. The sun appears to them in the same way. The same view, or scene, is presented to them both.

In fact, we often speak in this way. Thus the account of a recent solar eclipse:[14] 'Only a thin crescent remains; white light is now completely obscured; the sky appears a deep blue, almost purple, and the landscape is a monochromatic green . . . there are the flashes of light on the disc's circumference and now the brilliant crescent to the left . . . .' Newton writes in a similar way in the *Opticks:* 'These Arcs at their first appearance were of a violet and blue Color, and between them were white Arcs of Circles, which . . . became a little tinged in their inward Limbs with red and yellow . . . .'[15] Every physicist employs the language of lines, color patches, appearances, shadows. In so far as two normal observers use this language of the same event, they begin from the same data: they are making the same observation. Differences

FIGURE 1

between them must arise in the interpretations they put on these data.

Thus, to summarize, saying that Kepler and Tycho see the same things at dawn just because their eyes are similarly affected is an elementary mistake. There is a difference between a physical state and a visual experience. Suppose, however, that it is argued as above—that they see the same things because they have the same sense-datum experience. Disparities in their accounts arise in *ex post facto* interpretations of what is seen, not in the fundamental visual data. If this is argued, further difficulties soon obtrude.

Normal retinas and cameras are impressed similarly by fig. 1.[16] Our visual sense-data will be the same too. If asked to draw what we see, most of us will set out a configuration like fig. 1.

Do we all see the same thing?[17] Some will see a perspex cube viewed from below. Others will see it from above. Still others will see it as a kind of polygonally cut gem. Some people see only crisscrossed lines in a plane. It may be seen as a block of ice, an aquarium, a wire frame for a kite—or any of a number of other things.

Do we, then, all see the same thing? If we do, how can these differences be accounted for?

Here the 'formula' re-enters: 'These are different *interpretations* of what all observers see in common. Retinal reactions to fig. 1 are virtually identical; so too are our visual sense-data, since our drawings of what we see will have the same content. There is no place in the seeing for these differences, so they must lie in the interpretations put on what we see.'

This sounds as if I do two things, not one, when I see boxes and bicycles. Do I put different interpretations on fig. 1 when I see it now as a box from below, and now as a cube from above? I am aware of no such thing. I mean no such thing when I report that the box's perspective has snapped back into the page.[18] If I do not mean this, then the concept of seeing which is natural in this connection does not designate two diaphanous components, one optical, the other interpretative. Fig. 1 is simply seen now as a box from below, now as a cube from above; one does not first soak up an optical pattern and then clamp an interpretation on it. Kepler and Tycho just see the sun. That is all. That is the way the concept of seeing works in this connection.

'But,' you say, 'seeing fig. 1 first as a box from below, then as a cube from above, involves interpreting the lines differently in each case.' Then for you and me to have a different interpretation of fig. 1 just *is* for us to see something different. This does not mean we see the same thing and then interpret it differently. When I suddenly exclaim 'Eureka—a box from above', I do not refer simply to a different interpretation. (Again, there is a logically prior sense in which seeing fig. 1 as from above and then as from below is seeing the same thing differently, i.e., being aware of the same diagram in different ways. We can refer just to this, but we need not. In this case we do not.)

Besides, the word 'interpretation' is occasionally useful. We know where it applies and where it does not. Thucydides presented the facts objectively; Herodotus put an interpretation on them. The word does not apply to everything—it has a meaning. Can interpreting always be going on when we see? Sometimes, perhaps, as when the hazy outline of an agricultural machine looms up on a foggy morning and, with effort, we finally identify it. Is this the 'interpretation' which is active when bicycles and boxes are clearly seen? Is it active when the perspective of fig. 1 snaps into reverse? There was a time when Herodotus was half-through with his interpretation of the Graeco-Persian wars. Could there be a time when one is half-through interpreting fig. 1 as a box from above, or as anything else?

'But the interpretation takes very little time—it is instantaneous.' Instantaneous interpretation hails from the Limbo that produced unsensed sensibilia, unconscious inference, incorrigible statements, negative facts and *Objektive*. These are ideas which philosophers force on the world to preserve some pet epistemological or metaphysical theory.

Only in contrast to 'Eureka' situations (like perspective reversals, where one cannot interpret the data) is it clear what is meant by saying that though Thucydides could have put an interpretation on history, he did not. Moreover, whether or not a historian is advancing an interpretation is an empirical question: we know what would count as evidence one way or the other. But whether we are employing an interpretation when we see fig. 1 in a certain way is not empirical. What could count as evidence? In no ordinary sense of 'interpret' do I interpret fig. 1 differently when its perspective reverses for me. If there is some extraordinary sense of word it is not clear, either in ordinary language, or in extraordinary (philosophical) language. To insist that different reactions to fig. 1 *must* lie in the interpretations put on a common visual experience is just to reiterate (without reasons) that the seeing of *x must* be the same for all observers looking at *x*.

'But "I see the figure as a box" means: I am having a particular visual experience which I always have when I interpret the figure as a box, or when I look at a box . . . .' ' . . . if I meant this, I ought to know it. I ought to be able to refer to the experience directly and not only indirectly . . . .'[19]

Ordinary accounts of the experiences appropriate to fig. 1 do not require visual grist going into an intellectual mill: theories and interpretations are 'there' in the seeing from the outset. How can interpretations 'be there' in the seeing? How is it possible to see an object according to an interpretation? 'The question represents it as a queer fact; as if something were being forced into a form it did not really fit. But no squeezing, no forcing took place here.'[20] . . .

A trained physicist could see one thing in fig. 2: an X-ray tube viewed from the cathode. Would Sir Lawrence Bragg and an Eskimo baby see the same thing when looking at an X-ray tube? Yes, and no. Yes—they are visually aware of the same object. No—the ways in which they are visually aware are profoundly different. Seeing is not only the having of a visual experience; it is also the way in which the visual experience is had.

At school the physicist had gazed at this glass-and-metal instrument. Returning now, after years in university and research, his eye lights upon the same object once again. Does he see the same thing now as he did then? Now he sees the instrument in terms of

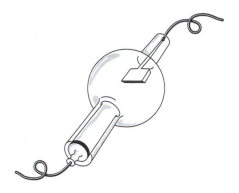

FIGURE 2

electrical circuit theory, thermodynamic theory, the theories of metal and glass structure, thermionic emission, optical transmission, refraction, diffraction, atomic theory, quantum theory and special relativity.

Contrast the freshman's view of college with that of his ancient tutor. Compare a man's first glance at the motor of his car with a similar glance ten exasperating years later.

'Granted, one learns all these things', it may be countered, 'but it all figures in the interpretation the physicist puts on what he sees. Though the layman sees exactly what the physicist sees, he cannot interpret it in the same way because he has not learned so much.'

Is the physicist doing more than just seeing? No; he does nothing over and above what the layman does when he sees an X-ray tube. What are you doing over and above reading these words? Are you interpreting marks on a page? When would this ever be a natural way of speaking? Would an infant see what you see here, when you see words and sentences and he sees but marks and lines? One does nothing beyond looking and seeing when one dodges bicycles, glances at a friend, or notices a cat in the garden.

'The physicist and the layman see the same thing', it is objected, 'but they do not make the same thing of it.' The layman can make nothing of it. Nor is that just a figure of speech. I can make nothing of the Arab word for *cat*, though my purely visual impressions may be indistinguishable from those of the Arab who can. I must learn Arabic before I can see what he sees. The layman must learn physics before he can see what the physicist sees.

If one must find a paradigm case of seeing it would be better to regard as such not the visual apprehension of color patches but things like seeing what time it is, seeing what key a piece of music is written in, and seeing whether a wound is septic.[21]

Pierre Duhem writes:

Enter a laboratory; approach the table crowded with an assortment of apparatus, an electric cell, silk-covered copper wire, small cups of mercury, spools, a mirror mounted on an iron bar; the experimenter is inserting into small openings the metal ends of ebony-headed pins; the iron oscillates, and the mirror attached to it throws a luminous band upon a celluloid scale; the forward-backward motion of this spot enables the physicist to observe the minute oscillations of the iron bar. But ask him what he is doing. Will he answer 'I am studying the oscillations of an iron bar which carries a mirror?' No, he will say that he is measuring the electric resistance of the spools. If you are astonished, if you ask him what his words mean, what relation they have with the phenomena he has been observing and which you have noted at the same time as he, he will answer that your question requires a long explanation and that you should take a course in electricity.[22]

The visitor must learn some physics before he can see what the physicist sees. Only then will the context throw into relief those features of the objects before him which the physicist sees as indicating resistance.

This obtains in all seeing. Attention is rarely directed to the space between the leaves of a tree, save when a Keats brings it to our notice.[23] (Consider also what was involved in Crusoe's seeing a vacant space in the sand as a footprint.) Our attention most naturally rests on objects and events which dominate the visual field. What a blooming, buzzing, undifferentiated confusion visual life would be if we all arose tomorrow without attention capable of dwelling only on what had heretofore been overlooked.[24]

The infant and the layman can see: they are not blind. But they cannot see what the physicist sees; they are blind to what he sees.[25] We may not hear that the oboe is out of tune, though this will be painfully obvious to the trained musician. (Who, incidentally, will not hear the tones and *interpret* them as being out of tune, but will simply hear the oboe to be out of tune.[26] We simply see what time it is; the surgeon simply sees a wound to be septic; the physicist sees the X-ray tube's anode overheating.) The elements of the visitor's visual field, though identical with those of the physicist, are not organized for him as for the physicist; the same lines, colors, shapes are apprehended by both, but not in the same way. There are indefinitely many ways in which a constellation of lines, shapes, patches, may be seen. *Why* a visual pattern is seen differently is a question for psychology, but *that* it may be seen differently is important in any examination of the concepts of seeing and observation. Here, as Wittgenstein might have said, the psychological is a symbol of the logical.

You see a bird, I see an antelope; the physicist sees an X-ray tube, the child a complicated lamp bulb; the microscopist sees coelenterate mesoglea, his new student sees only a gooey, formless stuff. Tycho and Simplicius see a mobile sun, Kepler and Galileo see a static sun.[27]

It may be objected, 'Everyone, whatever his state of knowledge, will see fig. 1 as a box or cube, viewed as from above or as from below.' True; almost everyone, child, layman, physicist, will see the figure as box-like one way or another. But could such observations be made by people ignorant of the construction of box-like objects? No. This objection only shows that most of us—the blind, babies, and dimwits excluded—have learned enough to be able to see this figure as a three-dimensional box. This reveals something about the sense in which Simplicius and Galileo do see the same thing (which I have never denied): they both see a brilliant heavenly body. The schoolboy and the physicist both see that the X-ray tube will smash if dropped. Examining how observers see different things in *x* marks something important about their seeing the same thing when looking at *x*. If seeing different things involves having different knowledge and theories about *x*, then perhaps the sense in which they see the same thing involves their sharing knowledge and theories about *x*. Bragg and the baby share no knowledge of X-ray tubes. They see the same thing only in that if they are looking at *x* they are both having some visual experience of it. Kepler and Tycho agree on

more: they see the same thing in a stronger sense. Their visual fields are organized in much the same way. Neither sees the sun about to break out in a grin, or about to crack into ice cubes. (The baby is not 'set' even against these eventualities.) Most people today see the same thing at dawn in an even stronger sense: we share much knowledge of the sun. Hence Tycho and Kepler see different things, and yet they see the same thing. That these things can be said depends on their knowledge, experience, and theories. . . .

The elements of [Kepler's and Tycho's] experiences are identical; but their conceptual organization is vastly different. Can their visual fields have a different organization? Then they can see different things in the east at dawn.

It is the sense in which Tycho and Kepler do not observe the same thing which must be grasped if one is to understand disagreements within microphysics. Fundamental physics is primarily a search for intelligibility—it is philosophy of matter. Only secondarily is it a search for objects and facts (though the two endeavors are as hand and glove). Microphysicists seek new modes of conceptual organization. If that can be done the finding of new entities will follow. Gold is rarely discovered by one who has not got the lay of the land.

To say that Tycho and Kepler, Simplicius and Galileo, Hooke and Newton, Priestley and Lavoisier, Soddy and Einstein, De Broglie and Born, Heisenberg and Bohm all make the same observations but use them differently is too easy.[28] It does not explain controversy in research science. Were there no sense in which they were different observations they could not be used differently. This may perplex some: that researchers sometimes do not appreciate data in the same way is a serious matter. It is important to realize, however, that sorting out differences about data, evidence, observation, may require more than simply gesturing at observable objects. It may require a comprehensive reappraisal of one's subject matter. This may be difficult, but it should not obscure the fact that nothing less than this may do.

There is a sense, then, in which seeing is a 'theory-laden' undertaking. Observation of *x* is shaped by prior knowledge of *x*. Another influence on observations rests in the language or notation used to express what we know, and without which there would be little we could recognize as knowledge.

NOTES

1.   Wär' nicht das Auge sonnenhaft,
     Die Sonne Könnt' es nie erblicken;
     Goethe, *Zahme Xenien* (Werke, Weimar, 1887–1918), Bk. 3, 1805.

2.   Cf. the papers by Baker and Gasatonby in *Nature*, 1949–present.

3.   This is not a *merely* conceptual matter, of course. Cf. Wittgenstein, *Philosophical Investigations* (Blackwell, Oxford, 1953), p. 196.

4.   a. G. Berkeley, *Essay Towards a New Theory of Vision* (in *Works,* vol. I (London, T. Nelson, 1948–56)), pp. 51 ff.

     b. James Mill, *Analysis of the Phenomena of the Human Mind* (Longmans, London, 1869), vol. I, p. 97.

     c. J. Sully, *Outlines of Psychology* (Appleton, New York, 1885).

     d. William James, *The Principles of Psychology* (Holt, New York, 1890–1905), vol. II, pp. 4, 78, 80 and 81; vol. I, p. 221.

     e. A. Schopenhauer, *Satz vom Grunde* (in *Sämmtliche Werke,* Liepzig, 1888), ch. IV.

     f. H. Spencer, *The Principles of Psychology* (Appleton, New York, 1897), vol. IV, chs. IX, X.

     g. E. von Hartmann, *Philosophy of the Unconscious* (K. Paul, London, 1931), B, chs. VII, VIII.

     h. W. M. Wundt, *Vorlesungen über die Menschen und Thierseele* (Voss, Hamburg, 1892), IV, XIII.

     i. H. L. F. von Helmholtz, *Handbuch der Physiologischen Optik* (Leipzig, 1867), pp. 430, 447.

     j. A. Binet, *La psychologie du raisonnement, recherches expérimentales par l'hypnotisme* (Alcan, Paris, 1886), chs. III, V.

     k. J. Grote, *Exploratorio Philosophica* (Cambridge, 1900), vol. II, pp. 201 ff.

     l. B. Russell, in *Mind* (1913), p. 76. *Mysticism and Logic* (Longmans, New York, 1918), p. 209. *The Problems of Philosophy* (Holt, New York, 1912), pp. 73, 92, 179, 203.

     m. Dawes Hicks, *Arist. Soc. Sup.* vol. II (1919), pp. 76–8.

     n. G. F. Stout, *A Manual of Psychology* (Clive, London, 1907, 2nd ed.), vol. II, 1 and 2, pp. 324, 561–4.

     o. A. C. Ewing, *Fundamental Questions of Philosophy* (New York, 1951), pp. 45 ff.

     p. G. W. Cunningham, *Problems of Philosophy* (Holt, New York, 1924), pp. 96–7.

5.   Galileo, *Dialogue Concerning the Two Chief World Systems* (California, 1953), 'The First Day', p. 33.

6.   '"Das ist doch kein Sehen!"—"Das ist doch ein Sehen!" Beide müssen sich begrifflich rechtfertigen lassen' (Wittgenstein, *Phil. Inv.* p. 203).

7.   Brain, *Recent Advances in Neurology* (with Strauss) (London, 1929), p. 88. Compare Helmholtz: 'The sensations are signs to our consciousness, and it is the task of our intelligence to learn to understand their meaning' (*Handbuch der Physiologischen Optik* (Leipzig, 1867), vol. III, p. 433.

     See also Husserl, 'Ideen zu einer Reinen Phaenomenologie', in *Jahrbuch für Philosophie,* vol. I (1913), pp. 75, 79, and Wagner's *Handwörterbuch der Physiologie,* vol. III, section I (1846), p. 183.

8.   Mann, *The Science of Seeing* (London, 1949), pp. 48–9. Arber, *The Mind and the Eye* (Cambridge, 1954). Compare Müller: 'In any field of vision, the retina sees only itself in its spatial extension during a state of affection. it perceives itself as . . . etc.' (*Zur vergleichenden Physiologie des Gesichtesinnes des Menschen und der Theire* (Leipzig, 1826), p. 54).

9.   Kolin: 'An astigmatic eye when looking at millimeter paper can accommodate to see sharply either the vertical lines or the horizontal lines' (*Physics* (New York, 1950), pp. 570 ff.).

10. Cf. Whewell, *Philosophy of Discovery* (London, 1860), 'The Parodoxes of Vision'.

11. Cf. e. g. J. Z. Young, *Doubt and Certainty in Science* (Oxford, 1951, The Reith Lectures), and Gray Walter's article in *Aspects of Form,* ed. by L. L. Whyte (London, 1953). Compare Newton: 'Do not the Rays of Light in falling upon the bottom of the Eye excite Vibrations in the Tunica Retina? Which Vibrations, being propogated along the solid Fibres of the Nerves into the Brain, cause the Sense of seeing' (*Optiks* (London, 1769), Bk. III, part I).

12. 'Rot und grün kann ich nur sehen, aber nicht hören' (Wittgenstein, *Phil. Inv.* p. 209).

13. Cf. 'An appearance is the same whenever the same eye is affected in the same way' (Lambert, *Photometria* (Berlin, 1760)); 'We are justified, when different perceptions offer themselves to us, to infer that the underlying real conditions are different' (Helmholtz, *Wissenschaftliche Abhandlungen* (Leipzig, 1882), vol. II, p. 656), and Hertz: 'We form for ourselves images or symbols of the external objects; the manner in which we form them is such that the logically necessary (*denknotwendigen*) consequences of the images in thought are invariably the images of materially necessary (*naturnotwendigen*) consequences of the

corresponding objects' (*Principles of Mechanics* (London, 1889), p. 1).

Broad and Price make depth a feature of the private visual pattern. However, Weyl (*Philosphy of Mathematics and Natural Science* (Princeton, 1949), p. 125) notes that a single eye perceives qualities spread out in a *two*-dimensional field, since the latter is dissected by any one-dimensional line running through it. But our conceptual difficulties remain even when Kepler and Tycho keep one eye closed.

Whether or not two observers are having the same visual sense-data reduces directly to the question of whether accurate pictures of the contents of their visual fields are identical, or differ in some detail. We can then discuss the publicly observable pictures which Tycho and Kepler draw of what they see, instead of those private, mysterious entities locked in their visual consciousness. The accurate picture and sense-datum must be identical; how could they differ?

14. From the B. B. C. report, 30 June 1954.

15. Newton, *Opticks,* Bk. II, part I. The writings of Claudius Ptolemy sometimes read like a phenomenalist's textbook. Cf. e.g. *The Almagest* (Venice, 1515), VI, section II, 'On the Directions in the Eclipses', 'When it touches the shadow's circle from within', 'When the circles touch each other from without'. Cf. also VII and VIII, IX (section 4). Ptolemy continually seeks to chart and predict 'the appearances'—the points of light on the celestial globe. *The Almagest* abandons any attempt to explain the machinery behind these appearances.

Cf. Pappus: 'The (circle) dividing the milk-white portion which owes its colour to the sun, and the portion which has the ashen colour natural to the moon itself is indistinguishable from a great circle' (*Mathematical Collection* (Hultsch, Berlin and Leipzig, 1864), pp. 554–60).

16. This famous illusion dates from 1832, when L. A. Necker, the Swiss naturalist, wrote a letter to Sir David Brewster describing how when certain rhomboidal crystals were viewed on end the perspective could shift in the way now familiar to us. Cf. *Phil. Mag.* III, no. I (1832), 329–37, especially p. 336. It is important to the present argument to note that this observational phenomenon began life not as a psychologist's trick, but at the very frontiers of observational science.

17. Wittgenstein answers: 'Denn wir sehen eben wirklich zwei verschiedene Tatsachen' (*Tractatus,* 5·5423).

18. 'Auf welche Vorgänge spiele ich an?' (Wittgenstein, *Phil. Inv.* p. 214).

19. *Ibid.* p. 194 (top).

20. *Ibid.* p. 200.

21. Often 'What do you see?' only poses the question 'Can you identify the object before you?'. This is calculated more to test one's knowledge than one's eyesight.

22. Duhem, *La théorie physique* (Paris, 1914), p. 218.

23. Chinese poets felt the significance of 'negative features' like the hollow of a clay vessel or the central vacancy of the hub of a wheel (cf. Waley, *Three Ways of Thought in Ancient China* (London, 1939), p. 155).

24. Infants are indiscriminate; they take in spaces, relations, objects, and events as being of equal value. They still must learn to organize their visual attention. The camera-clarity of their visual reactions is not by itself sufficient to differentiate elements in their visual fields. Contrast Mr. W. H. Auden who recently said of the poet that he is 'bombarded by a stream of varied sensations which would drive him mad if he took them all in. It is impossible to guess how much energy we have to spend every day in not-seeing, not-hearing, not-smelling, not-reacting.'

25. Cf. 'He was blind to the *expression* of a face. Would his eyesight on that account be defective?' (Wittgenstein, *Phil. Inv.* p. 210) and 'Because they seeing see not; and hearing they hear not, neither do they understand' (Matt. xiii. 10—13).

26. 'Es hört doch jeder nur, was er versteht' (Goethe, *Maxims* (*Werke,* Weimar, 1887–1918)).

27. Against this Professor H. H. Price has argued: 'Surely it appears to both of them to be rising, to be moving upwards, across the horizon. . . they both see a moving sun: they both see a round bright body which appears to be rising.' Philip Frank retorts: 'Our sense observation shows only that in the morning the distance between horizon and the sun is increasing, but it does not tell us whether the sun is ascending or the horizon is descending. . . ' (*Modern Science and its Philosophy* (Harvard, 1949), p. 231). Precisely. For Galileo and Kepler the horizon drops; for Simplicius and Tycho the sun rises. This is the difference Price misses, and which is central to this essay.

28. This parallels the too-easy epistemological doctrine that all normal observers see the same things in *x*, but interpret them differently.

25

THOMAS S. KUHN

The Structure of Scientific Revolutions

THE ROUTE TO NORMAL SCIENCE

In this essay, 'normal science' means research firmly based upon one or more past scientific achievements, achievements that some particular scientific community acknowledges for a time as supplying the foundation for its further practice. Today such achievements are recounted, though seldom in their original form, by science textbooks, elementary and advanced. These textbooks expound the body of accepted theory, illustrate many or all of its successful applications, and compare these applications with exemplary observations and experiments. Before such books became popular early in the nineteenth century (and until even more recently in the newly matured sciences), many of the famous classics of science fulfilled a similar function. Aristotle's *Physica*, Ptolemy's *Almagest*, Newton's *Principia* and *Opticks*, Franklin's *Electricity*, Lavoisier's *Chemistry*, and Lyell's *Geology*—these and many other works served for a time implicitly to define the legitimate problems and methods of a research field for succeeding generations of practitioners. They were able to do so because they shared two essential characteristics. Their achievement was sufficiently unprecedented to attract an enduring group of adherents away from competing modes of scientific activity. Simultaneously, it was sufficiently open-ended to leave all sorts of problems for the redefined group of practitioners to resolve.

Achievements that share these two characteristics I shall henceforth refer to as 'paradigms,' a term that relates closely to 'normal science.' By choosing it, I mean to suggest that some accepted examples of actual scientific practice—examples which include law, theory, application, and instrumentation together—provide models from which spring particular coherent traditions of scientific research. These are the traditions which the historian describes under such rubrics as 'Ptolemaic astronomy' (or 'Copernican'), 'Aristotelian dynamics' (or 'Newtonian'), 'corpuscular optics' (or 'wave optics'), and so on. The study of paradigms, including many that are far more specialized than those named illustratively above, is what mainly prepares the student for membership in the particular scientific community with which he will later practice. Because he there joins men who learned the bases of their field from the same concrete models, his subsequent practice will seldom evoke overt disagreement over fundamentals. Men whose research is based on shared paradigms are committed to the same rules and standards for scientific practice. That commitment and the apparent consensus it produces are prerequisites for normal science, i.e., for the genesis and continuation of a particular research tradition.

Because in this essay the concept of a paradigm will often substitute for a variety of familiar notions, more will need to be said about the reasons for its introduction. Why is the concrete scientific achievement, as a locus of professional commitment, prior to the various concepts, laws, theories, and points of view that may be abstracted from it? In what sense is the shared paradigm a fundamental unit for the student of scientific development, a unit that cannot be fully reduced to logically atomic components

The Structure of Scientific Revolutions (Chicago: University of Chicago Press, 1970). Reprinted with permission from The University of Chicago Press.

which might function in its stead? Answers to these questions and to others like them will prove basic to an understanding both of normal science and of the associated concept of paradigms. That more abstract discussion will depend, however, upon a previous exposure to examples of normal science or of paradigms in operation. In particular, both these related concepts will be clarified by noting that there can be a sort of scientific research without paradigms, or at least without any so unequivocal and so binding as the ones named above. Acquisition of a paradigm and of the more esoteric type of research it permits is a sign of maturity in the development of any given scientific field.

If the historian traces the scientific knowledge of any selected group of related phenomena backward in time, he is likely to encounter some minor variant of a pattern here illustrated from the history of physical optics. Today's physics textbooks tell the student that light is photons, i.e., quantum-mechanical entities that exhibit some characteristics of waves and some of particles. Research proceeds accordingly, or rather according to the more elaborate and mathematical characterization from which the usual verbalization is derived. That characterization of light is, however, scarcely half a century old. Before it was developed by Planck, Einstein, and others early in this century, physics texts taught that light was transverse wave motion, a conception rooted in a paradigm that derived ultimately from the optical writings of Young and Fresnel in the early nineteenth century. Nor was the wave theory the first to be embraced by almost all practitioners of optical science. During the eighteenth century the paradigm for this field was provided by Newton's *Opticks*, which taught that light was material corpuscles. At that time physicists sought evidence, as the early wave theorists had not, of the pressure exerted by light particles impinging on solid bodies.[1]

These transformations of the paradigms of physical optics are scientific revolutions, and the successive transition from one paradigm to another via revolution is the usual developmental pattern of mature science. It is not, however, the pattern characteristic of the period before Newton's work, and that is the contrast that concerns us here. No period between remote antiquity and the end of the seventeenth century exhibited a single generally accepted view about the nature of light. Instead there were a number of competing schools and subschools, most of them espousing one variant or another of Epicurean, Aristotelian, or Platonic theory. One group took light to be particles emanating from material bodies; for another it was a modification of the medium that intervened between the body and the eye; still another explained light in terms of an interaction of the medium with an emanation from the eye; and there were other combinations and modifications besides. Each of the corresponding schools derived strength from its relation to some particular metaphysic, and each emphasized, as paradigmatic observations, the particular cluster of optical phenomena that its own theory could do most to explain. Other observations were dealt with by *ad hoc* elaborations, or they remained as outstanding problems for further research.[2]

At various times all these schools made significant contributions to the body of concepts, phenomena, and techniques from which Newton drew the first nearly uniformly accepted paradigm for physical optics. Any definition of the scientist that excludes at least the more creative members of these various schools will exclude their modern successors as well. Those men were scientists. Yet anyone examining a survey of physical optics before Newton may well conclude that, though the field's practitioners were scientists, the net result of their activity was something less than science. Being able to take no common body of belief for granted, each writer on physical optics felt forced to build his field anew from its foundations. In doing so, his choice of supporting observation and experiment was relatively free, for there was no standard set of methods or of phenomena that every optical writer felt forced to employ and explain. Under these circumstances, the dialogue of the resulting books was often directed as much to the members of other schools as it was to nature. That pattern is not unfamiliar in a number of creative fields today, nor is it incompatible with significant discovery and invention. It is not, however, the pattern of development that physical optics acquired after Newton and that other natural sciences make familiar today. . . .

NORMAL SCIENCE AS PUZZLE-SOLVING

Perhaps the most striking feature of the normal research problems we have just encountered is how little they aim to produce major novelties, conceptual or phenomenal. Sometimes, as in a wave-length measurement, everything but the most esoteric detail of the result is known in advance, and the typical latitude of expectation is only somewhat wider. Coulomb's measurements need not, perhaps, have fitted an inverse square law; the men who worked on heating by compression were often prepared for any one of several results. Yet even in cases like these the range of anticipated, and thus of assimilable, results is always small compared with the range that imagination can conceive. And the project whose outcome does not fall in that narrower range is usually just a research failure, one which reflects not on nature but on the scientist.

In the eighteenth century, for example, little attention was paid to the experiments that measured electrical attraction with devices like the pan balance. Because they yielded neither consistent nor simple results, they could not be used to articulate the paradigm from which they derived. Therefore, they remained *mere* facts, unrelated and unrelatable to the continuing progress of electrical research. Only in retrospect, possessed of a subsequent paradigm, can we see what characteristics of electrical phenomena they display. Coulomb and his contemporaries, of course, also possessed this later paradigm or one that, when applied to the problem of attraction, yielded the same expectations. That is why Coulomb was able to design apparatus that gave a result assimilable by paradigm articulation. But it is also why that result surprised no one and why several of Coulomb's contemporaries had been able to predict it in advance. Even the project whose goal is paradigm articulation does not aim at the *unexpected* novelty.

But if the aim of normal science is not major substantive novelties—if failure to come near the anticipated result is usually failure as a scientist—then why are these problems undertaken at all? Part of the answer has already been developed. To scientists, at least, the results gained in normal research are significant because they add to the scope and precision with which the paradigm can be applied. That answer, however, cannot account for the enthusiasm and devotion that scientists display for the problems of normal research. No one devotes years to, say, the development of a better spectrometer or the production of an improved solution to the problem of vibrating strings simply because of the importance of the information that will be obtained. The data to be gained by computing ephemerides or by further measurements with an existing instrument are often just as significant, but those activities are regularly spurned by scientists because they are so largely repetitions of procedures that have been carried through before. That rejection provides a clue to the fascination of the normal research problem. Though its outcome can be anticipated, often in detail so great that what remains to be known is itself uninteresting, the way to achieve that outcome remains very much in doubt. Bringing a normal research problem to a conclusion is achieving the anticipated in a new way, and it requires the solution of all sorts of complex instrumental, conceptual, and mathematical puzzles. The man who succeeds proves himself an expert puzzle-solver, and the challenge of the puzzle is an important part of what usually drives him on.

The terms 'puzzle' and 'puzzle-solver' highlight several of the themes that have become increasingly prominent in the preceding pages. Puzzles are, in the entirely standard meaning here employed, that special category of problems that can serve to test ingenuity or skill in solution. Dictionary illustrations are 'jigsaw puzzle' and 'crossword puzzle,' and it is the characteristics that these share with the problems of normal science that we now need to isolate. One of them has just been mentioned. It is no criterion of goodness in a puzzle that its outcome be intrinsically interesting or important. On the contrary, the really pressing problems, e.g., a cure for cancer or the design of a lasting peace, are often not puzzles at all, largely because they may not have any solution. Consider the jigsaw puzzle whose pieces are selected at random from each of two different puzzle boxes. Since that problem is likely to defy (though it might not) even the most ingenious of men, it cannot serve as a test of skill in solution. In any usual sense it is not a puzzle at all. Though intrinsic value is no criterion for a puzzle, the assured existence of a solution is.

We have already seen, however, that one of the things a scientific community acquires with a paradigm is a criterion for choosing problems that, while the paradigm is taken for granted, can be assumed to have solutions. To a great extent these are the only problems that the community will admit as scientific or encourage its members to undertake. Other problems, including many that had previously been standard, are rejected as metaphysical, as the concern of another discipline, or sometimes as just too problematic to be worth the time. A paradigm can, for that matter, even insulate the community from those socially important problems that are not reducible to the puzzle form, because they cannot be stated in terms of the conceptual and instrumental tools the paradigm supplies. Such problems can be a distraction, a lesson brilliantly illustrated by several facets of seventeenth-century Baconianism and by some of the contemporary social sciences. One of the reasons why normal science seems to progress so rapidly is that its practitioners concentrate on problems that only their own lack of ingenuity should keep them from solving. . . .

THE NATURE AND NECESSITY OF SCIENTIFIC REVOLUTIONS

These remarks permit us at last to consider the problems that provide this essay with its title. What are scientific revolutions, and what is their function in scientific development? Much of the answer to these questions has been anticipated in earlier sections. In particular, the preceding discussion has indicated that scientific revolutions are here taken to be those non-cumulative developmental episodes in which an older paradigm is replaced in whole or in part by an incompatible new one. There is more to be said, however, and an essential part of it can be introduced by asking one further question. Why should a change of paradigm be called a revolution? In the face of the vast and essential differences between political and scientific development, what parallelism can justify the metaphor that finds revolutions in both?

One aspect of the parallelism must already be apparent. Political revolutions are inaugurated by a growing sense, often restricted to a segment of the political community, that existing institutions have ceased adequately to meet the problems posed by an environment that they have in part created. In much the same way, scientific revolutions are inaugurated by a growing sense, again often restricted to a narrow subdivision of the scientific community, that an existing paradigm has ceased to function adequately in the exploration of an aspect of nature to which that paradigm itself had previously led the way. In both political and scientific development the sense of malfunction that can lead to crisis is prerequisite to revolution. Furthermore, though it admittedly strains the metaphor, that parallelism holds not only for the major paradigm changes, like those attributable to Copernicus and Lavoisier, but also for the far smaller ones associated with the assimilation of a new sort of phenomenon, like oxygen or X-rays. Scientific revolutions . . . need seem revolutionary only to those whose paradigms are affected by them. To outsiders they may, like the Balkan revolutions of the early twentieth century, seem normal parts of the developmental process. Astronomers, for example, could accept X-rays as a mere addition to knowledge, for their paradigms were unaffected by the existence of the new radiation. But for men like Kelvin, Crookes, and Roentgen, whose research dealt with radiation theory or with cathode ray tubes, the emergence of X-rays necessarily violated one paradigm as it created another. That is why these rays could be discovered only through something's first going wrong with normal research.

This genetic aspect of the parallel between political and scientific development should no longer be open to doubt. The parallel has, however, a second and more profound aspect upon which the significance of the first depends. Political revolutions aim to change political institutions in ways that those institutions themselves prohibit. Their success therefore necessitates the partial relinquishment of one set of institutions in favor of another, and in the interim, society is not fully governed by institutions at all. Initially it is crisis alone that attenuates the role of political institutions as we have already seen it attenuate the role of paradigms. In increasing numbers individuals become increasingly estranged from political life and behave more and more eccentrically within it. Then, as the crisis deepens, many of these individuals commit themselves to some concrete proposal for the reconstruction of society in a new institutional framework. At that point the society is divided into competing camps or parties, one seeking to defend the old institutional constellation, the others seeking to institute some new one. And, once that polarization has occurred, *political recourse fails*. Because they differ about the institutional matrix within which political change is to be achieved and evaluated, because they acknowledge no supra-institutional framework for the adjudication of revolutionary difference, the parties to a revolutionary conflict must finally resort to the techniques of mass persuasion, often including force. Though revolutions have had a vital role in the evolution of political institutions, that role depends upon their being partially extrapolitical or extrainstitutional events.

The remainder of this essay aims to demonstrate that the historical study of paradigm change reveals very similar characteristics in the evolution of the sciences. Like the choice between competing political institutions, that between competing paradigms proves to be a choice between incompatible modes of community life. Because it has that character, the choice is not and cannot be determined merely by the evaluative procedures characteristic of normal science, for these depend in part upon a particular paradigm, and that paradigm is at issue. When paradigms enter, as they must, into a debate about paradigm choice, their role is necessarily circular. Each group uses its own paradigm to argue in that paradigm's defense.

The resulting circularity does not, of course, make the arguments wrong or even ineffectual. The man who premises a paradigm when arguing in its defense can nonetheless provide a clear exhibit of what scientific practice will be like for those who adopt the new view of nature. That exhibit can be immensely persuasive, often compellingly so. Yet, whatever its force, the status of the circular argument is only that of persuasion. It cannot be made logically or even probabilistically compelling for those who refuse to step into the circle. The premises and values shared by the two parties to a debate over paradigms are not sufficiently extensive for that. As in political revolutions, so in paradigm choice—there is no standard higher than the assent of the relevant community. To discover how scientific revolutions are effected, we shall therefore have to examine not only the impact of nature and of logic, but also the techniques of persuasive argumentation effective within the quite special groups that constitute the community of scientists. . . .

REVOLUTIONS AS CHANGES OF WORLD VIEW

Examining the record of past research from the vantage of contemporary historiography, the historian of science may be tempted to exclaim that when paradigms change, the world itself changes with them. Led by a new paradigm, scientists adopt new instruments and look in new places. Even more important, during revolutions scientists see new and different things when looking with familiar instruments in places they have looked before. It is rather as if the professional community had been suddenly transported to another planet where familiar objects are seen in a different light and are joined by unfamiliar ones as well. Of course, nothing of quite that sort does occur: there is no geographical transplantation; outside the laboratory everyday affairs usually continue as before. Nevertheless, paradigm changes do cause scientists to see the world of their research-engagement differently. In so far as their only recourse to that world is through what they see and do, we may want to say that after a revolution scientists are responding to a different world.

It is as elementary prototypes for these transformations of the scientist's world that the familiar demonstrations of a switch in visual gestalt prove so suggestive. What were ducks in the scientist's world before the revolution are rabbits afterwards. The man who first saw the exterior of the box from above later sees its interior from below. Transformations like these, though usually more gradual and almost always irreversible, are common concomitants of scientific training. Looking at a contour map, the student sees lines on paper, the cartographer a picture of a terrain. Looking at a bubble-chamber photograph, the student sees confused and broken lines, the physicist a record of familiar subnuclear events. Only after a number of such transformations of vision does the student become an inhabitant of the scientist's world, seeing what the scientist sees and responding as the scientist does. The world that the student then enters is not, however, fixed once and for all by the nature of the environment, on the one hand, and of science, on the other. Rather, it is determined jointly by the environment and the particular normal-scientific tradition that the student has been trained to pursue. Therefore, at times of revolution, when the normal-scientific tradition changes, the scientist's perception of his environment must be re-educated—in some familiar situations he must learn to see a new gestalt. After he has done so the world of his research will seem, here and there, incommensurable with the one he had inhabited before. That is another reason why schools guided by different paradigms are always slightly at cross-purposes.

In their most usual form, of course, gestalt experiments illustrate only the nature of perceptual transformations. They tell us nothing about the role of paradigms or of previously assimilated experience in the process of perception. But on that point there is a rich body of psychological literature, much of it stemming from the pioneering work of the Hanover Institute. An experimental subject who puts on goggles fitted with inverting lenses initially sees the entire world upside down. At the start his perceptual apparatus functions as it had been trained to function in the absence of the goggles, and the result is extreme disorientation, an acute personal crisis. But after the subject has begun to learn to deal with his new world, his entire visual field flips over, usually after an intervening period in which vision is simply confused. Thereafter, objects are again seen as they had been before the goggles were put on. The assimilation of a previously anomalous visual field has reacted upon and changed the field itself.[3] Literally as well as metaphorically, the man accustomed to inverting lenses has undergone a revolutionary transformation of vision. . . .

Still other experiments demonstrate that the perceived size, color, and so on, of experimentally displayed objects also varies with the subject's previous training and experience.[4] Surveying the rich experimental literature from which these examples are drawn makes one suspect that something like a par-

adigm is prerequisite to perception itself. What a man sees depends both upon what he looks at and also upon what his previous visual-conceptual experience has taught him to see. In the absence of such training there can only be, in William James's phrase, "a bloomin' buzzin' confusion."

In recent years several of those concerned with the history of science have found the sorts of experiments described above immensely suggestive. N. R. Hanson, in particular, has used gestalt demonstrations to elaborate some of the same consequences of scientific belief that concern me here.[5] Other colleagues have repeatedly noted that history of science would make better and more coherent sense if one could suppose that scientists occasionally experienced shifts of perception like those described above. Yet, though psychological experiments are suggestive, they cannot, in the nature of the case, be more than that. They do display characteristics of perception that *could* be central to scientific development, but they do not demonstrate that the careful and controlled observation exercised by the research scientist at all partakes of those characteristics. Furthermore, the very nature of these experiments makes any direct demonstration of that point impossible. If historical example is to make these psychological experiments seem relevant, we must first notice the sorts of evidence that we may and may not expect history to provide.

The subject of a gestalt demonstration knows that his perception has shifted because he can make it shift back and forth repeatedly while he holds the same book or piece of paper in his hands. Aware that nothing in his environment has changed, he directs his attention increasingly not to the figure (duck or rabbit) but to the lines on the paper he is looking at. Ultimately he may even learn to see those lines without seeing either of the figures, and he may then say (what he could not legitimately have said earlier) that it is these lines that he really sees but that he sees them alternately *as* a duck and *as* a rabbit. . . . Unless there were an external standard with respect to which a switch of vision could be demonstrated, no conclusion about alternate perceptual possibilities could be drawn.

With scientific observation, however, the situation is exactly reversed. The scientist can have no recourse above or beyond what he sees with his eyes and instruments. If there were some higher authority by recourse to which his vision might be shown to have shifted, then that authority would itself become the source of his data, and the behavior of his vision would become a source of problems (as that of the experimental subject is for the psychologist). The same sorts of problems would arise if the scientist could switch back and forth like the subject of the gestalt experiments. The period during which light was "sometimes a wave and sometimes a particle" was a period of crisis—a period when something was wrong—and it ended only with the development of wave mechanics and the realization that light was a self-consistent entity different from both waves and particles. In the sciences, therefore, if perceptual switches accompany paradigm changes, we may not expect scientists to attest to these changes directly. Looking at the moon, the convert to Copernicanism does not say, "I used to see a planet, but now I see a satellite." That locution would imply a sense in which the Ptolemaic system had once been correct. Instead, a convert to the new astronomy says, "I once took the moon to be (or saw the moon as) a planet, but I was mistaken." That sort of statement does recur in the aftermath of scientific revolutions. If it ordinarily disguises a shift of scientific vision or some other mental transformation with the same effect, we may not expect direct testimony about that shift. Rather we must look for indirect and behavioral evidence that the scientist with a new paradigm sees differently from the way he had seen before.

Let us then return to the data and ask what sorts of transformations in the scientist's world the historian who believes in such changes can discover. Sir William Herschel's discovery of Uranus provides a first example. . . . On at least seventeen different occasions between 1690 and 1781, a number of astronomers, including several of Europe's most eminent observers, had seen a star in positions that we now suppose must have been occupied at the time by Uranus. One of the best observers in this group had actually seen the star on four successive nights in 1769 without noting the motion that could have suggested another identification. Herschel, when he first observed the same object twelve years later, did so with a much improved telescope of his own manufacture. As a result, he was able to notice an apparent disk-size that was at least unusual for

stars. Something was awry, and he therefore postponed identification pending further scrutiny. That scrutiny disclosed Uranus' motion among the stars, and Herschel therefore announced that he had seen a new comet! Only several months later, after fruitless attempts to fit the observed motion to a cometary orbit, did Lexell suggest that the orbit was probably planetary.[6] When that suggestion was accepted, there were several fewer stars and one more planet in the world of the professional astronomer. A celestial body that had been observed off and on for almost a century was seen differently after 1781 because it could no longer be fitted to the perceptual categories (star or comet) provided by the paradigm that had previously prevailed.

The shift of vision that enabled astronomers to see Uranus, the planet, does not, however, seem to have affected only the perception of that previously observed object. Its consequences were more far-reaching. Probably, though the evidence is equivocal, the minor paradigm change forced by Herschel helped to prepare astronomers for the rapid discovery, after 1801, of the numerous minor planets or asteroids. Because of their small size, these did not display the anomalous magnification that had alerted Herschel. Nevertheless, astronomers prepared to find additional planets were able, with standard instruments, to identify twenty of them in the first fifty years of the nineteenth century.[7] The history of astronomy provides many other examples of paradigm-induced changes in scientific perception, some of them even less equivocal. Can it conceivably be an accident, for example, that Western astronomers first saw change in the previously immutable heavens during the half-century after Copernicus' new paradigm was first proposed? The Chinese, whose cosmological beliefs did not preclude celestial change, had recorded the appearance of many new stars in the heavens at a much earlier date. Also, even without the aid of a telescope, the Chinese had systematically recorded the appearance of sunspots centuries before these were seen by Galileo and his contemporaries.[8] Nor were sunspots and a new star the only examples of celestial change to emerge in the heavens of Western astronomy immediately after Copernicus. Using traditional instruments, some as simple as a piece of thread, late sixteenth-century astronomers repeatedly discovered that comets wandered at will through the space previously reserved for the immutable planets and stars.[9] The very ease and rapidity with which astronomers saw new things when looking at old objects with old instruments may make us wish to say that, after Copernicus, astronomers lived in a different world. In any case, their research responded as though that were the case.

NOTES

1. Joseph Priestley, *The History and Present State of Discoveries Relating to Vision, Light, and Colours* (London, 1772), pp. 385–90.
2. Vasco Ronchi, *Histoire de la lumière,* trans. Jean Taton (Paris, 1956), chaps i–iv.
3. The original experiments were by George M. Stratton, "Vision without Inversion of the Retinal Image," *Psychological Review,* IV (1897), 341–60, 463–81. A more up-to-date review is provided by Harvey A. Carr, *An Introduction to Space Perception* (New York, 1935), pp. 18–57.
4. For examples, see Albert H. Hastorf, "The Influence of Suggestion on the Relationship between Stimulus Size and Perceived Distance," *Journal of Psychology,* XXIX (1950), 195–217; and Jerome S. Bruner, Leo Postman, and John Rodrigues, "Expectations and the Perception of Color," *American Journal of Psychology,* LXIV (1951), 216-27.
5. N. R. Hanson, *Patterns of Discovery* (Cambridge, 1958), chap. i.
6. Peter Doig, *A Concise History of Astronomy* (London, 1950), pp. 115–16.
7. Rudolph Wolf, *Geschichte der Astronomie* (Munich, 1877), pp. 513–15, 683–93. Notice particularly how difficult Wolf's account makes it to explain these discoveries as a consequence of Bode's Law.
8. Joseph Needham, *Science and Civilization in China,* III (Cambridge, 1959), 423–29, 434–36.
9. T. S. Kuhn, *The Copernican Revolution* (Cambridge, Mass., 1957), pp. 206–9.

26

LARRY LAUDAN

A Problem-Solving Approach to Scientific Progress

DESIDERATA

Studies of the historical development of science have made it clear that any normative model of scientific rationality which is to have the resources to show that science has been largely a rational enterprise must come to terms with certain persistent features of scientific change. To be specific, we may conclude from the existing historical evidence that:

1. Theory transitions are generally noncumulative, i.e., neither the logical nor empirical content (nor even the 'confirmed consequences') of earlier theories is wholly preserved when those theories are supplanted by newer ones.

2. Theories are generally not rejected simply because they have anomalies nor are they

Beyond Positivism and Relativism (Boulder, CO: Westview Press, 1996), pp. 77–87. Reprinted with permission from the publisher.

generally accepted simply because they are empirically confirmed.

3. Changes in, and debates about, scientific theories often turn on conceptual issues rather than on questions of empirical support.

4. The specific and 'local' principles of scientific rationality which scientists utilize in evaluating theories are not permanently fixed, but have altered significantly through the course of science.

5. There is a broad spectrum of cognitive stances which scientists take toward theories, including accepting, rejecting, pursuing, and entertaining. Any theory of rationality which discusses only the first two will be incapable of addressing itself to the vast majority of situations confronting scientists.

6. There is a range of levels of generality of scientific theories, from laws at the one end to broad conceptual frameworks at the other. Principles of testing, comparison, and evaluation of theories seem to vary significantly from level to level.

7. Given the notorious difficulties with notions of 'approximate truth'—at both the semantic and epistemic levels—it is implausible that characterizations of scientific progress which view evolution toward greater truthlikeness as the central aim of science will allow one to represent science as a rational activity.

8. The coexistence of rival theories is the rule rather than the exception, so that theory evaluation is primarily a comparative affair.

The challenge to which this chapter is addressed is whether there can be a normatively viable philosophy of science which finds a place for most or all of these features of science *wie es eigentlich gewesen ist.*

THE AIM OF SCIENCE

To ask if scientific knowledge shows cognitive progress is to ask whether science through time brings us closer to achieving our cognitive aims or goals. Depending upon our choice of cognitive aims, one and the same temporal sequence of theories may

be progressive or nonprogressive. Accordingly, the stipulative task of specifying the aims of science is more than an academic exercise. Throughout history, there has been a tendency to characterize the aims of science in terms of such transcendental properties as truth or apodictic certainty. So conceived, science emerges as nonprogressive since we evidently have no way of ascertaining whether our theories are more truthlike or more nearly certain than they formerly were. We do not yet have a satisfactory semantic characterization of truthlikeness, let alone any epistemic account of when it would be legitimate to judge one theory to be more nearly true than another.[1] Only by setting goals for science which are in principle achievable, and which are such that we can tell whether we are achieving (or moving closer to achieving) them, can we even hope to be able to make a positive claim about the progressive character of science. There are many nontranscendent immanent goals in terms of which we might attempt to characterize science; we could view science as aiming at well-tested theories, theories which predict novel facts, theories which 'save the phenomena,' or theories which have practical applications. My own proposal, more general than these, is that the aim of science is to secure theories with a high problem-solving effectiveness. From this perspective, *science progresses just in case successive theories solve more problems than their predecessors.*

The merits of this proposal are twofold: (1) it captures much that has been implicit all along in discussions of the growth of science; and (2) it assumes a goal which (unlike truth) is not intrinsically transcendent and hence closed to epistemic access. The object of this chapter is to spell out this proposal in some detail and to examine some of the consequences that a problem-solving model of scientific progress has for our understanding of the scientific enterprise.

KINDS OF PROBLEM–SOLVING: A TAXONOMY

Despite the prevalent talk about problem-solving among scientists and philosophers, there is little agreement about what counts as a problem, what kinds of problems there are, and what constitutes a solution to a problem. To begin with, I suggest that we separate *empirical* from *conceptual* problems.

At the empirical level, I distinguish between potential problems, solved problems, and anomalous problems. 'Potential problems' constitute what we take to be the case about the world, but for which there is as yet no explanation. 'Solved' or 'actual' problems are that class of putatively germane claims about the world which have been solved by some viable theory or other. 'Anomalous problems' are actual problems which rival theories solve but which are not solved by the theory in question. It is important to note that, according to this analysis, unsolved or potential problems need not be anomalies. A problem is only anomalous for some theory if that problem has been solved by a viable rival. Thus, a prima facie falsifying instance for a theory, T, may not be an anomalous problem (specifically, when no other theory has solved it); and an instance which does not falsify T may nonetheless be anomalous for T (if T does not solve it and one of T's rivals does).

In addition to empirical problems, theories may be confronted by *conceptual* problems. Such problems arise for a theory, T, in any of the following circumstances:

1. when T is internally inconsistent or the theoretical mechanisms it postulates are ambiguous;

2. when T makes assumptions about the world that run counter to other theories or to prevailing metaphysical assumptions, or when T makes claims about the world which cannot be warranted by prevailing epistemic and methodological doctrines;

3. when T violates principles of the research tradition of which it is a part (to be discussed below);

4. when T fails to utilize concepts from other, more general theories to which it should be logically subordinate.

Conceptual problems, like anomalous empirical problems, indicate liabilities in our theories (i.e., partial failures on their part to serve all the functions for which we have designed them).

Running through much of the history of the philosophy of science is a tension between coherentist and correspondentist accounts of scientific knowledge. Coherentists stress the need for appropriate types of conceptual linkages between our beliefs, while correspondentists emphasize the grounding of beliefs in the world. Each account typically makes only minimal concessions to the other. (Correspondentists, for instance, will usually grant that theories should minimally cohere in the sense of being consistent with our other beliefs.) Neither side, however, has been willing to grant that a *broad range* of both empirical and conceptual checks are of equal importance in theory-testing. The problem-solving model, on the other hand, explicitly acknowledges that both concerns are co-present. Empirical and conceptual problems represent respectively the correspondentist and coherentist constraints which we place on our theories. The latter show up in the demand that conceptual difficulties (whose nature will be discussed below) should be minimized; the former are contained in the dual demands that a theory should solve a maximal number of empirical problems, while generating a minimal number of anomalies. Where most empiricists and pragmatic philosophers have assigned a subordinate role to conceptual factors in theory appraisal (essentially allowing such factors to come into play only in the choice between theories possessing equivalent empirical support), the problem-solving model argues that the elimination of conceptual difficulties is as much constitutive of progress as increasing empirical support. Indeed, on this model, it is *possible* that a change from an empirically well-supported theory to a less well-supported one could be progressive, provided that the latter resolved significant conceptual difficulties confronting the former.

The centrality of conceptual concerns here represents a significant departure from earlier empiricist philosophers of science. Many types of conceptual difficulties that theories regularly confront have been given little or no role to play by these philosophers in their models of scientific change. Even those like Popper who have paid lip service to the heuristic role of metaphysics in science leave no scope for rational conflicts between a theory and prevailing views about scientific methodology. This is because they have assumed that the meta-scientific evaluative criteria which scientists use for assessing theories are immutable and uncontroversial.

Why do most models of science fail at this central juncture? In assessing prior developments, they quite properly attend carefully to what evidence a former scientist had and to his substantive beliefs about the world, but they also assume without argument that earlier scientists adhered to our views about the rules of theory evaluation. Extensive scholarship on this matter makes it vividly clear that the views of the scientific community about how to test theories and about what counts as evidence have changed dramatically through history. (This should not be surprising, since we are as capable of learning more about how to do science as we are of learning more about how the world works.) The fact that the evaluative strategies of scientists of earlier eras are different from our strategies makes it quixotic to suppose that we can assess the rationality of their science by ignoring completely their views about how theories should be evaluated. Short of invoking Hegel's 'cunning of reason' or Marx's 'false consciousness,' it is anachronistic to judge the rationality of the work of an Archimedes, a Newton, or an Einstein by asking whether it accords with the contemporary methodology of a Popper or a Lakatos. The views of former scientists about how theories should be evaluated must enter into judgments about how rational those scientists were in testing their theories in the ways that they did. The problem-solving model brings such factors into play through the inclusion of conceptual problems, one species of which arises when a theory conflicts with a prevailing epistemology. Models of science which did not include a scientist's theory of evidence in a rational account of his actions and beliefs are necessarily defective.

I have talked of problems, but what of solutions? In the simplest cases, a theory solves an *empirical* problem when it entails, along with appropriate initial and boundary conditions, a statement of the problem. A theory solves or eliminates a *conceptual* problem when it fails to exhibit a conceptual difficulty of its predecessor. It is important to note that, on this account, *many different theories may solve the same* (empirical or conceptual) *problem*. The worth of a theory will depend *inter alia* on how many problems it solves. Unlike most models of explanation which insist that a theory does not really explain anything unless it is the best theory (or possesses a high degree of confirmation), the problem-solving approach allows a problem solution to be credited to a theory, independent of how well established the theory is, just so long as the theory stands in a certain formal relation to (a statement of) the problem. Some of the familiar paradoxes of confirmation are avoided by the correlative demand that theories must minimize conceptual difficulties; because standard theories of support leave no scope for the broad range of coherentist considerations sketched above, their deductivistic models of inductive support lead to many conundrums which the present approach readily avoids.

PROGRESS WITHOUT CUMULATIVE RETENTION

Virtually all models of scientific progress and rationality (with the exception of certain inductive logics which are otherwise flawed) have insisted on wholesale retention of content or success in every progressive-theory transition. According to some well-known models, earlier theories are required to be contained in, or limiting cases of, later theories; while in others, the empirical content or confirmed consequences of earlier theories are required to be subsets of the content or consequence classes of the new theories. Such models are appealing in that they make theory choice straightforward. If a new theory can do everything its predecessor could and more besides, then the new theory is clearly superior. Unfortunately, history teaches us that theories rarely if ever stand in this relation to one another, and recent conceptual analysis even suggests that theories could not possibly exhibit such relations under normal circumstances.

What is required, if we are to rescue the notion of scientific progress, is a breaking of the link between cumulative retention and progress, so as to allow for the possibility of progress even when there are explanatory losses as well as gains. Specifically, we must work out some machinery for setting off gains against losses. This is a much more complicated affair than simple cumulative retention and we are not close to having a fully developed account of it. But the outlines of such an account can be perceived. Cost-benefit analysis is a tool developed especially to handle such a situation. Within a problem-solving model, such analysis proceeds as follows: for every theory, assess the number and the weight of the empirical problems it is known to solve; similarly, assess the number and weight of its empirical anomalies; finally, assess the number and centrality of its conceptual difficulties or problems. Constructing appropriate scales, our principle of progress tells us to prefer that theory which comes closest to solving the largest number of important empirical problems while generating the smallest number of significant anomalies and conceptual problems.

Whether the details of such a model can be refined is still unclear. But the attractiveness of the general program should be obvious; for what it in principle allows us to do is to talk about rational and progressive theory change in the absence of cumulative retention of content. The technical obstacles confronting such an approach are, of course, enormous. It presumes that problems can be individuated and counted. How to do that is still not completely clear; but then *every* theory of empirical support requires us to be able to identify and to individuate the confirming and disconfirming instances which our theories possess. More problematic is the idea of weighting the importance of the problems, solved and unsolved. I discuss some of the factors that influence weighting in *Progress and Its Problems,* but do not pretend to have more than the outlines of a satisfactory account.

THE SPECTRUM OF COGNITIVE MODALITIES

Most methodologies of sciences have assumed that cognitive stands scientists adopt toward theories are exhausted by the oppositions between 'belief' and 'disbelief' or, more programmatically, 'acceptance' and 'rejection'. Even a superficial scrutiny of science reveals, however, that there is a much wider range of cognitive attitudes which should be included in our account. Many, if not most, theories deal with ideal cases. Scientists neither believe such theories nor accept them as true. But neither does 'disbelief' or 'rejection' correctly characterize scientists' attitudes toward such theories. Moreover, scientists often claim that a theory, even if unacceptable, deserves investigation, or warrants further elaboration. The logic of acceptance and rejection is simply too re-

strictive to represent this range of cognitive attitudes. Unless we are prepared to say that such attitudes are beyond rational analysis—in which case most of science is nonrational—we need an account of evidential support which will permit us to say when theories are worthy of further investigation and elaboration. My view is that this continuum of attitudes between acceptance and rejection can be seen to be functions of the relative problem-solving progress (and the rate of progress) of our theories. A highly progressive theory may not yet be worthy of acceptance but its progress may well warrant further pursuit. A theory with a high initial rate of progress may deserve to be entertained even if its net problem-solving effectiveness—compared to some of its older and better-established rivals—is unsatisfactory. Measures of a theory's progress show promise for rationalizing this important range of scientific judgments.

THEORIES AND RESEARCH TRADITIONS

Logical empiricists performed a useful service when they developed their account of the structure of a scientific theory. Theories of the type they discussed—consisting of a network of statements which, in conjunction with initial conditions, lead to explanations and predictions of specific phenomena—do come close to capturing the character of those frameworks which are typically tested by scientific experiments. But limiting our attention to theories so conceived prevents our saying very much about enduring, long-standing commitments which are so central a feature of scientific research. There are significant family resemblances between certain theories which mark them off as a group from others. Theories represent exemplifications of more fundamental views about the world, and the manner in which theories are modified and changed only makes sense when seen against the backdrop of those more fundamental commitments. I call the cluster of beliefs which constitute such fundamental views 'research traditions'. Generally, these consist of at least two components: (1) a set of beliefs about what sorts of entities and processes make up the domain of inquiry; and (2) a set of epistemic and methodological norms about how the domain is to be investigated, how theories are to be tested, how data are to be collected, and the like.

Research traditions are not directly testable, both because their ontologies are too general to yield specific predictions and because their methodological components, being rules or norms, are not straight-forwardly testable assertions about matters of fact. Associated with any active research tradition is a family of theories. Some of these theories, for instance, those applying the research tradition to different parts of the domain, will be mutually consistent while other theories, for instance, those which are rival theories within the research tradition, will not. What all the theories have in common is that they share the ontology of the parent research tradition and can be tested and evaluated using its methodological norms.

Research traditions serve several specific functions. Among others: (1) they indicate what assumptions can be regarded as uncontroversial 'background knowledge' to all the scientists working in that tradition; (2) they help to identify those portions of a theory that are in difficulty and should be modified or amended; (3) they establish rules for the collection of data and for the testing of theories; (4) they pose conceptual problems for any theory in the tradition which violates the ontological and epistemic claims of the parent tradition.

ADEQUACY AND PROMISE

Compared to single theories, research traditions tend to be enduring entities. Where theories may be abandoned and replaced very frequently, research traditions are usually long-lived, since they can obviously survive the demise of any of their subordinate theories. Research traditions are the units which endure through theory change and which establish, along with solved empirical problems, much of what continuity there is in the history of science. But even research traditions can be overthrown. To understand how, we must bring the machinery of problem-solving assessment into the picture.

Corresponding to the idealized modalities of acceptance and pursuit are two features of theories,

both related to problem-solving efficiency. Both of these features can be explained in terms of the problem-solving effectiveness of a theory, which is itself a function of the number and importance of the empirical problems a theory has solved and of the anomalies and conceptual problems which confront it. One theory is more adequate (i.e., more acceptable) than a rival just in case the former has exhibited a greater problem-solving effectiveness than the latter. One research tradition is more adequate than another just in case the ensemble of theories which characterize it at a given time are more adequate than the theories making up any rival research tradition.

If our only goal was that of deciding which theory or research tradition solved the largest number of problems, these tools would be sufficient. But there is a *prospective* as well as a retrospective element in scientific evaluation. Our hope is to move to theories which can solve more problems, including potential empirical problems, than we are now able to deal with. We seek theories which promise fertility in extending the range of what we can now explain and predict. The fact that one theory (or research tradition) is now the most adequate is not irrelevant to, but neither is it sufficient grounds for, judgments about promise or fertility. New theories and research traditions are rarely likely to have managed to achieve a degree of problem-solving effectiveness as high as that of old, well-established theories. How are we to judge when such novel approaches are worth taking seriously? A natural suggestion involves assessing the progress or rate of progress of such theories and research traditions. That progress in defined as the difference between the problem-solving effectiveness of the research tradition in its latest form and its effectiveness at an earlier period. The rate of progress is a measure of how quickly a research tradition has made whatever progress it exhibits.

Obviously, one research tradition may be less adequate than a rival, and yet more progressive. Acknowledging this fact, one might propose that highly progressive theories should be explored and pursued whereas only the most adequate theories should be accepted. Traditional philosophies of science (e.g., Carnap's, Popper's) and some more recent ones (e.g., Lakatos's) share the view that both adequacy and promise are to be assessed by the same measure. The approach outlined here has the advantage of acknowledging that we evaluate scientific ideas with different measures which are appropriate to those different ends. How progressive a research tradition is and how rapidly it has progressed are different, if equally relevant, questions from asking how well supported the research tradition is.

PATTERNS OF SCIENTIFIC CHANGE

According to Thomas Kuhn's influential view, science can be periodized into a series of epochs, the boundaries between which are called scientific revolutions. During periods of normal science, one paradigm reigns supreme. Raising fundamental conceptual concerns or identifying anomalies for the prevailing doctrine of actively developing alternative 'paradigms' are, in Kuhn's view, disallowed by the scientific community, which has a very low tolerance for rival points of view. The problem-solving model gives rise to a very different picture of the scientific enterprise. It suggests that the coexistence of rival research traditions is the rule rather than the exception. It stresses the centrality of debates about conceptual foundations and argues that the neglect of conceptual issues (a neglect which Kuhn sees as central to the 'normal' progress of science) is undesirable. That the actual development of science is closer to the picture of permanent coexistence of rivals and the omnipresence of conceptual debate than to the picture of normal science seems clear. It is difficult, for instance, to find any lengthy period in the history of any science in the last 300 years when the Kuhnian picture of 'normal science' prevails. What seems to be far more common is for scientific disciplines to involve a variety of co-present research approaches (traditions). At any given time, one or other of these may have the competitive edge, but there is a continuous and persistent struggle taking place, with partisans of one view or another pointing to the empirical and conceptual weaknesses of rival points of view and to the problem-solving progressiveness of their own approach. Dialectical confrontations are essential to the growth and improvement of scientific knowledge; like nature, science is red in tooth and claw.

SCIENCE AND THE NONSCIENCES

The approach taken here suggests that there is no fundamental difference in kind between scientific and other forms of intellectual inquiry. All seek to make sense of the world and of our experience. All theories, scientific and otherwise, are subject alike to empirical and conceptual constraints. Those disciplines that we call the 'sciences' are generally more progressive than the 'nonsciences'; indeed, it may be that we call them 'sciences' simply because they are more progressive rather than because of any methodological or substantive traits they possess in common. If so, such differences as there are turn out to be differences of degree rather than of kind. Similar aims, and similar evaluative procedures, operate across the spectrum of intellectual disciplines. It is true, of course, that *some* of the 'sciences' utilize vigorous testing procedures which do not find a place in the nonsciences; but such testing procedures cannot be constitutive of science since many 'sciences' do not utilize them.

The quest for a specifically scientific form of knowledge, or for a demarcation criterion between science and nonscience, has been an unqualified failure. There is apparently no epistemic feature or set of such features which all and only the 'sciences' exhibit. Our aim should be, rather, to distinguish reliable and well-tested claims to knowledge from bogus ones. The problem-solving model purports to provide the machinery to do this, but it does not assume that the distinction between warranted and unwarranted knowledge claims simply maps onto the science/non-science dichotomy. It is time we abandoned that lingering 'scientistic' prejudice which holds that 'the sciences' and sound knowledge are coextensive; they are not. Given that, our central concern should be with distinguishing theories of broad and demonstrable problem-solving scope from theories which do not have this property—regardless of whether the theories in question fall in areas of physics, literary theory, philosophy, or common sense.

THE COMPARATIVE NATURE OF THEORY EVALUATION

Philosophers of science have generally sought to characterize a set of epistemic and pragmatic features which were such that, if a theory possessed those features, it could be judged as satisfactory or acceptable independently of a knowledge of its rivals. Thus, inductivists maintained that once a theory passed a certain threshold of confirmation, it was acceptable; Popper often maintained that if a theory made surprising predictions, it had 'proved its mettle'. The approach taken here relativizes the acceptability of a theory to its competition. The fact that a theory has a high problem-solving effectiveness or is highly progressive warrants no judgments about the worth of the theory. Only when we compare its effectiveness and progress to that of its extant rivals are we in a position to offer any advice about which theories should be accepted, pursued, or entertained.

CONCLUSION

Judging this sketch of a problem-solving model of science against the desiderata discussed at the beginning of the chapter, it is clear that the model allows for the possibility that a theory may be acceptable even when it does not preserve cumulativity (specifically if the problem-solving effectiveness of the new exceeds the old). The model allows a rational role for controversies about the conceptual credentials of a theory; such controversies may even lead to progressive conceptual clarifications of our basic assumptions. By bringing the epistemic assumptions of a scientist's research tradition into the calculation of the adequacy of a theory, the model leaves scope for changing local principles of rationality in the development of science. Broadening the spectrum of cognitive modalities beyond acceptance and rejection is effected by the distinction between a theory's effectiveness, its progress, and its rate of progress. The model explains how it may be rational for scientists to accept theories confronted by anomalies and why scientists are sometimes loathe to accept certain prima facie well-confirmed theories. Through its characterization of the aims of science, the model avoids attributing transcendent or unachievable goals to science. Finally, the model rationalizes the ongoing coexistence of rival theories, showing why theoretical pluralism contributes to scientific progress.

None of this establishes that the problem-solving approach is a viable model of progress and rationality. What can be said, however, is that the model can accommodate as rational a number of persistent features of scientific development which prevailing accounts of science view as intrinsically irrational. To that degree, it promises to be able to explain why science works as well as it does.

PART 5 SUGGESTIONS FOR FURTHER READING

Brown, Harold I. *Perception, Theory, and Commitment*. Chicago: University of Chicago Press, 1977.

Grandy, R., ed. *Theories and Observation in Science*. Englewood Cliffs, NJ: Prentice-Hall, 1973.

Gutting, Gary, ed. *Paradigms and Revolutions*. Notre Dame, IN: University of Notre Dame Press, 1980.

Hanson, N. R., *Patterns of Discovery*. New York: Cambridge University Press, 1958.

Kordig, Carl R. "The Theory-ladenness of Observation." *Review of Metaphysics* 24 (1971): 450–455.

———. *The Justification of Scientific Change*. Dordrecht: D. Reidel, 1971.

Kukla, Andre. "The Theory-Observation Distinction." *Philosophical Review* 105 (1996): 173–230.

Scheffler, Isreal. *Science and Subjectivity*. Indianapolis: Bobbs-Merrill, 1967.

PART 6

SCIENCE AND OBJECTIVITY
The Science Wars

The positivists drew a distinction between the context of discovery and the context of justification. The process by which theories are arrived at is not one that can be analyzed in logical terms. Often they are the result of flashes of insight or intuition that resist any sort of formalization. But regardless of where a theory comes from, once it's before us, we can subject it to rational scrutiny. There are procedures we can follow, such as putting it to the test, which can determine whether we are justified in believing it.

Kuhn's claim that the choice among paradigms is not rationally analyzable—that it's more like a political decision or a religious conversion—led a number of thinkers to question the distinction between the context of discovery and the context of justification.

When we ask why someone believes something, we can be asking two very different things: (1) what caused him or her to believe it or (2) what justifies him or her in believing it. For example, if someone asked you why you believe in God and your response was because your parents told you that God exists, you would be giving an answer to the first sort of question—you would have identified the cause of your belief. But this sort of response doesn't answer the second type of question. From the fact that your parents told you that God exists, it doesn't follow that God exists or even that it is probable that God exists. After all, they probably also told you that Santa Claus and the tooth fairy exist, and we all agree that we are not justified in believing in them.

Some sociologists, however, claim that to answer the first sort of question is to answer the second sort of question—that there is no more to justifying a belief than identifying its cause. Latour and Woolgar, for example, say that rational considerations of the sort traditionally studied by philosophers play little or no role in determining what scientists believe. It is all a matter of politics and peer pressure. To prove their point, they decided to undertake a sociological investigation of a scientific research center. They recorded the interactions among scientists in the same way that sociologists record the interactions among primitive peoples. What they found was that scientific discoveries are the result of a complex negotiation process that is influenced by all sorts of social variables. From this they concluded that scientific reality is constructed by scientists in the laboratory. In particular, they claim that the chemical structure of thyrotropin-releasing factor (TRF) was socially constructed.

Thomas Cole finds this conclusion unwarranted because Latour and Woolgar have not shown that the outcome of the scientific investigation would have been any different had any of the social factors involved been different. Scientists certainly engage in discussions with one another, and what they say to one another is influenced by their status, seniority, gender, and so on. But what Latour and Woolgar must show to make their point is that if these factors had been different, the result of the investigation would have been different. And this, Cole claims, they have failed to show. Social factors may influence how long it takes to reach a conclusion, but there is no evidence that it affects the conclusion reached.

Sandra Harding attacks the objectivity of science from a different angle. She claims that Francis Bacon, one of the leading figures of the scientific revolution of the sixteenth century, used rape and torture metaphors to make the experimental method more attractive to male researchers. Just as Newton's mechanistic metaphors served to make machines seem more natural, Harding suggests that Bacon's torture and rape metaphors served to make torture and rape seem more natural. In fact, this implicit legitimization of rape and torture may have been responsible for some of the negative effects of applied science, such as pollution, alienation, and war.

Alan Soble disputes Harding's interpretation of Bacon. One of the offending passage is this:

> For you have but to hound nature in her wanderings, and you will be able when you like to lead and drive her afterwards to the same place. . . . Neither ought a man to make scruple of entering and penetrating into those holes and corners when the inquisition of truth is his whole object.

Soble points out that someone who had no scruples about penetrating holes could be a rapist, but he could also be a proctologist or a billiard player. In any event, the sexist interpretation is not the only plausible one. Another possibility is that Bacon is comparing science to a fox hunt. On this reading, nature is like a fox that can be hounded back into her hole. If you want to catch the fox, however, you can't have any scruples about reaching down into the hole and pulling her out. Given that organized fox hunting began in England at about the same time that Bacon was writing this passage, this interpretation is not entirely without merit.

Bacon had many reasons for extolling the experimental method. Chief among them was its potential to improve the human condition. Bacon even wrote a utopian novel, *The New Atlantis,* describing an ideal society guided by science. Soble doubts that comparisons to rape and torture would have made it more appealing, and these metaphors are not consistent with Bacon's vision of improving the human condition through science.

Whether or not Bacon used misogynistic metaphors to advance the cause of science, the fact remains that the scientific establishment has not treated women or women's issues with the respect they deserve. Janet Sayers describes some of the feminist critiques that have been leveled against the scientific establishment: It has discriminated against women in both hiring and promotion, it has neglected the study of female biology in favor of male biology, and it has interpreted various findings in ways that are detrimental to women. These criticisms are based on the traditional standard of objectivity. The unfair treatment of women shows that the scientific establishment has not been living up to its own standard.

An inquiry has traditionally been considered to be objective if it is not contaminated by social practices. Sayers argues, however, that this very conception of objectivity contributes to sexism in science. Scientific practice has been shaped by social factors. Specifically, it has been affected by the separation of production from reproduction brought about by the industrial revolution. Only by recognizing the social determinant of scientific practice can science hope to become truly objective.

Just as Sayers contends that we need to rethink the traditional notion of objectivity, other feminists claim that we need to rethink the traditional notion of knowledge itself. They claim that by focusing on women's experience, we can arrive at an entirely new conception of knowledge. Janet Richards views such a project as fundamentally misguided. It may well be the case that what scientists have considered to be knowledge is not really knowledge, but this doesn't mean that scientists' conception of knowledge is mistaken; they may have simply misapplied the concept. From the fact that some people take fool's gold for gold, it doesn't follow that anything is wrong with the concept of gold. What's more, you can maintain that a knowledge claim is mistaken only if you assume a particular theory of knowledge. But if the theory of knowledge you assume is the one you are arguing for, you are guilty of begging the question, for you have assumed what you are trying to prove. To one who has not already accepted your position, your argument would carry no weight. So feminism, Richards concludes, can give us no reason to alter our fundamental theory of knowledge. Feminism is essentially derivative; all of its results presuppose and thus require the traditional theory of knowledge.

27

BRUNO LATOUR AND STEVE WOOLGAR

The Social Construction of Scientific Facts

THE CONSTRUCTION AND DISMANTLING OF FACTS IN CONVERSATION

One way to examine the microprocesses of fact construction in science is by looking at conversation and discussions between members of the laboratory. For various reasons, it was not possible to tape-record discussions in the laboratory. For a total of twenty-five discussions, however, notes were compiled which include records of the timing, gestures, and intonation. A number of informal discussions, including snatches of conversations at benches, in the lobby, and at lunch, were similarly noted. Tape recorders could not be used, so these notes lack the precision necessary for "conversational analysis." Even in their somewhat crude or "tidied" state, however, these discussion notes provide a useful opportunity for a close analysis of the construction of facts.

We began by considering three short excerpts from an informal discussion in order to illustrate some of the ways in which arguments are constantly modified, reinforced, or negated during ordinary interaction in the laboratory. The conversation took place between Wilson, Flower, and Smith in the lobby. Smith was on the point of leaving when Wilson began to talk about an experiment he had done some days previously:

a. *Wilson (to Flower)*: You know how difficult this ACTH assay is, for the lower amount. . . . I

was thinking, well, for fifteen years I have wasted my money on his assay . . . Dietrich had calculated an ideal curve. Last time he made a mistake, because if you look at the real data, each time ACTH goes down Endorphin goes down, each time ACTH goes up Endorphin goes up. So we are going to calculate the fit between the two curves. Snoopy did it, it's 0.8.

Flower: Wooh!

Wilson: And we are going to do it with the means, which is perfectly legal. It will be, I am sure, 0.9.

Wilson and Flower then began to discuss a paper they were writing for *Science*. As Smith again started to leave, however, Wilson turned to him:

b. *Wilson (to Smith)*: By the way, I saw on the computer yesterday a 93% (match between) hemoglobin . . . or yeast?! . . .

(to Flower): You know what we are talking about? Our friend Brunick yesterday announced at the Endocrine Society Meeting that he had an amino acid analysis for GRF. You know what happened with his GRF? Smith had a computer program to look at homologies and found a 98% homology with hemoglobin, and I don't know what . . . yeast floating in the air . . . .

Flower: That's a case for concern.

Wilson (laughing): Depends on who you are. . . .

In the first excerpt, the notion that ACTH and endorphin were the same was reinforced by the

Laboratory Life: The Social Construction of Scientific Facts (Beverly Hills: Sage Publications, 1979), pp. 154–158, 235–250, 243. Copyright © 1979 Sage Publications, Inc. Reprinted by permission of the publisher.

suggestion of a probable improvement in the fit between two curves. As a result, Smith and Flower were persuaded that the operation met the desired professional standards. In the second excerpt, however, a colleague's claim was dismissed by showing an almost perfect fit between GRF, an important and long sought-after releasing factor, and a piece of hemoglobin, a relatively trivial protein. The dismissal effect is heightened by the creation of a link between his recent claim and the well-known blunder which the same colleague had committed a few years earlier (cf., Wynne, 1976: 327). Brunick had then claimed to have found a very important releasing factor, which later turned out to be a piece of hemoglobin. Brunick's recent claim was severely jeopardized by reference to this past incident. Flower's subsequent comment ("that's a case for concern") triggered a response which can be taken as indicating Wilson's high regard for his own professional standards compared to those of Brunick.

Smith left when Wilson suggested returning to discussion of the *Science* article. Wilson showed Flower a new mapping of the pituitary vascular system which had been sent to him by a European scientist. There then ensued a discussion of the map.

c. *Wilson*:   Anyway, the question for this paper is what I said in one of the versions that there was *no evidence* that there was any psychobehavioral effect of these peptides injected I.V. . . . Can we write that down?

Flower: That's a *practical* question . . . what do *we accept* as a negative answer? [Flower mentioned a paper which reported the use of an 'enormous' amount of peptides with a positive result.]

Wilson: That *much*?

Flower: Yes, so it depends on the peptides . . . but it is very important to do . . .

Wilson: I will give you the peptides, yes we have to do it . . . but I'd like to read the paper. . . .

Flower: You know it's the one where . . . .

Wilson: Oh, I have it, OK.

Flower: The threshold is 1 µg. . . . OK, if we want to inject 100 rats (we need at least a few micrograms) . . . it's a practical issue.

Unlike previous excerpts, this last sequence shows Wilson asking a series of questions. Wilson and Flower can be thought to have roughly the same academic status, even though Flower is about ten years younger than Wilson. They are both heads of laboratories and members of the National Academy of Science. However, Flower is an expert in the psychobehavioral effects of neurotransmitters whereas Wilson is a newcomer to this field. Wilson therefore needs the benefit of Flower's expertise in writing the collaborative paper (drafts of which had already been prepared at the time of the above conversation). More specifically, Wilson wants to know the basis for the claim that the peptides have no activity when injected intravenously (I.V.), so that they can counter any possible objections to their argument. At first sight, a Popperian might be delighted by Flower's response. It is clear, however, that the question does not simply hinge on the presence or absence of evidence. Rather Flower's comment shows that it depends on *what they choose to accept as* negative evidence. For him, the issue is a practical question. Flower and Wilson follow this exchange with a discussion of the amount of peptides they require to investigate the presence of psychobehavioral effects. Wilson had manufactured these rare and expensive peptides in his own laboratory. So the question for Flower is what quantity of peptides Wilson is willing to provide. The discussion between them thus entails a complex negotiation about what constitutes a legitimate quantity of peptides. Wilson has control over the availability of the substances; Flower has the necessary expertise to determine the amounts of these substances. At the same time, a claim has been made in the literature which could make it necessary to consider using an "enormous" quantity of peptides. In the light of this claim, Wilson's denial that intravenous injection gives a behavioral effect is weakened. On the other hand, Wilson argues that the amount of peptides used in the earlier work is ridiculous because it is far in excess of anything on a physiological scale. Nevertheless, Wilson agreed to give the peptides to Flower and to carry out the investigation with the amount of peptides used by the other researcher. They decided that this was the only way that Wilson's contention could be supported. Significantly, this experiment was planned after Wilson's contention had already been drafted.

Given the context of these discussions, it becomes clear that negotiation between Flower and Wilson does not depend solely on their evaluation of the epistemological basis for their work. In other words, although an idealized view of scientific activity might portray participants assessing the importance of a particular investigation for the extension of knowledge, the above excerpts show that entirely different considerations are involved. When, for example, Flower says, "it is very important to do . . .," it is possible to envisage a range of alternative responses about the relative importance of the uses of peptides. In fact, Wilson's reply ("I will give you the peptides") indicates that Wilson hears Flower's utterance as a request for peptides. Instead of simply asking for them, Flower casts his request in terms of the importance of the investigation. In other words, epistemological or evaluative formulations of scientific activity are being made to do the work of social negotiation.

A single discussion, occupying no more than a few minutes, can thus comprise a series of complex negotiations. The contention that ACTH and endorphin had some common relation was reinforced, Brunick's recent claim was degraded, and work was planned to enhance the resistance to attacks on Wilson's contention about the lack of psychobehavioral effects of certain peptides. These, then, are the results of just some mircoprocesses of fact construction which take place continually throughout the laboratory. Indeed, the encounter reported above is typical of hundreds of similar exchanges. In the course of these exchanges beliefs are changed, statements are enhanced or discredited, and reputations and alliances between researchers are modified. For our present purposes, the most important characteristic of these kinds of exchange is that they are devoid of statements which are "objective" in the sense that they escape the influence of negotiation between participants. Moreover, there is no indication that such exchanges comprise a kind of reasoning process which is markedly different from those

characteristic of exchanges in nonscientific settings. Indeed, for an observer, any presupposed difference between the quality of "scientific" and "common-sense" exchanges soon disappears. If, as this suggests, there are similarities between conversational exchanges in the laboratory and those which take place outside, it is possible that differences between scientific and common sense activity are best characterized by features other than differences in reasoning processes. . . .

CREATING A LABORATORY: THE MAIN ELEMENTS OF OUR ARGUMENT

The first concept used in our argument is that of *construction* (Knorr, in press). Construction refers to the slow, practical craftwork by which inscriptions are superimposed and accounts backed up or dismissed. It thus underscores our contention that the difference between object and subject or the difference between facts and artifacts should not be the starting point of the study of scientific activity; rather, it is through practical operations that a statement can be transformed into an object or a fact into an artifact. For instance, we followed the collective construction of a chemical structure, and showed how, after eight years of bringing inscription devices to bear on the purified brain extracts, the statement stabilized sufficiently to enable it to switch into another network. It was not simply that TRF was conditioned by social forces; rather it was constructed by and constituted through microsocial phenomena. We showed how statements are constantly modalized and demodalized in the course of conversations at the laboratory bench. Argument between scientists transforms some statements into figments of one's subjective imagination, and others into facts of nature. The constant fluctuation of statements' facticity allowed us approximately to describe the different stages in the construction of facts, as if a laboratory was a factory where facts were produced on an assembly line. The demystification of the difference between facts and artifacts was necessary for our discussion of the way in which the term fact can simultaneously mean what is fabricated and what is not fabricated. By observing artifact construction, we showed that reality was the consequence of the settlement of a dispute rather than its cause. Although obvious, this point has been overlooked by many analysts of science, who have taken the difference between fact and artifact as given and miss the process whereby laboratory scientists strive to *make* it a given.[1]

The second main concept which we have used constantly, is that of *agonistic* (Lyotard, 1975). If facts are constructed through operations designed to effect the dropping of modalities which qualify a given statement, and, more importantly, if reality is the consequence rather than the cause of this construction, this means that a scientist's activity is directed, not toward "reality," but toward these operations on statements. The sum total of these operations is the agonistic field. The notion of agonistic contrasts significantly with the view that scientists are somehow concerned with "nature." Indeed, we have avoided using nature throughout our argument, except in showing that one of its current components, namely the structure of TRF, has been created and incorporated in our view of the body. Nature is a usable concept only as a by-product of agonistic activity.[2] It does not help explain scientists' behavior. An advantage of the notion of agonistic is that it both incorporates many characteristics of social conflict (such as disputes, forces, and alliance) and explains phenomena hitherto described in epistemological terms (such as proof, fact, and validity). Once it is realized that scientists' actions are oriented toward the agonistic field, there is little to be gained by maintaining the distinction between the "politics" of science and its "truth;" the same "political" qualities are necessary both to make a point and to out-maneuver a competitor.

An agonistic field is in many ways similar to any other political field of contention. Papers are launched which transform statement types. But the many positions which already make up the field influence the likelihood that a given argument will have an effect. An operation may or may not be successful depending on the number of people in the field, the unexpectedness of the point, the personality and institutional attachment of the authors, the stakes,[3] and the style of the paper. This is why scientific fields do not display the orderly pattern with which some analysts of science like to contrast the disorderly tremors of political life. The field of neuroendocrinology thus comprises a multitude of claims and many substances exist only locally. For example, MSH releasing factor exists only in Louisiana, Argentina, and one place in Canada, and in one other in France; most of the associated literature was considered meaningless by our informants.[4] The negotiations as to what counts as a proof or what constitutes a good assay are no more or less disorderly than any argument between lawyers or politicians.[5]

Our use of agonistic is not meant to imply any especially wicked or dishonest character attribute of scientists. Although scientists' interaction can appear antagonistic, it is never concerned solely with psychological or personal evaluations of competitors. The solidity of the argument is always central to the dispute. But the constructed character of this solidity means that the agonistic necessarily plays a part in deciding which argument is the more persuasive. Neither agonistic nor construction have been used in our argument as a way of undermining the solidity of scientific facts; the reason for our nonrelativist use of these terms will be clear in our discussion of the third main concept used in our argument.

We have insisted on the importance of the material elements of the laboratory in the production of facts. For instance, we demonstrated how the very existence of the objects of study depended on the accumulation inside the laboratory walls of what Bachelard has called "phenomenotechnique." But this allows us only to describe the equipment of the group at one point in time. At some earlier point, each item of equipment had been a contentious set of arguments in a neighboring discipline. Consequently, one cannot take for granted the difference between "material" equipment and "intellectual" components of laboratory activity: the same set of intellectual components can be shown to become incorporated as a piece of furniture a few years later. In the same way, the long and controversial construction of TRF was eventually superceded by the appearance of TRF as a noncontroversial material component in other assays. Similarly, we briefly indicated how investments made within the laboratory were eventually realized in clinical studies and in drug industries. In order to emphasize the importance of the time dimension, we shall refer to the above process as *materialization* or *reification* (Sartre, 1943). Once a statement stabilizes in the agonistic field, it is reified and becomes part of the tacit skills or material equipment of another laboratory.[6] We shall return later to this point.

The fourth concept upon which we have drawn is that of *credibility* (Bourdieu, 1977). We used credibility to define the various investments made by scientists and the conversions between different aspects of the laboratory. Credibility facilitates the synthesis of economic notions (such as money, budget, and payoff) and epistemological notions (such as certitude, doubt, and proof). Moreover, it emphasizes that information is *costly*. The cost-benefit analysis applies to the type of inscription devices to be employed, the career of scientists concerned, the decisions taken by funding agencies, as well as to the nature of the data, the form of paper, the type of journal, and to readers' possible objections. The cost itself varies according to the previous investments in terms of money, time, and energy already made.[7] The notion of credibility permits the linking of a string of concepts, such accreditation, credentials, and credit to beliefs ("credo," "credible") and to accounts ("being accountable," "counts," and "credit accounts"). This provides the observer with an homogeneous view of fact construction and blurs arbitrary divisions between economic, epistemological, and psychological factors.[8]

The fifth concept used in our argument, albeit somewhat programmatically, is that of *circumstances* (Serres, 1977). Circumstances (that which stands around) have generally been considered irrelevant to the practice of science.[9] Our argument could be summarized as an attempt to demonstrate their relevance. Our claim is not just that TRF is surrounded, influenced by, in part depends on, or is also caused by circumstances; rather, we argue that science is entirely fabricated out of circumstance; moreover, it is precisely through specific localized practices that science appears to escape all circumstances. Although this has already been demonstrated by some sociologists (for example, Collins, 1974; Knorr, 1978; Woolgar, 1976), the concept of circumstances has also been developed from a philosophical perspective by (Serres (1977). . . .

The sixth and final concept upon which we have drawn is *noise* (or, more exactly, the ratio of signal to noise), which is borrowed from information theory (Brillouin, 1962). Its application to an understanding of scientific activity is not new (Brillouin, 1964; Singh, 1966; Atlan, 1972), but our usage is very metaphorical. We have not, for example, attempted to calculate the signal to noise ratio produced by the laboratory. But we have retained the central idea that information is measured against a background of equally probable events, or as Singh (1966) puts it:

> We measure the information content of a message in any given ensemble of messages by the logarithm of the probability of its occurrence. This way of defining information has an earlier precedent in statistical mechanics where the measure of entropy is identical in form with that of information (Singh, 1966: 73).

The concept of noise fits closely with our observations of participants busily reading the written tracts of inscription devices. The notion of equally probable alternatives also allowed us to describe the final construction of TRF: the import of mass spectrometry delimited the number of probable statements. The notion of demand, which allowed us to develop the idea of a market for information and to permit the operation of the credibility cycle, was based on the premise that any decrease in the noise of one participant's operation enhances the ability of another participant to decrease noise elsewhere.

The result of the *construction* of a fact is that it appears unconstructed by anyone; the result of rhetorical *persuasion* in the agonistic field is that participants are convinced that they have not been convinced; the result of *materialization* is that people can swear that material considerations are only minor components of the "thought process"; the result of the investments of credibility, is that participants can claim that economics and beliefs are in no way related to the solidity of science; as to the *circumstances*, they simply vanish from accounts, being better left to political analysis than to an appreciation of the hard and solid world of facts! Although it is unclear whether this kind of inversion is peculiar to science,[10] it is so important that we have devoted much of our argument to specifying and describing the very moment at which inversion occurs. . . .

The portrayal resulting from the above combination of concepts used throughout our argument has one central feature: the set of statements considered too costly to modify constitute what is referred to as reality. Scientific activity is not "about nature," it is a fierce fight to *construct* reality. The *laboratory* is the workplace and the set of productive forces, which makes construction possible. Every time a statement stabilizes, it is reintroduced into the laboratory (in the guise of a machine, inscription device, skill, routine, prejudice, deduction, program, and so on), and it is used to increase the difference between statements. The cost of challenging the reified statement is impossibly high. Reality is secreted.[11]

NOTES

1. This point has been made frequently by Bachelard (for example, 1934; 1953). However, his interest in demonstrating the "mediations" in scientific work was never extended. His "rational materialism," as he put it, was more often than not the basis for distinguishing between science and "prescientific" ideas. His exclusive interest in "la coupure épistémologique" prevented him from undertaking sociological investigations of science, even though many of his remarks about science make better sense when set within a sociological framework.

2. From the outset, the observer was struck by the almost absurd contrast between the mass of the apparatus and the minute quantities of processed brain extract. The interaction between scientific "minds" and "nature" could not adequately account for this contrast.

3. In a different context, the importance of the stakes may vary. For example, the importance of somatostatin for the treatment of diabetes ensures that each of the group's articles is carefully checked. In the case of endorphine, by contrast, any article (no matter what the wildness of its conjectures) will initially be accepted as fact.

4. On his first day in the laboratory, the observer was greeted with a maxim which was constantly repeated to him in one or another modified form throughout his time in the field: "The truth of the matter is that 99.9% (90%) of the literature is meaningless (crap)."

5. We base this argument on several conversational exchanges which took place between lawyers and scientists. Unfortunately, we are not able to make explicit use of this material here.

6. It is crucial to our argument that anything can be reified, no matter how mythical, absurd, whimsical, or logical it might seem either before or after the event. Callon (1978), for example, has shown how technical apparatus can incorporate the outcome of totally absurd decisions. Once reified, however, these decisions take the role of premise in subsequent logical arguments. In more philosophical terms, one cannot understand science by accepting the Hegelian argument that "real is rational."

7. But for a few pages in Lacan (1966) and some indirect hints by Young (n.d.), a psychoanalytic understanding of these kinds of energy investments is as yet undeveloped.

8. For example, Machlup (1962) and Rescher (1978) have attempted to understand the information market in economic terms. However, their approach extends rather than transforms the central notion of economic investment. By contrast, Bourdieu (1977) and Foucault (1978) have outlined a general framework for a political economy of truth (or of credit) which subsumes monetary economics as one particular form of investment.

9. The philosophical enterprise can be characterized as an attempt to eliminate any trace of circumstances. Thus, the task of Socrates in Plato's *Apology of Socrates* is to eliminate circumstances included in the definition of activity provided by the artist, the lawyer, and so on. Such elimination is the price which has to be paid in order to establish the existence of an "idea." Sohn Rethel (1975) has argued that such philosophical operations were essential for the development of science and economics. It could be argued, therefore, that the task of reconstructing circumstances is fundamentally hampered by the legacies of a philosophical tradition.

10. Barthes argues that this kind of transformation is typical of modern economics. It is thus possible that there is some useful similarity between Marx's (1867) notion of fetishism and the notion of scientific facts. (Both fact and fetish share a common etymological origin.) In both cases, a complex variety of processes come into play whereby participants forget that what is "out there" is the product of their own "alienated" work.

11. If reality means anything, it is that which "resists" (from the Latin "res"—thing) the pressure of a force. The argument between realists and relativists is exacerbated by the absence of an adequate definition of reality. It is possible that the following is sufficient: that which cannot be changed at will is what counts as real.

REFERENCES

Atlan, H. (1972) *L'organisation biologique et la théorie de l'information*. Paris: Hermann.

Bachelard, G. (1934) *Le nouvel esprit scientifique*. Paris: P.U.F.

——— (1953) *Le matérialisme rationnel*. Paris: P.U.F.

Bourdieu, P. (1972) *Esquisse d'une théorie de la pratique*. Genève: Droz.

——— (1975a) "Le couturier et sa griffe." *Actes de la Recherche en Sciences Sociales* 1 (1).

——— (1975b) "The specificity of the scientific field and the social conditions of the progress of reason." *Social Science Information* 14 (6): 19–47.

——— (1977) "La production de la croyance: contribution à une économie des biens symbolique." Actes de la Recherche en Sciences Sociales 13:3–43

Brillouin, L. (1962) *Science and Information Theory*. New York: Academic Press.

——— (1964) *Scientific Uncertainty and Information*. New York: Academic Press.

Callon, M. (1978) *De problèmes en problèmes: itinéraires d'un laboratoire universitaire saisi par l'aventure technologique*. Paris: Cordes.

Collins, H. M. (1974) "The T.E.A. set: tacit knowledge and scientific networks." *Science Studies* 4:165–186.

Foucault, M. (1978) "Vérité et pouvoir." *L'arc* 70.

Knorr, K. (1978) "Producing and reproducing knowledge: descriptive or constructive." *Social Science Information* 16(6):669–696.

——— (1978) "From scenes to scripts: on the relationships between research and publication in science." Paper presented at the American Sociological Association meeting.

——— (1979) "The research process: tinkering towards success or approximation of truth." *Theory and Society* 8 (3):347–376.

Lacan, J. (1966) *Les écrits*. Chapter: "La science et la vérité," pp. 865–879. Paris: Le Seuil.

Lyotard, J. F. (1975, 1976) *Lessons on Sophists*. San Diego: University of California.

Machlup, F. (1962) *The Production and Distribution of Knowledge*. Princeton, N.J.: Princeton University Press.

Marx, K. (1970) *Feuerbach: Opposition of the Naturalistic and Idealistic Outlook*. New York: Beckman.

———. (1977) *The Capital,* Vol. 1. New York: Random House.

Rescher, N. (1978) *Scientific Progress: A Philosophical Essay on the Economics of Research in Natural Science*. Oxford: Blackwell.

Sartre, J. P. (1943) *L'Etre et le Néant*. Paris: Gallimard.

Serres, M. (1972) *L'interférence,* Hermes II. Paris: Ed. de Minuit.

——— (1977a) *La distribution,* Hermes IV. Paris: Ed. de Minuit.

——— (1977b) *La naissance de la physique dans la texte de Lucrèce: fleuves et turbulences*. Paris: Ed. de Minuit.

Singh, J. (1966) *Information Theory, Language and Cybernetics*. New York: Dover.

Sohn, Rethel, A. (1975) "Science as alienated consciousness." *Radical Science Journal* 2/3: 65–101.

Woolgar, S. W. (1976a) "Writing an intellectual history of scientific development: the use of discovery accounts." *Social Studies of Science* 6:395–422.

——— (1976b) "Problems and possibilities in the sociological analysis of scientists' accounts." Paper presented at 4S/ISA Conference on the Sociology of Science. Cornell, New York, Nov. 4–6.

Wynne, B. (1976) "C.G. Barkla and the J phenomenon: a case study in the treatment of deviance in physics." *Social Studies of Science* 6:307–347.

Young, B. (n. d.) "Science is social relations" (mimeographed).

28

STEPHEN COLE

Voodoo Sociology: Recent Developments in the Sociology of Science

Up until the 1970s, sociologists of science did not examine the actual cognitive content of scientific ideas, as they believed that these were ultimately determined by nature and not a product of social processes and variables. Beginning in the late 1970s and early 1980s, a group of European sociologists, adopting a relativist epistemological position, began to challenge this view. At first they called themselves "relativist-constructivists" and later, more simply, "social constructivists." Their numbers were few, but within the short time span of roughly one decade, this group has come to completely dominate the sociology of science and the interdisciplinary field called the social studies of science. Although some like to deny this dominance because ideologically they do not like to see themselves as the power elite, their control of all the major associations and specialty journals is clear to anyone participating in the field. This dominance may easily be seen in the recently published *Handbook of Science and Technology Studies*,[1] published by the Society for the Social Study of Science. Virtually all the contributors are either constructivists or political allies of the constructivists. Tom Gieryn, a former student of Merton, but now a convert to constructivism, in his contribution to this handbook claims as an aside, "If science studies has *now convinced everybody* that scientific facts are only contingently credi-

ble and claims about nature are only as good as their local performance. . . ."[2]

It is very important to point out that social constructivism is not simply an intellectual movement, a way of looking at science, but it is an interest group that tries to monopolize rewards for its members or fellow travelers and exclude from any recognition those who question any of its dogma. In fact, the editors of the *Handbook* discussed above refused to include a chapter on the important topic of social stratification in science (in the most prestigious general journals of sociology like the *American Sociological Review*, more articles have been published on this topic than any other), claiming at first that they could find no one willing to write such a chapter. When I volunteered, an invitation to contribute was never forthcoming.

In the leading monographs that established the social constructivists—the two most important being the laboratory studies by Bruno Latour and Steve Woolgar, *Laboratory Life*, and Karin Knorr-Cetina's *Manufacturing Knowledge*, published in 1981—the epistemological position taken was highly relativistic. Scientific facts were not constrained by nature but were socially constructed or made up in the laboratory by the scientists. Harry M. Collins, another leading constructivist, stated that the *"natural world has a small or non-existent role in the construction of scientific knowledge."*[3] Some constructivists, including Collins, later claimed that he "really didn't mean this"; it was just being used in a "polemic" or a "programmatic" statement or to de-

The Flight from Science and Reason, Paul R. Gross, Norman Levitt, and Martin W. Lewis, eds. (Baltimore: Johns Hopkins University Press, 1996), pp. 274–287. Reprinted by permission of the publisher.

scribe "methodological relativism" as opposed to "epistemological relativism." Well that is fine, because if constructivists do not *really* subscribe to epistemological relativism, then there is nothing very radical about their work and no reason why it cannot be integrated with other nonrelativistic work in the social studies of science. But clearly he and they did and do really mean it, as it is the core belief of the entire program.

Some of the constructivists, influenced by symbolic interactionism and ethnomethodology, argued that what is a scientific fact was a result of social negotiations among scientists in the laboratory. Others, with a more neo-Marxist orientation, argued that social and economic interests determined the content of scientific idea. Thus, David Bloor argued[4] that Boyle's law was influenced by his conservative political beliefs and his desire to maintain the status quo in order to protect his vast Irish land holdings. This argument was strikingly similar to that made in the 1930s about Newton's work by the Soviet sociologist of science Boris Hessen.[5]

Where did these views come from? Although similar movements were occurring in other disciplines at the same time, literary theory, for example, there is little evidence that the constructivists in science were directly influenced by developments in other fields.[6] The single most important influence on the development of social constructivism was clearly the classic book by Thomas S. Kuhn, *The Structure of Scientific Revolutions*. In light of the growing popularity of relativistic views in the more humanistic areas of the social sciences (not the dominant quantitative areas that were then and still are today committed to an old-fashioned positivism), Kuhn's book seemed to give warrant to the view that social consensus determined "nature" rather than nature determining scientific consensus.

Politics also played a role in the emergence of constructivism. There were many young politically left people who entered sociology in the 1960s, a substantial portion being neo-Marxist or sympathetic to such views, who considered the then dominant approach to sociology—the functionalist view of Robert K. Merton and his students—to be politically conservative and welcomed any vehicle that allowed them to attack this group.

As the relativist-constructivist approach to science began to gain steam in the early 1980s, it began to draw some serious critical attention. Perhaps the earliest people to attack the constructivist approach were philosophers of science, with Larry Laudan,[7] Ronald Giere,[8] and David Hull[9] being prominent among them. With a group of eminent philosophers of science as contributors to this volume, I will not attempt to discuss the objections that philosophers had to the constructivist approach. Historians also began to raise questions about the accuracy of the constructivist portrayal of the development of science. Among the critics were Stephen Brush,[10] Martin Rudwick,[11] Peter Galison,[12] and ultimately Kuhn himself. In Kuhn's 1992 Rothschild Lecture he stated:

> "the strong program" [another term for the relativist-constructivist approach] has been widely understood as claiming that power and interest are all there are. Nature itself, whatever that may be, has seemed to have no part in the development of beliefs about it. Talk of evidence, or the rationality of claims drawn from it, and of the truth or probability of those claims has been seen as simply the rhetoric behind which the victorious party cloaks its power. What passes for scientific knowledge becomes, then, simply the belief of the winners. *I am among those who have found the claims of the strong program absurd: an example of deconstruction gone mad.*[13]

It is interesting that many members of the constructivist school do not see some of the scholars mentioned above, such as Rudwick and Galison, as being opposed to their position. In fact these two are frequently positively cited by constructivists. This is because all of the people I have mentioned have rejected the stereotyped view of positivism that the constructivists have set up as a straw man and therefore on some issues can be seen as having the same views as the constructivists. Virtually everyone writing about science today, the constructivists and their critics (including myself), reject the overly rationalized and idealized view of science that was prominent prior to the 1970s. But although scholars like Galison and Giere will reject the same stereotyped positivism that Kuhn rejected, they are also just as opposed to the relativism that is at the heart of the constructivist program.

Although virtually all of the leaders of the social constructivist school are or claim to be sociologists, criticisms by sociologists have been almost nonexistent. In 1982 Thomas Gieryn, a Merton student, published a criticism of the relativist-constructivists, arguing (I believe incorrectly) that all the major insights of the constructivists had been made by Merton.[14] Soon after, Gieryn, sensing the increasing power of the constructivists, had a conversion and today is proudly displayed as a member of that school as converts were displayed during the cold war. In fact, it was not until 1992 with the publication by Harvard University Press of my book *Making Science: Between Nature and Society* that there was any significant critique by a sociologist of the constructivist school.

Why were sociologists of science so late in coming to the party? First, as I pointed out above, many sociologists disliked the so called "Mertonians" for political reasons and were glad to see the rise of any school that would challenge Merton's perceived control of the specialty. Although Merton is widely respected by nonsociologists and sociologists in many countries abroad,[15] many of his ideas are currently "out of fashion."

Second, many correctly perceived the constructivist approach as an attack on the natural sciences and were pleased to see these sciences, which have long lorded it over the social sciences, knocked off their pedestal. The most eminent leader of the constructivists, Bruno Latour, readily admits that the delegitimization of the natural sciences was one of their goals. For example, in a recent debate with Collins, Callon and Latour argue that:

> The field of science studies has been engaged in a moral struggle to strip science of its extravagant claim to authority. Any move that waffles on this issue appears unethical, since it could also help scientists and engineers to reclaim this special authority which science studies has had so much trouble undermining.[16]

And in the same piece they go on to say: "we wish to attack scientists' hegemony on the definition of nature, we have never wished to accept the essential source of their power: that is the very distribution between what is natural and what is social and the fixed allocation of ontological status that goes with it."[17] And a few pages later when they list their goals, the first listed is "disputing scientists' hegemony."[18]

Third, most American sociologists of science lacked the philosophical training of the European constructivists and either did not really understand much of the constructivist work or were afraid to engage in a battle that would necessarily involve philosophical argument—an area where they felt distinctly disadvantaged. It is my bet that more than fifty percent of those citing the work of Latour could not give a coherent explanation of that work if asked to do so. People have jumped on a bandwagon that they do not really understand.

Fourth, Merton himself has a well-known dislike for controversy and has never made a public statement on the social constructivist school. Thus, the "Mertonians" were left without a leader in the debate. And finally, up until the last five years, the constructivists had very little influence in the United States and thus could safely be ignored as a European phenomenon that had no effect on the ability of nonconstructivist sociologists of science (a group dwindling in size all along) to do their quantitative more traditional work and have that work published in mainstream journals where the rewards for sociologists lie. American sociologists of science, for the most part, considered themselves to be primarily interested in sociology, with science used as a research site. The European constructivists were primarily interested in science and in many (not all) cases were abysmally ignorant of more general sociological issues and concerns.

Let me now briefly summarize my critique of the social constructivist approach to science. If a sociologist wants to show that social variables influence the cognitive content of science, she must be careful in specifying exactly what about the cognitive content is being influenced. There are three different ways in which social factors could influence cognitive content. The first has been called the foci of attention or what problems scientists choose to study. There is no question that problem choice is influenced at least to some significant extent by social factors. This was illustrated well by Merton in his classic study of seventeenth-century English science[19] in which he showed how practical military and economic problems of the day played a strong role in determining what problems scientists attempted to solve. A second way to look at how social factors influence the cognitive content is to look at the rate of advance. How does the social organization of science and the society in which it is embedded influence the rate at which problems will be solved? The centers of scientific advance have changed over time; how can we explain this? This was also analyzed by Merton in his study of seventeenth-century science when he asked the question of what influenced the dispersion of talent within a society. The question has also significantly been addressed by the late Joseph Ben-David[20] and Derek de Solla Price.[21] Tom Phelan, a former student of mine, and I are currently investigating this problem using nation states as the unit of analysis and the number of highly cited papers written as a measure of the dependent variable.

The third way to look at cognitive content is to look at the actual substance of solutions to specific scientific problems. For example, in *Laboratory Life,* Latour and Woolgar try to show how the discovery of the chemical sequence of thyrotropin releasing factor (TRF) made by R. Guillemin and A. V. Schally was socially constructed. The scientific community came to believe that TRF was made up of the sequence Pyro-Glu-His-Pro-NH_2 rather than some other sequence. It is this latter sense of cognitive content that the social constructivists are interested in. And they claim that since science is not constrained by nature, the solution to the chemical structure of TRF could have been different and that the specialty of neuroendocrinology would have progressed to the same degree or perhaps an even greater degree had some other structure been identified.

I argue in my book that there is not one single example in the *entire constructivist literature* that supports this view of science. In order to demonstrate the credibility of their view, one must show how a specific social variable influences a specific cognitive content. In all of their work there is always at least one or another crucial piece missing. In some of their work they illustrate well how social processes influence the doing of science; but they fail to show how they have had a significant effect

on what I call a knowledge outcome or a piece of science that has come to be accepted as true by the scientific community and thereby entered the core knowledge of that discipline.

In order to demonstrate this, it is necessary to do a very close reading of the texts produced by the constructivists. First, I will give an example of how they discuss the social processes influencing scientists as they go about their work; but fail to show how these processes influence any scientific outcome. In *Laboratory Life,* probably the most heavily cited and influential work done by constructivists, Latour and Woolgar present a long description of the social negotiation taking place between two scientists, Wilson and Flower.[22] They succeed in showing that scientists engage in social negotiations about their work in the same way as any other people negotiate about other aspects of their interaction. But what they have decisively failed to show is how this social negotiation influenced any aspect of science. *They have in effect "black boxed" the science; the very thing they so indignantly accused the "Mertonians" of doing.*

This is not just one example. Much of the work in which the constructivists talk about how scientists engage in social negotiations or are "human" fails to show how such negotiation of "humanness" influences any piece of communally held science in any way. Knorr-Cetina does the same thing when she describes social negotiation between a Watkins and a Dietrich[23]—but she never says how it influenced the scientific outcome. Also, in her famous analysis of how scientists negotiated fifteen different versions of a paper they are writing, she does not say whether it made any difference. Was the last published draft a significantly different piece of science than the first draft? And even if it were, if no one paid attention to this particular paper (if, in other words, it was a trivial piece of science she was analyzing), what difference would all the negotiation make for the scientific community?

In other cases, constructivists analyze scientific conflicts but fail to show how the resolution of these conflicts was influenced by social variables. It is a sad commentary on the constructivist program that the example they like to cite most frequently is the now well-known work by Andrew Pickering, *Constructing Quarks.* In this book and several earlier papers, Pickering discusses the debate theoretical high-energy physicists had over two theories—one was called "charm" and the other "color." The former theory won out; but Pickering fails to show that this was a result of social factors. His own analysis leads to the opposite conclusion, that it was data from experiments that led to the resolution of the "conflict." And research recently done by some students of mine in an undergraduate seminar suggests that there may never have been a "conflict" at all. There were almost no citations to the "color" theory other than by its two proponents. If there ever really was a conflict, we would expect to see some citation to both theories and then, when the conflict was resolved, a drop in citation to the "loser." Whom, we may ask, ever believed the "color" theory, or did Pickering manufacture this "conflict" because it would be an easy one to deal with? The primary point, however, is that if there was or was not a conflict, it was resolved, as Pickering himself points out, by empirical evidence.

In what the constructivists would call the naive and outdated language of positivist sociology, they either fail to have an adequate dependent variable, an adequate independent variable, or to demonstrate the link between these variables. Even the constructivists frequently show that they are aware that their theoretical approach cannot explain what they empirically observe. For example, consider another case from *Laboratory Life*. Latour and Woolgar tell us that Schally was about to publish the formula that eventually turned out to "be" TRF; but he believed in Guillemin's work more than his own and held back in publication. Guillemin was at that time arguing that TRF might not even be a peptide. Essentially what Latour and Woolgar are saying is that Guillemin's authority was so great that it served to delay the "discovery" of TRF by several years. But what they decisively fail to say is that the structure of TRF would have been anything other than Pyro-Glu-His-Pro-NH$_2$ whether or not this discovery would have been made several years earlier or at the time that it was. In other words, they have sneakily changed the dependent variable to be the rate of advance rather than the cognitive content of knowledge.

As a general rule, readers of the work of the social constructivists should always ask (1) have they identified a real social independent variable? and (2)

have they shown that it has influenced the actual cognitive content of some piece of science rather than the foci of attention or the rate of advance? By influence, we mean that the cognitive content (as it was accepted by the scientific community) turned out one way rather than another *because* of some social process. Of course, processes like those described by Latour and Woolgar affect how long it takes for a particular discovery to be made; but they have failed to give a single example where the social processes influence the content of such a discovery.

A frequently used constructivist rhetorical trick is to argue that it is impossible to separate the technical from the social; that all science is inherently social. This turns their entire argument into a tautology. If science is inherently social, this means that the technical aspects of scientific discoveries by necessity must be determined socially. This is indeed the question we are examining—the extent to which the technical aspects of science have social determinants. If we take as an assumption one answer to the question we are researching, then we might as well all pack up our bags and go home because there would be nothing more to research.

If there is only one correct outcome or solution to a scientific problem, like the structure of TRF, *before the actual discovery is made,* then this gives no room for social factors to influence it. If the constructivist position is correct, this means that it would have been possible that some other structure of TRF could have been accepted as fact and that the discipline in both its pure and applied aspects would have proceeded with just as great success. But even Latour and Woolgar show their skepticism about such a belief. In discussing the story of TRF, they point out that at one time it looked like the scientists might have been forced to abandon the program by inability to obtain enough research material:

It was then feasible that partially purified fractions would be continued to be used in the study of modes of action, that localization and classical physiology could have continued, and that Guillemin would merely have lost a few years in working up a blind alley. TRF would have attained a status similar to GRF or CRF, each of which refers to some activity in the bioassay, *the precise chemical structure of which had not yet been constructed.*[24]

Note again that they are not saying that the chemical structure of TRF would have been any different; but only that the problem would not have been solved.

They fail to explain why some local "productions" (as researchers like Knorr-Cetina like to call the results of laboratory science) are successful in the larger community and others are not. Latour's discussion of strategies and power clearly fails to explain cases like DNA in which the discovery was accepted into the core almost overnight. Its authors were both unknown, and their opponents were leaders of the field. What made opponents such as Pauling, Wilkens, and Chargaff enroll in the Watson and Crick bandwagon?

My book is full of many examples, based upon detailed readings of other constructivist texts, that show how in each and every case they fail to do what they claim. An examination of what happened to this book is good evidence of how the constructivists treat criticism. First, all reviews by constructivists were harshly negative, including one by Shapin in *Science* and one by Pickering in the *Times Literary Supplement*. Fuller actually wrote two negative reviews in two different journals. All the reviews of the book in mainstream American sociology journals that I have seen were moderately to strongly positive, including an extremely positive review by Mary Frank Fox in *Contemporary Sociology*. But the most noticeable aspect of the constructivists' reaction to the book was to ignore it. Where they have to review it, they will give it a good bashing, but where they have any control, they feel the best course of action is to keep the book unknown. Thus the book has gone unreviewed in the two leading specialty journals in the field, the *Social Studies of Science* and *Science, Technology, and Human Values*. It is quite probable that considerably more than half of the members of the Society for the Social Studies of Science do not even know of the book's existence.

Besides out and out distortion of what is said in the book and using the review as an occasion for general "Mertonian" bashing, the most frequent tactic taken is what I call the "we never said that (or meant that)" tactic. Most of the constructivist leaders criticized in my book are not stupid; far from it.

They know that what they say cannot hold water and, when pushed to its real foundations, is logically absurd. Therefore, the only way to defend themselves is to say that they never said what I said they said (or if they said it they did not mean it). There are two answers to this rebuttal. First, sit down with my book and their texts and read both closely and then determine whether they did or did not say what I said they said. For example, a recent review[25] says that my criticism "relies on misreadings" of the constructivist works. "He insists that the constructivist research program is premised on denial that the realities of nature play any part in scientists' deliberations, whereas his antagonists merely presume that these realities cannot be abstracted from the theories and technologies that frame them."[26] Statements such as these make me wonder how the reviewer ever got out of the eighth grade. Can she or can she not read the direct quote from Collins above in this article? Or does she want to accept the claim that he did not really mean it? To the extent that constructivists do not "really mean" their statements of relativism, then there is nothing contradictory between their beliefs and my own.

Shapin is another one who is notorious for putting a sugar coating on the constructivist pill.[27] For example, in a recent piece on the sociology of scientific knowledge, Shapin describes Latour's work as showing that there were more "politics" within the walls of scientific workplaces than there were outside and that to secure the support of other scientists for a scientific claim was a thoroughly social process.[28] There is no "Mertonian" sociologist of science who would disagree with such conclusions. In the long review article, he fails to deal at all with the relativism that is at the core of the constructivist program. Shapin himself, who used to write polemics supporting the constructivist program,[29] is now doing fairly traditional social history of science on topics that Mertonians studied more than twenty years ago. In his most recent book,[30] he essentially asks what social processes were involved in establishing authority in seventeenth-century science. He emphasizes the importance of the "gentleman" as an individual who could be trusted. There is no discussion, anywhere in the book, of how any social process influenced the content of science. The science is virtually black boxed throughout the book. He is arguing against the rationalistic view of science that rejects trust and authority as mechanisms that influence belief as opposed to direct observation. Consider the following quote:

> According to the classical view of the history and philosophy of science, consensus is determined by the empirical phenomena themselves. Theories supported by empirical observation would become part of the consensus, theories at odds with observable "facts" would be discarded. . . . Once we accept the notion that consensus does not automatically spring from nature, we are forced to pay more attention to the sociological processes through which consensus is developed, maintained, and eventually shifted. One of the primary mechanisms through which consensus is maintained is the practice of vesting authority in elites.

There is one problem here; this quote is not from Shapin's book. Rather it is from a book published in 1973 by Jonathan and Stephen Cole, entitled *Social Stratification in Science*.[31] This latter book is not cited by Shapin. Now the point is not to say that we have priority on this matter. Clearly we never did the type of detailed and admirable historical work conducted by Shapin to show how the mechanisms of authority may have developed in a particular society at a particular time. It is, however, to point out that without the relativism, there is no great gap between contemporary work done in the social studies of science and the past work of the misguided "Mertonians."

What is the current state of constructivist sociology of science? It is a field that is in intellectual disarray but stronger than ever in its political control of organizations, journals, and science studies programs within universities. After the constructivists' unexpected and amazingly easy takeover of the sociology of science, when it was no longer fun to flog the "Mertonian" bad boys, they fell out among themselves and split into a bunch of warring clans.

One problem always faced by the constructivists was that of reflexivity. If in the natural sciences facts were not based upon empirical evidence from the external world, then how could they be said to be so based in the sociology of science? Why should any-

body bother reading the works of Latour et al. if they only represented an attempt to push a point of view by power? Some constructivists such as Woolgar and Ashmore and to some extent Mulkay went off in the direction of taking their own work as the subject of analysis.[32] In the mean time Bruno Latour, who had become the demigod of the constructivist movement, and his French sidekick Michel Callon, began to recognize the problems inherent in their relativist position and turned on a dime and claimed they are not now and never have been relativists. Instead, he and Callon are now developing what they call actor-actant network theory. The most interesting part of this work is that what used to be considered the object of study, quarks, for example, now become equal to the humans. Scientific, and indeed social, outcomes are a result of interaction among a network of scientists, practitioners, other people, and things.

In a famous paper by Callon in which he analyzes some applied science on scallop fishing,[33] he concludes his story by saying that the reason that the experiment failed was because the scallops would not cooperate. In a vicious and amusing polemic between Latour and Callon on one side and Harry Collins and Steven Yearley on the other, Collins correctly accuses Latour of abandoning the relativist-constructivist program. As Collins and Yearley say: "The crucial final quotation [in Callon's article on scallops] is: 'To establish that larvae anchor, the complicity of the scallops is needed as much as that of the fishermen.'"[34] It does not take very much insight to see that this is a nifty way to bring nature back into the analysis. Latour and Callon have no answer to this, although they do successfully make some criticisms of Collins's work which privileges sociology while trying to attack the privileging of the natural sciences. For example:

> That scallops do not interfere at all in the debate among scientists striving to make scallops interfere in their debates—is not only counterintuitive but empirically stifling. *It is indeed this absurd position that has made the whole field of SSK* [social study of scientific knowledge—another term for constructivism] *look ridiculous* and lend itself to the "mere social" interpretation.[35]

An intelligent reader of this polemic between Collins and Yearley on the one side and Callon and Latour on the other can do nothing other than agree with both sides. Their work if taken seriously is nothing other than absurd or voodoo sociology. The sociology of science with its many potentially interesting questions has gotten lost in a tautological mess of philosophical arguments that, as many of even the constructivists have now seen, lead nowhere.[36]

Latour's latest book[37] is an obscure philosophical essay (which he proudly proclaims has no examples because of his "Gallic" tradition) that does not answer the pressing problems facing the sociology of science today and goes little beyond Callon and Latour[38] in developing actor-actant network theory. In both pieces, Latour argues that modernists have tried to locate things along a nature-society polarity and this has made it impossible for them to deal with "hybrids" or with phenomena that move back and forth between the nature end of the continuum and the social end of the continuum as they develop. Thus, he would argue there is no way to deal with a phenomenon like TRF, which is social as the scientists struggle to define it and becomes a result of nature once it has been defined. Latour suggests throwing in another dimension, which is time. But this does not solve any of the problems because it does not tell us why certain objects become stabilized and others do not and why some become stabilized at the nature end of the pole and others at the social. It is perhaps because of the deep obscurity of Latour's latest book that his followers are defining it as his "best contribution yet."

I find it exceedingly strange when my book can be criticized for addressing itself to the following type of questions:

Can the rate of scientific innovation in a nation be counted a linear function of the sheer number of scientists it manages to sustain (and, by implication, is scientific advance somehow linked to national development)? Can non-scientists whose decisions affect the direction of research be provided with quantitative measures of the quality of scientific work? Do universalistic criteria inform routine procedures for evaluating proposed projects, so that the most deserving applicants are funded? Do status differentials within disciplinary communities impede the free flow of communication so necessary to the growth of knowledge, possibly to a degree that precludes recognition of outstanding work done by persons who toil in professional obscurity? Do scientists' creative powers diminish as they age, so that an individual of relatively advanced age should, *ceteris paribus,* be presumed an unworthy recipient of support? And, even if there is no association between creativity and age, should younger grant applicants receive preferential consideration, lest the scientific community become moribund through failure to promote young talent (and the population as a whole thereby cease to enjoy the fruits of scientific progress)?[39]

Are not these indeed the type of questions that sociologists of science should be interested in and representative of a type of work that has literally been wiped out by the dominance of the new social constructivists?

In my book, *Making Science,* I call for a rapprochement between the social constructivists and the more traditional sociologists of science. Much of the work of the constructivists has been useful in pointing out that science is not the type of rational endeavor as depicted in the introductory philosophy of science chapters in science courses. It clearly is not easy to determine what is true or false, worthwhile or trivial. In my own work, I make a distinction between the frontier and the core. The core consists of a small set of theories, analytic techniques, and facts that represent the given at any particular point in time. Core knowledge is that which is accepted by the scientific community as being both true and important. The other component of knowledge, the research frontier, consists of all the work currently being produced by all active researchers in a given discipline. The research frontier is where all new knowledge is produced. On the research frontier, science is characterized by a lack of consensus. In fact I have shown that at the frontier, there is no more consensus in the natural sciences than there is in the social sciences. The lack of consensus is so great that whether or not one receives a National Science Foundation grant is fifty percent a result of the quality of the proposal and fifty percent a result of luck—which reviewers out of the pool of eligibles happen to be sent your proposal.[40]

Clearly social factors play an important role in the evaluation of new knowledge; but so does evidence obtained from the natural world. The sociologist of science should study how social factors and such evidence interact in the evaluation process. This is perhaps the most crucial question of the discipline. And if one abandons the frequently programmatic and polemic relativism of the constructivists, their work is of use in answering this question. In fact, I like to think of myself as a realist-constructivist. Yes science is socially constructed, but yes how it is constructed is to various degrees and extent constrained by nature.

My work in the sociology of science has led me to strongly reject the conclusion that the natural sciences are entirely socially constructed; but my life in the social sciences has made me more amenable to the possibility that these sciences may indeed be entirely socially constructed. Ideology, power, and network ties seem to determine what social scientists believe; evidence is frequently entirely ignored. I have recently begun to address this problem in an article entitled, "Why Sociology Doesn't Make Progress like the Natural Sciences." That social science is completely or almost completely socially constructed helps explain how the social constructivist view of science could have become so powerful in the absence of any good supporting evidence and in the face of such devastating empirical critiques as those found in books like Peter Galison's *How Experiments End.*

NOTES

1. S. Jasonoff, G. D. Markle, J. C. Peterson & T. Pinch, eds., *Handbook of Science and Technology Studies.*
2. Ibid., p. 440 (emphasis added).
3. Harry M. Collins, "Stages in the Empirical Program of Relativism," p. 3 (emphasis added).
4. David Bloor, "Durkheim and Mauss Revisited: Classification and the Sociology of Knowledge."
5. Boris Hessen, "The Social and Economic Roots of Newton's *Principia.*"
6. For a critique of the social constructivists and a discussion of similar attacks made in other fields of scholarship see Paul Gross & Norman Levitt,

Higher Superstition: The Academic Left and Its Quarrels with Science.
7. Larry Laudan, *Science and Values: The Aims of Science and Their Role in Scientific Debate; Relativism and Science.*
8. Ronald Giere, *Explaining Science.*
9. David Hull, *Science as Process.*
10. Stephen Brush, "Should the History of Science Be Rated X?"
11. Martin Rudwick, *The Great Devonian Controversy: The Shaping of Scientific Knowledge among Gentlemanly Specialists.*
12. Peter Galison, *How Experiments End.*
13. Thomas S. Kuhn, *Rothschild Lecture*, pp. 8–9 (emphasis added).
14. Thomas Gieryn, "Relativist/Constructivist Programs in the Sociology of Science: Redundancy and Retreat."
15. So much so, that he probably would have to be considered the most eminent living sociologist.
16. M. Callon & B. Latour, "Don't Throw the Baby Out with the Bath School! A Reply to Collins and Yearley," p. 346.
17. Ibid., p. 348.
18. Ibid., p. 351.
19. R. K. Merton, *Science, Technology, and Society in Seventeenth-Century England.*
20. Joseph Ben-David, "Scientific Productivity and Academic Organization in Nineteenth-Century Medicine."
21. Derek de Solla Price, *Little Science, Big Science . . . and Beyond.* For a review of literature on the rate of scientific advance see Stephen Cole, *Making Science: Between Nature and Society,* ch. 9.
22. Bruno Latour & Steve Woolgar, *Laboratory Life: The Construction of Scientific Facts,* pp. 154–158.
23. Karin Knorr-Cetina, *The Manufacture of Knowledge: An Essay on the Constructivist and Contextual Nature of Science.*
24. Latour & Woolgar, *Laboratory Life,* p.128 (emphasis added).
25. H. Kuklick, "Mind over Matter?"
26. Ibid., p. 370.
27. Unfortunately, because as the deadline for this article approached and passed my reply to Shapin's review of my book in *Science* remained somewhere in transit between Stony Brook and Queensland, I will be unable to include here details of his misinterpretations. But certainly one of the most important is to say that constructivists do not really mean their relativist manifestos.

28. S. Shapin, "Here and Everywhere: Sociology of Scientific Knowledge."

29. S. Shapin, "History of Science and Its Sociological Reconstructions."

30. S. Shapin, *A Social History of Truth: Civility and Science in Seventeenth-Century England.*

31. Jonathan Cole & Stephen Cole, *Social Stratification in Science*, pp. 77–78.

32. See Andrew Pickering, ed., *Science as Practice and Culture;* S. Shapin's review article on the sociology of scientific knowledge in *Annual Reviews of Sociology;* S. Jasonoff, G. D. Markle, J. C. Peterson and T. Pinch, eds., *Handbook of Science and Technology Studies* for discussions of recent trends in social constructivism.

33. Michel Callon, "Some Elements of a Sociology of Translation: Domestication of the Scallops and the Fishermen of St. Brieux Bay."

34. Harry M. Collins & Steven Yearley, "Epistemological Chicken," p. 314.

35. Michel Callon & Bruno Latour, "Don't Throw the Baby Out," p. 353 (emphasis added).

36. Andrew Pickering, *Science as Practice and Culture.*

37. Bruno Latour, *We Have Never Been Modern* (originally published in French in 1991).

38. Callon & Latour, "Don't Throw the Baby Out."

39. H. Kuklick, "Mind over Matter?" p. 369.

40. Stephen Cole, *Making Science: Between Nature and Society,* ch. 4.

REFERENCES

Ben-David, Joseph. "Scientific Productivity and Academic Organization in Nineteenth-Century Medicine." *American Sociological Review* 25 (1960): 828–843.

Bloor, David. "Durkheim and Mauss Revisited: Classification and the Sociology of Knowledge." *Studies in History and Philosophy of Science* 13 (1982): 267–297.

Brush, Stephen. "Should the History of Science Be Rated X?" *Science* 183 (1974): 1164–1172.

Callon, Michel. "Some Elements of a Sociology of Translation: Domestication of the Scallops and the Fisherman of St. Brieux Bay." In *Power, Action, and Belief: A New Sociology of Knowledge?*, edited by John Law, pp. 196–229. London: Routledge and Kegan Paul, 1986.

Callon, Michel & Bruno Latour. "Don't Throw the Baby Out with the Bath School! A Reply to Collins and Yearley." *In Science as Practice and Culture,* edited by A. Pickering. Chicago, IL: University of Chicago Press, 1992.

Cole, Jonathan & Stephen Cole. *Social Stratification in Science.* Chicago IL: University of Chicago Press, 1973.

Cole, Stephen. *Making Science: Between Nature and Society.* Cambridge, MA: Harvard University Press, 1992.

———. "Why Sociology Doesn't Make Progress Like the Natural Sciences." *Sociological Forum* 9 (1994): 133–154.

Collins, Harry M. "Stages in the Empirical Program of Relativism." *Social Studies of Science* 12 (1981): 3–10.

Fox, Mary Frank. "Realism, Social Constructivism and Outcomes in Science." *Contemporary Sociology* 22 (1993): 481–483.

Galison, Peter. *How Experiments End.* Chicago, IL: University of Chicago Press, 1987.

Giere, Ronald N. *Explaining Science.* Chicago, IL: University of Chicago Press, 1988.

Gieryn, Thomas. "Relativist/Constructivist Programs in the Sociology of Science: Redundance and Retreat." *Social Studies of Science* 12 (1982): 279–297.

Gross, Paul R. & Normal Levitt. *Higher Superstition: The Academic Left and Its Quarrels with Science.* Baltimore, MD: Johns Hopkins University Press, 1994.

Hessen, Boris. "The Social and Economic Roots of Newton's *Principia.*" In *Science at the Crossroads: Papers Presented to the International Congress of the History of Science and Technology.* London: Frank Cass, 1971.

Hull, David. *Science as Process.* Chicago, IL: University of Chicago Press, 1988.

Jasonoff, S., G. D. Markle, J. C. Peterson & T. Pinch, eds. *Handbook of Science and Technology Studies.* Thousand Oaks, CA: Sage Publications, 1995.

Knorr-Cetina, Karin. *The Manufacture of Knowledge: An Essay on the Constructivist and Contextual Nature of Science.* New York, NY: Pergamon Press, 1981.

Kuhn, Thomas S. Rothschild Lecture. 1992.

———. *The Structure of Scientific Revolutions.* Chicago, IL: University of Chicago Press, 1970.

Kuklick, H. "Mind over Matter?" *Historical Studies in the Physical and Biological Sciences* 25 (1995): 361–378.

Laudan, Larry. *Relativism and Science.* Chicago, IL: University of Chicago Press, 1990.

———. *Science and Values: The Aims of Science and Their Role in Scientific Debate.* Berkeley, CA: University of California Press, 1984.

Latour, Bruno. *We Have Never Been Modern.* Cambridge, MA: Harvard University Press, 1993.

Latour, Bruno & Steve Woolgar. *Laboratory Life: The Construction of Scientific Facts.* Princeton, NJ: Princeton University Press, 1986.

Merton, R. K. *Science, Technology, and Society in Seventeenth-Century England.* New York, NY: Howard Fertig, 1970.

Pickering, Andrew. *Constructing Quarks: A sociological History of Particle Physics.* Edinburgh: Edinburgh University Press, 1984.

—— *Science as Practice and Culture.* Chicago, IL: University of Chicago Press, 1992.

Price, Derek de Solla. *Little Science, Big Science . . . and Beyond.* New York, NY: Columbia University Press, 1986.

Rudwick, Martin J. S. *The Great Devonian Controversy: The Shaping of Scientific Knowledge among Gentlemanly Specialists.* Chicago, IL: University of Chicago Press, 1985.

Shapin, S. "Here and Everywhere: Sociology of Scientific Knowledge." In *Annual Review of Sociology,* edited by John Hagan & Karen S. Cook, pp. 289–321. Palo Alto, CA: Annual Reviews, Inc., 1995.

—— "History of Science and Its Sociological Reconstructions." *History of Science* 20 (1982): 157–211.

—— *A Social History of Truth: Civility and Science in Seventeenth-Century England.* Chicago, IL: University of Chicago Press, 1994.

29

SANDRA HARDING

Should the History and Philosophy of Science Be X-Rated?

This question[1] is only slightly antic once we look at the metaphors and models of gender politics with which scientists and philosophers of science have explained how we all should think about nature and inquiry. Examples of gender symbolization generally occur in the margins, in the asides, of texts—in those places where speakers reveal the assumptions they think they do not need to defend, beliefs they expect to share with their audiences. We will see assumptions that the audiences for these texts are men, that scientists and philosophers are men, and that the best scientific activity and philosophic thinking about science are to be modeled on men's most misogynous relationships to women—rape, torture, choosing "mistresses," thinking of mature women as good for nothing but mothering. Let us look first at some striking examples from the history of science, and then examine some comments by contemporary scientists and philosophers.

HISTORICAL IMAGES

Contemporary science presents its conceptions of nature and inquiry as truths discovered at the birth of modern science—as objective, universally valid reflections of *the* way nature is and *the* way to arrive at mirrorlike descriptions and explanations. But historians point out that conceptions of nature and inquiry have changed over time, and that they have been highly influenced by the political strategies used in historically identifiable battles between the genders. Gender politics has provided resources for the advancement of science, and science has provided resources for the advancement of masculine domination. I raised this issue earlier in asking whether it could possibly be reasonable to regard as a pure coincidence the development of sexology hot on the heels of the nineteenth-century women's movement.

We should note at the start that there are a number of problems with these historical studies. One origin of these problems is the mystifying philosophy of social science directing them, especially the misleading understandings of the complete "life history" of the role of metaphor in scientific explanation. Another origin is the inadequacy of histories which say little about social relations between the genders, let alone about how changes in these relations were experienced, perceived, and responded to by the culture in general, including the scientific thinkers of the day. We can see that the five substantive problems with the conceptual schemes of the social sciences pointed out by feminist critics infest the source materials available to historians today. In spite of such shortcomings, these studies greatly advance our understanding of science's place in its social worlds.

One phenomenon feminist historians have focused on is the rape and torture metaphors in the writings of Sir Francis Bacon and others (e.g., Machiavelli) enthusiastic about the new scientific method. Traditional historians and philosophers have said that these metaphors are irrelevant to the *real* meanings and referents of scientific concepts held by those who used them and by the public for whom they wrote. But when it comes to regarding nature as a machine, they have quite a different analysis: here, we are told, the metaphor provides the interpretations of Newton's mathematical laws: it directs inquirers to fruitful ways to apply his theory and suggests the appropriate methods of inquiry and the kind of metaphysics the new theory supports.[2] But if we are to believe that mechanistic metaphors were a fundamental component of the explanations the new science provided, why should we believe that the gender metaphors were not? A consistent analysis would lead to the conclusion that understanding nature as a woman indifferent to or even welcoming rape was equally fundamental to the interpretations of these new conceptions of nature and inquiry. Presumably these metaphors, too, had fruitful pragmatic, methodological, and metaphysical consequences for science. In that case, why is it not as illuminating and honest to refer to Newton's laws as "Newton's rape manual" as it is to call them "Newton's mechanics"?

We can now see that metaphors of gender politics were used to make morally and politically attractive the new conceptions of nature and inquiry required by experimental method and the emerging technologies of the period. The organicist conception of nature popular in the medieval period—nature as alive, as part of God's domain—was appropriate neither for the new experimental methods of science nor for the new technological applications of the results of inquiry. Carolyn Merchant identifies five changes in social thought and experience in Europe during the fifteenth to seventeenth centuries that contributed to the distinctive gender symbolism of the subsequent scientific world view.[3]

First of all, when Copernican theory replaced the earth-centered universe with a sun-centered universe, it also replaced a woman-centered universe with a man-centered one. For Renaissance and earlier thought within an organic conception of nature, the sun was associated with manliness and the earth with two opposing aspects of womanliness. Nature, and especially the earth, was identified on the one hand with a nurturing mother—"a kindly, beneficent female who provided for the needs of mankind in an ordered, planned universe"—and on the other with the "wild and uncontrollable [female] nature that could render violence, storms, droughts, and general chaos" (p. 2). In the new Copernican theory, the womanly earth, which had been God's special creation for man's nurturance, became just one tiny, externally moved planet circling in an insignificant orbit around the masculine sun.

Second, for the Platonic organicism, active power in the universe was associated with the alive, nurturing mother earth; for the Aristotelian organicism, activity was associated with masculinity and passivity with womanliness. Central to Aristotle's biological theory, this association was revived in sixteenth-century views of the cosmos, where "the marriage and impregnation of the 'material' female earth by the higher 'immaterial' celestial masculine heavens was a stock description of biological generation in nature." Copernicus himself draws on this metaphor: "Meanwhile, the earth conceives by the sun and becomes pregnant with annual offspring" (p. 7). Resistance to this shift in the social meaning of womanliness is evident in the sixteenth-century conflicts over whether it was morally proper to treat mother earth in the new ways called for by such commercial activities as mining. But as the experience of "violating the body" of earth became increasingly more common during the rise of modern science and its technologies, the moral sanctions against such activities provided by the older organic view slowly died away. Simultaneously, a criterion for distinguishing the animate from the inanimate was being created. (This distinction is a theoretical construct of modern science, not an observational given familiar to people before the emergence of science. And, as we shall see, it is one that increasingly ceases to reflect "common sense.") Thus a "womanly" earth must be only passive, inert matter and indifferent to explorations and exploitations of her insides.

Third, the new universe that science disclosed was one in which change—associated with "corruption," decay, and disorder—occurred not just on earth, as the Ptolemaic "two-world view" held, but also throughout the heavens. For Renaissance and Elizabethan writers, these discoveries of change in the heavens suggested that nature's order might break down, leaving man's fate in chaos (p. 128). Thinkers of the period consistently perceived unruly, wild nature as rising up against man's attempts to control his fate. Machiavelli appealed to sexual metaphors in his proposition that the potential violence of fate could be mastered: "Fortune is a woman and it is necessary if you wish to master her to conquer her by force; and it can be seen that she lets herself be overcome by the bold rather than by those who proceed coldly, and therefore like a woman, she is always a friend to the young because they are less cautious, fiercer, and master her with greater audacity" (p. 130).

Fourth, man's fate seemed difficult to control because of disorder not only in the physical universe but also in social life. The breakdown of the ancient order of feudal society brought the experience of widespread social disorder during the period in which the scientific world view was developing. Par-

ticularly interesting is the possibility that women's increased visibility in public life during this period was perceived as threatening deep and widespread changes in social relations between the genders. Women were active in the Protestant reform movements of northern Europe, and Elizabeth I occupied England's throne for an unprecedentedly long reign. Prepared by the organic view's association of wild and violent nature with one aspect of the womanly, and by the absence of clear distinctions between the physical and the social, the Renaissance imagination required no great leap to associate all disorder, natural and social, with women. By the end of the fifteenth century, this association had been fully articulated in the witchcraft doctrines. To women was attributed a "method of revenge and control that could be used by persons both physically and socially powerless in a world believed by nearly everyone to be animate and organismic" (p. 140).

Fifth, the political and legal metaphors of scientific method originated at least in part in the witchcraft trials of Bacon's day. Bacon's mentor was James I of England, a strong supporter of antifeminist and antiwitchcraft legislation in both England and Scotland. An obsessive focus in the interrogations of alleged witches was their sexual practices, the purpose of various tortures being to reveal whether they had "carnally known" the Devil. In a passage addressed to his monarch, Bacon uses bold sexual imagery to explain key features of the experimental method as the inquisition of nature: "For you have but to follow and as it were hound nature in her wanderings, and you will be able when you like to lead and drive her afterward to the same place again. . . . Neither ought a man to make scruple of entering and penetrating into those holes and corners, when the inquisition of truth is his whole object—as your majesty has shown in your own example" (p. 168). It might not be immediately obvious to the modern reader that this is Bacon's way of explaining the necessity of aggressive and controlled experiments in order to make the results of research replicable!

As I indicated earlier, this kind of analysis raises a number of problems and challenges. . . . There does, however, appear to be reason to be concerned about the intellectual, moral, and political structures of modern science when we think about how, from its very beginning, misogynous and defensive gender politics and the abstraction we think of as scientific method have provided resources for each other. The severe testing of hypotheses through controlled manipulations of nature, and the necessity of such controlled manipulations if experiments are to be repeatable, are here formulated by the father of scientific method in clearly sexist metaphors. Both nature and inquiry appear conceptualized in ways modeled on rape and torture—on men's most violent and misogynous relationships to women—and this modeling is advanced as a reason to value science. It is certainly difficult to imagine women as an enthusiastic audience for these interpretations of the new scientific method.

If appeal to gender politics provides resources for science, does appeal to science provide resources for gender politics? Do not metaphors illuminate in both directions? As nature came to seem more like a machine, did not machines come to seem more natural? As nature came to seem more like a woman whom it is appropriate to rape and torture than like a nurturing mother, did rape and torture come to seem a more natural relation of men to women? Could the uses of science to create ecological disaster, support militarism, turn human labor into physically and mentally mutilating work, develop ways of controlling "others"—the colonized, women, the poor—be just misuses of applied science? Or does this kind of conceptualization of the character and purposes of experimental method ensure that what is called bad science or misused science will be a distinctively masculinist science-as-usual? Institutions, like individuals, often act out the repressed and unresolved dilemmas of their infancies. To what extent is the insistence by science today on a value-neutral, dispassionate objectivity in the service of progressive social relations an attempt by a guilty conscience to resolve some of these early but still living dilemmas?

NOTES

1. My apologies to Stephen Brush, whose paper "Should the History of Science Be X-Rated?" in *Science* 183 (no. 4130) (1974) did not deal with the gender behavior of scientists (or philosophers).
2. See, e.g., the philosophers and scientists criticized in Hesse (1966).
3. Merchant (1980). Subsequent page references to this work (and the authors cited within it) appear in the text.

REFERENCES

Hesse, Mary. 1966. *Models and Analogies in Science.* Notre Dame, Ind.: University of Notre Dame Press.
Merchant, Carolyn. 1980. *The Death of Nature.* New York: Harper & Row.

30

ALAN SOBLE

In Defense of Bacon

What a man had rather were true he more readily believes. Therefore he rejects difficult things from impatience of research.

NOVUM ORGANUM, BOOK I, APH. 49

I. SCIENCE AND RAPE

In an article printed in the august pages of *The New York Times,* Sandra Harding (1989) introduced to the paper's readers one of the more shocking ideas to emerge from feminist science studies:

> Carolyn Merchant, who wrote a book called "Death of Nature," and Evelyn Keller's collection of papers called "Reflections on Gender & Science" talk about the important role that sexual metaphors played in the development of modern science. They see these notions of dominating mother nature by the good husband scientist. If we put it in the most blatant feminist terms used today, we'd talk about marital rape, the husband as scientist forcing nature to his wishes.*

Harding asserts elsewhere, too, that sexual metaphors played an important role in the development of science (e.g., 1986, 112, 113, 116; 1991, 43, 267). But here she understates the point by referring to "marital rape," and so does not convey it in the *most* blatant terms, because her own way of making the point is usually to talk about rape and torture in the same breath, not mentioning marriage. (I do *not* mean that marital rape is less vicious or more excusable than nonmarital rape. But the connotations of rape adjoined to torture are stronger than those of marital rape.) For example, Harding (1986) refers to "the rape and torture metaphors in the writings of Sir Francis Bacon and others (e.g., Machiavelli) enthusiastic about the new scientific method" (113). By associating rape metaphors with science, Harding is trying to accomplish what metaphor itself does; she wants the unsavory connotations of rape to spill over, with full moral condemnation, onto science:

> Understanding nature as a woman indifferent to or even welcoming rape was . . . fundamental to the interpretations of these new conceptions of nature and inquiry. . . . There does . . . appear to be reason to be concerned about the intellectual, moral, and political structures of modern science when we think about how, from its very beginning, misogynous and defensive gender politics and the abstraction we think of as scientific method have provided resources for each other (113,116).

I dare not hazard a guess as to how many people read Harding's article in the *Times;* how many clipped out this scandalous bit of bad publicity for science and put it on the icebox; or how many still have some vague idea tying science to rape. But the belief that vicious sexual metaphors were and are important in science has gained some currency in the academy.[1] This is unfortunate, not only for the

Philosophy of the Social Sciences 25 (1995), pp. 192–200. © 1995 by Sage Publications, Inc. Reprinted by permission of the publisher.

reputations of those who engage in or extol science, but also for our understanding of its history.

II. CONTEMPORARY SEXUAL METAPHORS

In *Whose Science? Whose Knowledge? (WS? WK?)*, Harding (1991) proposes that we abolish the "sexist and misogynistic metaphors" that have "infused" science and replace them with "positive images of strong, independent women," metaphors based on "womanliness" and "female eroticism woman-designed for women" (267, 301). Harding defends her proposal by claiming that "the prevalence of such alternative metaphors" would lead to "less partial and distorted descriptions and explanations" and would "foster the growth of knowledge":

> If they were to excite people's imaginations in the way that rape, torture, and other misogynistic metaphors have apparently energized generations of male science enthusiasts, there is no doubt that thought would move in new and fruitful directions. (267)

What are the misogynistic metaphors that "energized" science? In a footnote, Harding sends us to Chapter 2 of *WS? WK?* There we find a section titled "The Sexual Meanings of Nature and Inquiry" (1991, 42–6), which contains *four* examples of metaphors in the writings of two philosophers (Francis Bacon, Paul Feyerabend), one scientist (Richard Feynman), and the unnamed preparers of a booklet published by the National Academy of Sciences.

In the passage from Feynman's Nobel Lecture quoted by Harding, which she interprets as an example of "thinking of mature women as good for nothing but mothering" (1986, 112), the physicist reminisces about a particular theory in physics as if it were a woman with whom he long ago fell in love, a woman who has become old, yet had been a good mother and left many children (1986, 120; 1991, 43–4). From the NAS booklet, Harding quotes: "The laws of nature are not . . . waiting to be plucked like fruit from a tree. They are hidden and unyielding, and the difficulties of grasping them add greatly to the satisfaction of success" (1991, 44). Here, says Harding, one can hear "restrained but clear echoes" of sexuality. Perhaps the metaphors used by Feynman and the NAS are sexual, but they are hardly misogynistic or vicious, and I wonder why Harding thinks they deserve to be put on display. In fact, Harding only claims that in Bacon, of her four examples, is there a rape metaphor. But let us examine her treatment of Feyerabend first, for there are significant connections between them.

Harding quotes the closing lines of a critique of Kuhn and Lakatos by Feyerabend, who closes his long technical paper with the joke that his view "changes science from a stern and demanding mistress into an attractive and yielding courtesan who tries to anticipate every wish of her lover. Of course, it is up to us to choose either a dragon or a pussycat for our company. I do not think I need to explain my own preferences" (1970, 229).[2] In *WS?WK?*, Harding exhibits, but barely comments on, this passage. Her gripe cannot be that Feyerabend (or Feynman, or the NAS) employed a *sexual* metaphor, for we know that in *WS?WK?*, Harding condones "alternative" sexual images reflecting "female eroticism woman-designed for women." Feyerabend's metaphor—science is a selfish shrew who exploits us *or* she is a prostitute who waits on us hand and foot—must therefore be the *wrong kind* of sexual metaphor, even if not of rape. Harding quotes the same passage in her earlier *The Science Question in Feminism (SQIF)*, giving it as an example of how gender is attributed to scientific inquiry (120). In her view, the passage conveys, as does Feynman's, a cultural image of "manliness." Feynman depicts "the good husband and father," while Feyerabend's idea of manliness, says Harding, is "the sexually competitive, locker-room jock" (1986, 120). Thus science, in Feyerabend's metaphor, is an accommodating, sexually passive woman, and the scientist and the philosopher of science are the jocks she sexually pleases. I do not see how portraying science as a courtesan implies that the men who visit her, scientists and philosophers, are locker-room jocks. The fancy word "courtesan," if it implies anything at all, vaguely alludes to a debonair and educated Hugh Hefner puffing on his pipe, not to a Terry Bradshaw swatting bare male butt with a wet towel.[3]

Harding concludes her discussion of Feyerabend by claiming that his metaphor, coming as it does strategically at the end of his paper, serves a pernicious purpose. He depicts "science and its theories" as "exploitable women," and the scientist as a masculine, manly man, in order to tell his (male) audience that his philosophical "proposal should be appreciated *because* it replicates gender politics" of the sort they find congenial (1986, 121). In *WS? WK?*, Harding similarly asserts that this metaphor is the way Feyerabend "recommended" his view (1991, 43). I agree that a woman reading Feyerabend's paper would probably not empathize with the metaphor, even if she fully concurred with the critique of Kuhn that precedes it; but she could, if she wished, ignore it as irrelevant to Feyerabend's arguments. (*Had* Bacon employed rape metaphors, Harding [1986] would be right that "it is . . . difficult to imagine women as an enthusiastic audience" [116]. Still, *had* there been any women in Bacon's audience, they could have disregarded his metaphors and accepted, or rejected, the rest on its own merits.) Further, asserting that the men in Feyerabend's audience would be, in part, persuaded by this appeal and that Feyerabend thought that he could seduce them with his "conscientious effort . . . at gender symbolism" (Harding, 1986, 120) is insulting to men. Some men readers undoubtedly prefer strict to submissive women. Would Feyerabend's contrary preference for kittens tend to undermine for them his critique of Kuhn because it does not match their own taste?

III. HARDING ON BACON

According to Harding (1991), vicious sexual metaphors were infused into modern science at its very beginning, were instrumental in its ascent, and eventually became "a substantive part of science" (44). She thinks Francis Bacon (1561–1626) was crucial in this process.[4] What Harding (1986) says about Feyerabend, that he hoped his view would "be appreciated *because* it replicates gender politics" (121), is what Harding (1991) claims about Bacon, although in more extreme terms: "Francis Bacon appealed to rape metaphors to persuade his audience that experimental method is a good thing" (43).

This is a *damning* criticism. Bacon is not depicted as a negligible Feyerabend making silly jokes

about science the old whore. Harding is claiming that Bacon drew an analogy between the experimental method and rape, and tried to gain advantage from it (see also 1986, 116), as Merchant had claimed before her that Bacon drew an analogy between the experimental method and torture (1980, 168, 172). Conjure up the image: Bacon wants to convince fellow scholars to study nature systematically, by using experimental methods that elicit changes in nature, rather than study nature by accumulating specimens and observing phenomena passively. In order to champion experimentalism, Bacon says to them: think of it as *rape;* think of it as forcing apart with your knees the slender thighs of an unwilling woman, pinning her under the weight of your body as she kicks and screams in your ears, grabbing her poor little jaw roughly with your fist to shut her mouth, and *trying* to thrust your penis into her dry vagina; *that is* what the experimental method is all about.

What did Bacon write to provoke Harding into this accusation? Here is the whole text she offers as evidence for her reading:

> For you have but to hound nature in her wanderings, and you will be able when you like to lead and drive her afterwards to the same place again. Neither ought a man to make scruple of entering and penetrating into those holes and corners when the inquisition of truth is his whole object. (1991, 43)[5]

I *suppose* that a man who made no scruple of penetrating holes (and *corners?*) might be a rapist, but he also might be a proctologist or billiard player. And to "hound" nature *could* be seen as raping her, but the spirited student who storms my office and too often sits down next to me in the cafeteria, hoping for some words of wisdom—no more than that—is *hounding* me. Why could that not be Bacon's point, that a student of nature must be willing to sit for prolonged periods on the floor in the hall, outside her office, waiting for her reluctant nod? It is unlikely, then, that the rape metaphor Harding perceives here is located *entirely* in "hound," unless she has in mind Robin Morgan's definition of rape (1977):

> *Rape exists any time sexual intercourse occurs when it has not been initiated by the woman, out of her*

own genuine affection and desire. . . . How many millions of times have women had sex "willingly" with men they didn't want to have sex with? . . . How many times have women wished just to sleep instead or read or watch the Late Show? . . . most of the decently married bedrooms across America are settings for nightly rape. (165–66; Harding's italics)

In Morgan's view, a man who pesters his wife for sex, when she prefers to watch TV, has committed rape if she caves in under his pressure. But Bacon's audience would never have recognized this prosaic sexual phenomenon as rape. Nor would most of us today—reasonable men and women alike—judge it to be rape. We could do so, of course, but that would require many other changes in our moral and legal concepts, probably more than a good Quinean could endure, and would make trivial the accusation that Bacon traded in rape metaphors.

Looking at Keller might help us discern a rationale behind Harding's reading of Bacon's *De Augmentis*. In her essay "Baconian Science," Keller (1985) quotes the first of the two sentences quoted by Harding, in order to illustrate her own claim that, even though, for Bacon, "Nature may be coy," she can still "be conquered":

> For you have but to follow and as it were hound nature in her wanderings, and you will be able, when you like, to lead and drive her afterwards to the same place again. (36)

What "leads to [this] conquest," in Keller's view, is "not simple violation, or rape, but forceful and aggressive seduction" (37). Now, the fact that Keller interprets this passage from *De Augmentis* as a rape-free zone does not mean there is no rape image there. So Harding, when reading Keller, might have concluded—with a slight push from Keller (1985), for whom "the distinction between rape and conquest sometimes seems too subtle" (37)—that Keller had been too cautious, that the "conquest" of nature Keller found in "hound" and "drive" together is more accurately described as rape. Harding's rape-interpretation of Bacon, taken as deriving from Keller, will then depend on (1) Keller's being right in finding even the *conquest* of nature in "hound" and "drive," a conquest that must be *sexual*,[6] and (2) eradicating the difference—also for

Bacon, since his mens rea is at stake—between rape and seduction. Bacon, of course, recognizes this difference, and advises that science would be more successful by patiently wooing nature than by raping her:

> Art . . . when it endeavors by much vexing of bodies to force Nature to its will and conquer and subdue her . . . rarely attains the particular end it aims at. . . . men being too intent upon their end . . . struggle with Nature than woo her embraces with due observation and attention. ("Erichthonius," Myth 20, *Wisdom of the Ancients;* Robertson, 843)

There is reason to think, then, that Keller is right not to perceive rape in Bacon, although the seduction, here at least, seems considerate and delicate, not "forceful and aggressive."[7]

Something else can be gleaned from Keller. Compare the one sentence Keller takes from Bacon's *De Augmentis* with the first sentence that Harding attributes to Bacon. The sentence as quoted by Keller correctly includes the words "follow and as it were" (*Works IV,* 296), which are missing from Harding. I think there is some difference between Bacon's nuanced "follow and as it were hound" nature and the crude "hound" her, which dilution makes a rape metaphor more difficult to discern. I do not want to make much of this error, despite the fact that students of Harding who read only *WS? WK?* will be misguided, because five years earlier, in *SQIF,* Harding quotes this *De Augmentis* passage twice, almost correctly.[8] Here is one instance:

> To say "nature is rapable"—or, in Bacon's words: "For you have but to follow and as it were hound nature in her wanderings, and you will be able when you like to lead and drive her afterward to the same place again. . . . Neither ought a man to make scruple of entering and penetrating into those holes and corners when the inquisition of truth is his whole object"—is to *recommend* that similar benefits can be gained from nature if it is conceptualized and treated like a woman resisting sexual advances. (1986, 237; ellipsis and italics are Harding's)

Harding seems not to see "follow and as it were" as a sturdy qualification. But "penetration" need not be taken as having "strong sexual implications" (contra Merchant 1980, 168); and even if "penetration" is sexual (was it *for Bacon?*), it does not per se entail rape. Perhaps Harding construes the *unscrupulous* (= immoral) penetration of holes to be an allusion to rape.[9] But this reading makes sense only by wrenching "scruple" out of context. Bacon's point, which he repeats elsewhere, is that any scientist determined to find the truth about nature should be prepared to get his hands dirty; when truth is the goal, *everything* must be investigated, even if, to prissy minds, the methods employed and the objects studied are foul. (Think about Freud or Kinsey justifying the study of sex.) Thus a few lines after "scruple" in *De Augmentis*, Bacon bemoans that "it is esteemed a kind of dishonor . . . for learned men to descend to inquiry or meditation upon matters mechanical"(*Works IV,* 296; see also *Advancement, Works III,* 332). Parts of *De Augmentis* (whose title begins *De Dignitate*) and *Novum Organum* are intended to establish the dignity, despite dirty hands, of engaging in science to improve the human condition (see also *Parasceve, Works IV,* 257–9). *Novum Organum* is especially clear on this. In one aphorism, Bacon condemns "an opinion . . . vain and hurtful; namely, that the dignity of the human mind is impaired by long and close intercourse with experiments" (*Works IV; Book I,* aph. 83;[10] see also *Cogitata et Visa,* in Farrington 1964, 82). Bacon returns to this theme later in *Novum Organum:*

And for things that are mean or even filthy, . . . such things, no less than the most splendid and costly, must be admitted into natural history. Nor is natural history polluted thereby; for the sun enters the sewer no less than the palace, yet takes no pollution. . . . For whatever deserves to exist deserves also to be known, for knowledge is the image of existence; and things mean and splendid exist alike. Moreover as from certain putrid substances—musk, for instance, and civet—the sweetest odors are sometimes generated, so too from mean and sordid instances there sometimes emanates excellent light and information. But enough and more than enough of this, such fastidiousness being merely childish and effeminate. (120; see also 121)

Bacon is, like Calvin, a rascal. He would much rather dissect bugs and chase snakes than play house or have an afternoon tea with Susie.

Bacon's two sentences from *De Augmentis* make yet another appearance in Harding's *SQIF:*

Bacon uses bold sexual imagery to explain key features of the experimental method as the inquisition of nature: "For you have but to follow and as it were hound nature in her wanderings, and you will be able when you like to lead and drive her afterward to the same place again. . . . Neither ought a man to make scruple of entering and penetrating into those holes and corners, when the inquisition of truth is his whole object. . . ." . . . this is Bacon's way of explaining the necessity of aggressive and controlled experiments in order to make the results of research replicable! (1986, 116; first ellipsis is Harding's)

But it is not obviously true that Bacon, with the phrase "to the same place," is referring to replicability. Nor is Keller's idea obviously correct, that Bacon here asserts that nature can be "conquered." William Leiss (1972) suggests, in *The Domination of Nature,* an alternative reading: "having discovered the course leading to the end result we are able to duplicate the process at will" (59). This reading makes sense, because, in the immediately preceding sentence, Bacon had written: "from the wonders of nature is the most clear and open passage to the wonders of art," and then by way of explaining or defending this idea, that nature teaches us how to fabricate artificial devices, Bacon now writes "*For you have but to follow.*" To learn how to achieve one of nature's effects (to use my own anachronistic example, the overcoming [*conquering*] of bacterial infection), we must study how nature accomplishes it (we *follow* nature by pestering her in the lab); once we discover Nature's Way (the various mechanisms of the immune system) we can then copy, modify, and rearrange its main ingredients (develop "artificial" devices, vaccinations, that elicit antibodies) to "lead" nature "to the same place again." Bacon is modifying a point appearing elsewhere in his writings (e.g., *A Description of the Intellectual Globe, Works V,* 507) and that he has made just a page before in *De Augmentis:*

The artificial does not differ from the natural in form or essence, but only in the efficient. . . . Nor matters it, provided things are put in the way to produce an effect, whether it be done by human means or otherwise. Gold is sometimes refined in the fire and sometimes found pure in the sands, nature having done the work for herself. So also the rainbow is made in the sky out of a dripping cloud; it is also made here below with a jet of water. Still therefore it is nature which governs everything. (*Works IV,* 294–95)

In the *Cogitata et Visa* (in Farrington 1964), Bacon goes so far as to say that phenomena found in nature (he praises silk spun by a worm), and from which we can learn, "are such as to elude and mock the imagination and thought of men" (96). Bacon's example in *De Augmentis* of obtaining a rainbow from a spray of water is serene, even lovely, and makes it improbable that he viewed experimental manipulations as nothing but mere acts of aggression. Furthermore, Bacon's affirmation that nature "governs everything"—the ways of nature are responsible even for the artificial rainbow we make with a spray of water—is reason to doubt that he conceives of the relationship between science and nature principally as that between man the master and dominated woman. At the very beginning and at the very end of the first book of *Novum Organum,* as well as in "The Plan" of *The Great Instauration* (*Works IV,* 32)—that is, often and in prominent places—Bacon writes that science is "the servant and interpreter of Nature" (*Novum Organum,* 1) and "Nature to be commanded must be obeyed" 3; see also 129).

Harding (1986) introduces the *De Augmentis* passage by saying that it contains "bold sexual imagery,"[11] but after the quote she escalates the charge: experimentalism, the "testing of hypotheses," is "here formulated by the father of scientific method in clearly sexist metaphors" (116). Harding straightaway takes the next step: "Both nature and inquiry appear conceptualized in ways modeled on rape and torture—on men's most violent and misogynous relationships to women—and this modeling is advanced as a reason to value science" (116). Of course, if the passage contains no rape metaphor, Harding's thesis (see also 1986, 237;

1991, 43) that Bacon employs a rape metaphor to recommend experimentalism falls apart. But even if the passage contains a rape metaphor, why think that Bacon uses it to promote experimentalism? Examine the two sentences Harding quotes from the 1623 *De Augmentis,* as they appeared in the 1605 English predecessor of this text, *The Advancement of Learning:*

> For it is no more but by following and as it were hounding Nature in her wanderings, to be able to lead her afterwards to the same place again. . . . Neither ought a man to make scruple of entering into these things for inquisition of truth. (*Works III,* 331)

The three purported offenders, "drive," "penetrate," and "holes," are missing. Which of these texts is more momentous for Bacon's program? In *Advancement,* written soon after the accession of James I, Bacon was surely trying to win over his king; when writing *De Augmentis,* the older Bacon had already been stripped of his official positions and was writing for posterity. In which situation should we expect Bacon to use harsh (or soft) language to do the persuading? This is treacherous terrain; at least, the contrast between the *Advancement* and *De Augmentis* should give us pause.

Furthermore, in *Novum Organum* (and elsewhere), Bacon argues on behalf of science in terms more likely to convince his audience: it will improve the human condition (81, 129; see *De Augmentis, Works IV,* 297), and so the works of science are works of love (*Valerius Terminus, Works III,* 221–2; "Preface" to *The Great Instauration, Works IV,* 21; see Farrington 1964, 28–9). At the end of the first book of *Novum Organum* (129), Bacon reminds his audience that science fulfills the Biblical command (*Gen.* 1:26) for humans to rule the universe (see Leiss 1972, 53). These themes in Bacon are typical and familiar.[12] Why conjecture that Bacon also appeals to a rape metaphor, as if that were the icing on the cake of his vindication of science? Perhaps Harding assumes that, from the fact that the text contains a rape metaphor, it follows that the metaphor must have been used to convince the audience to embrace his philosophy. It would be a mistake to conclude, however, solely from the presence of a rape metaphor, that it is intended to have a specific perlocutionary force. This mistake is similar to one made about *erotica*, namely, arguing solely from the presence in a text of a photographic or linguistic depiction of a certain sexual act that the text recommends or endorses the depicted act.[13]

My mentioning *erotica* here is not inappropriate. Harding (1986) titles the section of *SQIF* in which she first quotes Bacon "Should the History and Philosophy of Science Be X-Rated?" "This question is only slightly antic," because (previewing her comments on Bacon and Feyerabend) "we will see assumptions that . . . the best scientific activity and philosophical thinking about science are to be modeled on men's most misogynous relationships to women" (112). That is, Harding thinks that science, its history and philosophy, *should* be rated "x"; it contains, in her view, explicit and nasty sex. But I cannot perceive, as she does, the sexually aggressive locker-room jock in Feyerabend's metaphor, nor can I perceive, as she does, the rape metaphor in Bacon. So perhaps it is Harding's story itself that should be rated "x": *she* injects sex where there is none to begin with. Consider the NAS metaphor: the laws of nature are "not waiting to be plucked like fruit from a tree [but] are hidden and unyielding." Harding finds here a "clear echo" of sexuality (1991, 44). But these few words can be read, without effort, as innocent and nonsexual; so it is Harding, like the person who feels squeamish at the sight of uncovered piano legs, who has infused the sex into them. Further, if Harding's own psychology is uncommonly sensitive to the nuances of language and so enables her to extract a rape metaphor out of "hound" and "holes," or if the metaphor is one that mostly or only women could sense, that would undermine Harding's claim that Bacon used a rape metaphor to persuade his audience.[14]

One more example: Bacon's portrayal of inquiry as a "disclosing of the secrets of nature" (*Works IV,* 296; see *Novum Organum,* 18, 89). We *could* construe the language of discovering, or uncovering, the hidden secrets of nature as alluding to a quest for carnal knowledge of a deeply concealed female sexuality that is not keen on being exposed (for whatever reason, be it prudish modesty, girlish self-doubt, lazy reluctance). We might interpret such language this way in order to suggest that this is the deeper, hidden meaning of the philosophical claim that underneath

the appearances of things and events are unobservable structures and forces about which we can have no direct knowledge and about which we will remain ignorant unless we diligently investigate, experimentally, their phenomenal manifestations. But the sexual metaphor, if we insist on digging it out, is tame; there is no rape and no need to compel or twist the metaphor, against its will, to *be* rape.

NOTES

1. See Bordo (1987, 107–8); Longino (1988, 563; 1990, 205); Nelson (1990, 213, 353 n. 136); Tuana (1990, 62)—all under the influence of Harding, Merchant, and Keller.

2. I quote from Feyerabend because Harding (1991) gives, as the last sentence, "I think I do not have to explain my own preferences" (43). In her 1986 book, Harding quotes it correctly (120).

3. Although Harding takes his metaphor as being about "science and its theories" (1986, 121), and Feyerabend agrees (1970, 229), it makes equal sense to read it as being about the nature of the reality that lies beyond science. We should not view Nature as a stern and demanding mistress, which it is for Popper and Lakatos: their Nature screams "False!" or "Incompatible!" when it does not like our scientific theories. Instead, Nature is an indulgent courtesan, one who lets us do *whatever we want*—in theory construction. It is Nature that whispers "Anything goes, big guy," not Science.

4. For an overview of Bacon's life and philosophy, see Thomas Macaulay's 1837 essay on Bacon and John Robertson's critical reply (1905, vii–xvi).

5. The two sentences are from Bacon's 1623 *Of the Dignity and Advancement of Learning, Works IV,* 296; hereafter, *De Augmentis.* Harding takes the passage not from Bacon but from Merchant (1980, 168), who takes it from *Works IV.*

6. Keller also argues (1978, 412–3, 429; 1985, 91) that the Baconian scientist is *asexual;* even though he dominates a female nature, the marriage they have is chaste, cold, distant, detached. This sits uneasily with Bacon's "conquest" nature being forceful sexual seduction, let alone rape.

7. Rossi suggests that Bacon, in "Erichthonius," expressed his view that to be successful with nature, science has to "humbly beg her assistance" (1968, 101; see also "humble respect," 105). Keller (1985, 37) quotes Rossi, but not *this* phrase, creating the im-

pression that he agrees with her "forceful and aggressive seduction" reading.

8. Harding quotes these two *De Augmentis* sentences three times in two books, always informing us that her source is Merchant's *The Death of Nature* (1980, 168). Merchant includes the five missing words. In addition to failing to mark an ellipsis in the first of Bacon's two sentences, Harding makes a second mistake in *WS?WK?*: ellipsis points belong *between* the two sentences, since Harding omits a large chunk. Any hint of rape created by the *juxtaposition* of these sentences in *WS?WK?* is therefore artificial. There are other errors (cf. *Works IV*, 296). In Merchant, we correctly find "these holes," while in Harding, "those." Merchant and Harding write "whole object," but both are wrong, "sole object' is correct. Merchant, and Harding in *SQIF*, gives "drive her afterward," but both are wrong; in *WS? WK?*, Harding got "afterwards" right.

9. See also Bordo's (1987) remark on this sentence from *De Augmentis*, apparently provoked by Keller and Merchant: it illustrates "the famous Baconian imagery of sexual assault" (107–8).

10. Henceforth I supply for *Novum Organum* only the *Book I* aphorism number.

11. This is reminiscent of Merchant on Bacon's experimentalism: "Here, in bold sexual imagery, is the key feature of the modern experimental method— constraint of nature in the laboratory" (1980, 171).

12. Bacon is no blind optimist: he recognizes that science done poorly will go wrong. See, for example, "Daedalus" (Myth 19) in *Wisdom of the Ancients* (Robertson, 842–43).

13. See my (1985), at 73–4.

14. Similarly, it will not help Harding to claim that her experiences and social location as feminist or woman grant her an epistemic advantage—in this case, they make her an especially perceptive reader of early seventeenth-century texts (see her 1991, 121–33, 150–51, and my [1992] and [1994]). Harding's reading of Bacon is a politically inspired reading that goes wrong, and so subverts her claim that feminist scholarship is better *because* it is deliberately political.

REFERENCES

Bacon, F. 1857–1874. *The works of Francis Bacon, vols I–XIV*. Edited by J. Spedding, R. L. Ellis, and D. D. Heath. London: Longman. Reprinted 1962–1963. Stuttgart: Verlag.

Bleier, R. 1984. *Science and gender*. New York: Pergamon.

Bordo, S. 1987. *The flight to objectivity*. Albany, NY: SUNY Press.

Boyd, R., P. Gasper, and J. D. Trout, eds. 1991. *The philosophy of science*. Cambridge: MIT Press.

Farrington, B. 1964. *The philosophy of Francis Bacon*. Liverpool: Liverpool University Press. Reprinted 1966. Chicago: University of Chicago Press.

Feyerabend, P. 1970. Consolations for the specialist. In *Criticism and the Growth of Knowledge*, edited by I. Lakatos and A. Musgrave, 197–230. Cambridge: Cambridge University Press.

Harding, S. 1986. *The science question in feminism*. Ithaca, NY: Cornell University Press.

———. 1989. Value-free research is a delusion. *New York Times*, October 22, E24.

———. 1991. *Whose science? Whose knowledge?* Ithaca, NY: Cornell University Press.

Harding, S., and J. F. O'Barr, eds. 1987. *Sex and scientific inquiry*. Chicago: University of Chicago Press.

Keller, E. F. 1978. Gender and science. *Psychoanalysis and Contemporary Thought* 1:409–33.

———. 1980. Baconian science: A hermaphroditic birth. *Philosophical Forum* 11:299–308.

———. 1982. Feminism and science. *Signs* 7:589–602.

———. 1985. *Reflections on gender and science*. New Haven, CT: Yale University Press.

Leiss, W. 1972. *The domination of nature*. New York: George Braziller.

Longino, H. 1988. Review essay. Science, objectivity, and feminist values. *Feminist Studies* 14:561–74.

———. 1990. *Science as social knowledge*. Princeton, NJ: Princeton University Press.

Macaulay, T. B. 1967. Francis Bacon. In *Critical and historical essays*, Vol. 2, 290–398. New York: Dutton.

Merchant, C. 1980. *The death of nature*. New York: Harper & Row.

Morgan, R. 1977. *Going too far*. New York: Random House.

Nelson, L. H. 1990. *Who knows. From Quine to a feminist empiricism*. Philadelphia: Temple University Press.

Robbins, R. H. 1959. *The Encyclopedia of witchcraft and demonology*. New York: Crown.

Robertson, J. M., ed. 1905. *The philosophical works of Francis Bacon*. London: Routledge.

Rossi, P. 1968. *Francis Bacon. From magic to science*. London: Routledge and Kegan Paul.

Soble, A. 1985. Pornography: Defamation and the endorsement of degradation. *Social Theory and Practice* 11:61–87.

———. 1992. Review of Harding's *Whose science? Whose knowledge?* In *International Studies in the Philosophy of Science* 6:159–62.

———. 1994. Gender, objectivity, and realism. *Monist* 77:509–30.

Tuana, N. 1990. Review of Harding and O'Barr's *Sex and scientific inquiry,* in *American Philosophical Association Newsletter on Feminism and Philosophy* 89:61–2.

31

JANET SAYERS

Feminism and Science

Women are participating today, perhaps more than ever before, in the movement for a new science that will meet the needs of all people—women and men alike. In the past the women's movement has contributed to this struggle primarily through asserting the ostensible old values of science against what was seen as their sexist abuse. It is now also increasingly challenging those values themselves, and arguing the need for them to be forged anew if science is to serve women equally with men. My concern here will be to describe this development in feminism's engagement with science.

The women's movement has, from its inception, taken issue with science in so far as it reflects and contributes to the maintenance of existing social inequalities between the sexes. At first this meant criticizing science solely in its own terms. This involved drawing attention to the way science departs from its own self-professed canons of neutrality in the sex discrimination it exercises against hiring and promoting women within its ranks. It also involved exposing the sexism that all too often governs the choice of research priorities, methods of investigation, and interpretation of results in science.

There are many examples of such departures from science's proclaimed social neutrality. As regards the sex discrimination exercised against women within the scientific profession there is, of course, the notorious case of Rosalind Franklin, whose contribution to the discovery of DNA was so singularly neglected by contrast with the attention accorded that of James Watson and Francis Crick.

This same sexism has also influenced the selection of research priorities within science. Thus, for example, the premise that women are naturally less capable of contributing to scientific research has led many investigators in the behavioral sciences to search for natural determinants of men's alleged superiority in scientific ability to the neglect of investigating sex similarities, or even the superiority of women as regards certain aspects of this ability. The search for a biological factor—a sex-linked gene, hormone, or brain difference, for instance—that might cause women's alleged natural inferior capacity for engaging in science continues today despite the repeated failure of any investigations to demonstrate that biology makes women less capable than men as regards those skills—visuospatial, analytical,

Science and Beyond, Steven Rose and Lisa Appignanesi, eds. (New York: Basil Blackwell, 1986), pp. 169–178. Reprinted with permission from the author.

or mathematical—associated with scientific ability (cf. Saraga and Griffiths, 1981; Sayers, 1982).

Likewise as regards research methodology. It has been shown that the presumption that men are naturally more aggressive and competitive than women has resulted in a focus on factors in male biology—androgens say—that might determine aggression to the neglect of equally studying factors in female biology—estrogens say—that in fact also contribute to the determination of this behavior (see Bleier, 1976). Overlooking the fact that such research thus breaches science's claim to be neutral and free from bias, the sociobiologist E. O. Wilson maintains that it provides scientific evidence of the justice of existing sexual inequalities within science's ranks. Citing this kind of research, he adds that science demonstrates that

> boys consistently show more mathematical and less verbal ability than girls on the average, and [that] they are more aggressive from the first hours of social play at age 2 to manhood (Wilson, 1975, p. 50).

On this basis he concludes

> Thus, even with identical education and equal access to all professions, men are likely to play a disproportionate role in political life, business, and science (Wilson, 1975, p. 50).

Scientists have also been guilty of partiality in the interpretation of their data. Faced with the repeated finding that, on average, girls are verbally superior to boys as regards age of first speech and as regards verbal fluency, scientists—psychologists in this case—have interpreted these data as justifying the current unequal division of child-care between the sexes. Women's greater verbal ability, it is argued, renders them better fitted for child-rearing, for teaching their children to speak than men (see, e.g., Gray and Buffery, 1971). But this is to neglect the fact that such data also put in doubt the wisdom of the present sexual division of child-care in so far as it results in women's under-representation 'in political life, business and science' for which their greater verbal abilities would also seem to fit them better than men!

It continues to be important to challenge science on its own ground, and to draw attention to those instances in which it departs from its canons of neutrality in its professional practice, as well as in its selection and interpretation of evidence. For it is precisely the claim of scientists to neutrality that gives such weight and authority to the conclusions they draw from their data as regards the legitimacy of existing sexual inequalities in society. On the other hand, feminists have become increasingly aware of the need also to reassess the criteria by which scientific neutrality is assessed. It is to a brief account of this development within feminism that I shall now turn.

As indicated above, the traditional test of the neutrality of scientific theory is that it be free from contamination by social practice, in this case by that of sexual discrimination in society. But this very separation of theory from practice is itself implicated in women's social subordination, and in the failure of science to meet women's needs.

I shall seek to explain this point by way of an example—that of the history of the social management of childbirth. As many feminists have pointed out, this development involved the appropriation by medicine of many of the tasks formerly performed by midwives. Legitimation of this appropriation was often couched in terms that characterized midwifery as based on ignorant, unskilled superstition and customary practice. This it was said made it inferior to medical obstetrics based on scientific theory, and thereby held to involve superior technological skill and expertise (see, e.g., Oakley, 1976).

Clearly science has served women in freeing them from the abuses resulting from the hocus pocus of superstition as it informed midwifery and other areas of social life. On the other hand, the celebration of scientific theory over practical experience—the practical experience of women as midwives and as mothers say—has also operated against women's interest in so far as it has resulted in this experience being neglected in selecting the priorities and objectives of obstetric medicine. Consequently, as many feminists have pointed out, obstetrics—like other areas of science—fails to meet women's needs and interests as well as it might.

Feminists have accordingly often argued that the division of theory from practice, of mental from manual labor, must be transcended if science is equally to serve all people—women as well as men. In this, feminism is at one with socialism in its argument that, since this division constitutes one of the major determinants of the alienation and exploitation of the working class under capitalism, science will not serve the ruled equally with the rulers until this division is transcended in science as in other areas of social life (Rose, 1983). Socialists, however, all too often forget that if science is even-handedly to meet the needs of all people it must pay attention to women's needs just as much as to those of men. This, as feminists note, means not only transcending the division of theory and practice that currently characterizes science. It also means transcending the associated division of production from reproduction.

What is the bearing of production and reproduction, and their divisions, on feminism and science? In brief, and at the risk of over-simplification, I should explain that these divisions—both actual and ideological—can be shown to have been a major source of women's current social subordination. In the past, production was often conducted in conjunction with reproduction on a family basis within the home. The development of the forces of production—made possible in large part by the scientific advances of the late eighteenth and early nineteenth centuries—involved the development of machinery that was also initially operated on a family basis either inside or outside the home. Increasingly, however, it came to be operated outside the home on an individual basis. Productive activity thereby came to be separated from the reproductive activities involved in looking after the day-to-day physical and emotional needs of children and adults—activities that continued to be performed, as they are today, on a primarily family basis within the home.

While women of the working classes continued to be involved in both spheres of activity, middle-class women were discouraged from involvement in production. Increasingly they came to be identified with reproduction—with bearing and rearing the heirs to family property. Their menfolk, by contrast, were increasingly identified with production. This reflected the fact that it was mostly men who owned and administered the means of production and the wealth generated by it. Since it was in this sphere that wealth—at least exchange value—was generated, so production came to be valorized over reproduction. Furthermore, as a result of the ideological as well as economic dominance of middle- over working-class

ideology, women of the working class, like women of the middle class, came to be identified with reproduction even though they were also involved in production; while men of all classes came to be identified with production (whether as workers or owners) and with the wealth it generates and the value thereby accorded it in society.

Since the division of production from reproduction has thus been a major source of women's work, needs, and interests being accorded less value, worth, and attention than those of men, so the women's movement has made the transcendence of this division one of its major aims. In the view of socialist feminism this entails seeking to integrate production with reproduction, not so much through domesticizing production as through socializing reproduction—namely child-care, education, health, and all those activities associated with the daily and generational reproduction of society's individual members.

What has all this to do with science? Scientific advance has been stimulated by, and has in turn stimulated, advance in the productive transformation of nature and its resources to meet human needs, which have in their turn thereby been advanced and changed. Science has thus been very much allied with production and with the dynamic progress of society that contrasts with the resistance to change that is so often justified by an appeal to naturalism. As production came to be divorced from reproduction, and as production came to be identified with men, reproduction with women, so science too came to be increasingly regarded as masculine and its object—nature—as feminine in that the reproductive activity of women increasingly came to be equated solely with its natural, biological aspects (aspects that many scientists said would be endangered were women to engage in science). This view of science's object as that of nature equated with femininity is nicely illustrated by the historian of science Ludmilla Jordanova, who describes in this context a 'statue in the Paris medical faculty of a young woman, her breasts bare, her head slightly bowed beneath the veil she is taking off, which bears the inscription "Nature unveils herself before Science"' (Jordanova, 1980a, p. 54).

Although, as Jordanova (1980b) points out, science involves both subjective feeling as well as objective rationality, these aspects of its activity have tended to be polarized: the former aspect is deemed antithetical to, and therefore absent from, science and its pursuit of objectivity and neutrality. This polarization, I would argue, is itself an effect of the division of production from reproduction. For along with the ideological consequence of a division which equated men with production, women with reproduction, went a bifurcation of theory and practice, objectivity and subjectivity, reason and emotion, and of individuality and mutuality even though they are all involved alike in production and reproduction. As a consequence, production, and the science allied to it, came to be viewed as the seat of theory and of objective, individualistic rationality, and reproduction as the seat of practical concern, and of subjective emotion and mutuality. As feminists have pointed out, this ideological divide has in turn had baneful effects on science and its progress. The physicist Evelyn Fox Keller (1982) shows how it has resulted in many scientists—women as well as men—focusing on the individual operation of the objects they study to the neglect of their mutual interdependence and interaction.

The achievement of a science that does justice to the nature it studies, and of a science that serves women equally with men, depends on transcending antitheses such as those between individuality and interdependence that are in large measure a result and reflection of the ideological effects of the division of production and reproduction. It therefore depends on the integration of production with reproduction for which feminism strives. The struggle for a feminist science is thus one with the struggle of the women's movement in general. But it also poses peculiar problems of its own, of which I shall specify three below.

In the first place it depends on women gaining control over science's means of production in order that these means might be directed to meeting women's needs equally with those of men. This entails seeking to change the conditions that render women only skivvies and handmaidens to science—whether as factory hands making scientific equipment, domestic servants (whether at home or at work as cleaners, cooks, etc. to scientists), or as clerks, secretaries, technicians, and research assistants in scientific laboratories. It means wresting the control of scientific production from men so as to

share equally with them in determining the priorities, objectives, and applications of science. Yet such is the association of science with men's interests that few women believe they have anything to gain from such struggle. This is the case despite the fact that science clearly shapes women's as well as men's activity in production and reproduction, in public as well as in private life (Haraway, 1985). Furthermore, even when girls and women recognize their interests to be bound up with science, and even when they seek to become scientists in their own right, they are often dissuaded from such participation on the ground that scientific and mathematical achievement depends on much more effort and hard work in the case of women than it does in the case of men (see, e.g., Parsons et al., 1982). This argument in turn reflects the prevalent view of science as men's natural preserve, one for which women are not naturally fitted and in which therefore their success depends on much more effort than it does for men.

A second obstacle to the struggle for a feminist science is that such a goal is often dismissed as a contradiction in terms. The very essence of science, it is argued, lies in its neutrality. It should not therefore be allied to any political and social persuasion, feminist or otherwise. However, as indicated above, the development of science has been inextricably bound up with society and with the development and divisions of its productive and reproductive activities. Furthermore, despite its claims to the contrary, scientific research is also very much governed by subjective hunch and emotional inclination. Science can only fully make good its claim to objectivity—a claim that goes hand-in-hand with its claim to neutrality—if it acknowledges all the objective conditions of its own existence. This includes recognizing its political, social, subjective, and emotional determinants, and the lack of objective truth to the polarization of subjectivity from objectivity—a polarization that is more emphasized in the ideological elaboration of the division of production from reproduction than is warranted by reality. As the psychologist Jean Piaget once pointed out:

> Objectivity consists in so fully realizing the countless intrusions of the self in everyday thought and the countless illusions which result— illusions of sense, language, point of view, value,

etc.—that the preliminary step to every judgment is the effort to exclude the intrusive self. . . . So long as thought has not become conscious of self, it is a prey to perpetual confusions between objective and subjective, between the real and the ostensible (cited by Keller, 1982, p. 594).

If science is to serve women it must recognize that this goal—that of meeting women's needs and interests as of society's in general—falls within its proper province and preserve. This means challenging the way in which its canons of objectivity and neutrality are used to mystify the actual relation of science to society (Fee, 1983). This is not to reject these canons in favor of a hopeless social relativism that would accord equal validity to sexism as to feminism in science. Quite the reverse! It is to argue the need to recognize the objective ways in which the sexism of our society affects its science both in its theory and in its practice, and it is to argue the need to acknowledge, in order to combat, the objective determinants of this sexism—namely the divisions both ideological and actual, of production and reproduction.

This brings me to a third, related obstacle to the development of a feminist science. Such is the force of the separation of production from reproduction, and of the related separation of science from society, that feminism is all too easily dismissed as peripheral to science, the main activity of which, as I have said, is conceptualized as essentially unrelated to social issues, feminist or otherwise. Moreover, even where it is acknowledged that science is related to society, it is more usually recognized to be related to production than it is to be related to reproduction. As a result, feminism and its concern with reproduction is still viewed as having little, if anything, to do with science.

Both the above ways in which feminism is marginalized with respect to science are thus related to the self-same divide of production from reproduction against which feminism strives. This marginalization is not due to any lack of numbers of women whose interests feminism claims to represent. It reflects women's lack of clout in a society where power, along with value, is accorded to social production and in which women, because of their role (actual or ascribed) in reproduction, enjoy relatively little status and authority.

For feminism to be given a proper hearing within science, let alone for it to achieve the transformation of science necessary to science's meeting women's needs equally with those of men, the division of production from reproduction must therefore be transcended. Until that happens feminism's engagement with science will, perforce, remain principally that of gadfly and irritant. As the biochemist and historian of science, Elizabeth Fee, puts it:

> At this historical moment, what we are developing is not a feminist science, but a feminist critique of existing science. It follows from what has been said about the relationship of science to society that we can expect a sexist society to develop a sexist science; equally, we can expect a feminist society to develop a feminist science. For us to imagine a feminist science in a feminist society is rather like asking a medieval peasant to imagine the theory of genetics or the production of a space capsule; our images are, at best, likely to be sketchy and insubstantial (Fee, 1983, p. 22).

This does not, however, entail that the struggle for a new, feminist science is utopian. The conditions for its existence are already in the process of being forged. Despite the divisions that exist between them, production and reproduction are indissolubly interlinked. Thus, for example, the hours of work men and women can devote to social production is conditioned by reproduction, by the provisions available, say, for the care of their children and other dependents while they work. Furthermore, production and reproduction are also in contradiction with each other. It is women's experience of this contradiction—their experience of the conflicting demands of work and home—that has constituted the impetus of the current women's movement. Consciousness of this contradiction has been heightened as an effect of women's increasing involvement in production—an involvement that is the result, in part, of scientific advance, of developments in contraceptive technology, for example, which have meant that women are not now removed from social production as they once were by the demands of constant childbearing. The conflicts between production and reproduction are thus even now operating to undermine and transform the divisions between them—divisions that I have been arguing constitute a primary source of the current failure of science and society to meet women's needs equally with those of men. In this lies cause for optimism. For it means that the conditions for the emergence of a new, feminist science—one that will equally serve women as well as men—are in the process of emergence not least because of the vigor with which the women's movement is now pursuing its cause.

REFERENCES

Bleier, R. (1976) 'Myths of the biological inferiority of women: an exploration of the sociology of biological research', *University of Michigan Papers in Women's Studies,* 2, 39–63.

Fee, E. (1983) 'Women's nature and scientific objectivity'. In *Woman's Nature: Rationalizations of Inequality,* M. Lowe and R. Hubbard (eds.), Pergamon Press.

Gray, J. A. and Buffery, A. W. H. (1971) 'Sex differences in emotional and cognitive behavior in mammals including man: adaptive and neural bases', *Acta Psychologica,* 35, 89–111.

Haraway, D. (1985) 'A manifesto for cyborgs: science, technology and socialist feminism in the 1980s', *Socialist Review,* 15(2), 65–107.

Jordanova, L. (1980a) 'Natural facts: a historical perspective on science and sexuality', In *Nature, Culture and Gender,* C. P. MacCormack and M. Strathern (eds.), Cambridge University Press.

———. (1980b) 'Romantic science? Michelet, morals, and nature', *British Journal for the History of Science,* 13, 44–50.

Keller, E. (1982) 'Feminism and science', *Signs,* 7(3), 589–602.

Oakley, A. (1976) 'Wisewoman and medicine man: changes in the management of childbirth'. In *The Rights and Wrongs of Women,* A. Oakley and J. Mitchell (eds.), Penguin.

Parsons, J. E., Adler, T. F. and Kaczala, C. M. (1982) 'Socialization of achievement attitudes and beliefs', *Child Development,* 53(2), 310–39.

Rose, H. (1983) 'Hand, brain and heart: a feminist epistemology for the natural sciences', *Signs,* 9 (1), 73–90.

Saraga, E. and Griffiths, D. (1981) 'Biological inevitabilities or political choices? The future for girls in science'. In *The Missing Half,* A. Kelly (ed.), Manchester University Press.

Sayers, J. (1982) *Biological Politics: Feminist and Anti-Feminist Perspectives,* Tavistock.

Wilson, E. O. (1975) 'Human decency is animal', *New York Times Magazine,* 12 October, 38–40, 42–6, 48, 50.

32

JANET RADCLIFFE RICHARDS

Why Feminist Epistemology Isn't

Twenty years ago, when feminism was younger and greener, it would have been much easier to set about discussing the movement from the point of view of flights from reason and science. To start with, there would have been relatively little dispute that such a flight was going on. Many feminists themselves claimed to be taking wing from all such male devices for the oppression of women, as well as from morality, which was another of them. And, furthermore, the position was a relatively easy one for the skeptical outsider to attack. Unless feminists could say such things as that the present treatment of women was morally wrong, or prevailing ideas about their nature false or unfounded, or traditional reasoning about their position confused or fallacious, it was difficult to see on what basis they could rest the feminist case. And of course as they did say such things, all the time, it was obvious that any systematic attempt to reject ethics and rationality was systematically undercut by feminists' own arguments.[1]

In these more sophisticated times, however, the issue is much more complicated. The language, at least, has changed, and few feminists now can be heard to say that reason or science should be abandoned altogether. What they say instead is that particular, traditional accounts of these things must go, to be replaced by new, feminist conceptions of them; and this makes matters very different.

Here, for instance, is Elizabeth Grosz in an essay on feminist epistemology, describing the work of another feminist, Luce Irigaray:[2]

> Irigaray's work thus remains indifferent to such traditional values as "truth" and "falsity" (where these are conceived as correspondence between propositions and reality), Aristotelian logic (the logic of the syllogism), and accounts of reason based upon them. This does not mean her work could be described as 'irrational,' 'illogical,' or 'false.' On the contrary, her work is quite logical, rational, and true in terms of quite *different criteria,* perspectives, and values than those dominant now. She both combats and constructs, strategically questioning phallocentric knowledges without trying to replace them with more inclusive or more neutral truths. Instead, she attempts to reveal a *politics* of truth, logic and reason.

Statements of this kind obviously make matters much more difficult for the critic who is concerned about flights from reason and science in general, and suspects contemporary feminism of being at the forefront of the rush. If feminists claim that all they are doing is offering new and improved conceptions of these things, it no longer seems possible to object in principle to the whole project. Investigating the

The Flight from Science and Reason, Paul R. Gross, Norman Levitt, and Martin W. Lewis, eds. *Annals of the New York Academy of Sciences* 775:385–412. Reprinted with permission.

This title was borrowed from Larry Alexander's "Fancy Theories of Interpretation Aren't," an uncompromising gem of analysis with which I am pleased to have even this tenuous association.

foundations of science, epistemology, logic, ethics, and all the rest is a perfectly respectable philosophical activity, in which feminists seem as entitled to join as anyone else.

Presumably, therefore, any complaint about irrationality must be directed to the particular conceptions of science and epistemology put forward under the name of feminism. But here things become extraordinarily difficult, because anyone of even vaguely familiar epistemological views who has tried to tackle feminist works in these areas will know how quickly there comes the sensation of being adrift in uncharted seas, with no familiar landmarks in sight. There is nothing so simple as particular claims that, by more conventional standards, seem mistaken. In so many ways that they defy representative quotation, much of what is said seems already so laden with revisionary theory that the innocent reader is left with the uncomfortable feeling of having missed the story so far. What are "knowledges"?—which presumably, if the term is not just a perverse neologism, are intended as something other than just different things that are known. What is it for them to have, or for that matter fail to have, such astonishing qualities as phallocentricity? What is the significance of their being said to be "produced,"[3] rather than acquired, or by that production's being "intersected by gender,"[4] or by women's asserting "'a right to know' independent of and autonomous from the methods and presumptions regulating the prevailing (patriarchal) forms of knowledge"?[5] What, even, is "feminist knowledge"?[6] To the uninitiated it soon begins to look as though it would take whole chapters or books to unpick and contend with the presuppositions of even a single sentence, even if there were any obvious way to set about doing it.

Now in a sense this cannot reasonably be complained about. Feminist epistemology is supposed to seem strange to the traditionalist; if it were not at odds with more familiar ideas, there would presumably be no point in doing it. But on the other hand if there is no common ground, how can the debate be tackled at all? The first difficulty seems to be to find a way of even approaching the problem.

This is the issue I want to address, and the point from which I propose to start is this. Feminist epis-temology may indeed look pretty bizarre, and completely detached from familiar ways of thinking, to anyone who plunges straight in; but its recommenders cannot intend there to be no connection at all with more familiar ground. The term is not supposed to be an arbitrary label adopted by people who go around wrenching words from their accustomed forms and contexts for the fun of it. Feminists who advocate feminist epistemology did, after all—or at least their foremothers did—grow up with the kind of epistemological view they now claim to have overthrown, so they must themselves have had reasons, presumably connected with their feminism, for rejecting it. The idea seems to be that once you have a proper feminist view of things, you begin to see the limitations of the "standard"[7] views you started with, and can take off in new and enlightened directions. And if so, it should be possible to explain to others the reasons for doing so.

We can, then, focus the problem by concentrating on the question of what arguments might persuade someone of such standard epistemological and scientific views to abandon them for the unfamiliar world of feminist epistemology; and, in particular, on the question of why a *feminist* should think of making the change, which—the idea seems pretty obviously to be—she should. This question can be regarded as the landmark to which it is always possible to turn when there is nothing else recognizable in sight.

Consider, then, the situation of someone whose epistemological views are still of a fairly commonsensical sort: someone who holds such unremarkable opinions as that there are some things we know, some we do not know, and others of which we are unsure; that different people know different things; that we often think we know what we turn out not to have known; and probably also—what Alcoff and Potter call a "philosophical myth"—that philosophical work can be good only "to the extent that its substantive, technical content is free of political influence."[8] And think of this standard epistemologist also as a feminist whom I shall, to avoid complications irrelevant to the matter in hand, take to be a woman—on the shores of this somewhat alarming sea, dipping in the occasional tentative toe and wondering what reasons the mermaids can offer for her abandoning the familiar landscape behind and casting off.

THE FOUNDATIONS OF FEMINISM

It looks as though the first problem must be to say what can be meant by the claim that the feminist on the shore is indeed a feminist, especially since, *ex hypothesi*, she has not yet embraced the various ideas and approaches that go by the name of feminist science and epistemology. It is well known, and notorious, that there is no generally accepted definition of feminism, and I certainly do not want to send things awry at the outset by adopting a prescriptive definition that other feminists would reject. Fortunately, however, there is no need to. It will be enough to take an absolutely minimal account, and say that whatever the details of what she thinks, a feminist is, at the very least, someone who thinks that *something* has been seriously wrong with the traditional position and treatment of women.

Now of course this could hardly be more minimal, since it allows for great differences between feminists in their ideas about what is wrong and how to put it right. Nevertheless it is enough for my purposes here, because if a feminist is someone who is making a complaint about the present state of things, there are various things that must be generally true about her starting position, no matter what the details are.

In the first place, she must obviously have a view about the way things *are*, or she could not think there was anything wrong with it; and she must also have some ideas about what possibilities there are for change, or she would not be able to say that things should be otherwise. She must, in other words, have a range of first-order beliefs about the world: the kind of belief that is supported by empirical, often scientific, investigation. Beliefs of this kind also imply that she has other beliefs about second-order questions of epistemology and scientific method, since in reaching conclusions about what to believe about what the world is like and how it works she has, however unconsciously, depended on assumptions about how these things can be found out, and how to distinguish knowledge from lesser things. These assumptions will become more explicit if any part of her feminism involves (as it is pretty well bound to) accusing the traditionalist opposition of prejudice, or of perpetrating or perpetuating false beliefs about women.

Similar points apply to questions of value. In order to make any complaint whatever about the way things are, a feminist must at least implicitly appeal to standards that determine when one state of affairs or kind of conduct is better or worse than another; and if her complaint takes a moral form rather than a simply self-interested one—if, like virtually all feminists, she expresses her complaints in terms of such things as injustice and oppression and entitlements to equality—she must be appealing to moral standards of good and bad or right and wrong, of which she thinks the present state of things falls short. And if she has such normative, first-order, standards, that in turn will imply something about her attitudes to the higher-order questions of metaethics, whether or not she thinks of them as such.

To say this is, once again, to say nothing at all about the content of such beliefs and standards. The claim so far is only that for feminism to get going at all, in any form, there must already be in place ideas about the way things are, and standards by which they are found wanting. Different people may have different beliefs and standards, and so reach quite different conclusions about what is wrong. However, since the specific problem being addressed here is of how someone could get from *familiar* ideas of science and epistemology to the kinds that are claimed as feminist, it is most useful to start by assuming fairly ordinary kinds of belief about both facts and values, and to consider how anyone starting from that kind of position can have been led to feminism at all.

This may in itself seem to present a problem; for how, it may be asked, can anyone be both a feminist—of any kind—and a holder of traditional views? If feminism is essentially a challenge to received beliefs and attitudes, as it is, its starting point must be the idea that these views are in some way wrong. There must therefore be *some* differences between what even the most cautious feminist accepts, and what is standardly believed by people who have not yet reached this degree of enlightenment.

And of course this is true; by its very nature, feminism must challenge some received ideas. But feminism began as a movement—as it probably does for most individuals—not with some sudden éclaircissement that led its supporters to reject all familiar standards, and to embrace instead new ones according to which prevailing ideas about women appeared as wrong from the foundations. Rather, it began in effect with the recognition that familiar ideas about women were *anomalous,* in that they could not be justified by the standards that holders of these ideas quite routinely accepted in other contexts. The original point was that traditional standards of evidence and argument in science and ethics *themselves* did not support traditional conclusions about women.

At the very simplest level, consider the early feminist challenges to received beliefs about the nature of women. Mill, for instance, pointed out that since women and men had since records began been placed in different social situations and given systematically different kinds of education (as everyone knew, since that was the status quo the insisters on women's difference were trying to defend), none of the observed psychological or intellectual differences between the sexes could reliably be attributed to nature.[9] This was an objection to established beliefs, but it involved the adoption of no new standards of epistemology or scientific procedure; prevailing beliefs were challenged by appeal to the standards that would be applied in any other scientific context. The feminist claim was essentially that there was what might be called a *sex-connected incoherence* in the current view of things. On the basis of the most fundamental current views about the nature of knowledge and standards of evidence, some less fundamental beliefs could be shown to be unfounded.

The same kind of thing happened with early feminist challenges to moral values. Of course some traditional moral values—about the propriety of women's remaining subservient to their husbands and away from public life—were challenged by feminists. But the original challenges were made not by reference to completely new standards of moral assessment that transmuted traditionalist right into feminist wrong, but by arguments showing that familiar general ideas of morality, such as most people professed most of the time, were incompatible with traditional ideas about the treatment of women. Even if it were assumed that most women were congenitally unsuited to the kind of occupa-

tion reserved for men (which anyway, from the previous argument, there was no adequate reason to assume), widely held views about open opportunities and letting people rise by their own efforts were incompatible with a wholesale exclusion of women that did not allow them even to try. Or even if it were conceded both that women were systematically inferior to men in strength and intellect, and that the weak needed protection, it still took a pretty remarkable twist of reasoning to reach the accepted conclusion that this provided a justification for making women weaker still, by placing them in social and legal subordination to men.[10]

So the arguments through which traditional feminism reached its first conclusions involved no departure from familiar standards of evidence and argument in ethics, epistemology, and science, but actually presupposed them. It was *by appeal* to these standards that the position of women was first claimed to be wrong. And notice that all arguments of this kind also depend on absolutely ordinary logic. It is *because* the traditional conclusions do not follow from the traditional premises, or *because* traditional beliefs are incompatible with traditional standards of assessment, that the challenge to the received view in its own terms is possible.[11]

Here, then, are the beginnings. Although feminism, as a critical movement, necessarily challenges parts of the status quo, it typically does so, at the outset, by appealing to other, more fundamental parts that it holds constant. Feminism as a movement started with the broad standards of moral and empirical investigation and argument that most other people accepted at the time, and the recognition that these could not support familiar, supposedly commonsensical, ideas about women and their position. Most individual feminists probably begin in more or less this way as well.

FEMINIST PROGRESS

The fact that feminism must start with appeals to existing standards, however, does not imply that it can never escape them. Any aspect of belief can be rethought at any time; and since even in its earliest stages feminism considerably affects the way the world appears to its converts, it is likely that once these first changes have been made, other adjustments will soon be found necessary. What must now be considered is how, starting with the kind of first- and second-order beliefs most people have, and having reached some kind of feminism on their basis, the new feminist might be led *by her feminism* to reject them, and eventually adopt radically different ideas about reason and science.

Consider then the situation of the novice feminist, who has come to recognize that traditional ideas about women cannot be justified by traditional standards. She will of course immediately recognize the need to work for political change, but that will not be all. An equally important consequence of her new view of the world will be an increasing awareness of new questions that need to be answered. Once what was previously accepted as knowledge has been thrown into doubt, the problem inevitably arises of what should now be put in its place; and, as happens with any new perspective or information, there also comes the recognition of questions that people simply never thought of asking before.

For instance, once feminists realize that positive obstacles have always been placed in the way of women's achievements, they may start wondering whether the attitudes that brought about this state of affairs could also have led to the overlooking of what women actually did do. They may start searching historical records for evidence that, in spite of the obstacles, women achieved a good deal more than was traditionally thought. Or if, having brought about apparently equal treatment of the sexes in some area, feminists find women still doing less well than men, they may suspect the existence of more subtle obstacles, and set up experiments to investigate that possibility. They may try such things as swapping round the names of men and women on academic articles, to see whether this affects readers' assessment of their merits,[12] or making controlled observations to see whether girls who are interested in science are actively discouraged by their teachers or other children.[13]

So feminism can open up a range of inquiries that would probably otherwise have remained closed, and any of which may lead to further changes in beliefs about what the world is like and how it works. And of course every new discovery will in turn have its own implications. Some will lead to an expanded political program. If, for instance, it turns out that people are unconsciously biased against women even when they think they are being impartial, feminism will have a different, and much more complex, problem on hand from the one that arises only from awareness of overt discrimination. Most discoveries will also themselves suggest further questions, which, again, would not have arisen but for the feminist awareness that started the inquiries in the first place.

It is, however, essential to see exactly how feminism connects with changing beliefs of these kinds. The sorts of investigation I have been discussing are of a kind that arise directly from feminism: without feminism to suggest where to look, nobody might have thought of launching them. And when they are finished, feminists may well find their feminist agenda widened: they may recognize more scope for feminist research and more need for feminist action. But the extent to which this happens *depends on how the inquiries turn out,* and that has nothing to do with feminism at all. In the kind of investigation considered so far, there is no point at which feminism provides the *justification* for any change of belief. Feminism does not determine what counts as a proper inquiry, or what counts as a result one way or the other, or what the results should be. In conducting inquiries of this kind, feminists are still using the standards of evidence and argument that brought them to feminism in the first place. If they come to the belief that girls are disregarded or discouraged in science classes, or that academic articles thought to be written by women are systematically underrated, that is because the evidence shows this to be so. It is not because feminist principles demand that it must be.

During these early stages of feminist progress, therefore, the situation is still essentially the same as at the beginning. The feminist's background beliefs and standards themselves provide the justification for her expanding feminism, not the other way round. And, for that reason, her conclusions should be demonstrable to any impartial investigator who shares her basic standards and will look. Even though whatever changes in belief result from these investigations would probably not have occurred but for feminism, the new beliefs are not feminist in the sense of there being any reason for a feminist to hold them that a non-feminist—someone who has not yet recognized that women have grounds for complaint—has not.

It is important to stress this, because although the point is simple and obvious once seen, it seems to be widely overlooked. If so, this is probably at least partly a result of the ambiguity of "because" between cause and justification. You may change your views *because of feminism* in the sense that you would not otherwise have embarked on the inquiries that led to those changes, but that does not mean you change them *because of feminism* in the sense that feminism provides the justification for the change. Various well-known advances in science were (reputedly) made because their begetters soaked in baths, or reclined in orchards, or dreamt of snakes with their tails in their mouths; but no one thinks these causes of inspiration provided any part of the justification for accepting the resulting theories of displacement or gravitation or benzene rings. Feminism may set investigators on the track of new discoveries, and the feminist program may expand as a result of them, but that provides no more reason to count the new *beliefs themselves* as feminist than to count the others as bathist or appleist or snakeist.

So, to relate all this again to the lingerer on the shores of feminist epistemology, we can see that her feminism may well have made considerable progress since it first began, but nevertheless the *justification* for any changes in her view of things has so far had nothing to do with her being a feminist. The novice has exactly the same kinds of reason for progress within feminism as she had for becoming a feminist in the first place. Her mind has been changed only because of what the evidence has shown, and this has involved an appeal to her old familiar standards of epistemology and scientific method.

INTERMEDIATE STANDARDS

One way of expressing all this is that the discussion so far has been of feminism only as an applied subject. Feminism and feminist progress have been shown as emerging from standards of science and rationality that are not themselves feminist. The view of feminists in the thick of feminist epistemology, however, is that this is only the beginning:

Feminism made its first incursions into philosophy in a movement from the margins to the center. Applied fields, most notably applied ethics, were the first areas in which feminist work was published. . . . But from the applied areas we moved into more central ones as we began to see the problems produced by androcentrism in aesthetics, ethics, philosophy of science, and, finally and fairly recently, the "core" areas of epistemology and metaphysics. . . . The work of feminist philosophers is in the process of producing a new configuration of the scope, contours, and problematics of philosophy *in its entirety* [original emphasis].[14]

And when feminism reaches this new ground, quite different questions seem to arise. Presumably nothing that has been said about the progress of feminism as an applied subject, within the traditional framework, can be assumed to apply to feminist challenges to that framework itself.

Nevertheless, it is important to have discussed the less radical kind of change, because it is essential for clarifying the general issue of feminist challenges to accepted standards. Standards come in hierarchies, and a good many changes can be made at superficial or intermediate levels long before the fundamentals of rationality, philosophy of science, and epistemology are even approached. And in particular, it is essential to recognize that changes in first-order beliefs—about what the world is like and how it works—always in themselves amount to potential changes in standards, because well-entrenched first-order beliefs are automatically and necessarily used as the basis for assessing others regarded as less well established.

This is a fundamental fact about the way we ordinarily reason—and must reason—that is obvious as soon as it is thought about. Consider, for instance, the Phoenicians in Herodotus, who returned from their voyage to circumnavigate Africa claiming that as they had sailed westward round what we know as the Cape, the sun had lain on their right, to the north. Since this account was incompatible with contemporary views about the relationship of the earth to the sun, Herodotus, not unreasonably, did not believe them. It seemed much more likely that returning travelers should spin fantasies than that the sun should change its course. Now, however, our changed beliefs about astronomy have changed the standards by which we assess the story; and we regard it not only as true, but also as providing the best possible evidence that the Phoenicians really had circumnavigated Africa.[15] This kind of thing happens in every aspect of life. In forensic medicine we decide guilt by reference to our fundamental beliefs about blood groups or DNA, whereas once we might have decided it by whether the accused sank or floated in water. If some fringe medicine makes claims that are incompatible with well-entrenched scientific theory, that theory will be used as proof that the fringe claims must be wrong, and any anecdotal evidence in their favor will be explained away in other ways.[16] But if some previously unknown causal mechanism is eventually found to exist (as was briefly claimed in the case of homeopathy a few years ago[17]), attitudes may change and the fringe be incorporated, wholly or at least in part, into the main stream.

How likely such changes are to happen will depend on how well entrenched, how comprehensive, and how vulnerable any particular range of beliefs is; and this makes it likely that feminism, once begun, will lead eventually to extensive changes of standards at this level. This is essentially Mill's point again. Understanding of the world comes through observation of its constituents under varying circumstances; and since ideas about women and men have developed while women have been seen only in rather limited situations, it is to be expected that proper investigation will dislodge many traditional beliefs connected with them. And when changes do occur—either positively, in the acceptance of new beliefs to replace or supplement the old, or negatively, in the recognition that old beliefs are insecure—standards for the assessment of other beliefs and ideas will necessarily change with them.

This is already familiar from more or less everyday life. If the idea is challenged that women's nature allows them to find happiness only through making husbands and children the main focus of their attention, this will result in fundamentally different approaches to the assessment of individual women who are chronically discontented about the course of their lives. If it is accepted that sexual abuse of girls by their fathers is widespread, allegations that this has happened will become more likely to lead to criminal investigations of fathers than psychological probings of women's oedipal delusions. But the same general point is also potentially relevant to many of the issues discussed in the broad context of feminist epistemology and science.

For instance, suppose feminists are right in claiming that many beliefs typically held by women—passed down the generations, perhaps—have traditionally been dismissed as old wives' tales. If this has happened because of conflict with entrenched scientific theories, then the dislodging of those theories—perhaps as a result of feminist inquiries—would remove the basis on which the women's beliefs had been dismissed; and if enough independent evidence could be accumulated in favor of the women's beliefs, they might even themselves become the basis for rejecting the established theories. Or perhaps—another feminist idea—women may have ways of investigating the world that have been disregarded because they are different from the ones currently regarded as paradigmatic of good scientific procedure. But if feminist-inspired investigation eventually showed not only that such female techniques did exist, but also that they were just as scientifically effective, by ordinary criteria for scientific success, as the ones currently used as the touchstone for the worth of research and the promise of researchers, that would support feminist demands for changes in the structure of the scientific establishment and the standards by which aspirant scientists were assessed.[18] Whether or not such changes turn out to be justified, there is no theoretical problem about the possibility that they might be.

Just because of this relevance to matters of science and knowledge, however, it is essential to stress again that when changes in standards of these kinds do occur, they have nothing to do with changing standards of epistemology or scientific method. Changes in first-order beliefs of the kinds just discussed provide the basis for corresponding changes only in *superficial* or *intermediate* standards for the assessment of knowledge claims. They still give no reason for changes in epistemology or fundamental attitudes to science. Quite the contrary, in fact, because whether or not the feminist-inspired research actually justifies the relevant changes in first-order beliefs *depends*, once again, on the acceptance of more fundamental standards of science and epistemology.

This point is of great importance to the traditional feminist on the epistemological shore, because unless she takes great care to distinguish these different levels of standard, she may well slip into feminist epistemology by accident. Inevitably, in the course of her developing feminism, she will encounter innumerable traditional knowledge claims she regards with suspicion; and it would be a serious mistake to respond by saying that if these were what traditional epistemology counted as knowledge, there must be something wrong with traditional epistemology. This would not only be much too precipitate; it would also give far too much credit to partriarchal man. It would by implication concede that whatever he had claimed as knowledge must, by traditional standards, really *be* knowledge, to be dealt with only by complicated revolutionary epistemologies through which it could be shrunk into mere phallocentric knowledge, or otherwise emasculated. It is usually much simpler—and, you would think, much more feminist—to start with the assumption that what has been *claimed* or *accepted* as knowledge may not be knowledge of any kind, even phallocentric, but, by patriarchal man's very own epistemological standards, plain ordinary (frequently patriarchal) *mistakes*.

It is difficult to say how much of the impulse towards feminist involvement in epistemology and other fundamental parts of philosophy and science arises from the blurring of this distinction. My suspicion is that a great deal of it does.[19] Fortunately, however, this is not a matter that needs to be investigated here, since it would make no difference to the arguments of this paper whether the answer were all or none. All that matters for my purposes is to make the distinction clear. This is why it has been important to discuss feminist progress in first-order knowledge, *within* traditional views of science and epistemology, and to show how this does in itself bring about changes in standards for the assessment of other knowledge claims, or of abilities to make advances in knowledge. Only when that issue is out of the way is it possible to make an uncluttered assault on the real question, of how the feminist on the shore would approach the problem of genuine epistemological change.

EPISTEMOLOGY PROPER

There is of course no problem of principle about the inquiring feminist's taking her feminist awareness into the study of epistemology: in this context as in others she may want to raise new questions, or check that female-connected ideas have not been overlooked or given inadequate consideration. There is also no problem about her considering new epistemological ideas, since that is something anyone can do at any time; and there is no reason, at least in advance of detailed consideration, to rule out the possibility of her deciding to abandon her old views for the kind she sees advocated by feminist epistemologists. The specific question here, however, is not of whether the move to these ideas can be justified at all, but of whether any part of the justification can be provided *by feminism*.

In the cases so far discussed, feminism has not been involved in the justification of new beliefs. All the changes in our feminist's view of the world have been justified in terms of the same epistemological and scientific standards as she appealed to when becoming a feminist in the first place. Those standards, however, are now themselves at issue. The question therefore arises again of whether her feminism—her commitment to the pursuit of proper treatment for women—gives her reasons to abandon her old ideas and change to new ones.

Something on these lines does seem to be widely implied, and not only in the term "feminist epistemology" itself. Alcoff and Potter, for instance, in the introduction to their anthology, say:

> The history of feminist epistemology itself is the history of clash between the feminist commitment to the struggles of women to have their understandings of the world legitimated, and the commitment of traditional philosophy to various accounts of knowledge—positivist, postpositivist, and others—that have consistently undermined women's claims to know.[20]

And later, in a comment on one of the essays in the collection, they refer to "Alcoff and Dalmiya's concern that traditional epistemology has reduced much of women's knowledge to the status of 'old wives' tales'."[21] Both these comments suggest that one purpose of feminist epistemology is to find an account of knowledge that would result in women's knowledge claims' being accepted rather than dismissed.

Alcoff and Potter also make broader claims about the aims of feminist epistemology:

> For feminists, the purpose of epistemology is not only to satisfy intellectual curiosity, but also to contribute to an emancipatory goal, the expansion of democracy in the production of knowledge. This goal requires that our epistemologies make it possible to see how knowledge is authorized and who is empowered by it.[22]

They also say that what the essays in their collection have in common is (nothing more than) "their commitment to unearth the politics of epistemology,"[23] and that "feminist work in philosophy is unashamedly a political intervention;"[24] and, referring to the work of one of their contributors, that "feminist epistemologies must be tested by their effects on . . . practical political struggles."[25] And another contributor says of one particular (non-feminist) account of epistemology:

> Critics must ask for whom this epistemology exists; whose interests it serves; and whose it neglects or suppresses in the process.[26]

All this suggests that there are feminist, emancipatory, oppression-resisting *reasons* for taking up these approaches to epistemology, and that the adequacy of any candidate theory must be judged by the extent to which it contributes to that emancipation.

So how should the feminist on the shore respond to this? Should she adopt this new approach to epistemology on the grounds that her present, traditional, theories are themselves part of the apparatus by which women have been oppressed? Consider first the suggestion that traditional epistemology has been responsible for the relegation of women's knowledge to the status of old wives' tales, and should be abandoned *for that reason*. Could she be led to the conclusions of feminist epistemology by this route?

The first point to notice here is that the claim that familiar ideas of epistemology have "consistently undermined women's claims to know"[27] cannot just be slipped into the argument as if it were obviously true. Since the challenge is to demonstrate to the traditional feminist the inadequacy of her present position, and show why she should adopt instead the ideas of feminist epistemology, this is something of whose truth she needs to be persuaded. She will need to be shown that there really are such substantial and systematic differences in the kinds of knowledge claim made by men and women, that the ones most often ruled out by her present standards are women's, that it is actually her epistemology, rather than her first-order beliefs, that leads to their being ruled out, and that the recommended change in epistemology would lead to their being recognized as knowledge.

These are not small matters; and if the traditional feminist is like most of the uninitiated, all this will present her would-be persuaders with considerable problems. Even the simplest part of the claim, about the sexes' making systematically different knowledge claims that are differently treated, may well strike her as implausible; and there is likely to be even more difficulty about the idea that whatever beliefs she does discount are the result of her traditional epistemology, since that is something she is likely not only to reject, but even to have difficulty in understanding.[28] So problems of this kind might well be enough on their own to put an end to any real prospect of converting the traditional feminist to feminist epistemology by an argument of this kind.

Even if there were some reasonable prospect of their being overcome, however, there would still remain a more fundamental and completely intractable problem. Suppose our feminist agreed to accept, at least for the sake of argument, that the traditional epistemological standards she still accepts really did discount a good deal of what women had traditionally claimed to know. How exactly is that supposed to justify the change to a new epistemology?

It is significant that in the quotations above Alcoff and Potter say first that it is women's *claims to know* that have been undermined by traditional epistemology, and later that it is women's *knowledge*. Now of course if our feminist's traditional theory actually ruled out women's *knowledge*, it would indeed be wrong and should be changed; that is analytically true. But the whole problem is that she does not yet accept that what she is dismissing *is* knowledge. What she is dismissing are knowledge *claims*, which, by her present standards, *really do* amount to nothing more than old wives' tales. To accept the crucial premise that these claims did represent genuine knowledge, she would *already have to accept* the new epistemology that the argument is supposed to be justifying. So she cannot be persuaded to change by an argument of this kind, because if the premise is taken to be about knowledge claims, it provides no reason at all for any change in epistemology; and if it is taken to be about knowledge, it is flagrantly question begging.

And, furthermore, *until* she has seen reason to change her epistemology, the still-traditional feminist must also conclude that it is the advocates of feminist epistemology who are treating women wrongly. Her feminist principles combine with her present epistemological views to suggest that the proper way to treat any women who make these misguided knowledge claims is to give them a proper education and bring them out of their ignorance. To offer them instead an epistemology that passes their ignorance off as knowledge is only to cheat them into collusion with their own deprivation, and this is obviously something she must regard as a scandalous perpetuation of the traditional wrongs of women, to be fought with all the feminist energy she can muster.

Of course this particular argument, about getting women's knowledge properly acknowledged, is only one among many possible lines of feminist argument to the conclusion that epistemological change is needed, and to show that this one does not work is still to allow for the possibility that others might. But in fact a version of the same problem arises whenever the emancipation of women is used as part of an argument for change in fundamental standards of epistemology or science, even when the substance is quite different.

Suppose, for instance, our still-traditional feminist were urged by feminist epistemologists to recognize that prevailing epistemological standards had been put in place by men, and that women could never be free from oppression as long as they were judged by male standards, epistemological or otherwise. She would, once again, have to be persuaded that there really were such differences between male and female standards (which would certainly present problems, since she herself accepts the ones

said to be male). She would also have to be persuaded that women's knowledge and abilities were bound to fare badly as long as male standards prevailed (which might be equally difficult, since what she has seen of the standards recommended by feminists is unlikely to make her yearn to be judged by them). But even if those difficulties could be overcome, the more fundamental problem would remain. The idea that any group's knowledge claims cannot be properly assessed by the standards of another group *is itself* the epistemological theory being advocated, *opposed* to the one the inquirer now holds, and therefore cannot be invoked as any part of an argument that her present view should change. And, again, from the point of view of her present ideas of epistemology, any move towards judging each group by its own standards would itself constitute a wrong to women, in inducing them to mistake their real deprivation for inappropriate attitudes of the privileged.

The same thing happens if it is claimed that knowledge is at root a matter of politics: that feminist epistemology is a matter of revealing "a *politics* of truth, logic, and reason," and that until this is understood women will be misled by standards that pretend to objectivity, but are really nothing but manifestations of unjust male power. Once again, the idea that knowledge is a matter of politics is *itself* the epistemological position being defended, and therefore the claim that women are wrongly treated by epistemologies that deny it cannot be used as part of the argument in its defense. And *until* the traditional feminist has been persuaded to change her mind, she will continue to think that attempting to persuade women of its truth is, once again, to delude them into thinking that nothing but politics is needed to transform their ignorance into knowledge, and so obstruct their acquisition of the real knowledge that is needed for effective political activity of any kind.

The problem cannot be escaped even by a retreat to the most blatantly political position of all: the argument that we must adopt the epistemological ideas claimed as feminist because until we do, things will be worse for women; that the progress of women depends on making this change. Even that, if taken as a defense of a serious epistemology (as opposed to one professed in public for political reasons but denied in private) presupposes the idea that epistemology is logically secondary to ethics, which is itself an epistemological theory.[29]

Because the root of this matter is a logical one, it makes no difference how many variations of detail are tried. The point is essentially the one that was made at the beginning of this piece, about the foundations of feminism, that criteria by which proper treatment and assessment can be recognized need to be in place before it can be said that the present state of things is falling short of them. The claim that some set of epistemological and scientific standards results in inappropriate treatment of women and their ideas cannot be used as an argument against those standards, because, necessarily, to accept those standards *is* to accept that women should be treated according to them. But anyone who is suspicious of such succinct generalities can easily test arguments individually by recalling the image of the feminist on the shore, who must be offered reasons to abandon her current views for those of feminist epistemology. It will be found for any argument she might be offered that if the reasons given for the recommended change from traditional to feminist epistemology are themselves *feminist*—if they depend in any way on the idea that current epistemology wrongs women, or is bad for women in any way—they will turn out to depend not only on highly contentious empirical premises, but also on revisionary epistemological claims that presuppose the conclusion and therefore beg the question.[30]

Now of course all this shows only that *feminist* justifications of what is claimed as feminist epistemology cannot work. It still leaves open the possibility that the feminist on the shore might find other, nonfeminist, reasons for changing her epistemological ideas, just as she earlier found non-feminist justifications for changes in various first-order beliefs about the world. Philosophers have for centuries been producing arguments about epistemology that have nothing to do with feminism, and some such non-feminist argument might persuade her to accept the approaches now claimed as feminist. Furthermore, if she did become convinced that these new epistemological ideas were right, *and* that women would do better under these than under the old ones, she would also—necessarily—conclude that women were wrongly treated by traditional epistemology, and might well see it as part of her feminist politics

to develop, and persuade others to adopt, the epistemology she now regarded as right.[31]

This is pretty obviously what has happened in the case of feminist epistemologists. They have been persuaded by particular approaches to epistemology—typically the kind that derive from ideas about the sociology of knowledge and science, and stress the idea that dominant groups set the standards—and have made these ideas the foundation both of their future inquiries and of the form their feminist politics takes. And that, as far as it goes, is fine in principle, but it must not be mistaken for there being any feminist reasons for accepting that, or any other, approach to epistemology.

So what all this means is that the situation is just the same for the feminist on the shores of feminist epistemology as it was in the early stages of her feminist inquiry. Feminism may, perhaps,[32] prompt her to raise particular epistemological questions, and if the answers to her questions lead her to epistemological change, that change will affect both her politics and the course of future inquiries. But still her feminism cannot itself be the determinant of the answers.

Feminism, in other words, can never escape its beginnings as an applied field.[33] Conclusions about what should be done by feminists for women—irrespective of whether they want what is just or right for women, or merely what is good for them—are *at all stages* essentially derivative, and dependent on more fundamental ideas. No beliefs about matters of fact, and no theories of epistemology or science, can be required by feminism, because feminist conclusions depend on them.

This means that to attach the label "feminist" to particular theories of epistemology or anything else is completely arbitrary.[34] In no sense that is not seriously misleading can there be any such thing as feminist epistemology.

THE FLIGHT FROM SCIENCE AND REASON

Now this must all be tied to the theme of this volume: the flight from science and reason.

The most immediately obvious way into the analysis of what goes by the name of feminist epistemology is to take the specific claims, presuppositions, and lines of argument claimed by their advocates as feminist, and subject them to critical analysis. That, however, is not what I have been doing here. This paper has been concerned only with the more fundamental problem of how feminism fits into these inquiries at all; and the essential conclusion is that although it certainly has a place, that place is limited. To try to go beyond it is to run into incoherences far more damaging to the idea of feminist epistemology than any criticisms of the details of its content.

To risk an analogy no doubt much too frivolous for such solemn matters, but salutary for just that reason, the place of feminism in scientific and philosophical inquiry has emerged as strikingly similar to that of James ("The Amazing") Randi[35] in the Uri Geller investigations. When scientists started investigating Geller's telepathic and spoon-bending exploits, they of course thought they were conducting a careful inquiry that eliminated the possibility of fraud. Scientists, however, do not know about conjuring. When Randi was brought in, his practiced eye went immediately to what had been made invisible to the lay observer, and the Geller tricks were exposed. And this did, in its modest way, affect the course of science. Anyone whose view of the world had been influenced by the Geller phenomena now had to eliminate these apparent data from their calculations, and rethink their view of the world—and perhaps even their ideas about scientific methodology—on a different basis. But even if Randi's contribution had been a thousand times greater—even if everyone engaged in the inquiries had been busily conjuring, and all the data had had to be scrapped—he still would have been doing nothing that could possibly be described as conjurist science. His contribution lay entirely within the familiar framework of scientific investigation. He put the scientists in the way of eliminating certain misleading information, but they themselves had to be able to confirm that the suspected tricks were actually going on: if Randi had claimed that the tricks must remain invisible to anyone who lacked his conjurist insights, nobody would have been in the least interested. And after this purge of misleading information, the scientists went on just as before, influenced by conjuring only to the extent of being aware of that kind of possibility for deception. It would have been absurd for Randi—at least qua conjurer—to say anything

about what data should be taken into consideration after the spurious ones had been eliminated, or which theories were most promising, or what direction future research should take, let alone for him to have offered anything claimed as conjurist approaches to science as a whole.

Notwithstanding obvious differences between the two cases, the contribution of feminism to academic inquiry has emerged as much the same in kind. Feminists come to academic inquiry of all sorts with a particular interest—and, after a while, with some accumulated expertise—that makes them look where others had not thought to look, and that frequently results in their discovering what had previously been unknown and finding anomalies where all had been presumed smooth. But all these discoveries must be visible to anyone who is willing to consider them; and once they have been made, the question of how science and philosophy should proceed is no more the concern of feminists than of everybody else. There can be no *feminist reasons* for adopting either first- or second-order beliefs of any kind.

The proposal of this paper is that this simple point provides the most effective way of coping with the phenomenon of so-called (as I must now insist) feminist epistemology. It combines two great advantages, of being relatively simple and easy to demonstrate, and of considerable power once established.

Consider first the simple aspects that make the point relatively easy to demonstrate.

First, the argument depends *not at all* on what the content of feminist epistemology is supposed to be. This is a great advantage, because anyone who takes on the details not only has an enormous task on hand, but also runs the perpetual risk of wrangles about misrepresentation. None of the foregoing arguments depends on the details of what any feminist thinks, so it makes no difference, for instance, whether or not what is claimed as feminist epistemology is accurately characterized by Alcoff and Potter as being specifically concerned to "unearth the politics of knowledge," or whether or not I am right in my speculation that much of the impetus to a feminist epistemology arises from confusions of level. The essential point is a logical one about the relationship of feminism to any theory of epistemology, or, for that matter, any other type of theory claimed as feminist.

Second, the claim is itself simple and straightforward, involving none of the appalling complications waiting to ensnare any critic brave or unwary enough to start from inside feminist epistemology and try to find a way out. Although developing and illustrating the argument may take some time in the first instance, all that is really involved is a simple logical point: that feminism can provide no justification for holding one theory rather than another. It is also relatively easy to demonstrate, since even if the general argument is thought to be in some way suspect, any particular argument that is attempted will provide an illustration. The important point to keep in view—the landmark, when everything else has vanished into the fog—is the question of why the feminist with still-conventional views of epistemology *should change* to the views claimed as feminist; and it can quickly be shown that if the argument offered has anything to do with her feminism, it will run into question begging or self-contradiction. She cannot be persuaded to change her epistemological views on the grounds that her present ones discount women's knowledge, for instance, since until she has changed those views she will not accept that what she is discounting *is* knowledge.

And finally (though Mill might have described this as "resembling those celebrations of royal clemency with which . . . the king of Lilliput prefaced his most sanguinary decrees"[36]), the argument has the advantage of being essentially mild and unprovocative, because it implies no criticism of the *substance* of what is claimed as feminist epistemology. *All* it does, as such, is insist that answers to questions of epistemology and science are presupposed in any arguments about the proper treatment of women, and therefore cannot themselves be required by feminism. It shows that if the feminist on the shore is to be persuaded to embrace the theories claimed as feminist she must be offered non-feminist reasons for doing so, but that does not imply that such reasons could not exist, or that the whole thing is nonsense.

So the case being presented here is relatively simple and relatively uncontentious, and all this sweetness and light may, perhaps, suggest that its implications cannot be very far reaching. It may even seem to leave the heart of the issue untouched, in allowing for the possibility that what goes by the name of feminist epistemology may be good epistemology even though not feminist. But although that may be technically true, the case argued here has direct implications that are almost frighteningly out of line with the prevailing culture of academic politics, and indirectly makes all the difference that any skeptic needs.

The most important direct implication is no doubt obvious. It is that a commitment to feminism—to righting the wrongs of women—gives *not the slightest* presumption in favor of any theory or set of beliefs that happens to have labeled itself feminist. Once it is recognized that such theories can have no feminist justification, and that the name is arbitrary and misleading, it becomes clear that they should all be treated exactly as if the name were not there. Decisions about their appropriateness for teaching in universities, or for publication by serious publishers, should positively not be distorted by the mistaken idea that women's past oppression must entitle whatever calls itself feminist to special consideration. And in fact the case is even stronger, because the argument applies equally to the question of whether some area of theory is worth even detailed preliminary study. If an initial skim of any part of the literature suggests that its content is weak or confused or misguided, then—in a world of far too many books, where deciding to read one means not reading others—even the most committed feminist can, with a limpid feminist conscience, decide to go no further. She may, of course, have made a mistake, but that is true of all the other books she has no time to read, and she would *certainly* be making a mistake if she allowed the spurious association with feminism to influence her decision.

The second heretical implication of these arguments is that no expertise whatever in these knowledge-connected subjects—epistemology and the sciences—comes of being a feminist. A feminist awareness that sex-connected anomalies may come up in particular areas, or that hitherto unnoticed questions may arise in them, *does not constitute expertise in these areas.* In fact it is rather the other way round. Until she has enough of a grip on a particular subject, a feminist cannot be adequately aware of the ways in which sex-connected anomalies may lurk within it, or where to look for undiscovered facts that might be of feminist significance. Once again the conjuring analogy is useful. If Randi had been summoned to seek out conjurist fraud among scientists dealing with esoteric parts of modern physics, he would have had to learn enough of the physics to see where lay the possibilities for deceit by conjuring. No matter how conscious a feminist may be of women's oppression, she will not spot subtle mistakes in patriarchal argument unless she has enough understanding of logic to understand how good arguments work, nor recognize inadequate evidence in any part of science unless she knows what adequate evidence looks like.[37]

And what this means, schematically speaking, is that neither science nor epistemology can be properly studied in women's studies departments, or taught by feminists qua feminist. These subjects must be taught—by all means in the company of people who are aware of the need to keep on the lookout for sex-connected anomalies—in departments of science and philosophy, by people who have enough background in the area to understand how to conduct inquiries and seek out anomalies. There is no problem of principle about this, because the arguments that show that no theories or beliefs can be feminist also show, by implication, that they cannot be patriarchal or phallocentric either; if they cannot be inherently emancipatory, neither can they be inherently oppressive.[38] And when such departments make new appointments, they are seriously misguided if they think that concerns for women oblige them to appoint specialists in what goes by the name of feminist epistemology or feminist anything else. They can advance the cause of women by looking for excellent scientists or philosophers who have some special awareness of the way feminist issues may arise in those areas, but that is quite a different matter.

Even this does not suggest that there is anything wrong with the content of what is claimed as feminist epistemology. All the theories and approaches misleadingly claimed as feminist could, in principle, survive the loss of the name, and be regarded as worth teaching and publishing whether thought of as feminist or not. But the association with feminism has provided a hothouse within which nonsense has had every chance to rampage, and the test for the theories claimed as feminist is to see how well they can survive a draft of cooler air.

It is obvious, in the first place, that the name of feminism—the apparent seal of authenticity—makes all the difference in the world to the way theories claimed as feminist are approached. Absorbers of feminist epistemology are not epistemological surfers who happened to be entranced in passing by the substance of these theories, only to find later that they were claimed as feminist. It is feminism that has drawn the crowds, and their attitudes to feminism will inevitably extend to what they see as feminist. Committed feminists will be more inclined to attribute obscurity to their own confusion than to confusion in what is claimed as feminist theory, and to accept on trust conclusions whose supporting arguments they have too little time or skill to assess. Fellow travelers will uneasily presume that there must be something in what is going on, and will be reluctant to resist as baffling appointments and courses of invisible merits proliferate around them. But take away the label, and feminists will see that they need positive reasons for venturing into these waters rather than excuses for staying out, and feminist sympathizers will be no more willing to endure the severe cognitive dissonance many of them now suffer than they would for the sake of flat earthers or crop circlers.

The protection provided by the name, furthermore, stretches even further when the related idea takes root that science, epistemology, or anything else should be approached through feminism, rather than the other way round. Just about anything can be made plausible to people who approach their subject from the far side of the relevant academic disciplines, unequipped with the techniques of detailed criticism that are the basis of all real progress in both science and philosophy, and in no position to identify as caricatures whatever silly or simplified ideas may be attributed to the opposition. Astrology can seem well founded to people whose scientific background is too vague to allow criticism of plausible generalities ("science has shown that the stars do have an influence on the earth"), but it cannot survive even a minimal acquaintance with post-Newtonian physics; the inconsistencies hidden by the generalities glare in the details. Creationism can flourish among people who start with the Bible, keep themselves entirely surrounded by creationists, and limit their acquaintance with paleontology and scientific method to selected odds and ends, but they could never have reached creationism from paleontology. Whether or not the epistemological ideas claimed as feminist have any merits, they can hardly fail to look plausible if approached from the direction of feminism, through sweeping ideas about the imposition of alien standards on an oppressed group ("men have had the power, and have used their patriarchal standards to dismiss women's knowledge," or whatever), and kept in a self-reinforcing huddle that sees all outside criticism as irrelevant because patriarchal. The test for the theories is to see to what extent they can survive an approach from the other direction, by people who are familiar with the relevant techniques of detailed criticism, and who do not have to rely for their understanding of "standard" epistemology on the rather surprising accounts that sometimes appear in the feminist literature[39]—many of which are quite enough on their own to make change seem a matter of urgency.

My own view is that not much would survive. If the whole idea of feminist epistemology rests on a mistake, that in itself bodes ill for the details, and enough has already been said in passing—for instance, about confusions between epistemological and first-order standards, knowledge and knowledge claims, and the politics of knowledge and the politics of epistemology—to suggest a range of serious problems. And to the extent that Alcoff and Potter are right in seeing feminist epistemology as "an unashamedly political intervention," there is also the fundamental logical problem of the idea that politics even can, let alone should, be at the root of things. Politics is a matter of manipulating other people to bring about a desired set of ends, and nobody can start making political calculations until they think they know more or less how the relevant parts of the world work. That not only means that first-order beliefs must precede politics; it also presupposes an epistemology that has nothing to do with politics. Anyone who tried to think seriously and in detail about how to go in for politics on the basis of an epistemology that took power to be at the root of what knowledge *actually was* would soon be stopped by dizziness.

However, to demonstrate this would take a good deal of detailed work that would go far beyond the intended scope of this paper, and might also overreach what was worth doing at all. Once so-called feminist epistemology is approached with the recognition that the name is spurious, and that writings presented under its name should be assessed as if the name were not there, it becomes clear that anything that looks like gobbledygook may reasonably be presumed such until shown otherwise, and disregarded accordingly.[40] The only reason for going into the details of anything that did look hopelessly unpromising would be to try to demonstrate to its practitioners the incoherences of their current position; but there would be no point in even that unless there were some reasonable prospect of success, and I doubt that anyone unpersuaded by the general argument given here would be more likely to be persuaded by the details. Since the deep irrationality of thinking that there can be any connection between feminism and particular theories of epistemology is the root of most of the other irrationalities, that is the one to concentrate on.

NOTES

1. See, e.g., Janet Radcliffe Richards, *The Sceptical Feminist*, ch. 1, passim.
2. "Bodies and Knowledges: Feminism and the Crisis of Reason," p. 209.
3. E.g., in Alcoff & Potter, eds., *Feminist Epistemologies*, Introduction, p. 13: "For feminists, the purpose of epistemology is not only to satisfy intellectual curiosity, but also to contribute to an emancipatory goal: the expansion of democracy in the production of knowledge." For convenience, most of the illustrative quotations in this paper will come from the Alcoff and Potter anthology. This seems a pretty comprehensive and representative collection, but nothing in the argument presented here depends on whether or not this is so.
4. E.g., Elizabeth Potter, "Gender and Epistemic Negotiation," p. 172: " . . . claims put forward by feminist scholars that gender strongly intersects the production of much of our knowledge. . . ."
5. Grosz, "Bodies and Knowledges," p. 187: " . . . if the body is an unacknowledged or an inadequately acknowledged condition of knowledges, and if the body is always sexually specific, concretely 'sexed,' this implies that the hegemony over knowledges that masculinity has thus far accomplished can be subverted, upset, or transformed through women's assertion of a 'right to know' independent of and autonomous from the

methods and presumptions regulating the prevailing (patriarchal) forms of knowledge." This passage is perhaps as good an illustration as any of the general point about uncharted seas. Readers of P. G. Wodehouse may find themselves reminded of Bertie Wooster's encounter with the improving literature prescribed by one of his passing financées: "I opened it [*Types of Ethical Theory*], and I give you my honest word this was what hit me. . . ." (from "Jeeves Takes Charge").

6. E.g., in Lynn Hankinson Nelson, "Epistemological Communities," p. 122: ". . . for more than a decade feminists have argued that a commitment to epistemological individualism would preclude reasonable explanations of feminist knowledge; such explanations . . . would need to incorporate the historically specific social and political relationships and situations, including gender and political advocacy, that have made feminist knowledge possible."

7. I do not want to concede at any point that there really is any such thing as "standard" epistemology (hence the distancing inverted commas), let alone that it has the kinds of characteristic that are sometimes claimed by feminist epistemologists; but for the limited purposes of this essay, and for the sake of argument, it will do no harm here to allow the point to pass.

8. Alcoff & Potter, Introduction, p. 13. To people who accept this myth, they say, "feminist work in philosophy is scandalous primarily because it is unashamedly political intervention."

9. J. S. Mill, *The Subjection of Women*, pp. 23–24.

10. See, e.g., Janet Radcliffe Richards, "Traditional Spheres and Traditionalist Logic," pp. 319–338.

11. There is no space to deal here with feminist challenges to logic, but the broad conclusions of this paper will be seen to apply to those as well. There will also be no further discussion of feminism and ethics, but for everything that is said here about epistemology, arguments about moral and other value judgments run in parallel.

12. P. Goldberg, "Are Women Prejudiced against Women?" pp. 28–30; quoted in Ann Oakley, *Subject Women*, p. 126.

13. See, e.g., Allison Kelly, ed., *The Missing Half*, passim.

14. Alcoff & Potter, *Feminist Epistemologies*, pp. 2–3.

15. Herodotus, *The Histories*, pp. 283–284.

16. See Wallace Sampson, "Antiscience Trends in the Rise of the 'Alternative Medicine' Movement"; and Gerald Weissmann, "'Sucking with Vampires':

The Medicine of Unreason," *The Flight from Science and Reason.*

17. Claims about the efficacy of homeopathic medicines have usually been rejected by scientists as obviously absurd, because the dilutions recommended are sometimes so extreme as to leave not a single molecule of the original drug. At one time some researchers claimed to have discovered that molecules could leave behind impressions of themselves in their absence, making it seem that there might after all be some causal mechanism by which the medicines might work. I believe this apparent finding has now itself been rejected, so the dilutions are back where they were.

18. Many feminists have taken up (without her full concurrence) Evelyn Fox Keller's work on Barbara McClintock, whose "feel for the organism" they claim as exemplifying women's approach to science, and as having been resisted by the scientific establishment. There are many possible grounds for controversy here, about whether the approach really is specifically female and whether it was really rejected by the establishment (see, e.g., Fox Keller, "The Gender/Science System"); but even if the claims were right, that would show the need for changes in standards only at an intermediate level, themselves justifiable in terms of more fundamental ideas about the nature of scientific success. These points are discussed further in Note 28.

19. This is what seems to be going on, for instance, where Lorraine Code (in "Taking Subjectivity into Account," pp. 15ff) is criticizing what she calls "S-knows-that-p" epistemologies, and in doing so mentions a sociologist's claims to have proved scientifically (and therefore to know) that "orientals as a group are more intelligent, more family-oriented, more law-abiding and less sexually promiscuous than whites, and that whites are superior to blacks in all the same respects." She gives reasons for doubting this claim, and then goes on to say:

> . . . the "Science has proved . . ." rhetoric derives from the sociopolitical influence of the philosophies of science that incorporate and are underwritten by S-knows-that-p epistemologies. . . . The implicit claim is that empirical inquiry is not only a neutral and impersonal process but also an inexorable one; it is compelling, even coercive, in what it turns up to the extent that a rational inquirer *cannot* withhold assent.

But nobody I have ever heard of holds epistemological views according to which if someone *claims* that science has proved this or that, a rational inquirer cannot withhold assent. That is what the claimant wants us to think, of course, but as rational inquirers we can, and frequently do, distinguish between S's claiming to know that something has been proved and S's actually knowing it. We typically challenge the claim—*within* the framework of familiar epistemology—precisely by casting doubt on the first-order knowledge claims produced as evidence. The fact that claimed proofs can be mistaken—which no non-lunatic epistemology could possibly deny—does not even begin to show that there is something fundamentally flawed about traditional ideas of propositional knowledge.

A more general indication that confusions of level may be a source of problems is the huge range of topics typically raised in feminist writings about feminist science and epistemology. These may have just about anything to do with women—or some individual woman—and any aspect of science or its applications, or anything whatever to do with knowledge. This does not matter in itself, but it does matter that there is usually no systematic discussion of how the different kinds and levels of discussion relate to each other. I have often been struck in practice, for instance, by the way feminists who are (quite reasonably) angry about the male takeover of obstetrics describe the insensitive use of gadgets as "subjecting women to male science," then go on to take this as indicating some kind of global, woman-oppressing maleness of every aspect of the scientific enterprise.

20. Alcoff & Potter, *Feminist Epistemologies,* Introduction, p. 2.
21. Ibid., p. 11.
22. Ibid., pp. 13–14.
23. Ibid., p. 3. Notice that this is one step further on than usual. Many epistemologies concern the idea of a politics of *knowledge;* the idea that there is a politics of epistemology suggests that there are political reasons for adopting one *epistemology* rather than another.
24. Ibid., p. 13.
25. Ibid., p. 14.
26. Lorraine Code, "Taking Subjectivity into Account," p. 23.
27. Note in passing, though I shall not go into them, the problems inherent in using empirical evidence based on the assumptions of one epistemology as part of the argument for establishing a quite different one, which may well undermine the original evidence. There is no difficulty in accepting that a

feminist may gradually change the epistemology she started with, but she cannot do so *and* keep earlier conclusions that were actually based on the rejected epistemology.

28. There is a real problem about the idea that it is epistemological standards, rather than particular first-order beliefs, that underlie the rejection of women's knowledge claims, which is difficult to explain because any case that fulfills the conditions seems bound to look absurd.

Consider again, for instance, the familiar feminist idea that much traditional knowledge of midwives has been dismissed as nonsense because it conflicts with established scientific views. Suppose that in some such case careful, feminist-inspired study revealed that a particular group of midwives had more success, in terms of well-being of mothers and their children, than some corresponding group of male doctors who based their practice on current scientific theories. Individual doctors or the scientific community might, perhaps, go on insisting that the midwives were simply ignorant and should be disregarded; but no standard *epistemology* would support such an attitude. Any reasonable scientist would take the midwives' success as evidence that they were on to something (though they might well be wrong about what it was) and that there must therefore be some inadequacy in the scientific theory. Such a case would therefore show no need for revisionary epistemology, but only for changes in ideas about which first-order claims to use as the standard for judging others.

To find a case that required genuine epistemological change to turn the midwives' ignorance into knowledge, it would be necessary to move to something much more bizarre, and postulate a situation where their practices were not only at odds with received scientific theory, but also *less successful* than those of the doctors, resulting in *worse* statistics of maternal and child welfare (because if they were successful, ordinary epistemology would admit that they were raising problems for the received theory), and where feminists would argue that we must change epistemological standards until *these* practices were counted as demonstrating knowledge. It is difficult to imagine either what such standards would be, or that any feminist would want to recommend any such thing. And unless the traditional feminist can be brought to understand how a change in epistemology might result in changes in the assessment of women's knowledge claims, she obviously cannot be persuaded to make the change for that reason.

Similar problems arise with feminist ideas about the need for radical change in fundamental approaches to science, to accommodate women's ways of setting about understanding the world. There is no problem in principle with the supposition that women might have systematically different ways of doing things, or that these ways might be systematically more successful than men's (though I know of no serious evidence that either of these is actually true), but to the extent that this is what is claimed by feminists, it does not call for any changes in fundamental conceptions of science. If women were successful in this way, ordinary standards of scientific success (such as reaching successful theories more quickly than men) would show this to be so. To show that more fundamental changes were needed in the criteria for scientific success it would be necessary to imagine women's being *unsuccessful* by current standards—having theories that tests kept showing were getting nowhere, making predictions that were usually unfulfilled, and so on—and then saying that scientific standards should be changed to count *this* as good science. It is, again, difficult to imagine either what such standards would be, or that any feminist would recommend them. All this provides further reason for suspecting that many feminist claims about the need for epistemological change may really be about the need for change in the first-order beliefs that provide intermediate standards for judgment.

29. And, of course, a pragmatically self-refuting one. If we can tell what is going to benefit women, we must think we know something about how the world works, and therefore must presuppose an epistemology other than the one we are supposed to be defending.

30. Or perhaps (though this possibility has not been discussed here) to depend on traditional epistemological claims that contradict the conclusion.

31. In fact it would be stretching things a bit to count this as part of *feminist* politics, since it would be for the benefit of *anyone* who was disadvantaged by the present sort. This aspect of the arbitrariness of counting a particular kind of epistemology as feminist is in effect noted by Alcoff & Potter (p. 4), though differently expressed and understood. There is no sign of their being aware of the other problems involved in claiming particular theories as feminist.

32. But that may be less likely than in the case of first-order inquiries.

33. See above, beginning of "Intermediate Standards."

34. It may be objected that "feminist" can legitimately be used to mean (more or less) "done in a characteristically female way," and that in this sense of the word it is not arbitrary to claim particular approaches as feminist. This sense of "feminist" is formally repudiated by most feminists, though it is obviously entrenched in popular usage. My own impression (which of course needs fuller justification) is that when the world is used to characterize first-order activities—scientific practice, normative ethics—it tends to mean "female," and when it characterizes second-order questions of metaethics, philosophy of science, and epistemology, it concerns getting recognition for what is supposedly female at the other level. "Feminist epistemology" never seems to mean "female epistemology." Even if it did, however, this would not help. There would, first, be the serious empirical question of whether women did really do things in the way claimed as feminist—which the many women who repudiate the ideas of feminist epistemology would say was an unfounded slur. But even if they did, to claim these ideas as *feminist,* rather than just female, is to imply an endorsement of them (the term "feminist" would not be used by a feminist who thought these female ideas nonsense). That presupposes an epistemology to supply the endorsement; and that, by the arguments presented here, cannot itself be feminist.

35. James Randi, *The Truth about Uri Geller.*

36. Mill, *The Subjection of Women,* p. 44.

37. It would be irrelevant here to ask, rhetorically, "by whose standards?" or otherwise raise questions about the standards of logic and scientific method being used. This argument is neutral between different possible standards. Resolve the fundamental problems of these matters any way you please, even reaching conclusions that are skeptical or relativist, and the argument about feminism goes through in the same way: until you understand how to apply whatever standards you do accept, you cannot see whether the treatment of women is wrong by those standards. (And anyone who does go for relativism or radical skepticism, or anything else too far from familiar standards, is likely to run into difficulties in finding anything wrong with the situation of women or anything else.)

38. There may of course be continuing anomalous *treatment* of women in such departments, or an

unwillingness to address woman-connected anomalies, and these are a continuing cause for feminist concern. But that must not be confused with there being anything essentially patriarchal about particular theories.

39. See, e.g., note 30, but there are many more. If feminists think the rest of the world goes around with epistemology like this, it is no wonder they think change is needed—though of course even if it is, that does not meant that it is needed for feminist reasons.

40. This is *not* to suggest that everything that appears under the heading of feminist epistemology is gobbledygook. The very fact—already mentioned—that there is such a confusion of levels, and so many different things going on, in what is claimed as feminist epistemology or science means that the quality may be thoroughly mixed. Although I do think there is not much hope for any of the genuinely epistemology claimed as feminist, much that appears in the collections may come into the category of genuinely useful feminist criticism.

REFERENCES

Alcoff, Linda & Elizabeth Potter, eds. *Feminist Epistemologies*. London: Routledge, 1993.

Alexander, Larry. "Fancy Theories of Interpretation Aren't." *Washington University Law Quarterly* 73, no. 3 (1995): 1081–1082.

Code, Lorraine. "Taking Subjectivity into Account." In *Feminist Epistemologies,* edited by Linda Alcoff & Elizabeth Potter. London: Routledge, 1993.

Fox Keller, Evelyn. "The Gender/Science System." *Hypatia* 2, no. 3 (Fall 1987).

Goldberg, P. "Are Women Prejudiced against Women?" *Transaction* 5, no. 5 (1968): 28–30.

Grosz, Elizabeth. "Bodies and Knowledges: Feminism and the Crisis of Reason." In *Feminist Epistemologies,* edited by Linda Alcoff & Elizabeth Potter. London: Routledge, 1993.

Herodotus. *The Histories.* Edited by A. R. Burn. Translated by Sélincourt. London: Penguin, 1972.

Kelly, Alison, ed. *The Missing Half.* Manchester: Manchester University Press, 1981.

Mill, J. S. *The Subjection of Women.* Edited by Susan M. Okin. Indianapolis, IN: Hackett, 1988.

Nelson, Lynn Hankinson. "Epistemological Communities." In *Feminist Epistemologies,* edited by Linda Alcoff & Elizabeth Potter. London: Routledge, 1993.

Oakley, Ann. *Subject Women.* Oxford: Martin Robertson, 1981.

Potter, Elizabeth. "Gender and Epistemic Negotiation." In *Feminist Epistemologies,* edited by Linda Alcoff & Elizabeth Potter. London: Routledge, 1993.

Radcliffe Richards, Janet. *The Sceptical Feminist.* 2nd ed. London: Penguin, 1994.

———. "Traditional Spheres and Traditional Logic." In *Empirical Logic and Public Debate, Essays in Honour of Else M. Barth,* edited by Erik C. W. Krabbe, Renée José Dalitz & Pier A. Smit, pp. 319–338. Poznan Studies in the Philosophy of the Sciences and the Humanities 35. Amsterdam: Rodopi, 1993. (Reprinted in *The Sceptical Feminist,* 2nd ed., pp. 353ff, passim.)

Randi, James. *The Truth about Uri Geller.* Buffalo, NY: Prometheus Books, 1982.

PART 6 SUGGESTIONS FOR FURTHER READING

Barnes, Barry. *Scientific Knowledge and Sociological Theory.* London: Routledge, 1974.

———. *T. S. Kuhn and Social Science.* New York: Columbia University Press, 1982.

Harding, Sandra. *Whose Science? Whose Knowledge?* Ithaca, NY: Cornell University Press, 1991.

Harding, Sandra, and Jean F. O'Barr, eds. *Sex and Scientific Inquiry.* Chicago: University of Chicago Press, 1987.

Keller, Evelyn Fox. *Gender and Science.* New Haven, CT: Yale University Press, 1985.

Laudan, Larry. *Science and Relativism.* Chicago: University of Chicago Press, 1990.

Longino, Helen. *Science as Social Knowledge: Values and Objectivity in Scientific Inquiry.* Princeton, NJ: Princeton University Press, 1990.

Martin, Jane R. "Ideological Critiques and the Philosophy of Science." *Philosophy of Science* 56 (1989): 1–22.

Meiland, Jack, and Michael Krausz, eds. *Relativism: Cognitive and Moral.* Notre Dame, IN: University of Notre Dame Press, 1982.

Mendelsohn, Everett, Peter Weingart, and Richard Whitley, eds. *The Social Production of Scientific Knowledge.* Dordrecht: D. Reidel, 1977.

Newton-Smith, W. H. *The Rationality of Science.* Boston: Routledge and Kegan Paul, 1981.

Rouse, Joseph. *Knowledge and Power.* Ithaca, NY: Cornell University Press, 1987.

Searle, John R. *The Construction of Social Reality.* New York: Free Press, 1995.

Sokal, Alan, and Jean Bricmont, eds. *Fashionable Nonsense: Postmodern Intellectuals' Abuse of Science.* New York: Picador, 1998.

PART 7

REALISM AND ANTIREALISM
Does Science Reveal Reality?

To explain phenomena, scientists often postulate the existence of theoretical entities. For example, to explain inherited characteristics, Gregor Mendel (1822–1884) postulated the existence of genes; to explain chemical reactions, John Dalton (1766–1844) postulated the existence of atoms; and to explain electric charge, Hendrik Lorentz (1853–1928) postulated the existence of electrons. These entities are theoretical because their existence cannot be directly verified through sense experience. Because empiricists want to admit the existence of only things that can be sensed, the question arises: what kinds of things are theoretical entities? Are they merely convenient fictions, like mythological creatures, that help us tell a good story? Or are they as real as the tangible objects we encounter every day? Those who take theoretical entities to be real are known as realists, whereas those who take them to be fictional are known as antirealists.

Grover Maxwell finds none of the arguments for antirealism convincing. Simply because an entity cannot be directly observed is no reason to believe that it is unreal; viruses can be observed only by means of a microscope, but that's no reason to think that they exist only in our minds. If only directly observable things are real, then the things we see through eyeglasses or windowpanes are not real. But that is obviously absurd. There is no nonarbitrary place to draw a line between the observable and the unobservable. Consequently, there is no good reason to believe that reality is limited to the observable.

Bas van Fraassen agrees that no nonarbitrary line can be drawn between the observable and the unobservable. But from this, he claims, it doesn't follow that there is no viable distinction between the observable and the unobservable. No nonarbitrary line can be drawn between a sapling and a tree, but that doesn't mean there is no viable distinction between a sapling and a tree. Even though the moons of Jupiter cannot be seen from Earth without the aid of a telescope, they are observable because they could be seen with the unaided eye from close up. The same cannot be said for subatomic particles, however; they are unobservable in a way that distant objects are not.

Are subatomic particles unreal then? Not necessarily. But van Fraassen says we don't have to believe in their existence because we can accept a theory without believing that the entities it postulates are real. "To accept a theory," he says, "is (for us) to believe that it is empirically adequate—that what the theory says *about what is observable* (by us) is true." In other words, an empirically adequate theory "saves the phenomena"; it explains why we observe what we observe. But an empirically adequate theory need not be true, for as Duhem taught us, theories are underdetermined by their data—for any set of data, any number of theories can be constructed to account for that data. We choose among competing theories by appealing to criteria such as simplicity, coherence, and explanatory power. According to van Fraassen, however, these criteria are not indicators of a theory's truth. Their function is purely pragmatic; they simply serve to identify the most useful theory.

Paul Churchland finds van Fraassen's remarks about observability and criteria of adequacy unconvincing. In the first place, he sees no difference between the unobservability of Jupiter's moons and the unobservability of subatomic particles. It is possible, in principle, to directly observe Jupiter's moons by changing our position relative to them. Similarly, it is possible, in principle, to directly observe subatomic particles by changing our size or the configuration of our senses relative to them. From a logical point of view, then, there is no more reason to be skeptical about the existence of subatomic particles than about the existence of Jupiter's moons.

Churchland also believes that van Fraassen is mistaken about the function of criteria of adequacy. Not only do they indicate the usefulness of a theory, they also indicate its truthfulness. To prove his point, he proposes a thought experiment: Suppose there existed a society of people who had no senses (and thus no sensations) but had microcomputers located on the tops of their skulls that used sensors to produce in the people a steady stream of beliefs about their environment. Because these people have no sensations, they have no empirical data on which to base their theories. As a result, whatever theories they develop cannot be empirically adequate. Nevertheless, Churchland maintains that these people, by using such criteria as simplicity, coherence, and explanatory unification, could develop a science as sophisticated and successful as our own. So these criteria must be at least as good an indication of truthfulness as empirical adequacy. To deny the reality of the things discussed by their science simply because those things are unobservable by us would be unjustified.

Many realists argue from the success of science to the reality of the entities it postulates. The best explanation of the fact that a theory makes successful predictions, they claim, is that the theory is true, or approximately true. As Hilary Putnam put it, realism "is the only philosophy that doesn't make the truth of science a miracle." Ian Hacking doesn't buy this argument. Many successful theories, such as Ptolemy's earth-centered theory of the solar system, have turned out to be false. Although Hacking is not a realist about scientific theories, he *is* a realist about scientific entities. What a theory says about things may be false. But that does not mean that the things themselves don't exist. As long as we can manipulate those things, we have reason to believe that they exist.

The purpose of experimentation, according to Hacking, is to interfere in the course of nature—to "twist the lion's tail," as Bacon put it. If our intervention has the desired effect, we have reason to believe that the things used to do the intervening exist. For example, if we can use electrons to investigate weak neutral currents, we have good reason to believe that electrons exist. We are justified in believing in the existence of macroscopic objects because we can manipulate them; we can affect them, and they can affect us. What goes for macroscopic objects, Hacking claims, goes for microscopic objects as well.

Like Hacking, Arthur Fine also rejects the success-of-science argument for scientific realism. What's more, he claims that the success of quantum mechanics and relativity theory shows that a commitment to realism is not essential to good science. Both of these theories were able to advance without their practitioners believing in the reality of the entities postulated.

Those who accept the reality of theoretical entities, Fine claims, do so on the same grounds that they accept the reality of nontheoretical entities: they trust the evidence of their senses and they trust the checks and balances inherent in the scientific method. So if scientists say that molecules, atoms, and subatomic particles exist, we have good reason to believe that these things exist. Fine calls this the "natural ontological attitude" (NOA). In this view, the findings of science are considered to be just as true as the findings of common sense. Fine believes that this attitude toward science can be shared by realists and antirealists alike.

What distinguishes the realist from the antirealist, Fine argues, is that they hold different theories about the nature of truth. For the realist, truth consists in correspondence to reality, whereas for the antirealist, truth consists in something else, such as coherence with other beliefs, pragmatic usefulness, or reduction to the observable. Fine suggests that we need not side with either the realist or the antirealist in this dispute; the natural ontological attitude is all that is necessary for an adequate philosophy of science.

James Robert Brown believes that any adequate account of theoretical entities must explain the success of science. Specifically, it must explain the following facts: "(1) that our current theories organize, unify and generally account for a wide variety of phenomena; (2) that theories have been getting better and better at this; they are progressing; and (3) that a significant number of their novel predictions are true." Van Fraassen's account is inadequate, Brown claims, because it cannot explain item (3). The fact that a theory is empirically adequate—that it organizes and unifies large parts of our experience—is no reason to believe that it will successfully predict new phenomena. Yet most good theories do just that. The best explanation of this fact, say the realists, is that these theories are true or approximately true. But the notion of approximate truth or verisimilitude is very difficult to make precise, as Brown's discussion of Newton-Smith's account demonstrates. What's more, false theories can make novel predictions. So making successful novel predictions is not a guarantee of truth.

Brown also argues that Fine's natural ontological attitude does not resolve or dissolve the dispute between the realists and the antirealists. To show the inadequacy of Fine's view, Brown draws an analogy between theology and science. Realists are like fundamentalists who believe that biblical language should be taken literally, whereas antirealists are like liberal theologians who believe that biblical language should be taken metaphorically. The dispute between the fundamentalists and the liberals cannot be settled by pointing out that they both accept biblical language. It can be settled only by deciding which interpretation of biblical language is the most plausible. Similarly, claims Brown, the dispute between the realists and the antirealists cannot be settled by pointing out that they both accept the claims of science. It can be settled only by determining which interpretation of scientific claims is the most plausible.

Brown argues that the plausibility of the realist position can be strengthened by considering a different sort of explanation. Explanations often seek to show why something is necessary. Many scientific explanations, for example, explain why something had to happen, given the initial conditions and the laws of nature. But explanations can also seek to show how something is possible. Darwin's theory of evolution, for example, explains how it's possible for creatures to be adapted to their environment even though they are not the products of conscious design. Realism, Brown claims, should be seen as providing a how-possible rather than a why-necessary explanation of the success of science. A theory's being true would explain how it's possible for it to make such predictions. Such explanations don't prove that successful scientific theories should be interpreted realistically, but they make it more plausible to do so.

33

GROVER MAXWELL

The Ontological Status of Theoretical Entities

That anyone today should seriously contend that the entities referred to by scientific theories are only convenient fictions, or that talk about such entities is translatable without remainder into talk about sense contents or everyday physical objects, or that such talk should be regarded as belonging to a mere calculating device and, thus, without cognitive content—such contentions strike me as so incongruous with the scientific and rational attitude and practice that I feel this paper *should* turn out to be a demolition of straw men. But the instrumentalist views of outstanding physicists such as Bohr and Heisenberg are too well known to be cited, and in a recent book of great competence, Professor Ernest Nagel concludes that "the opposition between [the realist and the instrumentalist] views [of theories] is a conflict over preferred modes of speech" and "the question as to which of them is the 'correct position' has only terminological interest."[1] The phoenix, it seems, will not be laid to rest.

The literature on the subject is, of course, voluminous, and a comprehensive treatment of the problem is far beyond the scope of one essay. I shall limit myself to a small number of constructive arguments (for a radically realistic interpretation of theories) and to a critical examination of some of the more crucial assumptions (sometimes tacit, sometimes explicit) that seem to have generated most of the problems in this area.[2]

Although this essay is not comprehensive, it aspires to be fairly self-contained. Let me, therefore, give a pseudohistorical introduction to the problem with a piece of science fiction (or fictional science).

In the days before the advent of microscopes, there lived a Pasteur-like scientist whom, following the usual custom, I shall call Jones. Reflecting on the fact that certain diseases seemed to be transmitted from one person to another by means of bodily contact or by contact with articles handled previously by an afflicted person, Jones began to speculate about the mechanism of the transmission. As a "heuristic crutch," he recalled that there is an obvious *observable* mechanism for transmission of certain afflictions (such as body lice), and he postulated that all, or most, infectious diseases were spread in a similar manner but that in most cases the corresponding "bugs" were too small to be seen and, possibly, that some of them lived inside the bodies of their hosts. Jones proceeded to develop his theory and to examine its testable consequences. Some of these seemed to be of great importance for preventing the spread of disease.

After years of struggle with incredulous recalcitrance, Jones managed to get some of his preventative measures adopted. Contact with or proximity to diseased persons was avoided when possible, and articles which they handled were "disinfected" (a word coined by Jones) either by means of high temperatures or by treating them with certain toxic preparations which Jones termed "disinfectants." The results were spectacular: within ten years the death rate had declined 40 percent. Jones and his theory received their well-deserved recognition.

However, the "crobes" (the theoretical term coined by Jones to refer to the disease-producing organisms) aroused considerable anxiety among

Scientific Explanation, Space and Time. Minnesota Studies in the Philosophy of Science, Vol. III, Herbert Feigl and Grover Maxwell, eds. (Minneapolis: University of Minnesota Press, 1962), pp. 3–15. Reprinted by permission of the publisher.

many of the philosophers and philosophically inclined scientists of the day. The expression of this anxiety usually began something like this: "In order to account for the facts, Jones must assume that his crobes are too small to be seen. Thus the very postulates of his theory preclude their being observed; they are *unobservable in principle*." (Recall that no one had envisaged such a thing as a microscope.) This common prefatory remark was then followed by a number of different "analyses" and "interpretations" of Jones' theory. According to one of these, the tiny organisms were merely convenient fictions—*façons de parler*—extremely useful as heuristic devices for facilitating (in the "context of discovery") the thinking of scientists but not to be taken seriously in the sphere of cognitive knowledge (in the "context of justification"). A closely related view was that Jones' theory was merely an instrument, useful for organizing observation statements and (thus) for producing desired results, and that, therefore, it made no more sense to ask what was the nature of the entities to which it referred than it did to ask what was the nature of the entities to which a hammer or any other tool referred.[3] "Yes," a philosopher might have said, "Jones' theoretical expressions are just meaningless sounds or marks on paper which, when correlated with observation sentences by appropriate syntactical rules, enable us to predict successfully and otherwise organize data in a convenient fashion." These philosophers called themselves "instrumentalists."

According to another view (which, however, soon became unfashionable), although expressions containing Jones' theoretical terms were genuine sentences, they were translatable without remainder into a set (perhaps infinite) of observation sentences. For example, 'There are crobes of disease X on this article' was said to translate into something like this: 'If a person handles this article without taking certain precautions, he will (probably) contract disease X; and if this article is first raised to a high temperature, then if a person handles it at any time afterward, before it comes into contact with another person with disease X, he will (probably) not contract disease X; and . . .'

Now virtually all who held any of the views so far noted granted, even insisted, that theories played a useful and legitimate role in the scientific enterprise. Their concern was the elimination of "pseudo problems" which might arise, say, when one began wondering about the "reality of supraempirical entities," etc. However, there was also a school of thought, founded by a psychologist named Pelter, which differed in an interesting manner from such positions as these. Its members held that while Jones' crobes might very well exist and enjoy "full-blown reality," they should not be the concern of medical research at all. They insisted that if Jones had employed the correct methodology, he would have discovered, even sooner and with much less effort, all of the observation laws relating to disease contraction, transmission, etc. without introducing superfluous links (the crobes) into the causal chain.

Now, lest any reader find himself waxing impatient, let me hasten to emphasize that this crude parody is not intended to convince anyone, or even to cast serious doubt upon sophisticated varieties of any of the reductionistic positions caricatured (some of them not too severely, I would contend) above. I am well aware that there are theoretical entities and theoretical entities, some of whose conceptual and theoretical statuses differ in important respects from Jones' crobes. (I shall discuss some of these later.) Allow me, then, to bring the Jonesean prelude to our examination of observability to a hasty conclusion.

Now Jones had the good fortune to live to see the invention of the compound microscope. His crobes were "observed" in great detail, and it became possible to identify the specific kind of *microbe* (for so they began to be called) which was responsible for each different disease. Some philosophers freely admitted error and were converted to realist positions concerning theories. Others resorted to subjective idealism or to a thoroughgoing phenomenalism, of which there were two principal varieties. According to one, the one "legitimate" observation language had for its descriptive terms only those which referred to sense data. The other maintained the stronger thesis that *all* "factual" statements were *translatable* without remainder into the sense-datum language. In either case, any two non-sense data (e.g., a theoretical entity and what would ordinarily be called an "observable physical object") had virtually the same status. Others contrived means of modifying their views much less drastically. One group maintained that Jones' crobes actually never had been unobservable in principle, for, they said, the theory did not imply the impossibility of finding a means (e.g., the microscope) of observing them. A more radical contention was that the crobes were not observed at all; it was argued that what was seen by means of the microscope was just a shadow or an image rather than a corporeal organism.

THE OBSERVATIONAL–THEORETICAL DICHOTOMY

Let us turn from these fictional philosophical positions and consider some of the actual ones to which they roughly correspond. Taking the last one first, it is interesting to note the following passage from Bergmann: "But it is only fair to point out that if this . . . methodological and terminological analysis [for the thesis that there are no atoms] . . . is strictly adhered to, even stars and microscopic objects are not physical things in a literal sense, but merely by courtesy of language and pictorial imagination. This might seem awkward. But when I look through a microscope, all I see is a patch of color which creeps through the field like a shadow over a wall. And a shadow, though real, is certainly not a physical thing."[4]

I should like to point out that it is also the case that if this analysis is strictly adhered to, we cannot observe physical things through opera glasses, or even through ordinary spectacles, and one begins to wonder about the status of what we see through an ordinary windowpane. And what about distortions due to temperature gradients—however small and, thus, always present—in the ambient air? It really *does* "seem awkward" to say that when people who wear glasses describe what they see they are talking about shadows, while those who employ unaided vision talk about physical things—or that when we look through a windowpane, we can only *infer* that it is raining, while if we raise the window, we may "observe directly" that it is. The point I am making is that there is, in principle, a continuous series beginning with looking through a vacuum and containing these as members: looking through a windowpane, looking through glasses, looking through binoculars, looking through a low-power microscope, looking through a high-power microscope, etc., in the order given. The important consequence is that, so far, we are left without criteria which would enable us to draw a non-arbitrary line between "observation" and "theory." Certainly, we will often find it convenient to draw such a to-some-extent-arbitrary line; but its position will vary widely from context to context. (For example, if we are determining the resolving characteristics of a certain microscope, we would certainly draw the line beyond ordinary spectacles, probably beyond simple magnifying glasses, and possibly beyond another microscope with a lower power of resolution.) But what ontological ice does a mere methodologically convenient observational-theoretical dichotomy cut? Does an entity attain physical thinghood and/or "real existence" in one context only to lose it in another? Or, we may ask, recalling the continuity from observable to unobservable, is what is seen through spectacles a "little bit less real" or does it "exist to a slightly less extent" than what is observed by unaided vision?[5]

However, it might be argued that things seen through spectacles and binoculars look like ordinary physical objects, while those seen through microscopes and telescopes look like shadows and patches of light. I can only reply that this does not seem to me to be the case, particularly when looking at the moon, or even Saturn, through a telescope or when looking at a small, though "directly observable," physical object through a low-power microscope. Thus, again, a continuity appears.

"But," it might be objected, "theory tells us that what we see by means of a microscope is a real image, which is certainly distinct from the object on the stage." Now first of all, it should be remarked that it seems odd that one who is espousing an austere empiricism which requires a sharp observational-language/theoretical-language distinction (and one in which the former language has a privileged status) should need a theory in order to tell him what is observable. But, letting this pass, what is to prevent us from saying that we still observe the object on the stage, even though a "real image" may be involved? Otherwise, we shall be strongly tempted by phenomenalistic demons, and at this point we are considering a physical-object observation language rather

arbitrary—random choice or personal whim rather than any reason or system based on than a sense-datum one. (Compare the traditional puzzles: Do I see one physical object or two when I punch my eyeball? Does one object split into two? Or do I see one object and one image? Etc.)

Another argument for the continuous transition from the observable to the unobservable (theoretical) may be adduced from theoretical considerations themselves. For example, contemporary valency theory tells us that there is a virtually continuous transition from very small molecules (such as those of hydrogen) through "medium-sized" ones (such as those of the fatty acids, polypeptides, proteins, and viruses) to extremely large ones (such as crystals of the salts, diamonds, and lumps of polymeric plastic). The molecules in the last-mentioned group are macro, "directly observable" physical objects but are, nevertheless, genuine, single molecules; on the other hand, those in the first mentioned group have the same perplexing properties as subatomic particles (de Broglie waves, Heisenberg indeterminacy, etc.). Are we to say that a large protein molecule (e.g., a virus) which can be "seen" only with an electron microscope is a little less real or exists to somewhat less an extent than does a molecule of a polymer which can be seen with an optical microscope? And does a hydrogen molecule partake of only an infinitesimal portion of existence or reality? Although there certainly *is* a continuous transition from observability to unobservability, any talk of such a continuity from full-blown existence to nonexistence is, clearly, nonsense.

Let us now consider the next to last modified position which was adopted by our fictional philosophers. According to them, it is only those entities which are *in principle* impossible to observe that present special problems. What kind of impossibility is meant here? Without going into a detailed discussion of the various types of impossibility, about which there is abundant literature with which the reader is no doubt familiar, I shall assume what usually seems to be granted by most philosophers who talk of entities which are unobservable in principle—i.e., that the theory(s) itself (coupled with a physiological theory of perception, I would add) entails that such entities are unobservable.

We should immediately note that if this analysis of the notion of unobservability (and, hence, of observability) is accepted, then its use as a means of delimiting the observation language seems to be precluded for those philosophers who regard theoretical expressions as elements of a calculating device—as meaningless strings of symbols. For suppose they wished to determine whether or not 'electron' was a theoretical term. First, they must see whether the theory entails the sentence 'Electrons are unobservable.' So far, so good, for their calculating devices are said to be able to select genuine sentences, provided they contain no theoretical terms. But what about the selected "sentence" itself? Suppose that 'electron' is an observation term. It follows that the expression is a genuine sentence and asserts that electrons are unobservable. But this entails that 'electron' is *not* an observation term. Thus if 'electron' is an observation term, then it is *not* an observation term. Therefore it is not an observation term. But then it follows that 'Electrons are unobservable' is not a genuine sentence and does not assert that electrons are unobservable, since it is a meaningless string of marks and does not assert anything whatever. Of course, it could be stipulated that when a theory "selects" a meaningless expression of the form 'Xs are unobservable,' then 'X' is to be taken as a theoretical term. But this seems rather arbitrary.

But, assuming that well-formed theoretical expressions are genuine sentences, what shall we say about unobservability in principle? I shall begin by putting my head on the block and argue that the present-day status of, say, electrons is in many ways similar to that of Jones' crobes before microscopes were invented. I am well aware of the numerous theoretical arguments for the impossibility of observing electrons. But suppose new entities are discovered which interact with electrons in such a mild manner that if an electron is, say, in an eigenstate of position, then, in certain circumstances, the interaction does not disturb it. Suppose also that a drug is discovered which vastly alters the human perceptual apparatus—perhaps even activates latent capacities so that a new sense modality emerges. Finally, suppose that in our altered state we are able to perceive (not necessarily visually) by means of these new entities in a manner roughly analogous to that by which we now see by means of photons. To make this a little more plausible, suppose that the energy eigenstates of the electrons in some of the compounds present in the relevant perceptual organ are such that even

"Real" based on direct observability

the weak interaction with the new entities alters them and also that the cross sections, relative to the new entities, of the electrons and other particles of the gases of the air are so small that the chance of any interaction here is negligible. Then we might be able to "observe directly" the position and possibly the approximate diameter and other properties of some electrons. It would follow, of course, that quantum theory would have to be altered in some respects, since the new entities do not conform to all its principles. But however improbable this may be, it does not, I maintain, involve any logical or conceptual absurdity. Furthermore, the modification necessary for the inclusion of the new entities would not necessarily change the meaning of the term 'electron.'[6]

Consider a somewhat less fantastic example, and one which does not involve any change in physical theory. Suppose a human mutant is born who is able to "observe" ultraviolet radiation, or even X rays, in the same was we "observe" visible light.

Now I think that it is extremely improbable that we will ever observe electrons directly (i.e., that it will ever be reasonable to assert that we have so observed them). But this is neither here nor there; it is not the purpose of this essay to predict the future development of scientific theories, and, hence, it is not its business to decide what actually is observable or what will become observable (in the more or less intuitive sense of 'observable' with which we are now working). After all, we are operating, here, under the assumption that it is theory, and thus science itself, which tells us what is or is not, in this sense, observable (the 'in principle' seems to have become superfluous). And this is the heart of the matter; for it follows that, at least for this sense of 'observable,' there are no a priori or philosophical criteria for separating the observable from the unobservable. By trying to show that we can talk about the *possibility* of observing electrons without committing logical or conceptual blunders, I have been trying to support the thesis that any (nonlogical) term is a *possible* candidate for an observation term.

There is another line which may be taken in regard to delimitation of the observation language. According to it, the proper term with which to work is not 'observ*able*' but, rather, 'observ*ed*.' There immediately comes to mind the tradition beginning with Locke and Hume (No idea without a preceding impression!), running through Logical Atomism and the Principle of Acquaintance, and ending (perhaps) in contemporary positivism. Since the numerous facets of this tradition have been extensively examined and criticized in the literature, I shall limit myself here to a few summary remarks.

Again, let us consider at this point only observation languages which contain ordinary physical-object terms (along with observation predicates, etc., of course). Now, according to this view, all descriptive terms of the observation language must refer to that which has been observed. How is this to be interpreted? Not too narrowly, presumably, otherwise each language user would have a different observation language. The name of my Aunt Mamie, of California, whom I have never seen, would not be in my observation language, nor would 'snow' be an observation term for many Floridians. One could, of course, set off the observation language by means of this awkward restriction, but then, obviously, not being the referent of an observation term would have no bearing on the ontological status of Aunt Mamie or that of snow.

Perhaps it is intended that the referents of observation terms must be members of a *kind*, some of whose members have been observed, or instances of a *property*, some of whose instances have been observed. But there are familiar difficulties here. For example, given any entity, we can always find a kind whose only member is the entity in question; and surely expressions such as 'men over 14 feet tall' should be counted as observational even though no instances of the "property" of being a man over 14 feet tall have been observed. It would seem that this approach must soon fall back upon some notion of simples or determinables vs. determinates. But is it thereby saved? If it is held that only those terms which refer to observed simples or observed determinates are observation terms, we need only remind ourselves of such instances as Hume's notorious missing shade of blue. And if it is contended that in order to be an observation term an expression must at least refer to an observed determinable, then we can always find such a determinable which is broad enough in scope to embrace any entity whatever. But even if these difficulties can be circumvented, we see (as we knew all along) that this approach leads inevitably into phenomenalism, which is a view with which we have not been concerning ourselves.

Now it is not the purpose of this essay to give a detailed critique of phenomenalism. For the most part, I simply assume that it is untenable, at least in any of its translatability varieties.[7] However, if there are any unreconstructed phenomenalists among the readers, my purpose, insofar as they are concerned, will have been largely achieved if they will grant what I suppose most of them would stoutly maintain anyway, i.e., that theoretical entities are no worse off than so-called observable physical objects.

Nevertheless, a few considerations concerning phenomenalism and related matters may cast some light upon the observational-theoretical dichotomy and, perhaps, upon the nature of the "observation language." As a preface, allow me some overdue remarks on the latter. Although I have contended that the line between the observable and the unobservable is diffuse, that it shifts from one scientific problem to another, and that it is constantly being pushed toward the "unobservable" end of the spectrum as we develop better means of observation—better instruments—it would, nevertheless, be fatuous to minimize the importance of the observation base, for it is absolutely necessary as a confirmation base for statements which do refer to entities which are unobservable at a given time. But we should take as its basis and its unit not the "observational term" but, rather, the quickly decidable sentence. (I am indebted to Feyerabend, *loc. cit.*, for this terminology.) A quickly decidable sentence (in the technical sense employed here) may be defined as a singular, non-analytic sentence such that a reliable, reasonably sophisticated language user can very quickly decide[8] whether to assert it or deny it when he is reporting on an occurrent situation. 'Observation term' may now be defined as a 'descriptive (nonlogical) term which may occur in a quickly decidable sentence,' and 'observation sentence' as a 'sentence whose only descriptive terms are observation terms.'

Returning to phenomenalism, let me emphasize that I am not among those philosophers who hold that there are no such things as sense contents (even sense data), nor do I believe that they play no important role in our perception of "reality." But the fact remains that the referents of most (not all) of the statements of the linguistic framework used in everyday life and in science are *not* sense contents but, rather, physical objects and other publicly observable entities. Except for pains, odors, "inner states," etc., *we do not usually observe sense contents;* and although there is good reason to believe that they play an indispensable role in observation, *we are usually not aware of them when we* (visually or tactilely) *observe physical objects.* For example, when I observe a distorted, obliquely reflected image in a mirror, I may seem to be seeing a baby elephant standing on its head; later I discover it is an image of Uncle Charles taking a nap with his mouth open and his hand in a peculiar position. Or, passing my neighbor's home at a high rate of speed, I observe that he is washing a car. If asked to report these observations I could quickly and easily report a baby elephant and a washing of a car; I probably would not, without subsequent observations, be able to report what colors, shapes, etc. (i.e., what sense data) were involved.

Two questions naturally arise at this point. How is it that we can (sometimes) quickly decide the truth or falsity of a pertinent observation sentence? and, What role do sense contents play in the appropriate tokening of such sentences? The heart of the matter is that these are primarily scientific-theoretical questions rather than "purely logical," "purely conceptual," or "purely epistemological." If theoretical physics, psychology, neurophysiology, etc., were sufficiently advanced, we could give satisfactory answers to these questions, using, in all likelihood, the physical-thing language as our observation language and *treating sensations, sense contents, sense data, and "inner states" as theoretical* (yes, theoretical!) *entities.*[9]

It is interesting and important to note that, even before we give completely satisfactory answers to the two questions considered above, we can, with due effort and reflection, train ourselves to "observe directly" what were once theoretical entities—the sense contents (color sensations, etc.)—involved in our perception of physical things. As has been pointed out before, we can also come to observe other kinds of entities which were once theoretical. Those which most readily come to mind involve the use of instruments as aids to observation. Indeed, using our painfully acquired theoretical knowledge of the world, we come to see that we "directly observe" many kinds of so-called theoretical things. After listening to a dull speech while sitting on a hard bench, we begin to become poignantly aware of the presence of a considerably strong gravitational field, and as Professor Feyerabend is fond of pointing out, if we were carrying a heavy suitcase in a changing gravitational field, we could observe the changes of the $G_{\mu v}$ of the metric tensor.

I conclude that our drawing of the observational-theoretical line at any given point is an accident and a function of our physiological makeup, our current state of knowledge, and the instruments we happen to have available and, therefore, that it has no ontological significance whatever.

NOTES

1. E. Nagel, *The Structure of Science* (New York: Harcourt, Brace, and World, 1961), Ch. 6.
2. For the genesis and part of the content of some of the ideas expressed herein, I am indebted to a number of sources; some of the more influential are H. Feigl, "Existential Hypotheses," *Philosophy of Science,* 17:35–62 (1950); P. K. Feyerabend, "An Attempt at a Realistic Interpretation of Experience," *Proceedings of the Aristotelian Society,* 58:144–170 (1958); N. R. Hanson, *Patterns of Discovery* (Cambridge: Cambridge University Press, 1958); E. Nagel, *loc. cit.*; Karl Popper, *The Logic of Scientific Discovery* (London: Hutchinson, 1959); M. Scriven, "Definitions, Explanations, and Theories," in *Minnesota Studies in the Philosophy of Science,* Vol. II, H. Feigl, M. Scriven, and G. Maxwell, eds. (Minneapolis: University of Minnesota Press, 1958); Wilfrid Sellars, "Empiricism and the Philosophy of Mind," in *Minnesota Studies in the Philosophy of Science,* Vol. I, H. Feigl and M. Scriven, eds. (Minneapolis: University of Minnesota Press, 1956), and "The Language of Theories," in *Current Issues in the Philosophy of Science,* H. Feigl and G. Maxwell, eds. (New York: Holt, Rinehart, and Winston, 1961).
3. I have borrowed the hammer analogy from E. Nagel, "Science and [Feigl's] Semantic Realism," *Philosophy of Science,* 17:174–181 (1950), but it should be pointed out that Professor Nagel makes it clear that he does not necessarily subscribe to the view which he is explaining.
4. G. Bergmann, "Outline of an Empiricist Philosophy of Physics," *American Journal of Physics,* 11:248–258; 335–342 (1943), reprinted in *Readings in the Philosophy of Science,* H. Feigl and M. Brodbeck, eds. (New York: Appleton-Century-Crofts, 1953), pp. 262–287.
5. I am not attributing to Professor Bergmann the absurd views suggested by these questions. He seems to take a sense-datum language as his observation language (the base of what he called "the empirical hierarchy"), and, in some ways, such a position is more difficult to refute than one which purports to take an "observable-physical-object" view. However, I believe that demolishing the straw men with which I am now dealing amounts to desirable preliminary "therapy." Some nonrealist interpretations of theories which embody the presupposition that the observable-theoretical distinction is sharp and ontologically crucial seem to me to entail positions which correspond to such straw men rather closely.
6. For arguments that it is possible to alter a theory without altering the meanings of its terms, see my "Meaning Postulates in Scientific Theories," in *Current Issues in the Philosophy of Science,* Feigl and Maxwell, eds.
7. The reader is no doubt familiar with the abundant literature concerned with this issue. See, for example, Sellars' "Empiricism and the Philosophy of Mind," which also contains references to other pertinent works.
8. We may say "noninferentially" decide, provided this is interpreted liberally enough to avoid starting the entire controversy about observability all over again.
9. Cf. Sellars, "Empiricism and the Philosophy of Mind." As Professor Sellars points out, this is the crux of the "other-minds" problem. Sensations and inner states (relative to an intersubjective observation language, I would add) are theoretical entities (and they "really exist") and *not* merely actual and/or possible behavior. Surely it is the unwillingness to countenance theoretical entities—the hope that every sentence is translatable not only into some observation language but into the physical-thing language—which is responsible for the "logical behaviorism" of the neo-Wittgensteinians.

34

BAS C. VAN FRAASSEN

Constructive Empiricism

ALTERNATIVES TO REALISM

Scientific realism is the position that scientific theory construction aims to give us a literally true story of what the world is like, and that acceptance of a scientific theory involves the belief that it is true. Accordingly, anti-realism is a position according to which the aim of science can well be served without giving such a literally true story, and acceptance of a theory may properly involve something less (or other) than belief that it is true.

What does a scientist do then, according to these different positions? According to the realist, when someone proposes a theory, he is asserting it to be true. But according to the anti-realist, the proposer does not assert the theory; *he displays it,* and claims certain virtues for it. These virtues may fall short of truth: empirical adequacy, perhaps; comprehensiveness, acceptability for various purposes. This will have to be spelled out, for the details here are not determined by the denial of realism. For now we must concentrate on the key notions that allow the generic division.

The idea of a literally true account has two aspects: the language is to be literally construed; and so construed, the account is true. This divides the anti-realists into two sorts. The first sort holds that science is or aims to be true, properly (but not literally) construed. The second holds that the language of science should be literally construed, but its theories need not be true to be good. The anti-realism I shall advocate belongs to the second sort.

It is not so easy to say what is meant by a literal construal. The idea comes perhaps from theology, where fundamentalists construe the Bible literally, and liberals have a variety of allegorical, metaphorical, and analogical interpretations, which 'demythologize.' The problem of explicating 'literal construal' belongs to the philosophy of language. . . . 'literal' does not mean 'truth-valued.' The term 'literal' is well enough understood for general philosophical use, but if we try to explicate it we find ourselves in the midst of the problem of giving an adequate account of natural language. It would be bad tactics to link an inquiry into science to a commitment to some solution to that problem. The following remarks . . . should fix the usage of 'literal' sufficiently for present purposes.

The decision to rule out all but literal construals of the language of science, rules out those forms of anti-realism known as *positivism* and *instrumentalism.* First, on a literal construal, the apparent statements of science really are statements, *capable of being true or false.* Secondly, although a literal construal can elaborate, it cannot change logical relationships. (It is possible to elaborate, for instance, by identifying what the terms designate. The 'reduction' of the language of phenomenological thermodynamics to that of statistical mechanics is like that: bodies of gas are identified as aggregates of molecules, temperature as mean kinetic energy, and so on.) On the positivists' interpretation of science, theoretical terms have meaning only through their connection with the observable. Hence they hold that two theories may in fact *say the same thing* although in form they contradict each other. (Perhaps the one says that all matter consists of atoms, while the other postulates instead a universal continuous medium; they will say the same thing nevertheless if they agree in their observable consequences, according to the positivists.) But two theories which contradict each other in such a way can 'really' be saying the same thing only if they are not literally construed. Most specifically, if a theory says that something exists, then a literal construal may elaborate on what that something is, but will not remove the implication of existence.

There have been many critiques of positivist interpretations of science, and there is no need to repeat them. I shall add some specific criticisms of the positivist approach in the next chapter.

CONSTRUCTIVE EMPIRICISM

To insist on a literal construal of the language of science is to rule out the construal of a theory as a metaphor or simile, or as intelligible only after it is 'demythologized' or subjected to some other sort of 'translation' that does not preserve logical form. If the theory's statements include 'There are electrons', then the theory says that there are electrons. If in addition they include 'Electrons are not planets', then the theory says, in part, that there are entities other than planets.

But this does not settle very much. It is often not at all obvious whether a theoretical term refers to a concrete entity or a mathematical entity. Perhaps one tenable interpretation of classical physics is that there are no concrete entities which are forces—that 'there are forces such that . . .' can always be understood as a mathematical statement asserting the existence of certain functions. That is debatable.

Not every philosophical position concerning science which insists on a literal construal of the language of science is a realist position. For this insistence relates not at all to our epistemic attitudes toward theories, nor to the aim we pursue in constructing theories, but only to the correct understanding of *what a theory says*. (The fundamentalist theist, the agnostic, and the atheist presumably agree with each other (though not with liberal theologians) in their understanding of the statement

that God, or gods, or angels exist.) After deciding that the language of science must be literally understood, we can still say that there is no need to believe good theories to be true, nor to believe *ipso facto* that the entities they postulate are real.

Science aims to give us theories which are empirically adequate; and acceptance of a theory involves as belief only that it is empirically adequate. This is the statement of the anti-realist position I advocate; I shall call it *constructive empiricism*.

This formulation is subject to the same qualifying remarks as that of scientific realism in Section 1.1 above. In addition it requires an explication of 'empirically adequate'. For now, I shall leave that with the preliminary explication that a theory is empirically adequate exactly if what it says about the observable things and events in this world, is true—exactly if it 'saves the phenomena'. A little more precisely: such a theory has at least one model that all the actual phenomena fit inside. I must emphasize that this refers to *all* the phenomena; these are not exhausted by those actually observed, nor even by those observed at some time, whether past, present, or future. The whole of the next chapter will be devoted to the explication of this term, which is intimately bound up with our conception of the structure of a scientific theory.

The distinction I have drawn between realism and anti-realism, in so far as it pertains to acceptance, concerns only how much belief is involved therein. Acceptance of theories (whether full, tentative, to a degree, etc.) is a phenomenon of scientific activity which clearly involves more than belief. One main reason for this is that we are never confronted with a complete theory. So if a scientist accepts a theory, he thereby involves himself in a certain sort of research program. That program could well be different from the one acceptance of another theory would have given him, even if those two (very incomplete) theories are equivalent to each other with respect to everything that is observable—in so far as they go.

Thus acceptance involves not only belief but a certain commitment. Even for those of us who are not working scientists, the acceptance involves a commitment to confront any future phenomena by means of the conceptual resources of this theory. It determines the terms in which we shall seek explanations. If the acceptance is at all strong, it is exhibited in the person's assumption of the role of explainer, in his willingness to answer questions *ex cathedra*. Even if you do not accept a theory, you can engage in discourse in a context in which language use is guided by that theory—but acceptance produces such contexts. There are similarities in all of this to ideological commitment. A commitment is of course not true or false: The confidence exhibited is that it will be *vindicated*.

This is a preliminary sketch of the *pragmatic* dimension of theory acceptance. Unlike the epistemic dimension, it does not figure overtly in the disagreement between realist and anti-realist. But because the amount of belief involved in acceptance is typically less according to anti-realists, they will tend to make more of the pragmatic aspects. It is as well to note here the important difference. Belief that a theory is true, or that it is empirically adequate, does not imply, and is not implied by, belief that full acceptance of the theory will be vindicated. To see this, you need only consider here a person who has quite definite beliefs about the future of the human race, or about the scientific community and the influences thereon and practical limitations we have. It might well be, for instance, that a theory which is empirically adequate will not combine easily with some other theories which we have accepted in fact, or that Armageddon will occur before we succeed. Whether belief that a theory is true, or that it is empirically adequate, can be equated with belief that acceptance of it would, under ideal research conditions, be vindicated in the long run, is another question. It seems to me an irrelevant question within philosophy of science, because an affirmative answer would not obliterate the distinction we have already established by the preceding remarks. (The question may also assume that counterfactual statements are objectively true or false, which I would deny.)

Although it seems to me that realists and anti-realists need not disagree about the pragmatic aspects of theory acceptance, I have mentioned it here because I think that typically they do. We shall find ourselves returning time and again, for example, to requests for explanation to which realists typically attach an objective validity which anti-realists cannot grant.

THE THEORY/OBSERVATION 'DICHOTOMY'

For good reasons, logical positivism dominated the philosophy of science for thirty years. In 1960, the first volume of *Minnesota Studies in the Philosophy of Science* published Rudolf Carnap's 'The Methodological Status of Theoretical Concepts', which is, in many ways, the culmination of the positivist program. It interprets science by relating it to an observation language (a postulated part of natural language which is devoid of theoretical terms). Two years later this article was followed in the same series by Grover Maxwell's 'The Ontological Status of Theoretical Entities', in title and theme a direct counter to Carnap's. This is the *locus classicus* for the new realists' contention that the theory/observation distinction cannot be drawn.

I shall examine some of Maxwell's points directly, but first a general remark about the issue. Such expressions as 'theoretical entity' and 'observable-theoretical dichotomy' are, on the face of it, examples of category mistakes. Terms or concepts are theoretical (introduced or adapted for the purposes of theory construction); entities are observable or unobservable. This may seem a little point, but it separates the discussion into two issues. Can we divide our language into a theoretical and non-theoretical part? On the other hand, can we classify objects and events into observable and unobservable ones?

Maxwell answers both questions in the negative, while not distinguishing them too carefully. On the first, where he can draw on well-known supportive essays by Wilfrid Sellars and Paul Feyerabend, I am in total agreement. All our language is thoroughly theory-infected. If we could cleanse our language of theory-laden terms, beginning with the recently introduced ones like 'VHF receiver', continuing through 'mass' and 'impulse' to 'element' and so on into the prehistory of language formation, we would end up with nothing useful. The way we talk, and scientists talk, is guided by the pictures provided by previously accepted theories. This is true also, as Duhem already emphasized, of experimental reports. Hygienic reconstructions of language such as the positivists envisaged are simply not on. . . .

But does this mean that we must be scientific realists? We surely have more tolerance of ambiguity than that. The fact that we let our language be guided by a given picture, at some point, does not show how much we believe about that picture. When we speak of the sun coming up in the morning and setting at night, we are guided by a picture now explicitly disavowed. When Milton wrote *Paradise Lost* he deliberately let the old geocentric astronomy guide his poem, although various remarks in passing clearly reveal his interest in the new astronomical discoveries and speculations of his time. These are extreme examples, but show that no immediate conclusions can be drawn from the theory-ladenness of our language.

However, Maxwell's main arguments are directed against the observable–unobservable distinction. Let us first be clear on what this distinction was supposed to be. The term 'observable' classifies putative entities (entities which may or may not exist). A flying horse is observable—that is why we are so sure that there aren't any—and the number seventeen is not. There is supposed to be a correlate classification of human acts: an unaided act of perception, for instance, is an observation. A calculation of the mass of a particle from the deflection of its trajectory in a known force field, is not an observation of that mass.

It is also important here not to confuse *observing* (an entity, such as a thing, event, or process) and *observing that* (something or other is the case). Suppose one of the Stone Age people recently found in the Philippines is shown a tennis ball or a car crash. From his behavior, we see that he has noticed them; for example, he picks up the ball and throws it. But he has not seen *that* it is a tennis ball, or *that* some event is a car crash, for he does not even have those concepts. He cannot get that information through perception; he would first have to learn a great deal. To say that he does not see the same things and events as we do, however, is just silly; it is a pun which trades on the ambiguity between seeing and seeing that. (The truth-conditions for our statement '*x* observes *that A*' must be such that what concepts *x* has, presumably related to the language *x* speaks if he is human, enter as a variable into the correct truth definition, in some way. To say that *x* observed the tennis ball, therefore, does not imply at all that *x*

observed that it was a tennis ball; that would require some conceptual awareness of the game of tennis.)

The arguments Maxwell gives about observability are of two sorts: one directed against the possibility of drawing such distinctions, the other against the importance that could attach to distinctions that can be drawn.

The first argument is from the continuum of cases that lie between direct observation and inference:

> there is, in principle, a continuous series beginning with looking through a vacuum and containing these as members: looking through a windowpane, looking through glasses, looking through binoculars, looking through a low-power microscope, looking through a high-power microscope, etc., in the order given. The important consequence is that, so far, we are left without criteria which would enable us to draw a non-arbitrary line between 'observation' and 'theory'.[1]

This continuous series of supposed acts of observation does not correspond directly to a continuum in what is supposed observable. For if something can be seen through a window, it can also be seen with the window raised. Similarly, the moons of Jupiter can be seen through a telescope; but they can also be seen without a telescope if you are close enough. That something is observable does not automatically imply that the conditions are right for observing it now. The principle is:

> X is observable if there are circumstances which are such that, if X is present to us under those circumstances, then we observe it.

This is not meant as a definition, but only as a rough guide to the avoidance of fallacies.

We may still be able to find a continuum in what is supposed detectable: perhaps some things can only be detected with the aid of an optical microscope, at least; perhaps some require an electron microscope, and so on. Maxwell's problem is: where shall we draw the line between what is observable and what is only detectable in some more roundabout way?

Granted that we cannot answer this question without arbitrariness, what follows? That 'observable' is a *vague predicate.* There are many puzzles about vague predicates, and many sophisms designed to show that, in the presence of vagueness, no distinction can be drawn at all. In Sextus Empiricus, we find the argument that incest is not immoral, for touching your mother's big toe with your little finger is not immoral, and all the rest differs only by degree. But predicates in natural language are almost all vague, and there is no problem in their use; only in formulating the logic that governs them.[2] A vague predicate is usable provided it has clear cases and clear counter-cases. Seeing with the unaided eye is a clear case of observation. Is Maxwell then perhaps challenging us to present a clear counter-case? Perhaps so, for he says 'I have been trying to support the thesis that any (non-logical) term is a *possible* candidate for an observation term.'

A look through a telescope at the moons of Jupiter seems to me a clear case of observation, since astronauts will no doubt be able to see them as well from close up. But the purported observation of microparticles in a cloud chamber seems to me a clearly different case—if our theory about what happens there is right. The theory says that if a charged particle traverses a chamber filled with saturated vapor, some atoms in the neighborhood of its path are ionized. If this vapor is decompressed, and hence becomes supersaturated, it condenses in droplets on the ions, thus marking the path of the particle. The resulting silver-gray line is similar (physically as well as in appearance) to the vapor trail left in the sky when a jet passes. Suppose I point to such a trail and say: 'Look, there is a jet!'; might you not say: 'I see the vapor trail, but where is the jet?' Then I would answer: 'Look just a bit ahead of the trail . . . there! Do you see it?' Now, in the case of the cloud chamber this response is not possible. So while the particle is detected by means of the cloud chamber, and the detection is based on observation, it is clearly not a case of the article's being observed.

As second argument, Maxwell directs our attention to the 'can' in 'what is observable is what can be observed.' An object might of course be temporarily unobservable—in a rather different sense: it cannot be observed in the circumstances in which it actually is at the moment, but could be observed if the circumstances were more favorable. In just the same way, I might be temporarily invulnerable or invisi-

ble. So we should concentrate on 'observable' *tout court,* or on (as he prefers to say) 'unobservable in principle'. This Maxwell explains as meaning that the relevant scientific theory *entails* that the entities cannot be observed in any circumstances. But this never happens, he says, because the different circumstances could be ones in which we have different sense organs—electron–microscope eyes, for instance.

This strikes me as a trick, a change in the subject of discussion. I have a mortar and pestle made of copper and weighing about a kilo. Should I call it breakable because a giant could break it? Should I call the Empire State Building portable? Is there no distinction between a portable and a console record player? The human organism is, from the point of view of physics, a certain kind of measuring apparatus. As such it has certain inherent limitations— which will be described in detail in the final physics and biology. It is these limitations to which the 'able' in 'observable' refers—our limitations, *qua* human beings.

As I mentioned, however, Maxwell's article also contains a different sort of argument: even if there is a feasible observable/unobservable distinction, this distinction has no importance. The point at issue for the realist is, after all, the reality of the entities postulated in science. Suppose that these entities could be classified into observables and others; what relevance should that have to the question of their existence?

Logically, none. For the term 'observable' classifies putative entities, and has logically nothing to do with existence. But Maxwell must have more in mind when he says: 'I conclude that the drawing of the observational–theoretical line at any given point is an accident and a function of our physiological make-up, . . . and, therefore, that it has no ontological significance whatever.'[3] No ontological significance if the question is only whether 'observable' and 'exists' imply each other—for they do not; but significance for the question of scientific realism?

Recall that I defined scientific realism in terms of the aim of science, and epistemic attitudes. The question is what aim scientific activity has, and how much we shall believe when we accept a scientific theory. What is the proper form of acceptance: belief that the theory, as a whole, is true; or something else? To this question, what is observable by us seems eminently relevant. Indeed, we may attempt an answer at this point: to accept a theory is (for us) to believe that it is empirically adequate—that what the theory says *about what is observable* (by us) is true.

It will be objected at once that, on this proposal, what the anti-realist decides to believe about the world will depend in part on what he believes to be his, or rather the epistemic community's, accessible range of evidence. At present, we count the human race as the epistemic community to which we belong; but this race may mutate, or that community may be increased by adding other animals (terrestrial or extra-terrestrial) through relevant ideological or moral decisions ('to count them as persons'). Hence the anti-realist would, on my proposal, have to accept conditions of the form

If the epistemic community changes in fashion *Y*, then my beliefs about the world will change in manner *Z*.

To see this as an objection to anti-realism is to voice the requirement that our epistemic policies should give the same results independent of our beliefs about the range of evidence accessible to us. That requirement seems to me in no way rationally compelling; it could be honored, I should think, only through a thoroughgoing skepticism or through a commitment to wholesale leaps of faith. But we cannot settle the major questions of epistemology *en passant* in philosophy of science; so I shall just conclude that it is, on the face of it, not irrational to commit oneself only to a search for theories that are empirically adequate, ones whose models fit the observable phenomena, while recognizing that what counts as an observable phenomenon is a function of what the epistemic community is (that *observable* is *observable-to-us*).

The notion of empirical adequacy in this answer will have to be spelled out very carefully if it is not to bite the dust among hackneyed objections. I shall

try to do so in the next chapter. But the point stands: even if observability has nothing to do with existence (is, indeed, too anthropocentric for that), it may still have much to do with the proper epistemic attitude to science.

NOTES

1. G. Maxwell, 'The Ontological Status of Theoretical Entities', *Minnesota Studies in Philosophy of Science*, III (1962), p. 7.
2. There is a great deal of recent work on the logic of vague predicates; especially important, to my mind, is that of Kit Fine ('Vagueness, Truth, and Logic', *Synthese,* 30 (1975), 265–300) and Hans Kamp. The latter is currently working on a new theory of vagueness that does justice to the 'vagueness of vagueness' and the context-dependence of standards of applicability for predicates.
3. Op. cit., p. 15. In the next chapter I shall discuss further how observability should be understood. At this point, however, I may be suspected of relying on modal distinctions which I criticize elsewhere. After all, I am making a distinction between human limitations and accidental factors. A certain apple was dropped into the sea in a bag of refuse, which sank; relative to that information it is necessary that no one ever observed the apple's core. That information, however, concerns an accident of history, and so it is not human limitations that rule out observation of the apple core. But unless I assert that some facts about humans are essential, or physically necessary, and others accidental, how can I make sense of this distinction? This question raises the difficulty of a philosophical retrenchment for modal language. This I believe to be possible through an ascent to pragmatics. In the present case, the answer would be, to speak very roughly, that the scientific theories we accept are a determining factor for the set of features of the human organism counted among the limitations to which we refer in using the term 'observable'. The issue of modality will occur explicitly again in the chapter on probability.

35

PAUL M. CHURCHLAND

The Anti–Realist Epistemology of van Fraassen's
The Scientific Image

At several points in the reading of van Fraassen's book, I feared I would no longer be a realist by the time I completed it. Fortunately, sheer doxastic inertia has allowed my convictions to survive its searching critique, at least temporarily, and as of today, van Fraassen and I still hold different views. I am a scientific realist, of unorthodox persuasion, and van Fraassen is a constructive empiricist, whose persuasions currently define the doctrine. I assert that global excellence of theory is the ultimate measure of truth and ontology at all levels of cognition, even at the observational level. Van Fraassen asserts that descriptive excellence at the observational level is the only genuine measure of any theory's truth, and that one's acceptance of a theory should create no ontological commitments whatever beyond the observational level.

Against his first claim I will maintain that observational excellence or 'empirical adequacy' is only one epistemic virtue among others, of equal or comparable importance. And against his second claim I will maintain that the ontological commitments of any theory are wholly blind to the idiosyncratic distinction between what is and what is not humanly observable, and so should be our own ontological commitments. Criticism will be directed primarily at van Fraassen's *selective* skepticism in favor of observable ontologies over unobservable ontologies: and against his view that the superempirical theoretical virtues (simplicity,

coherence, explanatory power) are merely pragmatic virtues, irrelevant to the estimate of a theory's truth. My aims are not merely critical, however. Scientific realism does need reworking, and there are good reasons for moving it in the direction of van Fraassen's constructive empiricism, as will be discussed in the closing section of this paper. But those reasons do not support the skeptical theses at issue.

. . .

Before pursuing our differences, it will prove useful to emphasize certain convictions we share. Van Fraassen is already a scientific realist in the minimal sense that he interprets theories literally and he concedes them a truth value. Further, we agree that the observable/unobservable distinction is entirely distinct from the non-theoretical/theoretical distinction, and we agree as well that all observation sentences are irredeemably laden with theory.

Additionally, I absolutely reject many sanguine assumptions common among realists. I do not believe that on the whole our beliefs must be at least roughly true; I do not believe that the terms of 'mature' sciences must typically refer to real things; and I very much doubt that the Reason of *homo sapiens,* even at its best and even if allowed infinite time, would eventually encompass all and/or only true statements.

This skepticism is born partly from a historical induction: so many past theories, rightly judged excellent at the time, have since proved to be false. And their current successors, though even better founded, seem but the next step in a probably end-

Pacific Philosophical Quarterly 63 (July 1982). Reprinted with permission from Blackwell Publishers.

less and not obviously convergent journey. (For a most thorough and insightful critique of typical realist theses, see the recent paper by Laudan [4].)

Evolutionary considerations also counsel a healthy skepticism. Human reason is a hierarchy of heuristics for seeking, recognizing, storing, and exploiting information. But those heuristics were invented at random, and they were selected for within a very narrow evolutionary environment, cosmologically speaking. It would be miraculous if human reason were completely free of false strategies and fundamental cognitive limitations, and doubly miraculous if the theories we accept failed to reflect those defects.

Thus some very realistic reasons for skepticism with respect to any theory. Why then am I still a scientific realist? Because these reasons fail to discriminate between the integrity of observables and the integrity of unobservables. If anything is compromised by these considerations, it is the integrity of theories generally. That is, of *cognition* generally. Since our observational concepts are just as theory-laden as any others, and since the integrity of those concepts is just as contingent on the integrity of the theories that embed them, our observational ontology is rendered *exactly as dubious* as our non-observational ontology.

This parity should not seem surprising. Our history reveals mistaken ontological commitments in both domains. For example, we have had occasion to banish phlogiston, caloric, and the luminiferous ether from our ontology, but we have also had occasion to banish witches, and the starry sphere that turns about us daily. And these latter items were as 'observable' as you please.

Since these skeptical considerations are indifferent to the distinction between what is and is not observable, they provide no reason for resisting a commitment to unobservable ontologies *while allowing* a commitment to observable ontologies. The latter appear as no better off than the former. For me then, the empirical success of a theory remains a reason for thinking the theory to be true, and for accepting its overall ontology. The inference from success to truth should no doubt be tempered by the skeptical considerations adduced, but the inference to *unobservable* ontologies is not rendered *selectively* dubious. Thus I remain a scientific realist.

My realism is highly circumspect, but the circumspection is uniform for unobservables and observables alike.

Perhaps I am wrong in this. Perhaps we should be selectively skeptical in the fashion van Fraassen recommends. Does he have other arguments for refusing factual belief and ontological commitment beyond the observational domain? Indeed he does. In fact, he does not appeal to historical induction or evolutionary humility at all. These are *my* reasons for skepticism (and they will remain, even if I manage to undermine van Fraassen's). They have been introduced here to show that, while there are some powerful reasons for skepticism, those reasons do not place unobservables at a selective disadvantage.

Very well, what are van Fraassen's reasons for skepticism? They are very interesting. To summarize quickly, he does a compelling job of deflating certain standard realist arguments (from Smart, Sellars, Salmon, Boyd, and others) to the effect that, given the aims of science, we have no alternative but to bring unobservables (not just into our calculations, but) into our literal ontology. He also argues rather compellingly that the superempirical virtues, such as simplicity and comprehensive explanatory power, are at bottom merely pragmatic virtues, having nothing essential to do with any theory's truth. This leaves only empirical adequacy as a genuine measure of any theory's truth. Roughly, a theory is empirically adequate if and only if everything it says about *observable* things is true. Empirical adequacy is thus a necessary condition on a theory's truth.

However, claims van Fraassen, the truth of any theory whose ontology includes unobservables is always radically underdetermined by its empirical adequacy, since a great many logically incompatible theories can all be empirically equivalent. Accordingly, the inference from empirical adequacy to truth now appears presumptuous in the extreme, especially since it has just been disconnected from additional selective criteria such as simplicity and explanatory power, criteria which might have reduced the arbitrariness of the particular inference drawn. Fortunately, says van Fraassen, we do not need to make such wanton inferences, since we can perfectly well understand science as an enterprise that never really draws them. Here we arrive at his positive conception of science as an enterprise whose sole intellectual aims are empirical adequacy and the satisfaction of certain human intellectual needs.

The central element in this argument is the claim that, in the case of a theory whose ontology includes unobservables, its empirical adequacy underdetermines its truth. (We should notice that in the case of a theory whose ontology is completely free of unobservables, its empirical adequacy does not underdetermine its truth: in that case, truth and empirical adequacy are obviously identical. Thus van Fraassen's *selective* skepticism with respect to unobservables.) That is, for any theory T inflated with unobservables, there will always be many other such theories incompatible with T, but empirically equivalent to it.

In my view, the notions of "empirical adequacy" and its cognate relative term "empirically equivalent" are extremely thorny notions of doubtful integrity. If we attempt to explicate a theory's "empirical content" in terms of the observation sentences it entails (or entails-if-conjoined-with available background information, or with possible future background information, or with possible future theories), we generate a variety of notions which are variously empty, context-relative, ill-defined, or incompatible with the claim of underdetermination. Van Fraassen is entirely aware of these difficulties and proposes to avoid them by giving the notions at issue a model-theoretic rather than a syntactic explication. I am unconvinced that this improves matters decisively (on this issue see Wilson [7]). But let me sidestep the issue for now, since the matter is difficult and there is a simpler objection to be voiced.

The empirical adequacy of any theory is itself something that is radically underdetermined by any evidence conceivably available to us. Recall that, for a theory to be empirically adequate, what it says about observable things must be true—*all* observable things, in the past, in the indefinite future, and in the most distant corners of the cosmos. But since any actual data possessed by us must be finite in its scope, it is plain that we here suffer an underdetermination problem no less serious than that claimed above. This is Hume's problem, and the lesson is that even observation-level theories suffer radical underdetermination by the evidence. Accordingly, theories about observables and theories about unobservables appear on a par again, so far as skepticism is concerned.

Van Fraassen thinks there is an important difference between the two cases, and one's first impulse is to agree with him. We are all willing to concede the existence of Hume's problem—the problem of justifying the inference to unobserv*ed* entities. But the inference to entities that are downright unobserv*able* appears as a different and additional problem.

I do not see that it is. Consider the different reasons why entities or processes may go unobserved by us. First, they may go unobserved because, relative to our natural sensory apparatus, they fail to enjoy an appropriate spatial or temporal *position*. They may exist in the Upper Jurassic Period, for example, or they may reside in the Andromeda Galaxy. Second, they may go unobserved because, relative to our natural sensory apparatus, they fail to enjoy the appropriate spatial or temporal *dimensions*. They may be too small, or too brief, or too large, or too protracted. Third, they may fail to enjoy the appropriate *energy*, being too feeble, or too powerful, to permit useful discrimination. Fourth and fifth, they may fail to have an appropriate *wavelength*, or an appropriate *mass*. Sixth, they may fail to 'feel' the relevant fundamental forces our sensory apparatus exploits, as with our inability to observe the background neutrino flux, despite the fact that its energy density exceeds that of light itself.

This list could be lengthened, but it is long enough to suggest that being spatially or temporally distant from our sensory apparatus is only one undistinguished way, one among many ways, in which an entity or process can fall outside the compass of human observation.

There is perhaps some point to calling a thing "observ*able*" if it fails only the first test (spatio-temporal proximity), and "unobservable" if it fails any of the others. But that is only because of the contingent practical fact that one generally has somewhat more *control* over the spatio-temporal perspective of one's sensory systems than one has over their size, or reaction time, or mass, or wavelength sensitivity, or chemical constitution. Had we been less mobile—rooted to the earth like Douglas Firs, say—yet been more voluntarily plastic in our sensory constitution, the distinction between the 'merely unobserved' and the 'downright unobservable' would have been very differently drawn. It may help to imagine here a suitably rooted arboreal philosopher named (what else?) Douglas van Fiirrsen, who, in his sedentary wisdom, urges an antirealist skepticism concerning the spatially very *distant* entities postulated by his fellow trees.

Admittedly, for any distant entity one can in principle always change the relative spatial position of one's sensory apparatus so that the entity is observed: one can go to it. But equally, for any microscopic entity one can in principle always change the relative spatial *size* or *configuration* of one's sensory apparatus so that the entity is observed. Physical law imposes certain limitations on such plasticity, but so also does physical law limit how far one can travel in a lifetime.

The point of all this is that there is no special or additional problem about inferences to the existence of entities commonly called "unobservables." Such entities are merely those that go unobserved by us for reasons *other* than their spatial or temporal distance from us. But whether the 'gap' to be bridged is spatio-temporal, or one of the many other gaps, the logical/epistemological problem is the same in all cases: ampliative inference and underdetermined hypotheses. I therefore fail to see how van Fraassen can justify tolerating an ampliative inference when it bridges a gap of spatial distance, while refusing to tolerate an ampliative inference when it bridges a gap of, for example, spatial size. Hume's problem and van Fraassen's problem (or Duhem's problem) collapse into one.

Van Fraassen attempts to meet such worries about the inescapable ubiquity of speculative activity by observing that ". . . it is not an epistemological principle that one may as well hang for a sheep as for a lamb" ([5], p. 72). Agreed. But it is a principle of *logic* that one may as well hang for a sheep as for a sheep, and van Fraassen's lamb (empirical adequacy) is just another sheep.

Let me summarize. As van Fraassen sets it up, and as the instrumentalists set it up before him, the realist looks more gullible than the non-realist, since the realist is willing to extend belief beyond the observable, while the non-realist insists on confining belief within that domain. I suggest, however, that it is really the non-realists who are being the more gullible in this matter, since they suppose that the epistemic situation of our beliefs about observables is in some way superior to that of our beliefs about unobservables. But in fact their epistemic situation is not superior. They are exactly as dubious as their non-observational cousins. Their casual history is different perhaps, but not their credibility.

Simply to hold *fewer* beliefs from a given set is of course to be less adventurous, but it is not necessarily to be applauded. One might decide to relinquish all one's beliefs save those about objects weighing less than 500 kg, and perhaps one would then be logically safer. But in the absence of some relevant epistemic difference between one's beliefs about such objects and one's beliefs about other objects, this is perversity, not parsimony.

. . .

Let me now try to address the question of whether the theoretical virtues such as simplicity, coherence, and explanatory power are *epistemic* virtues genuinely relevant to the estimate of a theory's truth, as tradition says, or merely *pragmatic* virtues, as van Fraassen urges. His view promotes empirical adequacy, or evidence of empirical adequacy, as the only genuine measure of a theory's truth, the other virtues (insofar as they are distinct from these) being cast as purely pragmatic virtues, to be valued only for the human needs they satisfy. Despite certain compelling features of the account of explanation that van Fraassen provides, I remain inclined towards the traditional view.

My reason is simplicity itself. Since there is no way of conceiving or representing "the empirical facts" that is completely independent of speculative assumptions, and since we will occasionally confront theoretical alternatives on a scale so comprehensive that we must also choose between competing modes of conceiving what the empirical facts before us *are*, then the epistemic choice between these global alternatives cannot be made by comparing the extent to which they are adequate to some common touchstone, 'the empirical facts.' In such a case, the choice must be made on the comparative global virtues of the two global alternatives, T_1-plus-the-observational-evidence-therein-construed, versus T_2-plus-the-observational-evidence-therein-(differently)-construed. That is, it must be made on superempirical grounds such as relative coherence, simplicity, and explanatory unity. In such cases, "empirical adequacy" becomes just one dimension of coherence.

Such cases as these are reminiscent of Carnap's 'external' questions, and it may be that van Fraassen, like Carnap, does not regard them as factual questions, but as essentially pragmatic questions. I would disagree, since I regard so-called 'external' questions as arrayed on a smooth continuum with 'internal' (i.e., factual) questions. The arguments are presented elsewhere ([1], sections 7 and 10), however, so I shall not repeat them here.

As I see it then, values such as ontological simplicity, coherence, and explanatory power are some of the brain's criteria for recognizing information, for distinguishing information from noise. And I think they are even more fundamental values than is 'empirical adequacy,' since collectively they can overthrow an entire conceptual framework for representing the empirical facts. Indeed, they even dictate how such a framework is constructed by the questing infant in the first place. One's observational taxonomy is not 'read off' the world directly; rather, one comes to it piecemeal, and by stages, and one settles on that taxonomy which finds the greatest coherence and simplicity in the world, and most and the simplest lawful connections.

I can bring together my protective concerns for unobservables and for the superempirical virtues by way of the following thought experiment. Consider a man for whom absolutely *nothing* is observable. All of his sensory modalities have been surgically destroyed, and he has no visual, tactile, or other sensory experience of any kind. Fortunately, he has mounted on top of his skull a microcomputer fitted out with a variety of environmentally-sensitive transducers. The computer is connected to his association cortex (or perhaps the frontal lobe, or Wernicke's area) in such a way as to cause in him a continuous string of singular beliefs about his local environment. These "intellectual intuitions" are not infallible, but let us suppose that they provide him with much the same information that our perceptual judgments provide us.

For such a person, or for a society of such persons, the *observable* world is an empty set. There is no question, therefore, of their evaluating any theory by reference to its 'empirical adequacy,' as characterized by van Fraassen (i.e., isomorphism between some observable features of the world and some 'empirical substructure' of one of the theory's models). But

such a society is still capable of science, I assert. They can invent theories, construct explanations of the facts-as-represented-in-past-spontaneous-beliefs, hazard predictions of the facts-as-represented-in-future-spontaneous-beliefs, and so forth. In principle, there is no reason they could not learn as much as we have. (cf. Feyerabend [3])

But it is plain in this case that the global virtues of simplicity, coherence, and explanatory unification are what *must* guide the continuing evolution of their collected beliefs. And it is plain as well that their ontology, whatever it is, must consist entirely of *un*observable entities. To invite a van Fraassenean disbelief in unobservable entities is in this case to invite the suspension of all beliefs beyond tautologies! Surely reason does not require them to be so abstemious.

It is time to consider the objection that those aspects of the world which are successfully monitored by the transducing microcomputer should count as 'observables' for the folk described, despite the lack of any appropriate field of internal sensory qualia to mediate the external circumstance and the internal judgment it causes. Their tables-and-chairs ontology, as expressed in their spontaneous judgments, could then be conceded legitimacy.

I will be the first to accept such an objection. But if we do accept it, then I do not see how we can justify van Fraassen's selective skepticism with respect to the wealth of 'unobservable' entities and properties reliably monitored by *our* transducing measuring instruments (electron microscopes, cloud chambers, chromatographs, etc.). The spontaneous singular judgments of the working scientist, at home in his theoretical vocabulary and deeply familiar with the measuring instruments to which his conceptual system is responding, are not worse off, causally or epistemologically, than the spontaneous singular judgments of our transducer-laden friends. If skepticism is to be put aside above, it must be put aside here as well.

My concluding thought experiment is a complement to the one just outlined. Consider some folk who observe, not less of the world than we do, but more of it. Suppose them able to observe a domain normally closed to us: the micro-world of virus particles, DNA strands, and large protein molecules. Specifically, suppose a race of humanoid creatures each of whom is born with an electron microscope permanently in place over his left 'eye.' The scope is biologically constituted, let us suppose, and it projects its image onto a human-style retina, with the rest of their neurophysiology paralleling our own.

Science tells us, and I take it that van Fraassen would agree, that virus particles, DNA strands, and most other objects of comparable dimensions count as observable entities for the humanoids described. The humanoids, at least, would be justified in so regarding them and in including them in their ontology.

But we humans may not include such entities in our ontology, according to van Fraassen's position, since they are not observable with our unaided perceptual apparatus. We may not include such entities in our ontology *even though we can construct and even if we do construct electron microscopes of identical function, place them over our left eyes, and enjoy exactly the same microexperience as the humanoids.*

The difficulty for van Fraassen's position, if I understand it correctly, is that his position requires that a humanoid and a scope-equipped human must embrace *different* epistemic attitudes towards the microworld, even though their causal connections to the world and their continuing experience of it be identical: the humanoid is required to be a realist with respect to the microworld, and the human is required to be an anti-realist (i.e., an agnostic) with respect to the microworld. But this distinction between what we and they may properly embrace as real seems to me to be highly arbitrary and radically under-motivated. For the only difference between the humanoid and a scope-equipped human lies in the *causal origins* of the transducing instruments feeding information into their respective brains. The humanoid's scope owes its existence to information coded in his genetic material. The human's scope owes its existence to information coded in his cortical material, or in technical libraries. I do not see why this should make any difference in their respective ontological commitments, whatever they are, and I must decline to embrace any philosophy of science which says that it must.

. . .

I now turn from critic of van Fraassen's position to advocate. One of the most central elements in his view seems to me to be well-motivated and urgently deserving of further development. As he explains in his introductory chapter, his aim is to reconceive the relation of theory to world, and the units of scientific cognition, and the virtue of those units when successful. He says,

> I use the adjective 'constructive' to indicate my view that scientific activity is one of construction rather than discovery: construction of models that must be adequate to the phenomena, and not discovery of truth concerning the unobservable. ([5], p. 5)

The traditional view of human knowledge is that the unit of cognition is the sentence or proposition, and the cognitive virtue of such units is truth. Van Fraassen rejects this overtly linguistic guise for his empiricism. He invites us to reconceive a theory as a set of models (rather than as a set of sentences), and he sees empirical adequacy (rather than truth) as the principal virtue of such units.

Though I reject his particular reconception, and the selective skepticism he draws from it, I think the move away from the traditional conception is entirely correct. The criticism to which I am inclined is that van Fraassen has not moved quite far enough. Specifically, if we are to reconsider truth as the aim or product of cognitive activity, I think we must reconsider its applicability right across the board, and not just in some arbitrarily or idiosyncratically segregated domain of 'unobservables.' That is, if we are to move away from the more naive formulations of scientific realism, we should move in the direction of *pragmatism* rather than in the direction of a positivistic instrumentalism. Let me elaborate.

When we consider the great variety of cognitively active creatures on this planet—sea slugs and octopi, bats, dolphins, and humans; and when we consider the ceaseless reconfiguration in which their brains or central ganglia engage—adjustments in the response potentials of single neurons made in the microsecond range, changes in the response characteristics of large systems of neurons made in the seconds-to-hours range, dendritic growth and new synaptic connections and the selective atrophy of old connections effected in the day-upwards range; then van Fraassen's term 'construction' begins to seem highly appropriate. There is endless construction and reconstruction, both functional and structural. Further, it is far from obvious that truth is either the primary aim or the principal product of this activity. Rather, its function would appear to be the ever more finely tuned administration of the organism's *behavior*. Natural selection does not care whether a brain has or tends toward true beliefs, so long as the organism reliably exhibits reproductively advantageous behavior. Plainly there is going to be *some* connection between the faithfulness of the brain's 'world-model' and the propriety of the organism's behavior. But just as plainly the connection is not going to be direct.

While we are considering cognitive activity in biological terms and in all branches of the phylogenetic tree, we should note that it is far from obvious that sentences or propositions or anything remotely like them constitute the basic elements of cognition in creatures generally. Indeed, as I have argued at length elsewhere ([1], chapter 5; [2]), it is highly unlikely that the sentential kinematics embraced by folk psychology and orthodox epistemology represents or captures the basic parameters of cognition and learning even in humans. That framework is part of a common-sense theory that threatens to be either superficial or false. If we are ever to understand the *dynamics* of cognitive activity, therefore, we may have to reconceive our basic unit of cognition as something other than the sentence or proposition, and reconceive its virtue as something other than truth.

Success of this sort on the descriptive/explanatory front would likely have normative consequences. Truth, as currently conceived, might cease to be an aim of science. Not because we had lowered our sights and reduced our epistemic standards, as van Fraassen's constructive empiricism would suggest, but because we had raised our sights, in pursuit of some epistemic goal even *more* worthy than truth. I cannot now elucidate such goals, but we should be sensible of their possible existence. The notion of "truth," after all, is but the central element in a normative *theory*, and *praxis* makes progress no less than *theoria*.

The notion of truth is suspect on purely metaphysical grounds anyway. It suggests straightaway the notion of the Complete and Final True Theory: at a minimum, the infinite set of all true sentences. Such a theory would be, by epistemic criteria, the best theory possible. But nothing whatever guarantees the existence of such a unique theory. Just as there is no largest positive integer, it may be that there is no best theory. It may be that for any theory whatsoever, there is always an even better theory, and so ad infinitum. If we were thus unable to speak of the set of all true sentences, what sense could we make of truth sentence-by-sentence?

These considerations do invite a 'constructive' conception of cognitive activity, one in which the notion of truth plays at best a highly derivative role. The formulation of such a conception, adequate to all of our epistemic criteria, is the outstanding task of epistemology. I do not think we will find that conception in a model-theoretic version of positivistic instrumentalism, nor do I think we will find it quickly. But the empirical brain begs unravelling, and we have plenty of time.

Finally, there is a question put to me by Stephen Stich. If ultimately my view is even more skeptical than van Fraassen's concerning the relevance or applicability of the notion of truth, why call it scientific *realism* at all? For at least two reasons. The term "realism" still marks the principal contrast with its traditional adversary, positivistic instrumentalism. Whatever the integrity of the notion of truth, theories about unobservables have just *as much* a claim to truth, epistemologically and metaphysically, as theories about observables. Second, I remain committed to the idea that there exists a world, independent of our cognition, with which we interact, and of which we construct representations: for varying purposes, with varying penetration, and with varying success. Lastly, our best and most penetrating grasp of the real is still held to reside in the representations provided by our best theories. Global excellence of theory remains the fundamental measure of rational ontology. And that has always been the central claim of scientific realism.

REFERENCES

[1] Churchland, Paul M. (1979), *Scientific Realism and the Plasticity of Mind* (Cambridge University Press).

[2] ——— (1981), "Eliminative Materialism and the Propositional Attitudes," *The Journal of Philosophy*, vol. 78, no. 2, February.

[3] Feyerabend, Paul K. (1969), "Science Without Experience," *The Journal of Philosophy*, vol. 66, no. 22.

[4] Laudan, Larry (1981), "A Confutation of Convergent Realism," *Philosophy of Science*, vol. 48, no. 1, March.

[5] van Fraassen, Bas C. (1980), *The Scientific Image* (Oxford University Press).

[6] ———— (1981), "Critical Notice of Paul Churchland: Scientific Realism and the Plasticity of Mind," *Canadian Journal of Philosophy*, vol. 11, no. 3, September.

[7] Wilson, Mark. (1980), "The Observational Uniqueness of Some Theories," *The Journal of Philosophy*, vol. 77, no. 4, April.

36

IAN HACKING

Experimentation and Scientific Realism

Experimental physics provides the strongest evidence for scientific realism. Entities that in principle cannot be observed are regularly manipulated to produce new phenomena and to investigate other aspects of nature. They are tools, instruments not for thinking but for doing.

The philosopher's standard "theoretical entity" is the electron. I shall illustrate how electrons have become experimental entities, or experimenter's entities. In the early stages of our discovery of an entity, we may test hypotheses about it. Then it is merely a hypothetical entity. Much later, if we come to understand some of its causal powers and to use it to build devices that achieve well understood effects in other parts of nature, then it assumes quite a different status.

Discussions about scientific realism or antirealism usually talk about theories, explanation and prediction. Debates at that level are necessarily inconclusive. Only at the level of experimental practice is scientific realism unavoidable. But this realism is not about theories and truth. The experimentalist need only be a realist about the entities used as tools.

A PLEA FOR EXPERIMENTS

No field in the philosophy of science is more systematically neglected than experiment. Our grade school teachers may have told us that scientific method is experimental method, but histories of science have become histories of theory. Experiments, the philosophers say, are of value only when they test theory. Experimental work, they imply, has no life of its own. So we lack even a terminology to describe the many varied roles of experiment. Nor has this one-sidedness done theory any good, for radically different types of theory are used to think about the same physical phenomenon (e.g., the magneto-optical effect). The philosophers of theory have not noticed this and so misreport even theoretical inquiry.[1]

Different sciences at different times exhibit different relationships between "theory" and "experiment." One chief role of experiment is the creation of phenomena. Experimenters bring into being phenomena that do not naturally exist in a pure state.

Philosophical Topics 13 (1983), pp. 71–87. Reprinted by permission of University of Arkansas Press.

These phenomena are the touchstones of physics, the keys to nature and the source of much modern technology. Many are what physicists after the 1870s began to call "effects": the photo-electric effect, the Compton effect, and so forth. A recent high-energy extension of the creation of phenomena is the creation of "events," to use the jargon of the trade. Most of the phenomena, effects and events created by the experimenter, are like plutonium: they do not exist in nature except possibly on vanishingly rare occasions.[2]

In this paper I leave aside questions of methodology, history, taxonomy and the purpose of experiment in natural science. I turn to the purely philosophical issue of scientific realism. Call it simply "realism" for short. There are two basic kinds: realism about entities and realism about theories. There is no agreement on the precise definition of either. Realism about theories says we try to form true theories about the world, about the inner constitution of matter and about the outer reaches of space. This realism gets its bite from optimism; we think we can do well in this project, and have already had partial success.

Realism about entities—and I include processes, states, waves, currents, interactions, fields, black holes and the like among entities—asserts the existence of at least some of the entities that are the stock in trade of physics.[3]

The two realisms may seem identical. If you believe a theory, do you not believe in the existence of the entities it speaks about? If you believe in some entities, must you not describe them in some theoretical way that you accept? This seeming identity is illusory. The vast majority of experimental physicists are realists about entities without a commitment to realism about theories. The experimenter is convinced of the existence of plenty of "inferred" and "unobservable" entities. But no one in the lab believes in the literal truth of present theories about those entities. Although various properties are confidently ascribed to electrons, most of these properties can be embedded in plenty of different inconsistent theories about which the experimenter is agnostic. Even people working on adjacent parts of the same large experiment will use different and mutually incompatible accounts of what an electron is. That is because different parts of the experiment will make different uses of electrons, and the models that are useful for making calculations about one use may be completely haywire for another use.

Do I describe a merely sociological fact about experimentalists? It is not surprising, it will be said, that these good practical people are realists. They need that for their own self-esteem. But the self-vindicating realism of experimenters shows nothing about what actually exists in the world. In reply I repeat the distinction between realism about entities and realism about theories and models. Anti-realism about models is perfectly coherent. Many research workers may in fact hope that their theories and even their mathematical models "aim at the truth," but they seldom suppose that any particular model is more than adequate for a purpose. By and large most experimenters seem to be instrumentalists about the models they use. The models are products of the intellect, tools for thinking and calculating. They are essential for writing up grant proposals to obtain further funding. They are rules of thumb used to get things done. Some experimenters are instrumentalists about theories and models, while some are not. That is a sociological fact. But experimenters are realists about the entities that they use in order to investigate other hypotheses or hypothetical entities. That is not a sociological fact. Their enterprise would be incoherent without it. But their enterprise is not incoherent. It persistently creates new phenomena that become regular technology. My task is to show that realism about entities is a necessary condition for the coherence of most experimentation in natural science.

OUR DEBT TO HILARY PUTNAM

It was once the accepted wisdom that a word like "electron" gets its meaning from its place in a network of sentences that state theoretical laws. Hence arose the infamous problems of incommensurability and theory change. For if a theory is modified, how could a word like "electron" retain its previous meaning? How could different theories about electrons be compared, since the very word "electron" would differ in meaning from theory to theory?

Putnam saves us from such questions by inventing a referential model of meaning. He says that meaning is a vector, refreshingly like a dictionary entry. First comes the syntactic marker (part of speech). Next the semantic marker (general category of thing signified by the word). Then the stereotype (clichés about the natural kind, standard examples of its use, and present day associations. The stereotype is subject to change as opinions about the kind are modified). Finally there is the actual reference of the word, the very stuff, or thing, it denotes if it denotes anything. (Evidently dictionaries cannot include this in their entry, but pictorial dictionaries do their best by inserting illustrations whenever possible.)[4]

Putnam thought we can often guess at entities that we do not literally point to. Our initial guesses may be jejune or inept, and not every naming of an invisible thing or stuff pans out. But when it does, and we frame better and better ideas, then Putnam says that although the stereotype changes, we refer to the same kind of thing or stuff all along. We and Dalton alike spoke about the same stuff when we spoke of (inorganic) acids. J. J. Thomson, Lorentz, Bohr and Millikan were, with their different theories and observations, speculating about the same kind of thing, the electron.

There is plenty of unimportant vagueness about when an entity has been successfully "dubbed," as Putnam puts it. "Electron" is the name suggested by G. Johnstone Stoney in 1891 as the name for a natural unit of electricity. He had drawn attention to this unit in 1874. The name was then applied in 1897 by J. J. Thomson to the subatomic particles of negative charge of which cathode rays consist. Was Johnstone Stoney referring to the electron? Putnam's account does not require an unequivocal answer. Standard physics books say that Thomson discovered the electron. For once I might back theory and say Lorentz beat him to it. What Thomson did was to measure the electron. He showed its mass is 1/1800 that of hydrogen. Hence it is natural to say that Lorentz merely postulated the particle of negative charge, while Thomson, determining its mass, showed that there is some such real stuff beaming off a hot cathode.

The stereotype of the electron has regularly changed, and we have at least two largely incompatible stereotypes, the electron as cloud and the electron as particle. One fundamental enrichment of the idea came in the 1920s. Electrons, it was found, have angular momentum, or "spin." Experimental work by Stern and Gerlach first indicated this, and then Goudsmit and Uhlenbeck provided the theoretical understanding of it in 1925. Whatever we

think about Johnstone Stoney, others—Lorentz, Bohr, Thomson and Goudsmit—were all finding out more about the same kind of thing, the electron.

We need not accept the fine points of Putnam's account of reference in order to thank him for providing a new way to talk about meaning. Serious discussions of inferred entities need no longer lock us into pseudo-problems of incommensurability and theory change. Twenty-five years ago the experimenter who believed that electrons exist, without giving much credence to any set of laws about electrons, would have been dismissed as philosophically incoherent. We now realize it was the philosophy that was wrong, not the experimenter. My own relationship to Putnam's account of meaning is like the experimenter's relationship to a theory. I don't literally believe Putnam, but I am happy to employ his account as an alternative to the unpalatable account in fashion some time ago.

Putnam's philosophy is always in flux. At the time of this writing, July 1981, he rejects any "metaphysical realism" but allows "internal realism."[5] The internal realist acts, in practical affairs, as if the entities occurring in his working theories did in fact exist. However, the direction of Putnam's metaphysical anti-realism is no longer scientific. It is not peculiarly about natural science. It is about chairs and livers too. He thinks that the world does not naturally break up into our classifications. He calls himself a transcendental idealist. I call him a transcendental nominalist. I use the word "nominalist" in the old-fashioned way, not meaning opposition to "abstract entities" like sets, but meaning the doctrine that there is no nonmental classification in nature that exists over and above our own human system of naming.

There might be two kinds of Putnamian internal realist—the instrumentalist and the scientific realist. The former is, in practical affairs where he uses his present scheme of concepts, a realist about livers and chairs, but he thinks that electrons are mental constructs only. The latter thinks that livers, chairs, and electrons are probably all in the same boat, that is, real at least within the present system of classification. I take Putnam to be an internal scientific realist rather than an internal instrumentalist. The fact that either doctrine is compatible with transcendental nominalism and internal realism shows that our question of scientific realism is almost entirely independent of Putnam's present philosophy.

INTERFERING

Francis Bacon, the first and almost last philosopher of experiments, knew it well: the experimenter sets out "to twist the lion's tail." Experimentation is interference in the course of nature; "nature under constraint and vexed; that is to say, when by art and the hand of man she is forced out of her natural state, and squeezed and molded."[6] The experimenter is convinced of the reality of entities some of whose causal properties are sufficiently well understood that they can be used to interfere *elsewhere* in nature. One is impressed by entities that one can use to test conjectures about other more hypothetical entities. In my example, one is sure of the electrons that are used to investigate weak neutral currents and neutral bosons. This should not be news, for why else are we (non-skeptics) sure of the reality of even macroscopic objects, but because of what we do with them, what we do to them, and what they do to us?

Interference and intervention are the stuff of reality. This is true, for example, at the borderline of observability. Too often philosophers imagine that microscopes carry conviction because they help us see better. But that is only part of the story. On the contrary, what counts is what we can do to a specimen under a microscope, and what we can see ourselves doing. We stain the specimen, slice it, inject it, irradiate it, fix it. We examine it using different kinds of microscopes that employ optical systems that rely on almost totally unrelated facts about light. Microscopes carry conviction because of the great array of interactions and interferences that are possible. When we see something that turns out not to be stable under such play, we call it an artifact and say it is not real.[7]

Likewise, as we move down in scale to the truly un-seeable, it is our power to use unobservable entities that make us believe they are there. Yet I blush over these words "see" and "observe." John Dewey would have said that a fascination with seeing-with-the-naked-eye is part of the Spectator Theory of Knowledge that has bedeviled philosophy from earliest times. But I don't think Plato or Locke or anyone before the nineteenth century was as obsessed with the sheer opacity of objects as we have been since. My own obsession with a technology that manipulates objects is, of course a twentieth-century counterpart to positivism and phenomenology. Their proper rebuttal is not a restriction to a narrower domain of reality, namely to what can be positivistically "seen" (with the eye), but an extension to other modes by which people can extend their consciousness.

MAKING

Even if experimenters are realists about entities, it does not follow that they are right. Perhaps it is a matter of psychology: the very skills that make for a great experimenter go with a certain cast of mind that objectifies whatever it thinks about. Yet this will not do. The experimenter cheerfully regards neutral bosons as merely hypothetical entities, while electrons are real. What is the difference?

There are an enormous number of ways to make instruments that rely on the causal properties of electrons in order to produce desired effects of unsurpassed precision. I shall illustrate this. The argument—it could be called the experimental argument for realism—is not that we infer the reality of electrons from our success. We do not make the instruments and then infer the reality of the electrons, as when we test a hypothesis, and then believe it because it passed the test. That gets the time-order wrong. By now we design apparatus relying on a modest number of home truths about electrons to produce some other phenomenon that we wish to investigate.

That may sound as if we believe in the electrons because we predict how our apparatus will behave. That too is misleading. We have a number of general ideas about how to prepare polarized electrons, say. We spend a lot of time building prototypes that don't work. We get rid of innumerable bugs. Often we have to give up and try another approach. Debugging is not a matter of theoretically explaining or predicting what is going wrong. It is partly a matter of getting rid of "noise" in the apparatus. "Noise" often means all the events that are not understood by any theory. The instrument must be able to isolate, physically, the properties of the entities that we wish to use, and damp down all the other effects that might get in our way. *We are completely convinced of the reality of electrons when we regularly set out to build—and often enough succeed in building—new kinds of devices that use various well understood causal properties of electrons to interfere in other more hypothetical parts of nature.*

It is not possible to grasp this without an example. Familiar historical examples have usually become encrusted by false theory-oriented philosophy or history. So I shall take something new. This is a polarizing electron gun whose acronym is PEGGY II. In 1978 it was used in a fundamental experiment that attracted attention even in *The New York Times*. . . .

PEGGY II

The basic idea began when C. Y. Prescott noticed, (by "chance"!) an article in an optics magazine about a crystalline substance called Gallium Arsenide. GaAs has a number of curious properties that make it important in laser technology. One of its quirks is that when it is struck by circularly polarized light of the right frequencies, it emits a lot of linearly polarized electrons. There is a good rough and ready quantum understanding of why this happens, and why half the emitted electrons will be polarized, $3/4$ polarized in one direction and $1/4$ polarized in the other.

PEGGY II uses this fact, plus the fact that GaAs emits lots of electrons due to features of its crystal structure. Then comes some engineering. It takes work to liberate an electron from a surface. We know that painting a surface with the right substance helps. In this case, a thin layer of Cesium and Oxygen is applied to the crystal. Moreover the less air pressure around the crystal, the more electrons will escape for a given amount of work. So the bombardment takes place in a good vacuum at the temperature of liquid nitrogen.

We need the right source of light. A laser with bursts of red light (7100 Ångstroms) is trained on the crystal. The light first goes through an ordinary polarizer, a very old-fashioned prism of calcite, or Iceland spar.[8] This gives longitudinally polarized light. We want circularly polarized light to hit the crystal. The polarized laser beam now goes through a cunning modern device, called a Pockel's cell. It electrically turns linearly polarized photons into circularly polarized ones. Being electric, it acts as a very fast switch. The direction of circular polarization depends on the direction of current in the cell. Hence the direction of polarization can be varied randomly. This is important, for we are trying to de-

tect a minute asymmetry between right- and left-handed polarization. Randomizing helps us guard against any systematic "drift" in the equipment.[9] The randomization is generated by a radioactive decay device, and a computer records the direction of polarization for each pulse.

A circularly polarized pulse hits the GaAs crystal, resulting in a pulse of linearly polarized electrons. A beam of such pulses is maneuvered by magnets into the accelerator for the next bit of the experiment. It passes though a device that checks on a proportion of polarization along the way. The remainder of the experiment requires other devices and detectors of comparable ingenuity, but let us stop at PEGGY II.

BUGS

Short descriptions make it all sound too easy, so let us pause to reflect on debugging. Many of the bugs are never understood. They are eliminated by trial and error. Let us illustrate three different kinds: (1) The essential technical imitations that in the end have to be factored into the analysis of error. (2) Simpler mechanical defects you never think of until they are forced on you. (3) Hunches about what might go wrong.

1. Laser beams are not as constant as science fiction teaches, and there is always an irremediable amount of "jitter" in the beam over any stretch of time.

2. At a more humdrum level the electrons from the GaAs crystal are back-scattered and go back along the same channel as the laser beam used to hit the crystal. Most of them are then deflected magnetically. But some get reflected from the laser apparatus and get back into the system. So you have to eliminate these new ambient electrons. This is done by crude mechanical means, making them focus just off the crystal and so wander away.

3. Good experimenters guard against the absurd. Suppose that dust particles on an experimental surface lie down flat when a polarized pulse hits it, and then stand on their heads when hit by a pulse polarized in the opposite direction. Might that have a systematic effect, given that we are detecting a minute asymmetry? One of the team thought of this in the middle of the night and came down next

morning frantically using antidust spray. They kept that up for a month, just in case.[10]

RESULTS

Some 10^{11} events were needed to obtain a result that could be recognized above systematic and statistical error. Although the idea of systematic error presents interesting conceptual problems, it seems to be unknown to philosophers. There were systematic uncertainties in the detection of right- and left-handed polarization, there was some jitter, and there were other problems about the parameters of the two kinds of beam. These errors were analyzed and linearly added to the statistical error. To a student of statistical inference, this is real seat-of-the-pants analysis with no rationale whatsoever. Be that as it may, thanks to PEGGY II the number of events was big enough to give a result that convinced the entire physics community.[11] Left-handed polarized electrons were scattered from deuterium slightly more frequently than right-handed electrons. This was the first convincing example of parity-violation in a weak neutral current interaction.

COMMENT

The making of PEGGY II was fairly non-theoretical. Nobody worked out in advance the polarizing properties of GaAs—that was found by a chance encounter with an unrelated experimental investigation. Although elementary quantum theory of crystals explains the polarization effect, it does not explain the properties of the actual crystal used. No one has been able to get a real crystal to polarize more than 37 percent of the electrons, although in principle 50 percent should be polarized.

Likewise, although we have a general picture of why layers of cesium and oxygen will "produce negative electron affinity," i.e., make it easier for electrons to escape, we have no quantitative understanding of why this increases efficiency to a score of 37 percent.

Nor was there any guarantee that the bits and pieces would fit together. To give an even more current illustration, future experimental work, briefly described later in this paper, makes us want even more electrons per pulse than PEGGY II could give. When the parity experiment was reported in *The New York Times,* a group at Bell Laboratories read the newspaper and saw what was going on. They had been constructing a crystal lattice for totally unrelated purposes. It uses layers of GaAs and a related aluminum compound. The structure of this lattice leads one to expect that virtually all the electrons emitted would be polarized. So we might be able to doubt the efficiency of PEGGY II. But . . . that nice idea has problems. The new lattice should also be coated in work-reducing paint. But the cesium oxygen stuff is applied at high temperature. Then the aluminum tends to ooze into the neighboring layer of GaAs, and the pretty artificial lattice becomes a bit uneven, limiting its fine polarized-electron-emitting properties. So perhaps this will never work.[12] The group are simultaneously reviving a souped-up new thermionic cathode to try to get more electrons. Maybe PEGGY II would have shared the same fate, never working, and thermionic devices would have stolen the show.

Note, incidentally, that the Bell people did not need to know a lot of weak neutral current theory to send along their sample lattice. They just read *The New York Times.*

MORAL

Once upon a time it made good sense to doubt that there are electrons. Even after Millikan had measured the charge on the electron, doubt made sense. Perhaps Millikan was engaging in "inference to the best explanation." The charges on his carefully selected oil drops were all small integral multiples of a least charge. He inferred that this is the real least charge in nature, and hence it is the charge on the electron, and hence there are electrons, particles of least charge. In Millikan's day most (but not all) physicists did become increasingly convinced by one or more theories about the electron. However it is always admissible, at least for philosophers, to treat inferences to the best explanation in a purely instrumental way, without any commitment to the existence of entities used in the explanation.[13] But it is now [more than eighty] years after Millikan, and we no longer have to infer from explanatory success. Prescott et al. don't explain phenomena with electrons. They know a great deal about how to use them.

The group of experimenters do not know what electrons are, exactly. Inevitably they think in terms of particles. There is also a cloud picture of an electron which helps us think of complex wavefunctions of electrons in a bound state. The angular momentum and spin vector of a cloud make little sense outside a mathematical formalism. A beam of polarized clouds is fantasy so no experimenter uses that model—not because of doubting its truth, but because other models help more with the calculations. Nobody thinks that electrons "really" are just little spinning orbs about which you could, with a small enough hand, wrap the fingers and find the direction of spin along the thumb. There is instead a family of causal properties in terms of which gifted experimenters describe and deploy electrons in order to investigate something else, e.g., weak neutral currents and neutral bosons. We know an enormous amount about the behavior of electrons. We also know what does not matter to electrons. Thus we know that bending a polarized electron beam in magnetic coils does not affect polarization in any significant way. We have hunches, too strong to ignore although too trivial to test independently: e.g., dust might dance under changes of directions of polarization. Those hunches are based on a hard-won sense of the kinds of things electrons are. It does not matter at all to this hunch whether electrons are clouds or particles.

The experimentalist does not believe in electrons because, in the words retrieved from medieval science by Duhem, they "save the phenomena." On the contrary, we believe in them because we use them to *create* new phenomena, such as the phenomenon of parity violation in weak neutral current interactions.

WHEN HYPOTHETICAL ENTITIES BECOME REAL

Note the complete contrast between electrons and neutral bosons. Nobody can yet manipulate a bunch of neutral bosons, if there are any. Even weak neutral currents are only just emerging from the mists of hypothesis. By 1980 a sufficient range of convincing experiments had made them the object of investigation. When might they lose their hypothetical status and become commonplace reality like electrons? When we use them to investigate something else.

I mentioned the desire to make a better gun than PEGGY II. Why? Because we now "know" that parity is violated in weak neutral interactions. Perhaps by an even more grotesque statistical analysis than that involved in the parity experiment, we can isolate just the weak interactions. That is, we have a lot of interactions, including, say, electromagnetic ones. We can censor these in various ways, but we can also statistically pick out a class of weak interactions as precisely those where parity is not conserved. This would possibly give us a road to quite deep investigations of matter and anti-matter. To do the statistics one needs even more electrons per pulse than PEGGY II could hope to generate. If such a project were to succeed, we should be beginning to use weak neutral currents as a manipulable tool for looking at something else. The next step towards a realism about such currents would have been made.

The message is general and could be extracted from almost any branch of physics. Dudley Shapere has recently used "observation" of the sun's hot core to illustrate how physicists employ the concept of observation. They collect neutrinos from the sun in an enormous disused underground mine that has been filled with the old cleaning fluid (i.e., Carbon Tetrachloride). We would know a lot about the inside of the sun if we knew how many solar neutrinos arrive on the earth. So these are captured in the cleaning fluid; a few will form a new radioactive nucleus. The number that do this can be counted. Although the extent of neutrino manipulation is much less than electron manipulation in the PEGGY II experiment, here we are plainly using neutrinos to investigate something else. Yet not many years ago, neutrinos were about as hypothetical as an entity could get. After 1946 it was realized that when mesons disintegrate, giving off, among other things, highly energized electrons, one needed an extra nonionizing particle to conserve momentum and energy. At that time this postulated "neutrino" was thoroughly hypothetical, but now it is routinely used to examine other things.

CHANGING TIMES

Although realisms and anti-realisms are part of the philosophy of science well back into Greek prehistory, our present versions mostly descend from debates about atomism at the end of the nineteenth century. Anti-realism about atoms was partly a matter of physics: the energeticists thought energy was at the bottom of everything, not tiny bits of matter. It also was connected with the positivism of Comte, Mach, Pearson and even J. S. Mill. Mill's young associate Alexander Bain states the point in a characteristic way, apt for 1870:

> Some hypotheses consist of assumptions as to the minute structure and operations of bodies. From the nature of the case these assumptions can never be proved by direct means. Their merit is their suitability to express phenomena. They are Representative Fictions.[14]

"All assertions as to the ultimate structure of the particles of matter," continues Bain, "are and ever must be hypothetical." . . . "The kinetic theory of heat," he says, "serves an important intellectual function." But we cannot hold it to be a true description of the world. It is a Representative Fiction.

Bain was surely right a century ago. Assumptions about the minute structure of matter could not be proved then. The only proof could be indirect, namely that hypotheses seemed to provide some explanation and helped make good predictions. Such inferences need never produce conviction in the philosopher inclined to instrumentalism or some other brand of idealism.

Indeed the situation is quite similar to seventeenth-century epistemology. At that time knowledge was thought of as correct representation. But then one could never get outside the representations to be sure that they corresponded to the world. Every test of a representation is just another representation. "Nothing is so much like an idea as an idea," as Bishop Berkeley had it. To attempt to argue for scientific realism at the level of theory, testing, explanation, predictive success, convergence of theories and so forth is to be locked into a world of representations. No wonder that scientific anti-realism is so permanently in the race. It is a variant on "The Spectator Theory of Knowledge."

Scientists, as opposed to philosophers, did in general become realists about atoms by 1910. Michael Gardner, in one of the finest studies of real-life scientific realism, details many of the factors that went into that change in climate of opinion.[15] Despite the changing climate, some variety of instrumentalism or fictionalism remained a strong philosophical alternative in 1910 and in 1930. That is what the history of philosophy teaches us. Its most recent lesson is Bas van Fraassen's *The Scientific Image,* whose "constructive empiricism" is another theory-oriented anti-realism. The lesson is: think about practice, not theory.

Anti-realism about atoms was very sensible when Bain wrote a century ago. Anti-realism about *any* sub-microscopic entities was a sound doctrine in those days. Things are different now. The "direct" proof of electrons and the like is our ability to manipulate them using well understood low-level causal properties. I do not of course claim that "reality" is constituted by human manipulability. We can, however, call something real, in the sense in which it matters to scientific realism, only when we understand quite well what its causal properties are. The best evidence for this kind of understanding is that we can set out, from scratch, to build machines that will work fairly reliably, taking advantage of this or that causal nexus. Hence, engineering, not theorizing, is the proof of scientific realism about entities.[16]

NOTES

1. C. W. F. Everitt and Ian Hacking, "Which Comes First, Theory or Experiment?"
2. Ian Hacking, "Spekulation, Berechnung und die Erschaffnung der Phänomenen," in *Versuchungen: Aufsätze zur Philosophie Paul Feyerabends,* (P. Duerr, ed.), Frankfurt, 1981, Bd 2, 126–58.
3. Nancy Cartwright makes a similar distinction in a sequence of papers, including "When Explanation Leads to Inference," in the present issue. She approaches realism from the top, distinguishing theoretical laws (which do not state the facts) from phenomenological laws (which do). She believes in some "theoretical" entities and rejects much theory on the basis of a subtle analysis of modeling in physics. I proceed in the opposite direction, from experimental practice. Both approaches share an interest in real-life physics as opposed to philosophical

fantasy science. My own approach owes an enormous amount to Cartwright's parallel developments, which have often preceded my own. My use of the two kinds of realism is a case in point.

4. Hilary Putnam, "How Not to Talk About Meaning," "The Meaning of 'Meaning'," and other papers in the *Mind, Language and Reality, Philosophical Papers,* Vol. 2. Cambridge, 1975.

5. These terms occur in, e.g., Hilary Putnam, *Meaning and the Moral Sciences,* London, 1978, 123–30.

6. Francis Bacon, *The Great Instauration,* in *The Philosophical Works of Francis Bacon* (J. M. Robertson, ed; Ellis and Spedding, Trans.), London, 1905, p. 252.

7. Ian Hacking, "Do We See Through a Microscope?" *Pacific Philosophical Quarterly,* winter 1981.

8. Iceland spar is an elegant example of how experimental phenomena persist even while theories about them undergo revolutions. Mariners brought calcite from Iceland to Scandinavia. Erasmus Batholinus experimented with it and wrote about it in 1609. When you look through these beautiful crystals you see double, thanks to the so-called ordinary and extraordinary rays. Calcite is a natural polarizer. It was our entry to polarized light which for 300 years was the chief route to improved theoretical and experimental understanding of light and then electromagnetism. The use of calcite in PEGGY II is a happy reminder of a great tradition.

9. It also turns GaAs, a $3/4$-$1/4$ left-/right-hand polarizer, into a 50-50 polarizer.

10. I owe these examples to conversation with Roger Miller of SLAC (Stanford Linear Accelerator Center).

11. The concept of a "convincing experiment" is fundamental. Peter Gallison has done important work on this idea, studying European and American experiments on weak neutral currents conducted during the 1970s.

12. I owe this information to Charles Sinclair of SLAC.

13. My attitude to "inference to the best explanation" is one of many learned from Cartwright. See, for example, her paper on this topic in this issue.

14. Alexander Bain, *Logic, Deductive and Inductive,* London and New York, 1870, p. 362.

15. Michael Gardner, "Realism and Instrumentalism in 19th-Century Atomism," *Philosophy of Science* 46, (1979), 1–34.

16. (Added in proof, February, 1983). As indicated in the text, this is a paper of July, 1981, and hence is out of date. For example, neutral bosons are described as purely hypothetical. Their status has changed since CERN announced on Jan. 23, 1983, that a group there had found W, the weak intermediary boson, in proton-antiproton decay at 540 GeV. These experimental issues are further discussed in my book, *Representing and Intervening* (Cambridge, 1983).

37

ARTHUR FINE

The Natural Ontological Attitude

Let us fix our attention out of ourselves as much as possible; let us chace our imagination to the heavens, or to the utmost limits of the universe; we never really advance a step beyond ourselves, nor can conceive any kind of existence, but those perceptions, which have appear'd in that narrow compass. This is the universe of the imagination, nor have we any idea but what is there produced.

—HUME, *TREATISE*, BOOK 1, PART II, SECTION VI

Realism is dead. Its death was announced by the neopositivists who realized that they could accept all the results of science, including all the members of the scientific zoo, and still declare that the questions raised by the existence claims of realism were mere pseudoquestions. Its death was hastened by the debates over the interpretation of quantum theory, where Bohr's nonrealist philosophy was seen to win out over Einstein's passionate realism. Its death was certified, finally, as the last two generations of physical scientists turned their backs on realism and have managed, nevertheless, to do science successfully without it. To be sure, some recent philosophical literature, and some of the best of it represented by contributors to this book, has appeared to pump up the ghostly shell and to give it new life. But I think these efforts will eventually be seen and understood as the first stage in the process of mourning, the stage of denial. This volume contains some further expressions of this denial. But I think we shall pass through this first stage and into that of acceptance, for realism is well and truly dead, and we have work to get on with, in identifying a suitable successor. To aid that work I want to do three things in this essay. First, I want to show that the arguments in favor of realism are not sound, and that they provide no rational support for belief in realism. Then, I want to recount the essential role of nonrealist attitudes for the development of science in this century, and thereby (I hope) to loosen the grip of the idea that only realism provides a progressive philosophy of science. Finally, I want to sketch out what seems to me a viable nonrealist position, one that is slowly gathering support and that seems a decent philosophy for postrealist times.[1]

ARGUMENTS FOR REALISM

Recent philosophical argument in support of realism tries to move from the success of the scientific enterprise to the necessity for a realist account of its practice. As I see it, the arguments here fall on two distinct levels. On the ground level, as it were, one attends to particular successes; such as novel, confirmed predictions, striking unifications of disparate-seeming phenomena (or fields), successful piggybacking from one theoretical model to another, and the like. Then, we are challenged to account for such success, and told that the best and, it is slyly suggested, perhaps, the *only* way of doing so is on a realist basis. I do not find the details of these ground-level arguments at all convincing. Larry Laudan has provided a forceful and detailed analysis which shows that not even with a lot of hand waving (to shield the gaps in the argument) and

Scientific Realism, J. Leplin, ed. (Berkeley: University of California Press, 1984), pp. 83–107. Copyright © 1984 The Regents of the University of California. Reprinted with permission of the publisher.

charity (to excuse them) can realism itself be used to explain the very successes to which it invites our attention.[2] But there is a second level of realist argument, the methodological level, that derives from Popper's attack on instrumentalism as inadequate to account for the details of his own, falsificationist methodology. Arguments on this methodological level have been skillfully developed by Richard Boyd,[3] and by one of the earlier Hilary Putnams.[4] These arguments focus on the methods embedded in scientific practice, methods teased out in ways that seem to me accurate and perceptive about ongoing science. We are then challenged to account for why these methods lead to scientific success and told that the best, and (again) perhaps, the only truly adequate way of explaining the matter is on the basis of realism.

I want to point out a deep and, I think, insurmountable problem with this entire strategy of defending realism, as I have laid it out above. . . .

Those suspicious of realism, from A. Osiander to H. Poincaré and P. Duhem to the 'constructive empiricism' of van Fraassen,[5] have been worried about the significance of the explanatory apparatus in scientific investigations. While they appreciate the systematization and coherence brought about by scientific explanation, they question whether acceptable explanations need to be true and, hence, whether the entities mentioned in explanatory principles need to exist.[6] Suppose they are right. Suppose, that is, that the usual explanation-inferring devices in scientific practice do not lead to principles that are reliably true (or nearly so), nor to entities whose existence (or near-existence) is reliable. In that case, the usual abductive methods that lead us to good explanations (even to 'the best explanation') cannot be counted on to yield results even approximately true. But the strategy that leads to realism, as I have indicated, is just such an ordinary sort of abductive inference. Hence, if the non-realist were correct in his doubts, then such an inference to realism as the best explanation (or the like), while possible, would be of no significance—exactly as in the case of a consistency proof using the methods of an inconsistent system. It seems, then, that Hilbert's maxim applies to the debate over realism: to argue for realism one must employ methods more stringent than those in ordinary scientific practice. In particular, one must not beg the question as to the sig-

nificance of explanatory hypotheses by assuming that they carry truth as well as explanatory efficacy.

There is a second way of seeing the same result. Notice that the issue over realism is precisely the issue as to whether we should believe in the reality of those individuals, properties, relations, processes, and so forth, used in well-supported explanatory hypotheses. Now what *is* the hypothesis of realism, as it arises as an explanation of scientific practice? It is just the hypothesis that our accepted scientific theories are approximately true, where "being approximately true" is taken to denote an extratheoretical relation between theories and the world. Thus, to address doubts over the reality of relations posited by explanatory hypotheses, the realist proceeds to introduce a further explanatory hypothesis (realism), itself positing such a relation (approximate truth). Surely anyone serious about the issue of realism, and with an open mind about it, would have to behave inconsistently if he were to accept the realist move as satisfactory.

Thus, both at the ground level and at the level of methodology, no support accrues to realism by showing that realism is a good hypothesis for explaining scientific practice. If we are open-minded about realism to begin with, then such a demonstration (even if successful) merely begs the question that we have left open ("need we take good explanatory hypotheses as true?"). Thus, Hilbert's maxim applies, and we must employ patterns of argument more stringent than the usual abductive ones. What might they be? Well, the obvious candidates are patterns of induction leading to empirical generalizations. But, to frame empirical generalizations, we must first have some observable connections between observables. For realism, this must connect theories with the world by way of approximate truth. But no such connections are observable and, hence, suitable as the basis for an inductive inference. I do not want to labor the points at issue here. They amount to the well-known idea that realism commits one to an unverifiable correspondence with the world. So far as I am aware, no recent defender of realism has tried to make a case based on a Hilbert strategy of using suitably stringent grounds and, given the problems over correspondence, it is probably just as well.

The strategy of arguments to realism as a good explanatory hypothesis, then, *cannot* (logically speaking) be effective for an open-minded nonbeliever. But what of the believer? Might he not, at least, show a kind of internal coherence about realism as an overriding philosophy of science, and should that not be of some solace, at least for the realist?[7] Recall, however, the analogue with consistency proofs for inconsistent systems. That sort of harmony should be of no solace to anyone. But for realism, I fear, the verdict is even harsher. For, so far as I can see, the arguments in question just do not work, and the reason for that has to do with the same question-begging procedures that I have already identified. . . .

REALISM AND PROGRESS

If we examine the two twentieth-century giants among physical theories, relativity and the quantum theory, we find a living refutation of the realist's claim that only his view of science explains its progress, and we find some curious twists and contrasts over realism as well. The theories of relativity are almost singlehandedly the work of Albert Einstein. Einstein's early positivism and his methodological debt to Mach (and Hume) leap right out of the pages of the 1905 paper on special relativity.[8] The same positivist strain is evident in the 1916 general relativity paper as well, where Einstein (in Section 3 of that paper) tries to justify his requirement of general covariance by means of a suspicious-looking verificationist argument which, he says, "takes away from space and time the last remnants of physical objectivity."[9] A study of his tortured path to general relativity (see here the brilliant work of John Earman, following on earlier hints by Banesh Hoffmann)[10] shows the repeated use of this Machist line, always used to deny that some concept has a real referent. Whatever other, competing strains there were in Einstein's philosophical orientation (and there certainly were others), it would be hard to deny the importance of this instrumentalist/positivist attitude in liberating Einstein from various realist commitments. Indeed, on another occasion, I would argue in detail that without the "freedom from reality" provided by his early reverence for Mach, a central tumbler necessary to unlock the secret of special relativity would never have fallen into place.[11] A few years after his work on general relativity, however, roughly around 1920, Einstein underwent a philosophical conversion, turning away from his positivist youth (he was forty-one in 1920) and becoming deeply committed to realism.[12] His subsequent battle with the quantum theory, for example, was fought much more over the issue of realism than it was over the issue of causality or determinism (as it is usually portrayed). In particular, following his conversion, Einstein wanted to claim genuine reality for the central theoretical entities of the general theory, the four-dimensional space-time manifold and associated tensor fields. This is a serious business, for if we grant his claim, then not only do space and time cease to be real but so do virtually all of the usual dynamical quantities.[13] Thus motion, as we understand it, itself ceases to be real. The current generation of philosophers of space and time (led by Howard Stein and John Earman) have followed Einstein's lead here. But, interestingly, not only do these ideas boggle the mind of the average man in the street (like you and me), they boggle most contemporary scientific minds as well.[14] That is, I believe the majority opinion among working, knowledgeable scientists is that general relativity provides a magnificent organizing tool for treating certain gravitational problems in astrophysics and cosmology. But few, I believe, give credence to the kind of realist existence and nonexistence claims that I have been mentioning. For relativistic physics, then, it appears that a nonrealist attitude was important in its development, that the founder nevertheless espoused a realist attitude to the finished product, but that most who actually use it think of the theory as a powerful instrument, rather than as expressing a "big truth."

With quantum theory, this sequence gets a twist. Heisenberg's seminal paper of 1925 is prefaced by the following abstract, announcing, in effect, his philosophical stance: "In this paper an attempt will be made to obtain bases for a quantum-theoretical mechanics based exclusively on relations between quantities observable in principle.[15] In the body of the paper, Heisenberg not only rejects any reference to unobservables; he also moves away from the very idea that one should try to form any picture of a reality underlying his mechanics. To be sure, E. Schrödinger, the second father of quantum theory, seems originally to have had a vague picture of an underlying wavelike reality for his own equation.

But he was quick to see the difficulties here and, just as quickly, although reluctantly, abandoned the attempt to interpolate any reference to reality.[16] These instrumentalist moves, away from a realist construal of the emerging quantum theory, were given particular force by Bohr's so-called "philosophy of complementarity"; and this nonrealist position was consolidated at the time of the famous Solvay conference, in October of 1927, and is firmly in place today. Such quantum nonrealism is part of what every graduate physicist learns and practices. It is the conceptual backdrop to all the brilliant successes in atomic, nuclear, and particle physics over the past fifty years. Physicists have learned to think about their theory in a highly nonrealist way, and doing just that has brought about the most marvelous predictive success in the history of science.

The war between Einstein, the realist, and Bohr, the nonrealist, over the interpretation of quantum theory was not, I believe, just a sideshow in physics, nor an idle intellectual exercise. It was an important endeavor undertaken by Bohr on behalf of the enterprise of physics as a progressive science. For Bohr believed (and this fear was shared by Heisenberg, A. Sommerfield, W. Pauli, and M. Born—and all the major players) that Einstein's realism, if taken seriously, would block the consolidation and articulation of the new physics and, thereby, stop the progress of science. They were afraid, in particular, that Einstein's realism would lead the next generation of the brightest and best students into scientific dead ends. Alfred Landé, for example, as a graduate student, was interested in spending some time in Berlin to sound out Einstein's ideas. His supervisor was Sommerfeld, and recalling this period, Landé writes

The more pragmatic Sommerfeld . . . warned his students, one of them this writer, not to spend too much time on the hopeless task of "explaining" the quantum but rather to accept it as fundamental and help work out its consequences.[17]

The task of "explaining" the quantum, of course, is the realist program for identifying a reality underlying the formulas of the theory and thereby explaining the predicative success of the formulas as approximately true descriptions of this reality. It is this program that I have criticized in the first part of this paper, and this same program that the builders of quantum theory saw as a scientific dead end. Einstein knew perfectly well that the issue was joined right here. In the summer of 1935, he wrote to Schrödinger,

The real problem is that physics is a kind of metaphysics; physics describes 'reality'. But we do not know what 'reality' is. We know it only through physical description. . . . But the Talmudic philosopher sniffs at 'reality', as at a frightening creature of the naive mind.[18]

By avoiding the bogey of an underlying reality, the "Talmudic" originators of quantum theory seem to have set subsequent generations on precisely the right path. Those inspired by realist ambitions have produced no predictively successful physics. Neither Einstein's conception of a unified field nor the ideas of the de Broglie group about pilot waves, nor the Bohm-inspired interest in hidden variables has made for scientific progress. To be sure, several philosophers of physics, including another Hilary Putnam, and myself, have fought a battle over the last decade to show that the quantum theory is at least consistent with some kind of underlying reality. I believe that Hilary has abandoned the cause, perhaps in part on account of the recent Bell-inequality problem over correlation experiments, a problem that van Fraassen calls "the charybdis of realism."[19] My own recent work in the area suggests that we may still be able to keep realism afloat in this whirlpool.[20] But the possibility (as I still see it) for a realist account of the quantum domain should not lead us away from appreciating the historical facts of the matter.

One can hardly doubt the importance of a nonrealist attitude for the development and practically infinite success of the quantum theory. Historical counterfactuals are always tricky, but the sterility of actual realist programs in this area at least suggests that Bohr and company were right in believing that the road to scientific progress here would have been blocked by realism. The founders of quantum theory never turned on the nonrealist attitude that served them so well. Perhaps that is because the central underlying theoretical device of quantum theory, the densities of a complex-valued and infinite-dimensional wave function, are even harder to take seriously than is the four-dimensional manifold of relativity. But now, there comes a most curious twist. For just as the practitioners of relativity, I have suggested, ignore the *realist* interpretation in favor of a more pragmatic attitude toward the space-time structure, the quantum physicists would appear to make a similar reversal and to forget their nonrealist history and allegiance when it comes time to talk about new discoveries.

Thus, anyone in the business will tell you about the exciting period, in the fall of 1974, when the particle group at Brookhaven, led by Samuel Ting, discovered the J particle, just as a Stanford team at the Stanford Linear Accelerator Center (SLAC), under Burton Richter, independently found a new particle they called "ψ". These turned out to be one and the same, the so-called ψ/J particle (Mass 3,098 MeV, Spin 1, Resonance 67 KeV, Strangeness 0). To explain this new entity, the theoreticians were led to introduce a new kind of quark, the so-called charmed quark. The ψ/J particle is then thought to be made up out of a charmed quark and an anticharmed quark, with their respective spins aligned. But if this is correct, then there ought to be other such pairs anti-aligned, or with variable spin alignments, and these ought to make up quite new observable particles. Such predictions from the charmed-quark model have turned out to be confirmed in various experiments.

In this example, I have been intentionally a bit more descriptive in order to convey the realist feel to the way scientists speak in this area. For I want to ask whether this is a return to realism or whether, instead, it can somehow be reconciled with a fundamentally nonrealist attitude.[21] I believe that the nonrealist option is correct, but I will not defend that answer here, however, because its defense involves the articulation of a compelling and viable form of nonrealism; and that is the task of the third (and final) section of this paper.

NONREALISM

Even if the realist happens to be a talented philosopher, I do not believe that, in his heart, he relies for his realism on the rather sophisticated form of abductive argument that I have examined and rejected in the first section of this paper, and which the history of twentieth-century physics shows to be falla-cious. Rather, if his heart is like mine (and I *do* believe in a common nature), then I suggest that a more simple and homely sort of argument is what grips him. It is this, and I will put it in the first person. I certainly trust the evidence of my senses, on the whole, with regard to the existence and features of everyday objects. And I have similar confidence in the system of "check, double-check, triple-check" of scientific investigation, as well as the other safeguards built into the institutions of science. So, if the scientists tell me that there really are molecules, and atoms, and ψ/J particles and, who knows, maybe even quarks, then so be it. I trust them and, thus, must accept that there really are such things, with their attendant properties and relations. Moreover, if the instrumentalist (or some other member of the species "non-realistica") comes along to say that these entities, and their attendants, are just fictions (or the like), then I see no more reason to believe him than to believe that *he is* a fiction, made up (somehow) to do a job on me; which I do not believe. It seems, then, that I had better be a realist. One can summarize this homely and compelling line as follows: it is possible to accept the evidence of one's senses and to accept, *in the same way*, the confirmed results of science only for a realist; hence, I should be one (and so should you!).

What is it to accept the evidence of one's senses and, *in the same way*, to accept confirmed scientific theories? It is to take them into one's life as true, with all that implies concerning adjusting one's behavior, practical and theoretical, to accommodate these truths. Now, of course, there are truths, and truths. Some are more central to us and our lives, some less so. I might be mistaken about anything, but were I mistaken about where I am right now, that might affect me more than would my perhaps mistaken belief in charmed quarks. Thus, it is compatible with the homely line of argument that some of the scientific beliefs that I hold are less central than some, for example, perceptual beliefs. Of course, were I deeply in the charmed-quark business, giving up that belief might be more difficult than giving up some at the perceptual level. (Thus we get the phenomenon of "seeing what you believe," as is well known to all thoughtful people.) When the homely line asks us, then, to accept the scientific results "in the same way" in which we accept the evidence of our senses, I take it that we are to accept them both as true. I take it that we are being asked not to distinguish between kinds of truth or modes of existence or the like, but only among truths themselves, in terms of centrality, degrees of belief, or such.

Let us suppose this understood. Now, do you think that Bohr, the archenemy of realism, could toe the homely line? Could Bohr, fighting for the sake of science (against Einstein's realism) have felt compelled either to give up the results of science, or else to assign to its "truths" some category different from the truths of everyday life? It seems unlikely. And thus, unless we uncharitably think Bohr inconsistent on this basic issue, we might well come to question whether there is any necessary connection moving us from accepting the results of science as true to being a realist.[22]

Let me use the term 'antirealist' to refer to any of the many different specific enemies of realism: the idealist, the instrumentalist, the phenomenalist, the empiricist (constructive or not), the conventionalist, the constructivist, the pragmatist, and so forth. Then, it seems to me that both the realist and the antirealist must toe what I have been calling "the homely line." That is, they must both accept the certified results of science as on par with more homely and familiarly supported claims. That is not to say that one party (or the other) cannot distinguish more from less well-confirmed claims at home or in science; nor that one cannot single out some particular mode of inference (such as inference to the best explanation) and worry over its reliability, both at home and away. It is just that one must maintain parity. Let us say, then, that both realist and antirealist accept the results of scientific investigations as 'true', on par with more homely truths. (I realize that some antirealists would rather use a different word, but no matter.) And call this acceptance or scientific truths the "core position."[23] What distinguishes realists from antirealists, then, is what they add onto this core position.

The antirealist may add onto the core position a particular analysis of the concept of truth, as in the pragmatic and instrumentalist and conventionalist conceptions of truth. Or the antirealist may add on a special analysis of concepts, as in idealism, constructivism, phenomenalism, and in some varieties of empiricism. These addenda will then issue in a special meaning, say, for existence statements. Or the antirealist may add on certain methodological strictures, pointing a wary finger at some particular inferential tool, or constructing his own account for some particular aspects of science (e.g., explanations or laws). Typically, the antirealist will make several such additions to the core.

What then of the realist, what does he add to his core acceptance of the results of science as really true? My colleague, Charles Chastain, suggested what I think is the most graphic way of stating the answer—namely, that what the realist adds on is a desk-thumping, foot-stamping shout of "Really!" So, when the realist and antirealist agree, say, that there really are electrons and that they really carry a unit negative charge and really do have a small mass (of about 9.1×10^{-28} grams), what the realist wants to add is the emphasis that all this is really so. "There really are electrons, really!" This typical realist emphasis serves both a negative and a positive function. Negatively, it is meant to deny the additions that the antirealist would make to that core acceptance which both parties share. The realist wants to deny, for example, the phenomenalistic reduction of concepts or the pragmatic conception of truth. The realist thinks that these addenda take away from the substantiality of the accepted claims to truth or existence. "No," says he, "they *really* exist, and not in just your diminished antirealist sense." Positively, the realist wants to explain the robust sense in which *he* takes these claims to truth or existence, namely, as claims about reality—what is really, really the case. The full-blown version of this involves the conception of truth as correspondence with the world, and the surrogate use of approximate truth as near-correspondence. We have already seen how these ideas of correspondence and approximate truth are supposed to explain what *makes* the truth *true* whereas, in fact, they function as mere trappings, that is, as superficial decorations that may well attract our attention but do not compel rational belief. Like the extra "really," they are an arresting foot-thump and, logically speaking, of no more force.

It seems to me that when we contrast the realist and the antirealist in terms of what they each want to add to the core position, a third alternative emerges—and an attractive one at that. It is the core position itself, *and all by itself.* If I am correct in thinking that, at heart, the grip of realism only extends to the homely connection of everyday truths with scientific truths, and that good sense dictates our acceptance of the one on the same basis as our acceptance of the other, than the homely line makes the core position, all by itself, a compelling one, one that we ought to take to heart. Let us try to do so, and to see whether it constitutes a philosophy, and an attitude toward science, that we can live by.

The core position is neither realist nor antirealist; it mediates between the two. It would be nice to have a name for this position, but it would be a shame to appropriate another "ism" on its behalf, for then it would appear to be just one of the many contenders for ontological allegiance. I think it is not just one of that crowd but rather, as the homely line behind it suggests, it is for commonsense epistemology—the natural ontological attitude. Thus, let me introduce the acronym *NOA* (pronounced as in "Noah"), for *natural ontological attitude,* and, henceforth, refer to the core position under that designation.

To begin showing how NOA makes for an adequate philosophical stance toward science, let us see what it has to say about ontology. When NOA counsels us to accept the results of science as true, I take it that we are to treat truth in the usual referential way, so that a sentence (or statement) is true just in case the entities referred to stand in the referred-to relations. Thus, NOA sanctions ordinary referential semantics and commits us, via truth, to the existence of the individuals, properties, relations, processes, and so forth referred to by the scientific statements that we accept as true. Our belief in their existence will be just as strong (or weak) as our belief in the truth of the bit of science involved, and degrees of belief here, presumably, will be tutored by ordinary relations of confirmation and evidential support, subject to the usual scientific canons. In taking this referential stance, NOA is not committed to the progressivism that seems inherent in realism. For the realist, as an article of faith, sees scientific success, over the long run, as bringing us closer to the truth. His whole explanatory enterprise, using approximate truth, forces his hand in this way. But, a "noaer" (pronounced as "knower") is not so committed. As a scientist, say, within the context of the tradition in which he works, the noaer, of course, will believe in the existence of those entities to which his theories refer. But should the tradition change, say in the manner of the conceptual revolutions that Kuhn dubs "paradigm shifts," then nothing in NOA dictates that the change be assimilated as being progressive, that is, as a change where we learn more accurately about *the same things.* NOA is perfectly consistent with the Kuhnian alternative, which construes such changes as wholesale changes of reference. Unlike the realist, adherents to NOA are free to examine the facts in cases of paradigm shift, and to see whether or not a convincing case for stability of reference across paradigms can be made without superimposing on these facts a realist-progressivist superstructure. I have argued elsewhere that if one makes oneself free, as NOA enables one to do, then the facts of the matter will not usually settle the case;[24] and that this is a good reason for thinking that cases of so-called "incommensurability" are, in fact, genuine cases where the question of stability of reference is indeterminate. NOA, I think, is the right philosophical position for such conclusions. It sanctions reference and existence claims, but it does not force the history of science into prefit molds.

So far I have managed to avoid what, for the realist, is the essential point, for what of the "external world"? How can I talk of reference and of existence claims unless I am talking about referring to things right out there in the world? And here, of course, the realist, again, wants to stamp his feet.[25] I think the problem that makes the realist want to stamp his feet, shouting "Really!" (and invoking the external world) has to do with the stance the realist tries to take vis-à-vis the game of science. The realist, as it were, tries to stand outside the arena watching the ongoing game and then tries to judge (from this external point of view) what the point is. It is, he says, *about* some area external to the game. The realist, I think, is fooling himself. For he cannot (really!) stand outside the arena, nor can he survey some area off the playing field and mark it out as what the game is about.

Let me try to address these two points. How are we to arrive at the judgment that, in addition to, say, having a rather small mass, electrons are objects "out there in the external world"? Certainly, we can stand off from the electron game and survey its claims, methods, predictive success, and so forth. But what stance could we take that would enable us to judge what the theory of electrons is *about,* other than agreeing that it is about electrons? It is not like matching a blueprint to a house being built, or a map route to a country road. For we are *in* the world, both physically and conceptually.[26] That is, *we* are among the objects of science, and the concepts and procedures that we use to make judgments of subject matter and correct application are themselves part of that same scientific world. Epistemologically, the situation is very much like the situation with regard to the justification of induction. For the problem of the external world (so-called) is how to satisfy the realist's demand that we justify the existence claims sanctioned by science (and, therefore, by NOA) as claims to the existence of entities "out there." In the case of induction, it is clear that only an inductive justification will do, and it is equally clear that no inductive justification will do at all. So too with the external world, for only ordinary scientific inferences to existence will do, and yet none of them satisfies the demand for showing that the existent is really "out there." I think we ought to follow Hume's prescription on induction, with regard to the external world. There is no possibility for justifying the kind of externality that realism requires, yet it may well be that, in fact, we cannot help yearning for just such a comforting grip on reality. I shall return to this theme at the close of the paper.

If I am right, then the realist is chasing a phantom, and we cannot actually do more, with regard to existence claims, than follow scientific practice, just as NOA suggests. . . .

Indeed, perhaps the greatest virtue of NOA is to call attention to just how minimal an adequate philosophy of science can be. (In this respect, NOA might be compared to the minimalist movement in art.) For example, NOA helps us to see that realism differs from various antirealisms in this way: realism adds an outer direction to NOA, that is, the external world and the correspondence relation of approximate truth; antirealisms (typically) add an inner direction, that is, human-oriented reductions of truth, or concepts, or explanations (as in my opening citation from Hume). NOA suggests that the legitimate features of these additions are already contained in the presumed equal status of everyday truths with scientific ones, and in our accepting them both as

truths. No other additions are legitimate, and none are required.

It will be apparent by now that a distinctive feature of NOA, one that separates it from similar views currently in the air, is NOA's stubborn refusal to amplify the concept of truth, by providing a theory or analysis (or even a metaphorical picture). Rather, NOA recognizes in "truth" a concept already in use and agrees to abide by the standard rules of usage. These rules involve a Davidsonian-Tarskian, referential semantics, and they support a thoroughly classical logic of inference. Thus NOA respects the customary "grammar" of 'truth' (and its cognates). Likewise, NOA respects the customary epistemology, which grounds judgments of truth in perceptual judgments and various confirmation relations. As with the use of other concepts, disagreements are bound to arise over what is true (for instance, as to whether inference to the best explanation is always truth-conferring). NOA pretends to no resources for settling these disputes, for NOA takes to heart the great lesson of twentieth-century analytic and Continental philosophy, namely, that there *are* no general methodological or philosophical resources for deciding such things. The mistake common to realism and all the antirealisms alike is their commitment to the existence of such nonexistent resources. If pressed to answer the question of what, then, does it *mean* to say that something is true (or to what does the truth of so-and-so commit one), NOA will reply by pointing out the logical relations engendered by the specific claim and by focusing, then, on the concrete historical circumstances that ground that particular judgment of truth. For, after all, there *is* nothing more to say.[27]

Because of its parsimony, I think the minimalist stance represented by NOA marks a revolutionary approach to understanding science. It is, I would suggest, as profound in its own way as was the revolution in our conception of morality, when we came to see that founding morality on God and His Order was *also* neither legitimate nor necessary. Just as the typical theological moralist of the eighteenth century would feel bereft to read, say, the pages of *Ethics,* so I think the realist must feel similarly when NOA removes that "correspondence to the external world" for which he so longs. I too have regret for that lost paradise, and too often slip into the realist fantasy. I use my understanding of twentieth-century physics to help me firm up my convictions about NOA, and I recall some words of Mach, which I offer as a comfort and as a closing. With reference to realism, Mach writes

> It has arisen in the process of immeasurable time without the intentional assistance of man. It is a product of nature, and preserved by nature. Everything that philosophy has accomplished . . . is, as compared with it, but an insignificant and ephemeral product of art. The fact is, every thinker, every philosopher, the moment he is forced to abandon his one-sided intellectual occupation . . . , immediately returns [to realism].
>
> Nor is it the purpose of these "introductory remarks" to discredit the standpoint [of realism]. The task which we have set ourselves is simply to show why and for what purpose we hold that standpoint during most of our lives, and why and for what purpose we are . . . obliged to abandon it.

These lines are taken from Mach's *The Analysis of Sensations* (Sec. 14). I recommend that book as effective realism-therapy, a therapy that works best (as Mach suggests) when accompanied by historicophysical investigations (real versions of the breakneck history of my second section, "Realism and Progress"). For a better philosophy, however, I recommend NOA.

NOTES

1. In the final section, I call this postrealism "NOA." Among recent views that relate to NOA, I would include Hilary Putnam's "internal realism," Richard Rorty's "epistemological behaviorism," the "semantic realism" espoused by Paul Horwich, parts of the "Mother Nature" story told by William Lycan, and the defense of common sense worked out by Joseph Pitt (as a way of reconciling W. Sellars's manifest and scientific images). For references, see Hilary Putnam, *Meaning and the Moral Sciences* (London: Routledge and Kegan Paul, 1978); Richard Rorty, *Philosophy and the Mirror of Nature* (Princeton: Princeton University Press, 1979); Paul Horwich,

"Three Forms of Realism," *Synthese* 51 (1982): 181–201; William G. Lycan, "Epistemic Value" (preprint, 1982); and Joseph C. Pitt, *Pictures, Images and Conceptual Change* (Dordrecht: D. Reidel, 1981). The reader will note that some of the above consider their views a species of realism, whereas others consider their views antirealist. As explained below, NOA marks the divide; hence its "postrealism."

2. Larry Laudan, "A Confutation of Convergent Realism," *Philosophy of Science* 48 (1981): 19–49.

3. Richard N. Boyd, "Scientific Realism and Naturalistic Epistemology," in *PSA* (1980), vol. 2, ed. P. D. Asquith and R. N. Giere (E. Lansing: Philosophy of Science Association, 1981), 613–662.

4. Hilary Putnam, "The Meaning of 'Meaning'," in *Language, Mind and Knowledge,* ed. K. Gunderson (Minneapolis: University of Minnesota Press, 1975), 131–193.

5. Bas C. van Fraassen, *The Scientific Image* (Oxford: The Clarendon Press, 1980). See especially pp. 97–101 for a discussion of the truth of explanatory theories. To see that the recent discussion of realism is joined right here, one should contrast van Fraassen with W. H. Newton-Smith, *The Rationality of Science* (London: Routledge and Kegan Paul, 1981), esp. chap. 8.

6. Nancy Cartwright's *How the Laws of Physics Lie* (Oxford: Oxford University Press, 1983) includes some marvelous essays on these issues.

7. Some realists may look for genuine support, and not just solace, in such a coherentist line. They may see in their realism a basis for general epistemology, philosophy of language, and so forth (as does Boyd, "Scientific Realism and Naturalistic Epistemology"). If they find in all this a coherent and comprehensive world view, then they might want to argue for their philosophy as Wilhelm Wien argued (in 1909) for special relativity, "What speaks for it most of all is the inner consistency which makes it possible to lay a foundation having no self-contradictions, one that applies to the totality of physical appearances." Quoted by Gerald Holton, "Einstein's Scientific Program: Formative Years" in *Some Strangeness in the Proportion,* ed. H. Woolf (Reading: Addison-Wesley, 1980), 58. Insofar as the realist moves away from the abductive defense of realism to seek support, instead, from the merits of a comprehensive philosophical system with a realist core, he marks as a failure the bulk of recent defenses of realism. Even so, he will not avoid the critique pursued in the text. For although my argument above has been directed, in particular, against the abductive strategy, it is

itself based on a more general maxim, namely, that the form of argument used to support realism must be more stringent than the form of argument embedded in the very scientific practice that realism itself is supposed to ground—on pain of begging the question. Just as the abductive strategy fails because it violates this maxim, so too would the coherentist strategy, should the realist turn from one to the other. For, as we see from the words of Wien, the same coherentist line that the realist would appropriate for his own support, is part of ordinary scientific practice in framing judgments about competing theories. It is, therefore, not a line of defense available to the realist. Moreover, just as the truth-bearing status of abduction is an issue dividing realists from various nonrealists, so too is the status of coherence-based inference. Turning from abduction to coherence, therefore, still leaves the realist begging the question. Thus, when we bring out into the open the character of arguments for realism, we see quite plainly that they do not work.

In support of realism there seem to be only those "reasons of the heart" which, as Pascal says, reason does not know. Indeed, I have long felt that belief in realism involves a profound leap of faith, not at all dissimilar from the faith that animates deep religious convictions. I would welcome engagement with realists on this understanding, just as I enjoy conversation on a similar basis with my religious friends. The dialogue will proceed more fruitfully, I think, when the realists finally stop pretending to a rational support for their faith, which they do not have. Then we can all enjoy their intricate and sometimes beautiful philosophical constructions (of, e.g., knowledge, or reference, etc.), even though, as nonbelievers, they may seem to us only wonderful castles in the air.

8. See Gerald Holton, "Mach, Einstein, and the Search for Reality," in his *Thematic Origins of Scientific Thought* (Cambridge: Harvard University Press, 1973), 219–259. I have tried to work out the precise role of this positivist methodology in my "The Young Einstein and the Old Einstein," in *Essays in Memory of Imré Lakatos,* ed. R. S. Cohen et al. (Dordrecht: D. Reidel, 1976), 145–159.

9. A. Einstein et al., *The Principle of Relativity,* trans. W. Perrett and G. B. Jeffrey (New York: Dover, 1952), 117.

10. John Earman et al., "Lost in the Tensors," *Studies in History and Philosophy of Science* 9 (1978): 251–278. The tortuous path detailed by Earman is sketched by B. Hoffmann, *Albert Einstein, Creator and Rebel* (New York: New American Library, 1972), 116–128. A nontechnical and illuminating account is given by John Stachel, "The Genesis of General Relativity," in *Einstein Symposium Berlin,* ed. H. Nelkowski et al. (Berlin: Springer-Verlag, 1980).

11. I have in mind the role played by the analysis of simultaneity in Einstein's path to special relativity. Despite the important study by Arthur Miller, *Albert Einstein's Special Theory of Relativity* (Reading: Addison-Wesley, 1981), and an imaginative pioneering work by John Earman (and collaborators), the details of which I have been forbidden to disclose, I think the role of positivist analysis in the 1905 paper has yet to be properly understood. My ideas here were sparked by Earman's playful reconstructions. So I cannot expose my ideas until John is ready to expose his.

12. Peter Barker, "Einstein's Later Philosophy of Science," in *After Einstein,* ed. P. Barker and C. G. Shugart (Memphis: Memphis State University Press, 1981), 133–146, is a nice telling of this story.

13. Roger Jones in "Realism about What?" [*Philosophy of Science* 58 (1991) 185–202] explains very nicely some of the difficulties here.

14. I think the ordinary, deflationist attitude of working scientists is much like that of Steven Weinberg, *Gravitation and Cosmology: Principles and Applications of the General Theory of Relativity* (New York: Wiley, 1972).

15. See B. L. van der Waerden, *Sources of Quantum Mechanics* (New York: Dover, 1967), 261.

16. See Linda Wessels, "Schrödinger's Route to Wave Mechanics," *Studies in History and Philosophy of Science* 10 (1979): 311–340.

17. A. Landé, "Albert Einstein and the Quantum Riddle," *American Journal of Physics* 42 (1974): 460.

18. Letter to Schrödinger, June 19, 1935. See my "Einstein's Critique of Quantum Theory: The Roots and Significance of EPR," in *After Einstein* (see n. 16), 147–158, for a fuller discussion of the contents of this letter.

19. Bas van Fraassen, "The Charybdis of Realism: Epistemological Implications of Bell's Inequality," *Synthese* 52 (1982): 25–38.

20. See my "Antinomies of Entanglement: The Puzzling Case of the Tangled Statistics," *Journal of Philosophy* 79, 12 (1982), for part of the discussion and for reference to other recent work.

21. The nonrealism that I attribute to students and practitioners of the quantum theory requires more discussion and distinguishing of cases and kinds

than I have room for here. It is certainly not the all-or-nothing affair I make it appear in the text. See Arthur Fine, "Is Scientific Realism Compatible with Quantum Physics?" in A. Fine, *The Shaky Game: Einstein, Realism, and the Quantum Theory.* 2d ed. (Chicago: University of Chicago Press, 1996): 151–171. My thanks to Paul Teller and James Cushing, each of whom saw the need for more discussion here.

22. I should be a little more careful about the historical Bohr than I am in the text. For Bohr himself would seem to have wanted to truncate the homely line somewhere between the domain of chairs and tables and atoms, whose existence he plainly accepted, and that of electrons, where he seems to have thought the question of existence (and of realism, more generally) was no longer well defined. An illuminating and provocative discussion of Bohr's attitude toward realism is given by Paul Teller, "The Projection Postulate and Bohr's Interpretation of Quantum Mechanics," pp. 201–223 n. 3. Thanks, again, to Paul for helping to keep me honest.

23. In this context, for example, van Fraassen's "constructive empiricism" would prefer the concept of empirical adequacy, reserving "truth" for an (unspecified) literal interpretation and believing in that truth only among observables. It is clear, nevertheless, that constructive empiricism follows the homely line and accepts the core position. Indeed, this seems to be its primary motivating rationale. If we reread constructive empiricism in our terminology, then, we would say that it accepts the core position but adds to it a construal of truth as empirical adequacy. Thus, it is antirealist, just as suggested in the next paragraph below. I might mention here that in this classification Putnam's internal realism also comes out as antirealist. For Putnam also accepts the core position, but he would add to it a Peircean construal of truth as ideal rational acceptance. This is a mistake, which I expect that Putnam will realize and correct in future writings. He is criticized for it, soundly I think, by Paul Horwich ("Three Forms of Realism") whose own "semantic realism" turns out, in my classification, to be neither realist nor antirealist. Indeed, Horwich's views are quite similar to what is called "NOA" below, and could easily be read as sketching the philosophy of language most compatible with NOA. Finally the "epistemological behaviorism" espoused by Rorty is a form of antirealism that seems to me very similar to Putnam's position, but achieving the core parity between science and common sense by means of an acceptance that is neither ideal nor especially rational, at least in the normative sense. (I beg the reader's indulgence over this summary treatment of complex and important positions. I have been responding to Nancy Cartwright's request to differentiate these recent views from NOA. So if the treatment above strikes you as insensitive, or boring, please blame Nancy.)

24. "How to Compare Theories: Reference and Change," *Nous* 9 (1975): 17–32.

25. In his remarks at the Greensboro conference, my commentator, John King, suggested a compelling reason to prefer NOA over realism; namely, because NOA is less percussive! My thanks to John for the nifty idea, as well as for other comments.

26. "There is, I think, no theory-independent way to reconstruct phrases like 'really true'; the notion of match between the ontology of a theory and its 'real' counterpart in nature now seems to me illusive in principle." T. S. Kuhn, "Postscript," in *The Structure of Scientific Revolutions*, 2d ed. (Chicago: University of Chicago Press, 1970), 206. The same passage is cited for rebuttal by W. H. Newton-Smith, in *The Rationality of Science*. But the "rebuttal" sketched there in chapter 8, sections 4 and 5, not only runs afoul of the objections stated here in my first section, it also fails to provide for the required theory-independence. For Newton-Smith's explication of verisimilitude (p. 204) makes explicit reference to some unspecified background theory. (He offers either current science or the Peircean limit as candidates.) But this is not to rebut Kuhn's challenge (and mine); it is to concede its force.

27. No doubt I am optimistic, for one can always think of more to say. In particular, one could try to fashion a general, descriptive framework for codifying and classifying such answers. Perhaps there would be something to be learned from such a descriptive, semantical framework. But what I am afraid of is that this enterprise, once launched, would lead to a proliferation of frameworks not so carefully descriptive. These would take on a life of their own, each pretending to ways (better than its rivals) to settle disputes over truth claims, or their import. What we need, however, is less bad philosophy, not more. So here, I believe, silence is indeed golden.

38

JAMES ROBERT BROWN

Explaining the Success of Science

Richard Rorty's *Philosophy and the Mirror of Nature* (1979) has done much to undermine a particular view of scientific knowledge and intellectual progress. More recently, he has pooh-poohed the very idea of 'explaining the success of science', and with it he has dismissed one of the stronger arguments for scientific realism. (The argument runs: our theories are successful and truth is the best explanation for this success; therefore, our theories are probably true.) '[W]e do not itch', says Rorty, 'for an explanation of the success of recent Western science any more than for the success of recent Western politics' (1987, 41). Written just before the collapse of the Berlin Wall, the dissolution of the Warsaw Pact, the break-up of the Soviet Union and the crushing of Iraq in the Gulf War, it is hard to imagine a less plausible sentiment.

Not only are we at present swamped with (usually silly and smug) analyses of 'why the West won', but if events should turn sour (as they often seem in danger of doing) we will be awash with explanations of the 'failure' of Western policies. And our concern is quite fitting. Knowing why particular political strategies worked (or failed) is of obvious vital interest. The same can be said for science. I'm happy to join Rorty in lumping science and politics together, but let's try to explain the successes (or failures) of both, rather than turn our backs on them.

Karl Popper has a completely different motivation, but he too has steadfastly held that the success of science is not to be explained—it's a miracle. '[N]o theory of knowledge', he says, 'should attempt

to explain why we are successful in our attempts to explain things' (1972, 23). And even though 'science has been miraculously successful,' as he puts it, '[t]his strange fact cannot be explained' (*ibid.*, 204). Consistency with his other views requires him, no doubt, to disavow any presupposition that a scientific theory is likely to be true. Yet explanations of the success of science often make that very assumption: a theory's *success* is explained by assuming that the theory is *true*. Hence Popper's quandary. But throwing up our hands in despair or embracing miracles seem neither the heroic nor the reasonable thing to do. I have nothing heroic to offer by way of accounting for the success of science either, but I shall try a moderately reasonable stab at it.

Before proceeding further, something should be said about the term 'success'. There are several ways in which science is an overachiever. Its technological accomplishments are undeniable: it is very handy for building bridges and curing diseases. It is also a glorious entertainer: many of us would rather curl up in bed with a good piece of popular physics than with any novel. And science has also been a great success at extracting tax dollars from us all. (I do not say that cynically; I would gladly pay more.)

By calling science successful I do not mean that everything that is called science is successful, only that many current theories are. And by calling these theories successful I chiefly mean that:

1. they are able to organize and unify a great variety of known phenomena;

2. this ability to systematize the empirical data is more extensive now than it was for previous theories; and

Smoke and Mirrors (New York: Routledge, 1994), pp. 3–26. Reprinted by permission of the publisher.

3. a statistically significant number of novel predictions pan out, i.e., our theories get more predictions right than mere guessing would allow.

This, I think, is roughly what is involved in the normal use of the phrase 'the success of science', and I simply follow tradition here. At any rate these are the senses of success that I shall be dealing with. Even though they are common ingredients, they are not, however, always clearly distinguished by writers on this topic.

MIRACLES, DARWIN AND 'THE TRUTH'

The thing to be explained is the success of science, and the way realists often explain this fact is by claiming that theories are true, or at least approximately true, and that any conclusion deduced from true premises must itself be true. So the assumption that theories are (approximately) true explains the success of those theories. Realism, as Hilary Putnam (1975) puts it, is the only explanation which does not make the success of science a miracle. J. J. C. Smart states the case this way:

> If the phenomenalist about theoretical entities is correct, we must believe in a *cosmic coincidence*. That is, if this is so, statements about electrons, etc., are of only instrumental value: they simply enable us to predict phenomena on the level of galvanometers and cloud chambers. They do nothing to remove the *surprising character* of these phenomena. . . . Is it not odd that the phenomena of the world should be such as to make a purely instrumental theory true? On the other hand, if we interpret a theory in a realist way, then we have no need for such a cosmic coincidence: it is not surprising that galvanometers and cloud chambers behave in the sort of way they do, for if there really are electrons, etc., this is just what we should expect. A lot of surprising facts no longer seem surprising. (1968, 39)

We can reconstruct the argument in this passage in a way that makes it seem quite reasonable and convincing.

1. Conclusion O (an observation statement) can be *deduced* from theory T.

2. O is seen to be the case.

3. If T is true then the argument for O is *sound* and so O *had* to be true.

4. If T is false then the argument for O is *merely valid* and the probability of the arbitrary consequence O being true is very small (i.e., it would be a miracle if O were true).

5. Therefore the argument for O is probably sound.

6. Therefore T is probably true. (This is, even T's theoretical statements are probably true.)

This argument uses the realist's explanation of the success of science to draw ontological morals. Let us contrast it with a rival 'Darwinian' view of the anti-realist Bas van Fraassen, perhaps the most influential of recent anti-realists, who gives such an account of the success of science in *The Scientific Image*.[1] The explanation goes something like this: just as there are a great many species struggling for existence, so too have a great many theories been proposed. But just as species which are not adapted to their environment become extinct, so too are theories which do not make true observational predictions dropped. The belief that our theories might be true, or even approximately true, is therefore an illusion. It is similar to the illusion that Darwin undermined, that species are evolving *towards some goal*. Van Fraassen writes:

> I can best make the point by contrasting two accounts of the mouse who runs from its enemy, the cat. St. Augustine . . . provided an intentional explanation: the mouse *perceives that* the cat is its enemy, hence the mouse runs. What is postulated here is the 'adequacy' of the mouse's thought to the order of nature: the relation of enmity is correctly reflected in his mind. But the Darwinist says: Do not ask why the *mouse* runs from its enemy. Species which did not cope with their natural enemies no longer exist. That is why there are only ones who do.

And so, he continues:

> In just the same way, I claim that the success of current scientific theories is no miracle. It is not

even surprising to the scientific (Darwinist) mind. For any scientific theory is born into a life of fierce competition, a jungle red in tooth and claw. Only the successful theories survive—the ones which *in fact* latched on to actual regularities in nature. (1980, 39f)

'Truth' plays no role at all in the success of science for the Darwinian anti-realist. Yet for the realist it is the central explanatory factor. So here we have two main contenders, but could either of these explanations of the success of science be right?

THE DARWINIAN ANSWER

I characterized the success of science as having three ingredients. Van Fraassen's Darwinian explanation seems to account for two of these features, but not the third. He has an apparently adequate answer to the questions why theories get so much right and why newer theories get more right than the ones we have tossed out. The simple answer is that we have tossed out any theory which did not organize, unify and generally get a lot right; and we have tossed out theories which have done less well, comparatively, than others.

However, the third question is still unanswered. Why do our theories make correct predictions more often than one could expect on the basis of mere chance? Here the Darwinian analogy breaks down since most species could not survive a radical change of environment, the analogue of a novel prediction.

There is also a more general problem with van Fraassen's Darwinian approach. It is a problem which stems from the empiricism of anti-realists. An implicit assumption is that rational choice and success go hand in hand. On this assumption it is not surprising that science is successful in the senses (1) and (2), since we choose theories, says the empiricist, on that very basis. This, I think, is not so. Success, as characterized by a van Fraassen–type anti-realist, is a totally empirical notion. But in reality theories are rationally evaluated on the basis of several other considerations besides empirical factors. I do not wish to argue here for any in particular, but let us suppose that conceptual, metaphysical

and aesthetic concerns play a role in actual theory choice. (Van Fraassen calls these 'pragmatics' and allows that they play a role.) Consequently, it is *not* a trivial analytic truth that the rational thing to believe is also the most successful (as success was characterized above). Anyone who is not an extreme empiricist must concede that it is quite *possible* that the most rationally acceptable theory is not the most successful theory.

So even the Darwinian answers to (1) and (2) which above I tentatively conceded to be adequate are, in fact, not adequate after all. And (3), of course, remains entirely unexplained. The Darwinian account, linked to an empiricist methodology, yields a plausible account of two of the three aspects of success, but unlinked from this untenable methodology it accounts for nothing.

REALISM AND REFERENCE

A belief common to scientific realists is that the succession of theories is getting closer to the truth. This belief may well be true (I hope it is), but it is often tied to a doctrine that says that the central terms of one theory refer to the same things as the central terms of its successor and predecessor theories. Moreover, the intuitive idea of getting-closer-to-the-truth will itself need fleshing out in the form of an explicit doctrine of verisimilitude. Unfortunately, there are terrible problems with both of these. Beliefs about the constancy of reference run afoul of the history of science, and the concept of verisimilitude is plagued with technical problems. Even a cursory glance at the past suggests that there is no royal road to the truth such as that implied by the convergence picture, and every explication of verisimilitude so far proposed has been a crashing failure. Let's look at things now in some detail. In the most quoted version of the realist's explanation of the success of science, Putnam writes:

> The positive argument for realism is that it is the only philosophy that doesn't make the success of science a miracle. That terms in mature theories typically refer (this formulation is due to Richard Boyd), that the theories accepted in a mature science are typically

approximately true, that the same term can refer to the same thing even when it occurs in different theories—these statements are viewed by the scientific realist not as necessary truths but as part of the only scientific explanation of the success of science, and hence as part of any adequate scientific description of science and its relations to its objects. (1975, 73)

In the next section I shall examine the idea that mature theories are 'typically approximately true' by looking at Newton-Smith's views, since they are much more developed than Putnam's. This section will be devoted solely to examining the claim that 'terms in mature theories typically refer'. Let us begin by looking at a very simple theory:

Alasdair loves Hegel.

For the sake of the argument, let us suppose that it is quite a successful theory (there were reports of his buying several works by Hegel, waxing eloquent about Hegel's logic, hanging a picture of Hegel on his office wall, etc.) and that all the terms in this simple theory refer. But is the fact that all the terms refer *sufficient* to explain why the theory is successful? The simplest consideration completely undermines this supposition. The following theory, we may suppose, is very unsuccessful:

Alasdair does *not* love Hegel.

Yet all the relevant terms just as surely refer.

Not all counter-examples are so artificial; historical illustrations of the problem abound. Consider the succession of atomistic theories; some were successful, but many were not. So clearly, having the term 'atom' in the theory does not lead to success even though (we believe) the term 'atom' refers.

Reference is not *sufficient* for success, but is it *necessary*? This, too, seems most unlikely. Phlogiston theories, caloric theories, ether theories and numerous others have all had a definite heyday; yet, by our present best guesses, the central terms of these theories do not refer. In the Putnam–Boyd explanation of the success of science there is a caveat. The term 'typically' is used: 'terms typically refer' and theories are 'typically approximately true'. This seems to leave one free to dismiss the occasional example such as phlogiston or caloric as a tolerable aberra-

tion. It would then appear to be a question of degree, and consequently the historical case for or against this sort of realism is going to be rather difficult to establish.

One could seriously doubt that the historical cases will come out the way Putnam and Boyd expect, i.e., with successful theories *typically* having terms which refer. But even if this should be the case with almost every theory, there still remains one great problem. A *single* example of a successful theory with at least one central term which does not refer must count as a miracle. Thus, the success of the caloric theory of heat, by the lights of Putnam and Boyd, must rank with the raising of Lazarus from the dead; and what Priestley achieved with his phlogiston theory was no less an amazing feat than if he had turned water into wine.

By weakening the claim to just saying that reference is *typical*, easy counter-examples drawn from the history of science might be avoided. But the cost is impossibly high—every atypical example is a miracle.

REALISM AND VERISIMILITUDE

It is time now to look at the other key idea in the Putnam–Boyd explanation of the success of science, the idea that theories are 'typically approximately true'. Unfortunately, neither Putnam nor Boyd has bothered to unpack this notion, so I shall examine the similar but rather more developed views of William Newton-Smith instead.

Newton-Smith's approach to verisimilitude is a 'transcendental' one as he puts it. He too is looking for an explanation of what he sees is an undeniable fact: *science has made progress*. And how has this remarkable achievement come about? His realist answer is disarmingly simple: if our theories were getting closer to the truth then this is exactly what we should expect (1981, 196).

To maintain a doctrine of increasing *verisimilitude*, or truth-likeness, is to maintain that the succession of past theories, up to the present, has been getting *closer to the truth*. There may be several respects in which later theories are better than earlier ones; they may be better predictors, more elegant, technologically more fruitful. But the one respect the realist cares about most is veracity; later theories, it is hoped and claimed, are better in this regard. Verisimilitude is an intuitive notion to which most people subscribe; but it is extremely problematic. The most famous instance of trying to come to grips with it, namely Popper's account (1972), is a clear-cut failure. And unless someone is able to successfully explicate the notion soon, it is likely to have the same fate as such other intuitive notions as 'neutral observation' and 'simplicity'—it will be tossed on the junk pile of history.

There is one virtue of Newton-Smith's account of verisimilitude which needs to be stressed. Constancy of reference across successive theories is not required. The kind of problems phlogiston, caloric and the ether present for the convergence account of Putnam and Boyd have no bearing on Newton-Smith's version. This is what makes his account interesting, initially promising and worthy of special attention.

Let me now focus on some of the details. What is required, as Newton-Smith sees it (1981, 198), is an analysis of the notion which will then justify the crucial premise in his argument. That is, he must show that, on unpacking, the concept of verisimilitude yields this: an increase in verisimilitude implies the likeliness of an increase in observational success. And he is quite right to worry about this, for in spite of its intuitive nature, we cannot count on the properties of *truth* carrying over to *truthlikeness*. The consequences of a true theory must be true, but the consequences of a theory which is approximately true need not themselves be approximately true.

Before getting to his analysis of verisimilitude, we need to set the stage with Newton-Smith's characterization of a few key notions. A *theory* is the deductive closure of the postulates and appropriate auxiliary hypotheses; an *observational consequence* is a conditional, $p \rightarrow q$, where p is a statement of the observable initial conditions and q the observable final conditions; the consequences of a theory must be *recursively enumerable* (i.e., mechanically producible in a sequence—Newton-Smith does not defend this dubious condition). A theory *decides* p if it implies either p or its negation. The *content* of a theory is a fairly technical notion, but we can say roughly that one theory has more content than another if it de-

cides more sentences. Since typically both will decide infinitely many sentences, some technical complications in the definition are required. Imagine two theories, T_1 and T_2, with their consequences recursively enumerated. The nth member of the sequence generated from T_1 either will or will not be decided by T_2. We are to determine which it is. (Given Church's theorem, this is not going to be mechanically possible.) This process is generalized and finally we are able to form the appropriate ratio from the sentences decided by the two theories. In this way Newton-Smith is able to define which theory has the greater content, and he is able to do so in a manner which seems to capture our intuitive requirements. Of course, the definition is based on an infinite sequence, but for practical purposes greater content could be determined after a large, but finite, number of sentences have been examined.

The last important notion is that of *relative truth*. Consider again the theories T_1 and T_2 with their consequences enumerated recursively. After n terms there will be a number of truths and a number of falsehoods for each. The ratio of these numbers is the *truth ratio*. We then pick a third theory, T_3, to appraise the truth values of the sentences in the sequence generated by T_1 and T_2. (T_3 could be either from a God's eye point of view or it could be our presently held theory.) Newton-Smith then defines T_2 as having *greater truth relative to* T_3 than T_1 has, if and only if the infinite sequence of ratios, which give the ratio of truths in T_1 to the truths in T_2 as judged by reference to T_3, has a limit greater than $1/2$. Now we come to the main idea:

> T_2 has greater *verisimilitude* than T_1 if and only if both:
> (1) the relative content of T_2 is equal to or greater than T_1.
> (2) T_2 has greater truth relative to T_3 than T_1.
> (1981, 204)

So the rough idea is this: to have more verisimilitude is, first, to say more about the world and, second, to say more true things in doing so. Does this solve the initial problem which was to show that greater verisimilitude implied the likelihood of greater observational success? The answer, says Newton-Smith, is yes. Here is his argument: pick an arbitrary sentence from T_2 which we shall assume has greater verisimilitude than T_1 according to our definition. The chances of it being true, since it came from T_2, are greater than the chances of some arbitrary sentence which comes from T_1 being true. And since the set of arbitrary sentences of T_2 includes the observational sentences it follows that T_2 will likely have more observational successes.

Newton-Smith's account of the notion of truth-likeness certainly has its attractions. It is not obviously plagued with the same problems which beset Popper's account; it is simple and elegant; and it satisfies several of our most basic intuitions about the concept. However, it still seems to be not entirely satisfactory, as a number of considerations show.

Is Newton-Smith's explanation good at accounting for all three senses of success? Not entirely. It is very good at accounting for (2) (theories explain more now than in the past). But it doesn't say why present theories get much right. It is perfectly compatible with Newton-Smith's theory that our present beliefs organize the data poorly, make few successful novel predictions and generally get very little right. His theory guarantees that our present scientific theories do better than our past theories. But there are important senses of success left unexplained.

Another problem that I see with Newton-Smith concerns a theory's content. Historical considerations make his requirement of increasing content in the definition of greater verisimilitude implausible. Any event in the history of science where the domain shrank—and there are several of them—will stand as a counter-example. Newton-Smith's requirement is that the later theory must have equal or greater content than the former. But this did not happen in the following example which most of us would probably consider a progressive move: once there were theories which combined astronomy and astrology together; then a transition was made to purely astronomical theories. The earlier theories which combined both astronomical and astrological claims obviously said more about the world, so the later astronomical theories had less content. However we characterize truth-likeness, it must be compatible with such domain shrinking transitions in the history of science. Newton-Smith's account is not.

IS HYPOTHETICO–DEDUCTIVISM THE PROBLEM?

What about the style of Newton-Smith's argument which links greater verisimilitude with the likelihood of greater observational success? Anti-realists often decry the hypothetico-deductive (H-D) form of inference. That is, they reject arguments which go:

Theory → Observation
Observation
∴ (Probably) Theory

Given that they find H-D arguments unconvincing (claiming that it is a simple fallacy of affirming the consequent), why should anti-realists be persuaded to become realists by an argument that goes: verisimilitude would explain greater observational success and there has been greater observational success; thus, there must be greater verisimilitude? The style is the same in both cases:

Greater verisimilitude → Greater observational success
Greater observational success
∴ (Probably) Greater verisimilitude

The anti-realist will simply say that the question has been begged. Some of us may like Newton-Smith's argument for verisimilitude and the realist approach in general, but then we *already* liked H-D inference. Laudan, however, gives voice to the anti-realist sentiment when he writes:

> Ever since antiquity critics of epistemic realism have based their skepticism upon a deep-rooted conviction that the fallacy of affirming the consequent is indeed fallacious. . . . Now enters the new breed of realist . . . who wants to argue that epistemic realism can reasonably be presumed to be true by virtue of the fact that it has true consequences. But this is a monumental case of begging the question. (1981, 45)

Can the blame for the failures to explain the success of science be pinned on H-D inference? At first glance the fight between realists and anti-realists over the success of science seems but a dressed up version of the old problem of induction. If there is no hope of solving that problem, then how can we hope to explain the success of science? The answer, I think, is that they are not really the same problem. If H-D reasoning were really the issue here it would be a problem for anti-realists, too. But van Fraassen, a paradigm anti-realist, relies on H-D inference regularly, as he must, for instance, in the following type of argument.

T is empirically adequate → Observation O
Observation O
∴ (Probably) T is empirically adequate

Van Fraassen wants to go as little beyond the observable evidence as he can, but he does take some risks. He resists inferences to the truth, but in accepting a theory as empirically adequate he recognizes the need for ampliative inference.

Similarly, Laudan, when he has on his historian's hat, says the shift to the H-D style of inference with Hartley and LeSage was a step forward in the history of methodology (see, for example, Laudan 1977). Before their work, the Newtonian tradition of doing science was associated with the famous dictum, *hypothesis non fingo;* theories were to be deduced from the phenomena. The introduction of H-D in the eighteenth century marked a definite advance, says Laudan, and most would concur (though some recent readings of Newton would dispute this—see Harper 1990).

Anti-realists such as van Fraassen and Laudan are not *skeptics* about induction. They need and use inductive inference as much as realists do. If realists are committing a fallacy at the meta-level of explaining science, then so is everyone else (except perhaps Popper) at the theory level of explaining the world. But to give up inductive inference entirely, which neither realists nor anti-realists wish to do, is just to stop doing science altogether.

There is, in fact, a range of possibilities here where one might be tempted to draw a line. Consider the following:

I. Evidence E is true

II. Theory T is empirically adequate

III. The entities T posits exist

IV. T is true

They are ordered in terms of decreasing probability, given evidence E. An inductive skeptic will, of course, accept E given E, but will go no further. Van Fraassen will accept the likes of II, given E, but resists III and IV. The niche between II and IV is interesting, though not common. Ian Hacking (1983) holds something like it with his realism based on experimental manipulation, and Nancy Cartwright (1983) believes that there are electrons but that the electron theory is false. Her half realist–half antirealist view is partly revealed in her provocative title *How the Laws of Physics Lie.*[2] The full-blooded realist is prepared in principle to accept IV. All of this makes it clear that there are anti-realist positions between full realism and inductive skepticism. The fight, contrary to Laudan, is not over the legitimacy of induction, but when and where to use it. Laudan is not alone; Arthur Fine picks up on the same point.

FINE'S ONTOLOGICAL ATTITUDE

Arthur Fine's 'The natural ontological attitude' has been an influential and much discussed paper since it appeared in the middle of the debate about scientific realism. NOA, as he calls it, simply accepts the assertions of science at face value. It is not a brand of realism; 'And not anti-realism either', as the title of a follow-up paper announces. Fine's idea would seem to be that the common-sense reading of scientific assertions is the right one. But if this is so, why then are realists being attacked by Fine? Isn't this exactly what realists hold? Clearly it is; realism has only become an explicit doctrine because of the attack on it by anti-realists. I take realism to be just a reflective attempt to defend the 'natural', unreflective, commonsensical reading of the assertions of science.

Perhaps all that NOA comes to is realism *without* a defense. In other words, any argument for NOA would perhaps be an argument for realism. It is hard to say, since NOA is not spelled out or directly argued for; Fine criticizes realists and anti-realists, then NOA 'wins' by default. I'm inclined to see NOA as less the formulation of an ontological point of view than the ventilation of impatience with a perennial philosophical problem.

For the most part the NOA paper is a sustained attack on realists. One of Fine's chief targets is the no-miracles argument. As we have just seen, anti-realists have strong doubts about the inference: T is a good explanation, therefore T is (probably) true. Given these reservations it would seem (for them) to be a case of begging the question when a realist says: approximate truth is a good explanation of success, so we should accept the truth of that explanation (i.e., our theories are approximately true). But, as was suggested above, there is another way to look at things. Realists could say they are simply modeling their argument on a form that any anti-realist (who is not a complete skeptic) would accept. Again, let's make this explicit:

I. T is empirically adequate → Observation O
    Observation O
    ∴ T is empirically adequate

II. T is approximately true → T is successful
    T is successful
    ∴ T is approximately true

(In realistic situations background assumptions will play a role, making inferences more complex and subtle than represented here. But we can ignore this since it is the similarity between I and II which is at issue.) Anti-realists make ampliative inferences of the same form as the success argument. Of course, it is not deductively valid, but if it is an inductive *fallacy* then everyone is making it all over the place. Fine, however, won't let all of us make it. He demands more stringent standards for philosophy than for science. Inference forms such as the above are legitimate, perhaps, when they concern empirical adequacy, but not when they concern truth. But why?

Hilbert's program serves as a model for Fine: mathematics is allowed infinitary methods, but meta-mathematics may only employ finitary techniques. 'Hilbert's maxim applies to the debate over realism: to argue for realism one must employ methods more stringent than those in ordinary scientific practice' (1986, 115).

It is an interesting analogy, but Fine misuses it. He says of Hilbert's program that anything short of

stricter standards is worthless. But this is a false dichotomy: a proof of the consistency of mathematics is either finitary *or* completely worthless. (By analogy, any argument is either a non-inductive argument for realism or else is totally illegitimate.) This overlooks Gentzen's non-finitistic proof of the consistence of arithmetic, which was a great achievement. True, his technique was as 'dubious' as the number theory he set out to legitimize, but the fact that he succeeded in making the whole fit together better than before must surely increase our confidence in that whole. Gentzen's proof does not increase our confidence greatly, but its impact is not negligible either. This, I think, is how we should understand explanations of the success of science which use approximate truth.

Let me put the Gentzen point the other way around. Suppose that he had proved something quite different, an extension of Gödel's incompleteness result, to the effect that there is no infinitary proof of the consistency of mathematics either. If Gentzen had actually proved this, our faith in the consistency of classical mathematics might reasonably decline. So if such a negative result could lead to such an attitude, then Gentzen's actual positive result must surely be taken as lending support to the belief that arithmetic is indeed consistent. Analogously, suppose we had some sort of meta-argument that there could not be a success of science argument concluding that our theories are (approximately) true. Our faith in realism would be shattered. So, given that we have such an argument, shouldn't our faith in realism be at least slightly reinforced?

Attempts such as Fine's to shrug off the realism–anti-realism debate are likely to be unsuccessful. Consider an analogous situation outside of science, say in theology. The analogue of NOA would be to take the Bible at face value. This we might imagine to be done by all until it is pointed out that certain geological and biological facts are incompatible with Genesis. Some conclude from this evidence that the Bible is false; these are the atheists. Fundamentalists hold to its truth, and deny the alleged facts of science. There is, of course, yet a third position which stems from the debate so far. It holds that the Bible is indeed true, but should not be read literally. Atheists and fundamentalists, in reaction to the non-literal reading of the Bible, become explicit realists in theology (where realism means the statements of theology are literally true or literally false); the so-called liberal theologians are the anti-realists. These rival philosophical views inevitably arise and displace the unreflective 'natural' reading of the Bible. NOA is the initial outlook—in science or theology—but it cannot be the final one.

WHY TRUTH MATTERS (A LITTLE)

It is now time to take stock. By explaining success, remember, there are three things to be accounted for: (1) that our current theories organize, unify and generally account for a wide variety of phenomena; (2) that theories have been getting better and better at this, they are progressing; and (3) that a significant number of their novel predictions are true. It is now time to stand back and see to where we have reached.

The debate concerning this attempt to account for the success of science is not just a re-enactment of the problem of induction. So there is perhaps some hope of coming up with an answer. Realist explanations of success may well beg the question against that age-old problem, but then we all (including the anti-realists) do that all the time. Induction, in principle, is not what is at issue here; rather it is a *particular* inference that is being debated.

Van Fraassen's account has no answer at all for (3), i.e., for the fact that theories make novel predictions which are found to be true. It has an explanation of (1), the significant degree of empirical adequacy, and (2), the increasing degree of empirical adequacy over time, but it can explain these only by linking rational theory choice to success *by definition*. Since this is methodologically implausible, even his explanations of (1) and (2) are thus not acceptable.

Let us turn now to the realist's account of things. Explaining the second aspect of success (theories are getting better) is probably the most popular approach. Leplin (1980) thinks it is the most promising, and Newton-Smith, as we saw earlier, builds his doctrine of verisimilitude around it. Actually, it may

be the least promising. Some realist explanations of this sense of the success of science quite explicitly need a theory of verisimilitude. However, none seems available. Newton-Smith was criticized above and other versions of the doctrine have not gone unscathed either. The historical record makes the prospects for one look rather dim; verisimilitude may have to go the way of say, 'simplicity'.

The third sense of success (novel predictions) seems also to be promising for the realist. Predictions about the future which turn out to be true are not just lucky guesses on the realist's account. These predictions are deduced from the truth, says the realist; so it is no wonder that the 'guesses' panned out. There is no rival explanation for this; the Darwinian explanation of van Fraassen didn't even try to account for it. In Laudan's very detailed attack on convergent realism (1981) there is very little mention of this sense of success. So it remains something to which the realist might point as a genuine accomplishment, something to which the anti-realist fails to do justice. But how strong is this? How much support does this give to the realist? Unfortunately, many theories now thought to be false made true novel predictions. Ptolemaic astronomy, for instance, predicted eclipses fairly accurately. And Fresnel rather surprisingly got right his prediction of a bright spot in the middle of a shadow cast by a disk. So being true is hardly necessary for making successful predictions.

It is hard to say why realist accounts of the success of science have gone wrong. Of course, one answer is that realism itself is wrong. But this is an answer we should be loath to accept; so before we do, let's explore at least one different kind of approach to the problem. What realists need, I suggest, is a different style of explanation entirely. I shall now try to spell this out, if only briefly. I stress the tentative, exploratory and sketchy nature of the proposal to follow; it is intended merely as a beginning.

The last four decades have seen considerable quarrelling over the form of a proper explanation. The dominant theory has been the so-called deductive-nomological or covering law model proposed by Hempel (see Hempel 1965). For probabilistic situations there is the so-called inductive-statistical account. Either way, on Hempel's view, an explanation is an argument. Given the explanans, the explanandum is shown to have been expected. (In the deductive case it is certainly expected and in the inductive case the explanandum is expected with high probability.) In short, an explanation is a sufficient or almost sufficient condition for what is being explained.

Here lies the difficulty. The preceding considerations show that truth is neither a necessary nor a sufficient condition for the success of science. It does not meet the Hempelian conditions at all. Since it is not even close to being sufficient in any probabilistic sense we cannot subsume it under the inductive-statistical version of the covering law model either. But the idea that it might have something to do with statistical considerations is, perhaps, an idea worth exploring.

Wesley Salmon proposed[3] an account of explanation which rivals the covering law account of Hempel. An explanation is not an argument for a conclusion; it is instead the marshalling of the statistically relevant facts which have a bearing on the outcome. His view was introduced to cope with examples such as this: 'Why does Jones have paresis?' Explanation: 'Because he had syphilis.' This seems intuitively like a good explanation, yet the outcome, Jones's paresis, is not likely at all. The chances of getting paresis are very small with syphilis, but larger than they would be without it. Having syphilis, says Salmon, is statistically relevant; that is why it explains Jones's paresis. (A is statistically relevant to B if and only if Prob(B, given A) ≠ Prob(B).)[4]

We know that false premises can yield true conclusions, so truth is not (logically speaking) necessary for success. The reason truth is not sufficient for success is because of the presence of auxiliary assumptions which are also at work in any explanation. However, even though truth is neither sufficient nor necessary for success, it is, I shall say following Salmon, statistically relevant. The truth matters to the outcome, though it only matters a little. But it is not any statistical account of explanation that I really care to embrace. Instead, I mention it here only as a kind of introduction to another explanatory form, one which I do want to adopt for explaining the success of science.

NARRATIVE EXPLANATIONS

Salmon's statistical relevance model is not the only challenger to the Hempelian account. Some philosophers of biology and other philosophers of history[5] have advocated a *narrative* style of explanation. An event or condition is explained by telling a story in which the thing to be explained is embedded. In this way the explanandum is said to be rendered 'intelligible'; from the story we see how the events in question are possible. It is often claimed that Darwinian evolution, for instance, is unable to satisfy the Hempelian form, but that it is explanatory nevertheless. It provides neither necessary nor sufficient conditions, but it succeeds in some sense or other in explaining things.

Consider some examples: Why does the giraffe have a long neck? Explanation: The ancestors of the modern giraffe fed on trees, and those with long necks were able to reach more when food was scarce (such as in the occasional drought). There would have been some survival value in having a long neck, so there was, consequently, differential selection in its favor.

Is this meant by the evolutionist to be true? Not with any degree of confidence. It is only meant to be an evolutionary *possibility*, one of the many courses (within the Darwinian framework) that nature *might* have taken. . . .

My suggestion is that realism works as an explanation for the success of science in the same way as the story about the possible history of giraffes explains how the long necks of current giraffes are possible in Darwin's theory. In each case they answer a 'How possible?' question—standard fare for any narrative explanation. . . . The point of these explanations is to let us see how the phenomenon in question could come about. And this is exactly what truth-as-the-explanation-for-the-success-of-science does. It lets us see how science could be successful.

In some respects narrative explanations are similar to statistical relevance explanations. Neither are guaranteed to provide necessary or sufficient conditions for what is being explained. What both do, however, is provide something which is relevant to the outcome. Yet, there is also a difference between them. The statistically relevant information in, for example, the Jones' paresis case is the *known* fact that Jones had syphilis. In typical narrative explanations the relevant fact in the explanation is not known to be true—it is conjectured. So, the realist has an explanation for the success of science: truth is the explanation and the style of the explanation is narrative. The truth is not known to obtain; it is conjectured. But even if it did obtain, success would not automatically follow. The presence of the truth, however, does make a difference; truth is relevant to the outcome.

NOTES

1. I'm going to give an explanation below which is similar to Darwin's in one respect, but it should not be confused with this explanation. Nor should this be confused with the Darwinian epistemology which will be the subject of Chapter 4.
2. The idea is that laws and theories generally are false, but the things they talk about (at least the ones with causal capacities, e.g., electrons, genes, etc.) are quite real and certainly exist, according to Cartwright.
3. See his 'Statistical explanation' (1971). More recent is his *Scientific Explanation and the Causal Structure of the World* (1984).
4. There are problems with this account; see, for example, the relevant discussion by Cartwright in *How the Laws of Physics Lie* (1983). Salmon has further fine tuned his view in *Scientific Explanation and the Causal Structure of the World* (1984).
5. For example, Goudge (1961) and Dray (1964).

REFERENCES

Cartwright, N. (1983) *How the Laws of Physics Lie*, Oxford: Oxford University Press.

Dray, W. (1964). *Philosophy of History*, Englewood Cliffs, NJ: Prentice-Hall.

Fine, A. (1986) *The Shaky Game, Chicago*, IL: Chicago University Press.

van Fraassen, B. (1980) *The Scientific Image*, Oxford: Oxford University Press.

Goudge, T. (1961) *The Ascent of Life*, London: George Allen & Unwin.

Hacking, I. (1983) *Representing and Intervening*, Cambridge: Cambridge University Press.

Harper, W. (1990) 'Newton's classical deductions from phenomena,' in A. Fine, M. Forbes, and L. Wessels (eds.) *PSA 1990*.

Hempel, C. (1965) *Aspects of Scientific Explanation*, New York: Free Press.

Laudan, L. (1977) 'Sources of modern methodology', reprinted in L. Laudan *Science and Hypothesis*, Dordrecht: Reidel, 1981.

Laudan, L. (1981) 'A confutation of convergent realism', *Philosophy of Science*.

Leplin, J. (1980) 'The historical objection to scientific realism', in P. Asquith and R. Giere (eds.) *PSA 1980*, vol. I.

Newton-Smith, W. H. (1981) *The Rationality of Science*, London: Routledge & Kegan Paul.

Popper, K. (1972) *Objective Knowledge*, Oxford: Oxford University Press.

Putnam, H. (1975) *Philosophical Papers*, vol. I, Cambridge: Cambridge University Press.

Rorty, R. (1979) *Philosophy and the Mirror of Nature*, Princeton, NJ: Princeton University Press.

Rorty, R. (1987) 'Science as solidarity', reprinted in R. Rorty *Objectivity, Relativism, and Truth*, Cambridge: Cambridge University Press, 1991.

Salmon, W. (1971) 'Statistical explanation', reprinted in W. Salmon (ed.) *Statistical Explanation and Statistical Relevance*, Pittsburgh, PA: Pittsburgh University Press.

Salmon, W. (1984) *Scientific Explanation and the Causal Structure of the World*, Princeton, NJ: Princeton University Press.

Smart, J. J. C. (1968) *Between Science and Philosophy*, New York: Random House.

PART 7 SUGGESTIONS FOR FURTHER READING

Cartwright, Nancy. *How the Laws of Physics Lie*. Oxford: Clarendon Press, 1983.

Churchland, Paul, and Clifford Hooker, eds. *Images of Science: Essays on Realism and Empiricism*. Chicago: University of Chicago Press, 1985.

Foss, Jeff. "On Accepting van Fraassen's Image of Science." *Philosophy of Science* 51 (1984): 79–92.

Hacking, Ian. *Representing and Intervening.* Cambridge: Cambridge University Press, 1983.

Hausman, Daniel M. "Constructive Empiricism Contested." *Pacific Philosophical Quarterly* 63 (1982): 21–28.

Kline, A. D., and C. Matheson. "How the Laws of Physics Don't Even Fib." In *PSA 1986,* vol. 1, A. Fine and P. Machamer, eds. East Lansing, MI: Philosophy of Science Association, 1986.

Laudan, Larry. *Science and Hypothesis.* Dordrecht: D. Reidel, 1981.

Leplin, Jarrett, ed. *Scientific Realism.* Berkeley: University of California Press, 1984.

McMullin, Ernan. "Selective Anti-Realism." *Philosophical Studies* 61 (1991): 97–108.

Reiner, Richard, and Robert Pierson. "Hacking's Experimental Realism: An Untenable Middle Ground." *Philosophy of Science* 62 (1995): 60–69.

Trigg, Roger. *Reality at Risk.* Totowa, NJ: Barnes and Noble, 1980.

PART 8

SCIENCE AND RELIGION
Reason versus Faith

Both science and religion offer explanations of various phenomena. Religious explanations differ from scientific ones, however, in that they must conform to holy scripture. This difference is often expressed by saying that religious explanations are based on faith whereas scientific explanations are based on fact. But faith, by definition, is belief that is unsupported by logic or evidence. So scientific explanations are often perceived as superior to religious ones.

Paul Feyerabend argues that this perception is mistaken. Scientific explanations, he claims, are just as ideologically tainted as any others. They are inherently superior only for those who have already accepted a certain ideology. This ideology includes the belief that science possesses a special method to distinguish fact from opinion. Feyerabend argues that science has no such method, and even if it did, there is no logical relationship between facts and the theories that scientists concoct to explain them. Facts do not determine which theories scientists accept, for any number of theories can be constructed to account for a particular set of facts. Nor do facts determine which theories scientists reject, because theories cannot be tested in isolation; it's always possible to hold onto a theory in the face of adverse evidence as long as one is willing to give up other beliefs. So how do scientists decide which theories to accept? By vote. Theories are accepted not because they are true but because they have widespread support. Because science does not have a corner on the truth, Feyerabend claims that the state should not force its citizens to study it. In a real democracy, people should be free to choose their epistemology (theory of knowledge) as well as their ideology. Consequently, there should be a separation of state and science as well as a separation of state and church.

For Richard Dawkins, the difference between religion and science could not be more striking. Science is based on observations, which can be independently verified, whereas religion is based on holy scripture, which must be accepted on faith. Those who believe on faith can't be reasoned with. As a result, disputes between people of different faiths often end in war.

Dawkins considers faith to be a disease—a virus of the mind that can be just as deadly as any biological virus. Science and reason are the only cure. To immunize people against the disease of faith, Dawkins recommends injecting science into religious education. By encouraging children to reflect critically on the deep questions of human existence, they may come to realize that science offers a view of the world that is just as emotionally satisfying and awe-inspiring as any offered by religion.

As a Reformed Christian, Alvin Plantinga believes scripture is "a special revelation from God himself, demanding our absolute trust and allegiance." But he also believes reason is "a God-given power by virtue of which we have knowledge of ourselves, our world, our past, logic and mathematics, right and wrong, and God himself." So what should a Reformed Christian do when faith and reason clash, as they appear to do in the creation-evolution controversy? Some have suggested that we give up reason, others that we give up faith. Both of these alternatives are problematic, however, because we can never be sure that the apparent conflict is real. The seeming inconsistency may be due to a faulty interpretation of scripture or to a mistaken view of natural law. Because we can't assume that our view of scripture or nature is infallible, the best way to resolve disputes concerning faith and reason is to examine the evidence on both sides and go with the stronger.

Plantinga thinks that the evidence for creation is stronger than that for evolution. He admits that evolution can explain a number of disparate facts, such as the similarity of structures found in different species and the progressive development found in the fossil record. But from the fact that such phenomena can be explained by evolution, it doesn't follow that they *must* be explained by it. The theory that different types of creatures were specially created by God can also explain these facts. What's more, given the difficulty that evolution has in accounting for the gaps in the fossil record as well as the improbability of complex structures evolving by chance, Plantinga finds the special creation theory more probable than evolution.

If, as the doctrine of *methodological naturalism* suggests, science can offer only natural explanations of phenomena, then supernatural explanations such as special creation can't be considered scientific. But Plantings sees no reason to limit science to natural explanations. Scientists should look for the best explanations, and if the best explanations invoke the supernatural, then so be it. What we need, Plantinga says, is a theistic science that takes the supernatural seriously.

Ernan McMullin argues that there can be no theistic science of the sort that Plantinga envisions—not because it invokes the supernatural, but because it lacks the "systematic observation, generalization, and testing of explanatory hypotheses characteristic of genuine science." Plantinga wants a theistic science based on Christianity. But suppose someone else wanted a theistic science based on gnosticism, which holds that the physical world and everything in it was specially created by the devil. How would we decide between the God hypothesis and devil hypothesis? What observations or experiments could be used to settle the issue? There seem to be none. Consequently, neither of these theories is scientific.

McMullin also disagrees with Plantinga's claim that special creation is more probable than evolution. Even if God intervened in human history, it doesn't follow that He intervened in natural history. The human situation is unique. What's more, it's entirely possible that God chose evolution to be the mechanism by which different creatures are brought into existence. An all-powerful and all-wise creator may well have wanted to unite all living things in this way. Both St. Augustine and St. Thomas Aquinas would be sympathetic to such a developmental view of creation, says McMullin. So even though God could have intervened in the course of nature, there is no reason to believe that He did.

Peter Atkins finds no reason to believe in anything supernatural. Natural explanations are inherently better than supernatural ones because they are simpler—they make fewer assumptions. One of the fundamental principles of theory construction is the principle of simplicity, also known as Occam's razor (named after the medieval logician William of Occam), which says: Do not multiply entities beyond necessity. In other words, if you can explain a phenomenon without assuming that something exists, do not assume that it exists. It follows from this principle that if everything can be explained in natural terms, there's no reason to believe in the supernatural. Atkins argues that, in fact, everything can be explained in natural terms. Science has never encountered a barrier it could not overcome, and there's no reason to believe that it ever will. Even the mysteries of the origin of the universe and the nature of consciousness can be explained scientifically. According to Atkins, then, belief in the supernatural is not only unnecessary; it is irrational.

Martin Gardner dubs the view that science can explain everything "the ultimate in hubris." Gardner believes there is much that we will never know, from the number of hairs on Plato's head to all the decimal digits of pi. Science, of course, does not aspire to give us that sort of knowledge; it tries to identify the fundamental laws that govern the universe. But even if we identified those laws, the question, Why those laws rather than some others? would still be with us. What's more, there may be concepts that we are unable to grasp. Our evolutionary ancestor, the chimpanzee, is incapable of understanding what's taught in grade school. Our evolutionary successors, or even advanced aliens, may very well have concepts that are beyond our ken. So a complete explanation of everything may forever elude us.

39

PAUL FEYERABEND

Science and Myth

Science is much closer to myth than a scientific philosophy is prepared to admit. It is one of the many forms of thought that have been developed by man, and not necessarily the best. It is conspicuous, noisy, and impudent, but it is inherently superior only for those who have already decided in favor of a certain ideology, or who have accepted it without ever having examined its advantages and its limits. And as the accepting and rejecting of ideologies should be left to the individual it follows that the separation of state and *church* must be complemented by the separation of state and *science,* that most recent, most aggressive, and most dogmatic religious institution. Such a separation may be our only chance to achieve a humanity we are capable of, but have never fully realized. . . .

The rise of modern science coincides with the suppression of non-Western tribes by Western invaders. The tribes are not only physically suppressed, they also lose their intellectual independence and are forced to adopt the bloodthirsty religion of brotherly love—Christianity. The most intelligent members get an extra bonus: they are introduced into the mysteries of Western Rationalism and its peak—Western Science. Occasionally this leads to an almost unbearable tension with tradition (Haiti). In most cases the tradition disappears without the trace of an argument; one simply becomes a slave both in body and in mind. Today this development is gradually reversed—with great reluctance, to be sure, but it is reversed. Freedom is regained, old traditions are rediscovered, both among the minorities in Western countries and among large populations in non-Western continents. *But science still reigns supreme.* It reigns supreme because its practitioners are *unable to understand,* and *unwilling to condone,* different ideologies, because they have the *power* to enforce their wishes, and because they *use* this power just as their ancestors used *their* power to force Christianity on the peoples they encountered during their conquests. Thus, while an American can now choose the religion he likes, he is still not permitted to demand that his children learn magic rather than science at school. There is a separation between state and church; there is no separation between state and science.

And yet science has no greater authority than any other form of life. Its aims are certainly not more important than are the aims that guide the lives in a religious community or in a tribe that is united by a myth. At any rate, [scientists] have no business restricting the lives, the thoughts, the education of the members of a free society where everyone should have a chance to make up his own mind and to live in accordance with the social beliefs he finds most acceptable. The separation between state and church must therefore be complemented by the separation between state and science.

We need not fear that such a separation will lead to a breakdown of technology. There will always be people who prefer being scientists to being the masters of their fate and who gladly submit to the meanest kind of (intellectual and institutional) slavery provided they are paid well and provided also there are some people around who examine their work and sing their praise. Greece developed and progressed because it could rely on the services of unwilling slaves. We shall develop and progress with the help of

Against Method (London: Verso, 1975), pp. 295–309. Reprinted with permission.

Shouldn't we take responsibility for insuring common members of our society don't live as idiots?

the numerous *willing* slaves in universities and laboratories who provide us with pills, gas, electricity, atom bombs, frozen dinners and, occasionally, with a few interesting fairy-tales. We shall treat these slaves well, we shall even listen to them, for they have occasionally some interesting stories to tell, but we shall *not* permit them to impose their ideology on our children in the guise of 'progressive' theories of education. We shall not permit them to teach the fancies of science as if they were the only factual statements in existence. This separation of science and state may be our only chance to overcome the hectic barbarism of our scientific-technical age and to achieve a humanity we are capable of, but have never fully realized.[1] Let us, therefore, . . . review the arguments that can be adduced for such a procedure.

The image of 20th-century science in the minds of scientists and laymen is determined by technological miracles such as color television, the moon shots, the infra-red oven, as well as by a somewhat vague but still quite influential rumor, or fairy-tale, concerning the manner in which these miracles are produced.

According to the fairy-tale the success of science is the result of a subtle, but carefully balanced combination of inventiveness and control. Scientists have *ideas*. And they have special *methods* for improving ideas. The theories of science have passed the test of method. They give a better account of the world than ideas which have not passed the test.

The fairy-tale explains why modern society treats science in a special way and why it grants it privileges not enjoyed by other institutions.

Ideally, the modern state is ideologically neutral. Religion, myth, prejudices *do* have an influence, but only in a roundabout way, through the medium of politically influential *parties*. Ideological principles *may* enter the governmental structure, but only via a majority vote, and after a lengthy discussion of possible consequences. In our schools the main religions are taught as *historical phenomena*. They are taught as parts of the truth only if the parents insist on a more direct mode of instruction. It is up to them to decide about the religious education of their children. The financial support of ideologies does not exceed the financial support granted to parties and to private groups. State and ideology, state and church, state and myth, are carefully separated.

State and science, however, work closely together. Immense sums are spent on the improvement of scientific ideas. Bastard subjects such as the philosophy of science, which have not a single discovery to their credit, profit from the boom of the sciences. Even human relations are dealt with in a scientific manner, as is shown by education programs, proposals for prison reform, army training, and so on. Almost all scientific subjects are compulsory subjects in our schools. While the parents of a six-year-old child can decide to have him instructed in the rudiments of Protestantism, or in the rudiments of the Jewish faith, or to omit religious instruction altogether, they do not have a similar freedom in the case of the sciences. Physics, astronomy, history *must* be learned. They cannot be replaced by magic, astrology, or by a study of legends. . . .

The reason for this special treatment of science is, of course, our little fairy-tale: if science has found a method that turns ideologically contaminated ideas into true and useful theories, then it is indeed not mere ideology, but an objective measure of all ideologies. It is then not subjected to the demand for a separation between state and ideology.

But the fairy-tale is false, as we have seen. There is no special method that guarantees success or makes it probable. Scientists do not solve problems because they possess a magic wand—methodology, or a theory of rationality—but because they have studied a problem for a long time, because they know the situation fairly well, because they are not too dumb (though that is rather doubtful nowadays when almost anyone can become a scientist), and because the excesses of one scientific school are almost always balanced by the excesses of some other school. (Besides, scientists only rarely solve their problems, they make lots of mistakes, and many of their solutions are quite useless.) Basically there is hardly any difference between the process that leads to the announcement of a new scientific law and the process preceding passage of a new law in society: one informs either all citizens or those immediately concerned, one collects 'facts' and prejudices, one discusses the matter, and one finally votes. But while a democracy makes some effort to *explain* the process so that everyone can understand it, scientists either *conceal* it, or *bend* it, to make it fit their sectarian interests.

No scientist will admit that voting plays a role in his subject. Facts, logic, and methodology alone decide—this is what the fairy-tale tells us. But how do facts decide? What is their function in the advancement of knowledge? We cannot *derive* our theories from them. We cannot give a *negative* criterion by saying, for example, that good theories are theories which can be refuted, but which are not yet contradicted by any fact. A principle of falsification that removes theories because they do not fit the facts would have to remove the whole of science (or it would have to admit that large parts of science are irrefutable). The hint that a good theory *explains more* than its rivals is not very realistic either. True: new theories often predict new things—but almost always at the expense of things already known. Turning to logic we realize that even the simplest demands *are not* satisfied in scientific practice, and *could not be* satisfied, because of the complexity of the material. The ideas which scientists use to present the known and to advance into the unknown are only rarely in agreement with the strict injunctions of logic or pure mathematics, and the attempt to make them conform would rob science of the elasticity without which progress cannot be achieved. We see: facts alone are not strong enough for making us accept, or reject, scientific theories, the range they leave to thought is *too wide;* logic and methodology eliminate too much, they are *too narrow.* In between these two extremes lies the ever-changing domain of human ideas and wishes. And a more detailed analysis of successful moves in the game of science ('successful' from the point of view of the scientists themselves) shows indeed that there is a wide range of freedom that *demands* a multiplicity of ideas and *permits* the application of democratic procedures (ballot-discussion-vote) but that is actually closed by power politics and propaganda. *This is where the fairy-tale of a special method assumes its decisive function.* It conceals the freedom of decision which creative scientists and the general public have even inside the most rigid and the most advanced parts of science by a recitation of 'objective' criteria, and it thus protects the big-shots (Nobel Prize winners; heads of laboratories, of organizations such as the AMA, of special schools; 'educators'; etc.) from the masses (laymen; experts in non-scientific fields; experts in other fields of science): only those citizens count who were subjected to the pressures of scientific institutions (they have undergone a long process of education), who succumbed to these pressures (they have passed their examinations), and who are now firmly convinced of the truth of the fairy-tale. This is how scientists have deceived themselves and everyone else about their business, but without any real disadvantage: they have more money, more authority, more sex appeal than they deserve, and the most stupid procedures and the most laughable results in their domain are surrounded with an aura of excellence. It is time to cut them down in size, and to give them a more modest position in society. . . .

Modern science . . . is not at all as difficult and as perfect as scientific propaganda wants us to believe. A subject such as medicine, or physics, or biology appears difficult only because it is taught badly, because the standard instructions are full of redundant material, and because they start too late in life. During the war, when the American Army needed physicians within a very short time, it was suddenly possible to reduce medical instruction to half a year (the corresponding instruction manuals have disappeared long ago, however. Science may be simplified during the war. In peacetime the prestige of science demands greater complication.) And how often does it not happen that the proud and conceited judgment of an expert is put in its proper place by a layman! Numerous inventors built 'impossible' machines. Lawyers show again and again that an expert does not know what he is talking about. Scientists, especially physicians, frequently come to different results so that it is up to the relatives of the sick person (or the inhabitants of a certain area) to decide *by vote* about the procedure to be adopted. How often is science improved, and turned into new directions by non-scientific influences! It is up to us, it is up to the citizens of a free society to either accept the chauvinism of science without contradiction or to overcome it by the counterforce of public action. Public action was used against science by the Communists in China in the fifties, and it was again used, under very different circumstances, by some opponents of evolution in California in the seventies. Let us follow their example and let us free society from the strangling hold of an ideologically petrified science just as our ancestors freed *us* from the strangling hold of the One True Religion!

NOTE

1. For the humanitarian deficiencies of science cf. 'Experts in a Free Society,' *The Critic,* November/December 1971, or the improved German version of this essay and of 'Towards a Humanitarian Science' in Part II of Vol. I of my *Ausgewählte Aufsätze.* Vieweg, 1974.

40

RICHARD DAWKINS

Is Science a Religion?

brain virus of faith

It is fashionable to wax apocalyptic about the threat to humanity posed by the AIDS virus, "mad cow" disease, and many others, but I think a case can be made that *faith* is one of the world's great evils, comparable to the smallpox virus but harder to eradicate.

Faith, being belief that isn't based on evidence, is the principle vice of any religion. And who, looking at Northern Ireland or the Middle East, can be confident that the brain virus of faith is not exceedingly dangerous? One of the stories told to young Muslim suicide bombers is that martyrdom is the quickest way to heaven—and not just heaven but a special part of heaven where they will receive their special reward of 72 virgin brides. It occurs to me that our best hope may be to provide a kind of "spiritual arms control": send in specially trained theologians to deescalate the going rate in virgins.

Given the dangers of faith—and considering the accomplishments of reason and observation in the activity called science—I find it ironic that, whenever I lecture publicly, there always seems to be someone who comes forward and says, "Of course, your science is just a religion like ours. Fundamentally, science just comes down to faith, doesn't it?"

Well, science is not religion and it doesn't just come down to faith. Although it has many of religion's virtues, it has none of its vices. Science is based upon verifiable evidence. Religious faith not only lacks evidence, its independence from evidence is its pride and joy, shouted from the rooftops. Why else would Christians wax critical of doubting Thomas? The other apostles are held up to us as exemplars of virtue because faith was enough for them. Doubting Thomas, on the other hand, required evidence. Perhaps he should be the patron saint of scientists.

One reason I receive the comment about science being a religion is because I believe in the fact of evolution. I even believe in it with passionate conviction. To some, this may superficially look like faith. But the evidence that makes me believe in evolution is not only overwhelmingly strong; it is freely available to anyone who takes the trouble to read up on it. Anyone can study the same evidence that I have and presumably come to the same conclusion. But if you have a belief that is based solely on faith, I can't examine your reasons. You can retreat behind the private wall of faith where I can't reach you.

Now in practice, of course, individual scientists do sometimes slip back into the vice of faith, and a few may believe so single-mindedly in a favorite theory that they occasionally falsify evidence. However, the fact that this sometimes happens doesn't alter

---

Transcript of a speech delivered to the American Humanist Association accepting the award of 1996 Humanist of the Year.

the principle that, when they do so, they do it with shame and not with pride. The method of science is so designed that it usually finds them out in the end.

Science is actually one of the most moral, one of the most honest disciplines around—because science would completely collapse if it weren't for a scrupulous adherence to honesty in the reporting of evidence. (As James Randi has pointed out, this is one reason why scientists are so often fooled by paranormal tricksters and why the debunking role is better played by professional conjurors; scientists just don't anticipate deliberate dishonesty as well.) There are other professions (no need to mention lawyers specifically) in which falsifying evidence or at least twisting it is precisely what people are paid for and get brownie points for doing.

Science, then, is free of the main vice of religion, which is faith. But, as I pointed out, science does have some of religion's virtues. Religion may aspire to provide its followers with various benefits—among them explanation, consolation, and uplift. Science, too, has something to offer in these areas.

Humans have a great hunger for explanation. It may be one of the main reasons why humanity so universally has religion, since religions do aspire to provide explanations. We come to our individual consciousness in a mysterious universe and long to understand it. Most religions offer a cosmology and a biology, a theory of life, a theory of origins, and reasons for existence. In doing so, they demonstrate that religion is, in a sense, science; it's just bad science. Don't fall for the argument that religion and science operate on separate dimensions and are concerned with quite separate sorts of questions. Religions have historically always attempted to answer the questions that properly belong to science. Thus religions should not be allowed now to retreat from the ground upon which they have traditionally attempted to fight. They do offer both a cosmology and a biology; however, in both cases it is false.

Consolation is harder for science to provide. Unlike Religion, science cannot offer the bereaved a glorious reunion with their loved ones in the hereafter. Those wronged on this earth cannot, on a scientific view, anticipate a sweet comeuppance for their tormentors in a life to come. It could be argued that, if the idea of an afterlife is an illusion (as I believe it is), the consolation it offers is hollow. But that's not nec-

essarily so; a false belief can be just as comforting as a true one, provided the believer never discovers its falsity. But if consolation comes that cheap, science can weigh in with other cheap palliatives, such as pain-killing drugs, whose comfort may or may not be illusory, but they do work.

Uplift, however, is where science really comes into its own. All the great religions have a place for awe, for ecstatic transport at the wonder and beauty of creation. And it's exactly this feeling of spine-shivering, breath-catching awe—almost worship—this flooding of the chest with ecstatic wonder, that modern science can provide. And it does so beyond the wildest dreams of saints and mystics. The fact that the supernatural has no place in our explanations, in our understanding of so much about the universe and life, doesn't diminish the awe. Quite the contrary. The merest glance through a microscope at the brain of an ant or through a telescope at a long-ago galaxy of a billion worlds is enough to render poky and parochial the very psalms of praise.

Now, as I say, when it is put to me that science or some particular part of science, like evolutionary theory, is just a religion like any other, I usually deny it with indignation. But I've begun to wonder whether perhaps that's the wrong tactic. Perhaps the right tactic is to accept the charge gratefully and demand equal time for science in religious education classes. And the more I think about it, the more I realize that an excellent case could be made for this. So I want to talk a little bit about religious education and the place that science might play in it.

I do feel very strongly about the way children are brought up. I'm not entirely familiar with the way things are in the United States, and what I say may have more relevance to the United Kingdom, where there is state-obliged, legally enforced religious instruction for all children. That's unconstitutional in the United States, but I presume that children are nevertheless given religious instruction in whatever particular religion their parents deem suitable.

Which brings me to my point about mental child abuse. In a 1995 issue of the *Independent*, one of London's leading newspapers, there was a photograph of a rather sweet and touching scene. It was Christmas time, and the picture showed three children dressed up as the three wise men for a nativity play. The accompanying story described one child as a Muslim, one as a Hindu, and one as a Christian. The supposedly sweet and touching point of the story was that they were all taking part in this nativity play.

What is not sweet and touching is that these children were all four years old. How can you possibly describe a child of four as a Muslim or a Christian or a Hindu or a Jew? Would you talk about a four-year-old economic monetarist? Would you talk about a four-year-old neo-isolationist or a four-year-old liberal Republican? There are opinions about the cosmos and the world that children, once grown, will presumably be in a position to evaluate for themselves. Religion is the one field in our culture about which it is absolutely accepted, without question—without even noticing how bizarre it is—that parents have a total and absolute say in what their children are going to be, how their children are going to be raised, what opinions their children are going to have about the cosmos, about life, about existence. Do you see what I mean about mental child abuse?

Looking now at the various things that religious education might be expected to accomplish, one of its aims could be to encourage children to reflect upon the deep questions of existence, to invite them to rise above the humdrum preoccupations of ordinary life and think *sub specie aleternitatis*.

Science can offer a vision of life and the universe which, as I've already remarked, for humbling poetic inspiration far outclasses any of the mutually contradictory faiths and disappointingly recent traditions of the world's religions.

For example, how could any child in a religious education class fail to be inspired if we could get across to them some inkling of the age of the universe? Suppose that, at the moment of Christ's death, the news of it had started traveling at the maximum possible speed around the universe outwards from the earth? How far would the terrible tidings have traveled by now? Following the theory of special relativity, the answer is that the news could not, under any circumstances whatever, have reached more than one-fiftieth of the way across one galaxy—not one-thousandth of the way to our nearest neighboring galaxy in the 100-million-galaxy-strong universe. The universe at large couldn't possibly be anything other than indifferent to

Christ, his birth, his passion, and his death. Even such momentous news as the origin of life on Earth could have traveled only across our little local cluster of galaxies. Yet so ancient was that event on our earthly time-scale that, if you span its age with your open arms, the whole of human history, the whole of human culture, would fall in the dust from your fingertip at a single stroke of a nail file.

The argument from design, an important part of the history of religion, wouldn't be ignored in my religious education classes, needless to say. The children would look at the spellbinding wonders of the living kingdoms and would consider Darwinism alongside the creationist alternatives and make up their own minds. I think the children would have no difficulty in making up their minds the right way if presented with the evidence. What worries me is not the question of equal time but that, as far as I can see, children in the United Kingdom and the United States are essentially given *no* time with evolution yet are taught creationism (whether at school, in church, or at home).

It would also be interesting to teach more than one theory of creation. The dominant one in this culture happens to be the Jewish creation myth, which is taken over from the Babylonian creation myth. There are, of course, lots and lots of others, and perhaps they should all be given equal time (except that wouldn't leave much time for studying anything else). I understand that there are Hindus who believe that the world was created in a cosmic butter churn and Nigerian peoples who believe that the world was created by God from the excrement of ants. Surely these stories have as much right to equal time as the Judeo-Christian myth of Adam and Eve.

So much for Genesis; now let's move on to the prophets. Halley's Comet will return without fail in the year 2062. Biblical or Delphic prophecies don't begin to aspire to such accuracy; astrologers and Nostradamians dare not commit themselves to factual prognostications but, rather, disguise their charlatanry in a smokescreen of vagueness. When comets have appeared in the past, they've often been taken as portents of disaster. Astrology has played an important part in various religious traditions, including Hinduism. The three wise men I mentioned earlier were said to have been led to the cradle of Jesus by a star. We might ask the children by what physical route do they imagine the alleged stellar influence on human affairs could travel.

Incidentally, there was a shocking program on the BBC radio around Christmas 1995 featuring an astronomer, a bishop, and a journalist who were sent off on an assignment to retrace the steps of the three wise men. Well, you could understand the participation of the bishop and the journalist (who happened to be a religious writer), but the astronomer was a supposedly respectable astronomy writer, and yet she went along with this! All along the route, she talked about the portents of when Saturn and Jupiter were in the ascendant up Uranus or whatever it was. She doesn't actually believe in astrology, but one of the problems is that our culture has been taught to become tolerant of it, even vaguely amused by it—so much so that even scientific people who don't believe in astrology sort of think it's a bit of harmless fun. I take astrology very seriously indeed: I think it's deeply pernicious because it undermines rationality, and I should like to see campaigns against it.

When the religious education class turns to ethics, I don't think science actually has a lot to say, and I would replace it with rational moral philosophy. Do the children think there are absolute standards of right and wrong? And if so, where do they come from? Can you make up good working principles of right and wrong, like "do as you would be done by" and "the greatest good for the greatest number" (whatever that is supposed to mean)? It's a rewarding question, whatever your personal morality, to ask as an evolutionist where morals come from; by what route has the human brain gained its tendency to have ethics and morals, a feeling of right and wrong?

Should we value human life above all other life? Is there a rigid wall to be built around the species *Homo sapiens,* or should we talk about whether there are other species which are entitled to our humanistic sympathies? Should we, for example, follow the right-to-life lobby, which is wholly preoccupied with *human* life, and value the life of a human fetus with the faculties of a worm over the life of a thinking and feeling chimpanzee? What is the basis of this fence we erect around *Homo sapiens*—even around a small piece of fetal tissue? (Not a very sound evolutionary idea when you think about it.) When, in our evolutionary descent from our common ancestor with chimpanzees, did the fence suddenly rear itself up?

Well, moving on, then, from morals to last things, to eschatology, we know from the second law of thermodynamics that all complexity, all life, all laughter, all sorrow, is hell-bent on leveling itself out into cold nothingness in the end. They—and we—can never be more than temporary, local buckings of the great universal slide into the abyss of uniformity.

We know that the universe is expanding and will probably expand forever, although it's possible it may contract again. We know that, whatever happens to the universe, the sun will engulf the earth in about 60 million centuries from now.

Time itself began at a certain moment, and time may end at a certain moment—or it may not. Time may come locally to an end in miniature crunches called black holes. The laws of the universe seem to be true all over the universe. Why is this? Might the laws change in these crunches? To be really speculative, time could begin again with new laws of physics, new physical constants. And it has even been suggested that there could be many universes, each one isolated so completely that, for it, the others don't exist. Then again, there might be a Darwinian selection among universes.

So science could give a good account of itself in religious education. But it wouldn't be enough. I believe that some familiarity with the King James versions of the Bible is important for anyone wanting to understand the allusions that appear in English literature. Together with Book of Common Prayer, the Bible gets 58 pages in the *Oxford Dictionary of Quotations.* Only Shakespeare has more. I do think that not having any kind of biblical education is unfortunate if children want to read English literature and understand the provenance of phrases like "through a glass darkly," "all flesh is as grass," "the race is not to the swift," "crying in the wilderness," "reaping the whirlwind," "amid the alien corn," "Eyeless in Gaza," "Job's comforters," and "the widow's mite."

I want to return now to the charge that science is just a faith. The more extreme version of this charge—and one that I often encounter as both a scientist and a rationalist—is an accusation of zealotry and bigotry in scientists themselves as great as that found in religious people. Sometimes there may be a little bit of justice in this accusation; but as zealous bigots, we scientists are mere amateurs at the game. We're content to *argue* with those who disagree with us. We don't kill them.

But I would want to deny even the lesser charge of purely verbal zealotry. There is a very, very important difference between feeling strongly, even passionately, about something because we have thought about and examined the evidence for it on the one hand, and feeling strongly about something because it has been internally revealed to us, or internally revealed to somebody else in history and subsequently hallowed by tradition. There's all the difference in the world between a belief that one is prepared to defend by quoting evidence and logic and a belief that is supported by nothing more than tradition, authority, or revelation.

41

ALVIN PLANTINGA

When Faith and Reason Clash: Evolution and the Bible

My question is simple: how shall we Christians deal with apparent conflicts, between faith and reason, between what we know as Christians and what we know in other ways, between teaching of the Bible and the teachings of science? As a special case, how shall we deal with apparent conflicts between what the Bible initially seems to tell us about the origin and development of life, and what contemporary science seems to tell us about it? Taken at face value, the Bible seems to teach that God created the world relatively recently, that he created life by way of several separate acts of creation, that in another separate act of creation, he created an original human pair, Adam and Eve, and that these our original parents disobeyed God, thereby bringing ruinous calamity on themselves, their posterity and the rest of creation.

According to contemporary science, on the other hand, the universe is exceedingly old—some 15 or 16 billion years or so, give or take a billion or two. The earth is much younger, maybe 4½ billion years old, but still hardly a spring chicken. Primitive life arose on earth perhaps 3½ billion years ago, by virtue of processes that are completely natural if so far not well understood; and subsequent forms of life developed from these aboriginal forms by way of natural processes, the most popular candidates being perhaps random genetic mutation and natural selection.

Now we Reformed Christians are wholly in earnest about the Bible. We are people of the Word; *Sola Scriptura* is our cry; we take Scripture to be a special revelation from God Himself, demanding our absolute trust and allegiance. But we are equally enthusiastic about *reason,* a God-given power by virtue of which we have knowledge of ourselves, our world, our past, logic and mathematics, right and wrong, and God himself; reason is one of the chief features of the image of God in us. And if we are enthusiastic about reason, we must also be enthusiastic about contemporary natural science, which is a powerful and vastly impressive manifestation of reason. So

Christian Scholar's Review 21, (1991), pp. 8–32. Copyright © 1991 by *Christian Scholar's Review.* Reprinted by permission.

this is my question: given our Reformed proclivities and this apparent conflict, what are we to do? How shall we think about this matter?

I. WHEN FAITH AND REASON CLASH

If the question is simple, the answer is enormously difficult. To think about it properly, one must obviously know a great deal of science. On the other hand, the question crucially involves both philosophy and theology: one must have a serious and penetrating grasp of the relevant theological and philosophical issues. And who among us can fill a bill like that? Certainly I can't. (And that, as my colleague Ralph McInerny once said in another connection, is no idle boast.) The scientists among us don't ordinarily have a sufficient grasp of the relevant philosophy and theology; the philosophers and theologians don't know enough science; consequently, hardly anyone is qualified to speak here with real authority. This must be one of those areas where fools rush in and angels fear to tread. Whether or not it is an area where angels fear to tread, it is obviously an area where fools rush in. I hope this essay isn't just one more confirmation of that dismal fact. . . .

As everyone knows, there are various intellectual or cognitive powers, belief-producing mechanisms or powers, various sources of belief and knowledge. For example, there are perception, memory, induction, and testimony, or what we learn from others. There is also reason, taken narrowly as the source of logic and mathematics, and reason taken more broadly as including perception, testimony and both inductive and deductive processes; it is reason taken this broader way that is the source of science. But the serious Christian will also take our grasp of Scripture to be a proper source of knowledge and justified belief. Just how does Scripture work as a source of proper belief? An answer as good as any I now was given by John Calvin and endorsed by the Belgic Confession: this is Calvin's doctrine of the Internal Testimony of the Holy Spirit. This is a fascinating and important contribution that doesn't get nearly the attention it deserves; but here I don't have time to go into the matter. Whatever the mechanism, the Lord speaks to us in Scripture.

And of course what the Lord proposes for our belief is indeed what we should believe. Here there will be enthusiastic agreement on all sides. Some conclude, however, that when there is a conflict between Scripture (or our grasp of it) and science, we must reject science; such conflict automatically shows science to be wrong, at least on the point in question. In the immortal words of the inspired Scottish bard William E. McGonagall, poet and tragedian,

When faith and reason clash,
Let reason go to smash.

But clearly this conclusion doesn't follow. *The Lord* can't make a mistake: fair enough; but *we* can. Our grasp of what the Lord proposes to teach us can be faulty and flawed in a thousand ways. This is obvious, if only because of the widespread disagreement among serious Christians as to just what it is the Lord *does* propose for our belief in one or another portion of Scripture. Scripture is indeed perspicuous: what it teaches with respect to the way of salvation is indeed such that she who runs may read. It is also clear, however, that serious, well-intentioned Christians can disagree as to what the teaching of Scripture, at one point or another, really is. Scripture is inerrant: the Lord makes no mistakes; what he proposes for our belief is what we ought to believe. Sadly enough, however, our grasp of what he proposes to teach is fallible. Hence we cannot simply identify the teaching of Scripture with our grasp of that teaching; we must ruefully bear in mind the possibility that we are mistaken. "He sets the Earth on its foundations; it can never be moved," says the Psalmist.[1] Some sixteenth-century Christians took the Lord to be teaching here that the earth neither rotates on its axis nor goes around the sun; and they were mistaken.

So we can't identify our understanding or grasp of the teaching of Scripture with the teaching of Scripture; hence we can't automatically assume that conflict between what we see as the teaching of Scripture, and what we seem to have learned in some other way must always be resolved in favor of the former. Sadly enough, we have no guarantee that on every point our grasp of what Scripture teaches is correct; hence it is possible that our grasp of the teaching of Scripture be corrected or improved by what we learn in some other way—by way of science, for example.

But neither, of course, can we identify either the current deliverances of reason or our best contemporary science (or philosophy, or history, or literary criticism, or intellectual efforts of any kind) with the truth. No doubt what reason, taken broadly, teaches is by and large reliable; this is, I should think, a consequence of the fact that we have been created in the image of God. Of course we must reckon with the fall and its noetic effects; but the sensible view here, overall, is that the deliverances of reason are for the most part reliable. Perhaps they are most reliable with respect to such common everyday judgments as that there are people here, that it is cold outside, that the pointer points to 4, that I had breakfast this morning, that 2 + 1 = 3, and so on; perhaps they are less reliable when it comes to matters near the limits of our abilities, as with certain questions in set theory, or in areas for which our faculties don't seem to be primarily designed, as perhaps in the world of quantum mechanics. By and large, however, and over enormous swatches of cognitive territory, reason is reliable.

Still, we can't simply embrace current science (or current anything else either) as the truth. We can't identify the teaching of Scripture with our grasp of it because serious and sensible Christians disagree as to what Scripture teaches; we can't identify the current teachings of science with truth, because the current teachings of science change. And they don't change just by the accumulation of new facts. A few years back, the dominant view among astronomers and cosmologists was that the universe is infinitely old; at present the prevailing opinion is that the universe began some 16 billion years ago; but now there are straws in the wind suggesting a step back towards the idea that there was no beginning.[2] Or think of the enormous changes from nineteenth- to twentieth-century physics. A prevailing attitude at the end of the nineteenth century was that physics was pretty well accomplished; there were a few loose ends here and there to tie up and a few mopping up operations left to do, but the fundamental lineaments and characteristics of physical reality had been described. And we all know that happened next.

As I said above, we can't automatically assume that when there is a conflict between science and our grasp of the teaching of Scripture, it is science that is wrong and must give way. But the same holds

vice versa; when there is a conflict between our grasp of the teaching of Scripture and current science, we can't assume that it is our interpretation of Scripture that is at fault. It *could* be that, but it doesn't *have* to be; it could be because of some mistake or flaw in current science. The attitude I mean to reject was expressed by a group of serious Christians as far back as 1832, when deep time was first being discovered; "If sound science appears to contradict the Bible," they said, "we may be sure that it is our interpretation of the Bible that is at fault."[3] To return to the great poet McGonagall.

When faith and reason clash,
'Tis faith must go to smash.

This attitude—the belief that when there is a conflict, the problem must inevitably lie with our interpretation of Scripture, so that the correct course is always to modify that understanding in such a way as to accommodate current science—is every bit as deplorable as the opposite error. No doubt science can correct our grasp of Scripture; but Scripture can also correct current science. If, for example, current science were to return to the view that the world has no beginning, and is infinitely old, then current science would be wrong.

So what, precisely, must we do in such a situation? Which do we go with: faith or reason? More exactly, which do we go with, our grasp of Scripture or current science? I don't know of any infallible rule, or even any pretty reliable general recipe. All we can do is weigh and evaluate the relative warrant, the relative backing or strength, of the conflicting teachings. We must do our best to apprehend both the teachings of Scripture and the deliverances of reason; in either case we will have much more warrant for some apparent teachings than for others. It may be hard to see just what the Lord proposes to teach us in the Song of Solomon or Old Testament genealogies; it is vastly easier to see what he proposes to teach us in the Gospel accounts of Christ's resurrection from the dead. On the other side, it is clear that among the deliverances of reason is the proposition that the earth is round rather than flat; it is enormously harder to be sure, however, that contemporary quantum mechanics, taken realistically, has things right.[4] We must make as careful an estimate as we can of the degrees of warrant of the conflicting doctrines; we may then make a judgment as to where the balance of probability lies, or alternatively, we may suspend judgment. After all, we don't *have* to have a view on all these matters.

Let me illustrate from the topic under discussion. Consider that list of apparent teachings of Genesis: that God has created the world, that the earth is young, that human beings and many different kinds of plants and animals were separately created, and that there was an original human pair whose sin has afflicted both human nature and some of the rest of the world. At least one of these claims—the claim that the universe is young—is very hard to square with a variety of types of scientific evidence: geological, paleontological, cosmological and so on. Nonetheless a sensible person might be convinced, after careful and prayerful study of the Scriptures, that what the Lord teaches there implies that this evidence is misleading and that as a matter of fact the earth really *is* very young. So far as I can see, there is nothing to rule this out as automatically pathological or irrational or irresponsible or stupid.

And of course this sort of view can be developed in more subtle and nuanced detail. For example, the above teachings may be graded with respect to the probability that they really are what the Lord intends us to learn from early Genesis. Most clear, perhaps, is that God created the world, so that it and everything in it depends upon him and neither it nor anything in it has existed for an infinite stretch of time. Next clearest, perhaps, is that there was an original human pair who sinned and through whose sinning disaster befell both man and nature; for this is attested to not only here but in many other places in Scripture. That humankind was separately created is perhaps less clearly taught; that many other kinds of living beings were separately created might be still less clearly taught; that the earth is young, still less clearly taught. One who accepted all of these theses ought to be much more confident of some than of others—both because of the scientific evidence against some of them, and because some are much more clearly the teachings of Scripture than others. I do not mean to endorse the view that all of these propositions are true: but it isn't just silly or irrational to do so. One need not be a fanatic, or a Flat Earther, or an ignorant Fundamentalist in order to hold it. In my judgment the view is mis-

taken, because I take the evidence for an old earth to be strong and the warrant for the view that the Lord teaches that the earth is young to be relatively weak. But these judgments are not simply *obvious,* or inevitable, or such that anyone with any sense will automatically be obliged to agree.

II. FAITH AND EVOLUTION

So I can properly correct my view as to what reason teaches by appealing to my understanding of Scripture; and I can properly correct my understanding of Scripture by appealing to the teachings of reason. It is of the first importance, however, that we correctly *identify* the relevant teachings of reason. Here I want to turn directly to the present problem, the apparent disparity between what Scripture and science teach us about the origin and development of life. Like any good Christian Reformed preacher, I have three points here. First, I shall argue that the theory of evolution is by no means religiously or theologically neutral. Second, I want to ask how we Christians should in fact think about evolution; how probable is it, all things considered, that the Grand Evolutionary Hypothesis is true? And third, I want to make a remark about how, as I see it, our intellectuals and academics should serve us, the Christian community, in this area.

A. Evolution Religiously Neutral?

According to a popular contemporary myth, science is a cool, reasoned, wholly dispassionate attempt to figure out the truth about ourselves and our world, entirely independent of religion, or ideology, or moral convictions, or theological commitments. I believe this is deeply mistaken. Following Augustine (and Abraham Kuyper, Herman Dooyeweerd, Harry Jellema, Henry Stob and other Reformed thinkers), I believe that there is conflict, a battle between the *Civitas Dei,* the City of God, and the City of the World. As a matter of fact, what we have, I think, is a three-way battle. On the one hand there is Perennial Naturalism, a view going back to the ancient world, a view according to which there is no God, nature is all there is, and mankind is to be understood as a part of nature. Second, there is what I shall call 'Enlightenment Humanism': we could also call it 'Enlightenment Subjectivism' or 'Enlightenment Antirealism': this way of thinking goes back substantially to the great eighteenth-century enlightenment philosopher Immanuel Kant. According to its central tenet, it is really we human beings, we men and women, who structure the world, who are responsible for its fundamental outline and lineaments. Naturally enough, a view as startling as this comes in several forms. According to Jean Paul Sartre and his existentialist friends, we do this world-structuring freely and individually; according to Ludwig Wittgenstein and his followers we do it communally and by way of language; according to Kant himself it is done by the transcendental ego which, oddly enough, is neither one nor many, being itself the source of the one-many structure of the world. So two of the parties to this three-way contest are Perennial Naturalism and Enlightenment Humanism; the third party, of course, is Christian theism. Of course there are many unthinking and ill-conceived combinations, much blurring of lines, many cross currents and eddies, many halfway houses, much halting between two opinions. Nevertheless I think these are the three basic contemporary Western ways of looking at reality, three basically *religious* ways of viewing ourselves and the world. The conflict is real, and of profound importance. The stakes, furthermore, are high; this is a battle for men's souls.

Now it would be excessively naive to think that contemporary science is religiously and theologically neutral, standing serenely above this battle and wholly irrelevant to it. Perhaps *parts* of science are like that: mathematics, for example, and perhaps physics, or parts of physics—although even in these areas there are connections.[5] Other parts are obviously and deeply involved in this battle: and the closer the science in question is to what is distinctively human, the deeper the involvement.

To turn to the bit of science in question, the theory of evolution plays a fascinating and crucial role in contemporary Western culture. The enormous controversy about it is what is most striking, a controversy that goes back to Darwin and continues full force today. Evolution is the regular subject of courtroom drama; one such trial—the spectacular Scopes trial of 1925—has been made the subject of an extremely popular film. Fundamentalists regard evolution as the work of the Devil. In academia, on the other hand, it is an idol of the contemporary tribe; it serves as a shibboleth, a litmus test distinguishing the ignorant and bigoted fundamentalist goats from the properly acculturated and scientifically receptive sheep. Apparently this litmus test extends far beyond the confines of this terrestrial globe: according to the Oxford biologist Richard Dawkins, "If superior creatures from space ever visit earth, the first question they will ask, in order to assess the level of our civilization, is: 'Have they discovered evolution yet?'" Indeed many of the experts—for example, Dawkins, William Provine, Stephen Gould—display a sort of revulsion at the very idea of special creation by God, as if this idea is not merely not good science, but somehow a bit obscene, or at least unseemly; it borders on the immoral; it is worthy of disdain and contempt. In some circles, confessing to finding evolution attractive will get you disapproval and ostracism and may lose you your job; in others, confessing doubts about evolution will have the same doleful effect. In Darwin's day, some suggested that it was all well and good to discuss evolution in the universities and among the *cognoscenti;* they thought *public* discussion unwise, however; for it would be a shame if the lower classes found out about it. Now, ironically enough, the shoe is sometimes on the other foot; it is the devotees of evolution who sometimes express the fear that public discussion of doubts and difficulties with evolution could have harmful political effects.[6]

So why all the furor? The answer is obvious: evolution has deep religious connections; deep connections with how we understand ourselves at the most fundamental level. Many evangelicals and fundamentalists see in it a threat to the faith, they don't want it taught to their children, at any rate as scientifically established fact, and they see acceptance of it as corroding proper acceptance of the Bible. On the other side, among the secularists, evolution functions as a *myth*, in a technical sense of that term: a shared way of understanding ourselves at the deep level of religion, a deep interpretation of ourselves to ourselves, a way of telling us why we are here, where we come from, and where we are going.

It was serving in this capacity when Richard Dawkins (according to Peter Medawar, "one of the most brilliant of the rising generation of biologists") leaned over and remarked to A. J. Ayer at one of those elegant, candle-lit, bibulous Oxford dinners that he couldn't imagine being an atheist before 1859 (the year Darwin's *Origin of Species* was published); "although atheism might have been logically tenable before Darwin," said he, "Darwin made it possible to be an intellectually fulfilled atheist."[7] (Let me recommend Dawkins' book to you: it is brilliantly written, unfailingly fascinating, and utterly wrongheaded. It was second on the British best-seller list for some considerable time, second only to Mamie Jenkins' *Hip and Thigh Diet*.) Dawkins goes on:

> All appearances to the contrary, the only watchmaker in nature is the blind forces of physics, albeit deployed in a very special way. A true watchmaker has foresight: he designs his cogs and springs, and plans their interconnections, with a future purpose in his mind's eye. Natural selection, the blind, unconscious automatic process which Darwin discovered, and which we now know is the explanation for the existence and apparently purposeful form of all life, has no purpose in mind. It has no mind and no mind's eye. It does not plan for the future. It has no vision, no foresight, no sight at all. If it can be said to play the role of watchmaker in nature, it is the *blind* watchmaker (p. 5).

Evolution was functioning in that same mythic capacity in the remark of the famous zoologist G. G. Simpson: after posing the question "What is man?" he answers, "The point I want to make now is that all attempts to answer that question before 1859 are worthless and that we will be better off if we ignore them completely."[8] Of course it also functions in that capacity in serving as a litmus test to distinguish the ignorant fundamentalists from the properly enlightened *cognoscenti;* it functions in the same way in many of the debates, in and out of the courts, as to whether it should be taught in the schools, whether other views should be given equal time, and the like. Thus Michael Ruse: "the fight against creationism is a fight for all knowledge, and that battle can be won if we all work to see that Darwinism, which has had a great past, has an even greater future."[9]

The essential point here is really Dawkins' point: Darwinism, the Grand Evolutionary Story, makes it possible to be an intellectually fulfilled atheist. What he means is simple enough. If you are Christian, or a theist of some other kind, you have a ready answer to the question, how did it all happen? How is it that there are all the kinds of floras and faunas we behold; how did they all get here? The answer, of course, is that they have been created by the Lord. But if you are not a believer in God, things are enormously more difficult. How did all these things get here? How did life get started and how did it come to assume its present multifarious forms? It seems monumentally implausible to think these forms just popped into existence; that goes contrary to all our experience. So how did it happen? Atheism and Secularism need an answer to this question. And the Grand Evolutionary Story gives the answer: somehow life arose from nonliving matter by way of purely natural means and in accord with the fundamental laws of physics; and once life started, all the vast profusion of contemporary plant and animal life arose from those early ancestors by way of common descent, driven by random variation and natural selection. I said earlier that we can't automatically identify the deliverances of reason with the teaching of current science because the teaching of current science keeps changing. Here we have another reason for resisting that identification: a good deal more than reason goes into the acceptance of such a theory as the Grand Evolutionary Story. For the nontheist, evolution is the only game in town; it is an essential part of any reasonably complete nontheistic way of thinking; hence the devotion to it, the suggestions that it shouldn't be discussed in public, and the venom, the theological odium with which dissent is greeted.

B. The Likelihood of Evolution

Of course the fact that evolution makes it possible to be a fulfilled atheist doesn't show either that the theory isn't true or that there isn't powerful evidence for it. Well then, how likely is it that this theory is true? Suppose we think about the question from an explicitly theistic and Christian perspective; but suppose we temporarily set to one side the evidence, whatever exactly it is, from early Genesis. From this perspective, how good is the evidence for the theory of evolution?

The first thing to see is that a number of *different* large-scale claims fall under this general rubric of evolution. First, there is the claim that the earth is very old, perhaps some 4.5 billion years old: The *Ancient Earth Thesis,* as we may call it. Second, there is the claim that life has progressed from relatively simple to relatively complex forms of life. In the beginning there was relatively simple unicellular life, perhaps of the sort represented by bacteria and blue green algae, or perhaps still simpler unknown forms of life. (Although bacteria are simple compared to some other living beings, they are in fact enormously complex creatures.) Then more complex unicellular life, then relatively simple multicellular life such as seagoing worms, coral, and jelly fish, then fish, then amphibia, then reptiles, birds, mammals, and finally, as the culmination of the whole process, human beings: the *Progress Thesis,* as we humans may like to call it (jelly fish might have a different view as to where the whole process culminates). Third, there is the *Common Ancestry Thesis:* that life originated at only one place on earth, all subsequent life being related by descent to those original living creatures—the claim that, as Stephen Gould puts it, there is a "tree of evolutionary descent linking all organisms by ties of genealogy."[10] According to the Common Ancestry Thesis, we are literally cousins of all living things—horses, oak trees and even poison ivy—distant cousins, no doubt, but still cousins. (This is much easier to imagine for some of us than for others.) Fourth, there is the claim that there is a (naturalistic) *explanation* of this development of life from simple to complex forms; call this thesis *Darwinism,* because according to the most popular and well-known suggestions, the evolutionary mechanism would be natural selection operating on random genetic mutation (due to copy error or ultra violet radiation or other causes); and this is similar to Darwin's proposals. Finally, there is the claim that life itself developed from non-living matter without any special creative activity of God but just by virtue of the ordinary laws of physics and chemistry: call this the *Naturalistic Origins Thesis.* These five theses are of course importantly different from each other. They are also logically independent in pairs, except for the third and fourth theses: the fourth entails the third, in that you can't sensibly propose a mechanism or an explanation for evolution without agreeing that evolution has indeed occurred. The combination of all five of these theses is what I have been calling 'The Grand Evolutionary Story'; the Common Ancestry Thesis together

with Darwinism (remember, Darwinism isn't the view that the mechanism driving evolution is just what Darwin says it is) is what one most naturally thinks of as the Theory of Evolution.

So how shall we think of these five theses? First, let me remind you once more that I am no expert in this area. And second, let me say that, as I see it, the empirical or scientific evidence for these five different claims differs enormously in quality and quantity. There is excellent evidence for an ancient earth: a whole series of interlocking different kinds of evidence, some of which is marshalled by Howard van Till in *The Fourth Day.* Given the strength of this evidence, one would need powerful evidence on the other side—from Scriptural considerations, say—in order to hold sensibly that the earth is young. There is less evidence, but still good evidence in the fossil record for the Progress Thesis, the claim that there were bacteria before fish, fish before reptiles, reptiles before mammals, and mice before men (or wombats before women, for the feminists in the crowd). The third and fourth theses, the Common Ancestry and Darwinian Theses, are what is commonly and popularly identified with evolution; I shall return to them in a moment. The fourth thesis, of course, is no more likely than the third, since it includes the third and proposes a mechanism to account for it. Finally, there is the fifth thesis, the Naturalistic Origins Thesis, the claim that life arose by naturalistic means. This seems to me to be for the most part mere arrogant bluster; given our present state of knowledge, I believe it is vastly less probable, on our present evidence, than is its denial. Darwin thought this claim very chancy; discoveries since Darwin and in particular recent discoveries in molecular biology make it much less likely than it was in Darwin's day. I can't summarize the evidence and the difficulties here.[11]

Now return to evolution more narrowly so-called: the Common Ancestry Thesis and the Darwinian Thesis. Contemporary intellectual orthodoxy is summarized by the 1979 edition of the *New Encyclopedia Britannica,* according to which "evolution is accepted by all biologists and natural selection is recognized as its cause. . . . Objections . . . have come from theological and, for a time, from political standpoints" (Vol. 7). It goes on to add that "Darwin did two things; he showed that evolution was in fact contradicting Scriptural legends of creation and that its cause, natural selection, was automatic, with no room for divine guidance or design." According to most of the experts, furthermore, evolution, taken as the Thesis of Common Ancestry, is not something about which there can be sensible difference of opinion. Here is a random selection of claims of certainty on the part of the experts. Evolution is certain, says Francisco J. Ayala, as certain as "the roundness of the earth, the motions of the planets, and the molecular constitution of matter."[12] According to Stephen J. Gould, evolution is an established fact, not a mere theory; and no sensible person who was acquainted with the evidence could demur.[13] According to Richard Dawkins, the theory of evolution is as certainly true as that the earth goes around the sun. This comparison with Copernicus apparently suggests itself to many; according to Philip Spieth, "A century and a quarter after the publication of the *Origin of Species,* biologists can say with confidence that universal genealogical relatedness is a conclusion of science that is as firmly established as the revolution of the earth about the sun."[14] Michael Ruse trumpets, or perhaps screams, that "evolution is Fact, FACT, **FACT!**" If you venture to suggest doubts about evolution, you are likely to be called ignorant or stupid or worse. In fact this isn't merely *likely;* you have already *been* so-called: in a recent review in the *New York Times,* Richard Dawkins claims that "It is absolutely safe to say that if you meet someone who claims not to believe in evolution, that person is ignorant, stupid or insane (or wicked, but I'd rather not consider that)." (Dawkins indulgently adds that "You are probably not stupid, insane or wicked, and ignorance is not a crime. . . . ")

Well then, how should a serious Christian think about the Common Ancestry and Darwinian Theses? The first and most obvious thing, of course, is that a Christian holds that all plants and animals, past as well as present, have been created by the Lord. Now suppose we set to one side what we take to be the best understanding of early Genesis. Then the next thing to see is that God could have accomplished this creating in a thousand different ways. It was entirely within his power to create life in a way corresponding to the Grand Evolutionary scenario: it was within his power to create matter and energy, as in the Big Bang, together with laws for its behav-

ior, in such a way that the outcome would be first, life's coming into existence three or four billion years ago, and then the various higher forms of life, culminating, as we like to think, in humankind. This is a semideistic view of God and his workings: he starts everything off and sits back to watch it develop. (One who held this view could also hold that God constantly *sustains* the world in existence—hence the view is only *semi*deistic—and even that any given causal transaction in the universe requires specific divine concurrent activity.)[15] On the other hand, of course, God could have done things very differently. He has created matter and energy with their tendencies to behave in certain ways—ways summed up in the laws of physics—but perhaps these laws are *not* such that given enough time, life would automatically arise. Perhaps he did something different and special in the creation of life. Perhaps he did something different and special in creating the various kinds of animals and plants. Perhaps he did something different and special in the creation of human beings. Perhaps in these cases his action with respect to what he has created was different from the ways in which he ordinarily treats them.

How shall we decide which of these is initially the more likely? This is not an easy question. It is important to remember, however, that the Lord has not merely left the Cosmos to develop according to an initial creation and an initial set of physical laws. According to Scripture, he has often intervened in the working of his cosmos. This isn't a good way of putting the matter (because of its deistic suggestions); it is better to say that he has often treated what he has created in a way different from the way in which he ordinarily treats it. There are miracles reported in Scripture, for example; and, towering above all, there is the unthinkable gift of salvation for humankind by way of the life, death, and resurrection of Jesus Christ, his son. According to Scripture, God has often treated what he has made in a way different from the way in which he ordinarily treats it; there is therefore no initial edge to the idea that he would be more likely to have created life in all its variety in the broadly deistic way. In fact it looks to me as if there is an initial probability on the other side; it is a bit more probable, before we look at the scientific evidence, that the Lord created life and some of its forms—in particular human life—specially.

From this perspective, then, how shall we evaluate the evidence for evolution? Despite the claims of Ayala, Dawkins, Gould, Simpson and the other experts, I think the evidence here has to be rated as ambiguous and inconclusive. The two hypotheses to be compared are (1) the claim that God has created us in such a way that (a) all of contemporary plants and animals are related by common ancestry, and (b) the mechanism driving evolution is natural selection working on random genetic variation and (2) the claim that God created mankind as well as many kinds of plants and animals separately and specially, in such a way that the thesis of common ancestry is false. Which of these is the more probable, given the empirical evidence and the theistic context? I think the second, the special creation thesis, is somewhat more probable with respect to the evidence (given theism) than the first.

There isn't the space, here, for more than the merest hand waving with respect to marshalling and evaluating the evidence. But according to Stephen Jay Gould, certainly a leading contemporary spokesman,

> our confidence that evolution occurred centers upon three general arguments. First, we have abundant, direct observational evidence of evolution in action, from both field and laboratory. This evidence ranges from countless experiments on change in nearly everything about fruit flies subjected to artificial selection in the laboratory to the famous populations of British moths that became black when industrial soot darkened the trees upon which the moths rest. . . . [16]

Second, Gould mentions homologies: "Why should a rat run, a bat fly, a porpoise swim, and I type this essay with structures built of the same bones," he asks, "unless we all inherited them from a common ancestor?" Third, he says, there is the fossil record:

> transitions are often found in the fossil record. Preserved transitions are not common, . . . but they are not entirely wanting. . . . For that matter, what better transitional form could we expect to find than the oldest human, *Australopithecus afrarensis*, with its apelike palate, its human upright stance, and a cranial capacity larger than any ape's of the same body size but a full 1000

cubic centimeters below ours? If God made each of the half-dozen human species discovered in ancient rocks, why did he create in an unbroken temporal sequence of progressively more modern features, increasing cranial capacity, reduced face and teeth, larger body size? Did he create to mimic evolution and test our faith thereby?[17]

Here we could add a couple of other commonly cited kinds of evidence: (a) we along with other animals display vestigial organs (appendix, coccyx, muscles that move ears and nose); it is suggested that the best explanation is evolution. (b) There is alleged evidence from biochemistry: according to the authors of a popular college textbook, "All organisms . . . employ DNA, and most use the citric acid cycle, cytochromes, and so forth. It seems inconceivable that the biochemistry of living things would be so similar if all life did not develop from a single common ancestral group."[18] There is also (c) the fact that human embryos during their development display some of the characteristics of simpler forms of life (for example, at a certain stage they display gill-like structures). Finally, (d) there is the fact that certain patterns of geographical distribution—that there are orchids and alligators only in the American south and in China, for example—are susceptible to a nice evolutionary explanation.

Suppose we briefly consider the last four first. The arguments from vestigial organs, geographical distribution and embryology are suggestive, but of course nowhere near conclusive. As for the similarity in biochemistry of all life, this is reasonably probable on the hypothesis of special creation, hence not much by way of evidence against it, hence not much by way of evidence for evolution.

Turning to the evidence Gould develops, it too is suggestive, but far from conclusive; some of it, furthermore, is seriously flawed. First, those famous British moths didn't produce a new species; there were both dark and light moths around before, the dark ones coming to predominate when the industrial revolution deposited a layer of soot on trees, making the light moths more visible to predators. More broadly, while there is wide agreement that there is such a thing as microevolution, the question is whether we can extrapolate to macroevolution, with the claim that enough microevolution can account for the enormous differences between, say, bacteria and human beings. There is some experiential reason to think not; there seems to be a sort of envelope of limited variability surrounding a species and its near relatives. Artificial selection can produce several different kinds of fruit flies and several different kinds of dogs, but, starting with fruit flies, what it produces is only more fruit flies. As plants or animals are bred in a certain direction, a sort of barrier is encountered; further selective breeding brings about sterility or a reversion to earlier forms. Partisans of evolution suggest that in nature, genetic mutation of one sort or another can appropriately augment the reservoir of genetic variation. That it can do so sufficiently, however, is not known; and the assertion that it does is a sort of Ptolemaic epicycle attaching to the theory.

Next, there is the argument from the fossil record; but as Gould himself points out, the fossil record shows very few transitional forms. "The extreme rarity of transitional forms in the fossil record," he says, "persists as the trade secret of paleontology. The evolutionary trees that adorn our textbooks have data only at the tips and nodes of their branches; the rest is inference, however reasonable, not the evidence of fossils."[19] Nearly all species appear for the first time in the fossil record fully formed, without the vast chains of intermediary forms evolution would suggest. Gradualistic evolutionists claim that the fossil record is woefully incomplete. Gould, Eldredge and others have a different response to this difficulty: punctuated equilibriumism, according to which long periods of evolutionary stasis are interrupted by relatively brief periods of very rapid evolution. This response helps the theory accommodate some of the fossil data, but at the cost of another Ptolemaic epicycle.[20] And still more epicycles are required to account for puzzling discoveries in molecular biology during the last twenty years.[21] And as for the argument from homologies, this too is suggestive but far from decisive. First, there are of course many examples of architectural similarity that are not attributed to common ancestry, as in the case of the Tasmanian wolf and the European wolf; the anatomical givens are by no means conclusive proof of common ancestry. And secondly, God created several different kinds of animals; what would prevent him from using similar structures?

But perhaps the most important difficulty lies in a slightly different direction. Consider the mammalian eye: a marvelous and highly complex instrument resembling a telescope of the highest quality, with a lens, an adjustable focus, a variable diaphragm for controlling the amount of light, and optical corrections for spherical and chromatic aberration. And here is the problem: how does the lens, for example, get developed by the proposed means—random genetic variation and natural selection—when at the same time there has to be development of the optic nerve, the relevant muscles, the retina, the rods and cones, and many other delicate and complicated structures, all of which have to be adjusted to each other in such a way that they can work together? Indeed, what is involved isn't, of course, just the eye; it is the whole visual system, including the relevant parts of the brain. Many different organs and suborgans have to be developed together, and it is hard to envisage a series of mutations which is such that each member of the series has adaptive value, is also a step on the way to the eye, and is such that the last member is an animal with such an eye.

We can consider the problem a bit more abstractly. Think of a sort of space, in which the points are organic forms (possible organisms) and in which neighboring forms are so related that one could have originated from the other with some minimum probability by way of random genetic mutation. Imagine starting with a population of animals without eyes, and trace through the space in question all the paths that lead from this form to forms with eyes. The chief problem is that the vast majority of these paths contain long sections with adjacent points such that there would be no adaptive advantage in going from one point to the next, so that, on Darwinian assumptions, none of them could be the path in fact taken. How could the eye have evolved in this way, so that each point on its path through that space would be adaptive and a step on the way to the eye? (Perhaps it is possible that some of these sections could be traversed by way of steps that were not adaptive and were fixed by genetic drift; but the probability of the population's crossing such stretches will be much less than that of its crossing a similar stretch where natural selection is operative.) Darwin himself wrote, "To suppose that the eye, with all its inimitable contrivances . . . could have been formed by natural selection seems absurd in the highest degree." "When I think of the eye, I shudder" he said (3–4). And the complexity of the eye is enormously greater than was known in Darwin's time.

We are never, of course, given the *actual* explanation of the evolution of the eye, the actual evolutionary history of the eye (or brain or hand or whatever). That would take the form: in that original population of eyeless life forms, genes A_1–A_n mutated (due to some perhaps unspecified cause), leading to some structural and functional change which was adaptively beneficial; the bearers of A_1–A_n thus had an advantage and came to dominate the population. Then genes B_1–B_n mutated in an individual or two, and the same thing happened again; then gene C_1–C_n, etc. Nor are we even given any possibilities of these sorts. (We couldn't be, since, for most genes, we don't know enough about their functions.) We are instead treated to broad brush scenarios at the macroscopic level: perhaps reptiles gradually developed feathers, and wings, and warm-bloodedness, and the other features of birds. We are given possible evolutionary histories, not of the detailed genetic sort mentioned above, but broad macroscopic scenarios: what Gould calls "just-so stories."

And the real problem is that we don't know how to evaluate these suggestions. To know how to do *that* (in the case of the eye, say), we should have to start with some population of animals without eyes; and then we should have to know the rate at which mutations occur for that population; the proportion of those mutations that are on one of those paths through that space to the condition of having eyes; the proportion of *those* that are adaptive, and, at each stage, given the sort of environment enjoyed by the organisms at that stage, the rate at which such adaptive modifications would have spread through the population in question. Then we'd have to compare our results with the time available to evaluate the probability of the suggestion in question. But we don't know what these rates and proportions are. No doubt we *can't* know what they are, given the scarcity of operable time-machines; still, the fact is we don't know them. And hence we don't really know whether evolution is so much as biologically possible: maybe there is no path through that space. It is *epistemically* possible that evolution has occurred—that is, we don't know that it hasn't; for all we know, it has. But

it doesn't follow that it is *biologically* possible. (Whether every even number is the sum of two primes is an open question; hence it is epistemically possible that every even number is the sum of two primes, and also epistemically possible that some even numbers are not the sum of two primes; but one or the other of those epistemic possibilities is in fact mathematically impossible.) Assuming that it *is* biologically possible, furthermore, we don't know that it is not prohibitively improbable (in the statistical sense), given the time available. But then (given the Christian faith and leaving to one side our evaluation of the evidence from early Genesis) the right attitude towards the claim of universal common descent is, I think, one of certain interested but wary skepticism. It is *possible* (epistemically possible) that this is how things happened; God could have done it that way; but the evidence is ambiguous. That it is *possible* is clear; that it *happened* is doubtful; that it is *certain,* however, is ridiculous.

But then what about all those exuberant cries of certainty from Gould, Ayala, Dawkins, Simpson and the other experts? What about those claims that evolution, universal common ancestry, is a rock-ribbed certainty, to be compared with the fact that the earth is round and goes around the sun? What we have here is at best enormous exaggeration. But then what accounts for the fact that these claims are made by such intelligent luminaries as the above? There are at least two reasons. First, there is the cultural and religious, the mythic function of the doctrine evolution helps make it possible to be an intellectually fulfilled atheist. From a naturalistic point of view, this is the only answer in sight to the question "How did it all happen? How did all this amazing profusion of life get here?" From a nontheistic point of view, the evolutionary hypothesis is the only game in town. According to the thesis of universal common descent, life arose in just one place, then there was constant development by way of evolutionary mechanisms from that time to the present, this resulting in the profusion of life we presently see. On the alternative hypothesis, different forms of life arose independently of each other; on that suggestion there would be many different genetic trees, the creatures adorning one of these trees genetically unrelated to those on another. From a nontheistic perspective, the first hypothesis will be by far the more probable, if only because of the extraordinary difficulty in seeing how life could arise even once by any ordinary mechanisms which operate today. That it should arise many different times and at different levels of complexity in this way, is quite incredible.

From a naturalist perspective, furthermore, many of the arguments for evolution are much more powerful than from a theistic perspective. (For example, *given* that life arose naturalistically, it is indeed significant that all life employs the same genetic code.) So from a naturalistic, nontheistic perspective the evolutionary hypothesis will be vastly more probable than alternatives. Many leaders in the field of evolutionary biologists, of course, *are* naturalists—Gould, Dawkins, and Stebbins, for example; and according to William Provine, "very few truly religious evolutionary biologists remain. Most are atheists, and many have been driven there by their understanding of the evolutionary process and other science."[22] If Provine is right or nearly right, it becomes easier to see why we hear this insistence that the evolutionary hypothesis is certain. It is also easy to see how this attitude is passed on to graduate students, and, indeed, how accepting the view that evolution is certain is itself adaptive for life in graduate school and academia generally.

There is a second and related circumstance at work here. We are sometimes told that natural science is *natural* science. So far it is hard to object: but how shall we take the term 'natural' here? It could mean that natural science is science devoted to the study of nature. Fair enough. But it is also taken to mean that natural science involves a *methodological naturalism* or provisional atheism:[23] no hypothesis according to which God has done this or that can qualify as a *scientific* hypothesis. It would be interesting to look into this matter: is there really any compelling or even decent reason for thus restricting our study of nature? But suppose we irenically concede, for the moment, that natural science doesn't or shouldn't invoke hypotheses essentially involving God. Suppose we restrict our explanatory materials to the ordinary laws of physics and chemistry; suppose we reject divine special creation or other hypotheses about God as *scientific* hypotheses. Perhaps indeed the Lord has engaged in special creation, so we say, but that he has (if he has) is not something with which natural science can deal. So far as natu-

ral science goes, therefore, an acceptable hypothesis must appeal only to the laws that govern the ordinary, day-to-day working of the cosmos. As natural scientists we must eschew the supernatural—although, of course, we don't mean for a moment to embrace naturalism.

Well, suppose we adopt this attitude. Then perhaps it looks as if by far the most probable of all the properly scientific hypotheses is that of evolution by common ancestry: it is hard to think of any other real possibility. The only alternatives, apparently, would be creatures popping into existence fully formed, and that is wholly contrary to our experience. Of all the scientifically acceptable explanatory hypotheses, therefore, evolution seems by far the most probable. But if this hypothesis is vastly more probable than any of its rivals, then it must be certain, or nearly so.

But to reason this way is to fall into confusion compounded. In the first place, we aren't just given that one or another of these hypotheses is in fact correct. Granted: if we *knew* that one or another of those scientifically acceptable hypotheses were in fact correct, then perhaps this one would be certain; but of course we don't know that. One real possibility is that we don't have a very good idea how it all happened, just as we may not have a very good idea as to what terrorist organization has perpetrated a particular bombing. And secondly, this reasoning involves a confusion between the claim that of all of those *scientifically* acceptable hypotheses, that of common ancestry is by far the most plausible, with the vastly more contentious claim that of all the acceptable hypotheses *whatever* (now placing no restrictions on their kind) this hypothesis is by far the most probable. Christians in particular ought to be alive to the vast difference between these claims; confounding them leads to nothing but confusion.

From a Christian perspective, it is dubious, with respect to our present evidence, that the Common Ancestry Thesis is true. No doubt there has been much by way of microevolution: Ridley's gulls are an interesting and dramatic case in point. But it isn't particularly likely, given the Christian faith and the biological evidence, that God created all the flora and fauna by way of some mechanism involving common ancestry. My main point, however, is that Ayala, Gould, Simpson, Stebbins and their coterie are wildly mistaken in claiming that the Grand Evolutionary Hypothesis is *certain*. And hence the source of this claim has to be looked for elsewhere than in sober scientific evidence.

So it could be that the best scientific hypothesis was evolution by common descent—i.e., of all the hypotheses that conform to methodological naturalism, it is the best. But of course what we really want to know is not which hypothesis is the best from some artificially adopted standpoint of naturalism, but what the best hypothesis is *overall*. We want to know what the *best* hypothesis is, not which of some limited class is best—particularly if the class in question specifically excludes what we hold to be the basic truth of the matter. It could be that the best scientific hypothesis (again supposing that a scientific hypothesis must be naturalistic in the above sense) isn't even a strong competitor in *that* derby.

Judgments here, of course, may differ widely between believers in God and non-believers in God. What for the former is at best a methodological restriction is for the latter the sober metaphysical truth; her naturalism is not merely provisional and methodological, but, as she sees it, settled and fundamental. But believers in God can see the matter differently. The believer in God, unlike her naturalist counterpart, is free to look at the evidence for the Grand Evolutionary Scheme and follow it where it leads, rejecting that scheme if the evidence is insufficient. She has a freedom not available to the naturalist. The latter accepts the Grand Evolutionary Scheme because from a naturalistic point of view this scheme is the only visible answer to the question *what is the explanation of the presence of all these marvelously multifarious forms of life?* The Christian, on the other hand, knows that creation is the Lord's; and she isn't blinkered by *a priori* dogmas as to how the Lord must have accomplished it. Perhaps it was by broadly evolutionary means, but then again perhaps not. At the moment, 'perhaps not' seems the better answer.

Returning to methodological naturalism, if indeed natural science is essentially restricted in this way, if such a restriction is a part of the very essence of science, then what we need here, of course, is not natural science, but a broader inquiry that can include *all* that we know, including the truths that God has created life on earth and could have done it in many different ways. "Unnatural Science," "Creation Science," "Theistic Science"—call it what you will: what we need when we want to know how to think about the origin and development of contemporary life is what is most plausible from a Christian point of view. What we need is a scientific account of life that isn't restricted by that methodological naturalism.

C. What Should Christian Intellectuals Tell the Rest of Us?

Alternatively, how can Christian intellectuals—scientists, philosophers, historians, literary and art critics, Christian thinkers of every sort—how can they best serve the Christian community in an area like this? How can they—and since we are they, how can we—best serve the Christian community, the Reformed community of which we are a part, and more importantly, the broader general Christian community? One thing our experts can do for us is help us avoid rejecting evolution for stupid reasons. The early literature of Creation-Science, so-called, is littered with arguments of that eminently rejectable sort. Here is such an argument. Considering the rate of human population growth over the last few centuries, the author points out that even on a most conservative estimate the human population of the earth doubles at least every 1000 years. Then if, as evolutionists claim, the first humans existed at least a million years ago, by now the human population would have doubled 1000 times. It seems hard to see how there could have been fewer than two original human beings, so at that rate, by the inexorable laws of mathematics, after only 60,000 years or so, there would have been something like 36 quintillion people, and by now there would have to be 2^{1000} human beings. 2^{1000} is a large number; it is more than 10^{300}, 1 with 300 zeros after it; if there were that many of us the whole universe would have to be packed solid with people. Since clearly it isn't, human beings couldn't have existed for as long as a million years; so the evolutionists are wrong. This is clearly a lousy argument; I leave as homework the problem of saying just where it goes wrong. There are many other bad arguments against evolution floating around, and it is worth our while to learn that these arguments are indeed bad. We shouldn't reject contemporary science unless we have to, and we shouldn't reject it for the wrong reasons. It is a good thing for our scientists to point out some of those wrong reasons.

But I'd like to suggest, with all the diffidence I can muster, that there is something better to do here—or at any rate something that should be done in addition to this. And the essence of the matter is fairly simple, despite the daunting complexity that arises when we descend to the nitty-gritty level where the real work has to be done. The first thing to see, as I said before, is that Christianity is indeed engaged in a conflict, a battle. There is indeed a battle between the Christian community and the forces of unbelief. This contest or battle rages in many areas of contemporary culture—the courts, in the so-called media and the like—but perhaps most particularly in academia. And the second thing to see is that important cultural forces such as science are not neutral with respect to this conflict—though of course certain parts of contemporary science and many contemporary scientists might very well be. It is of the first importance that we discern in detail just *how* contemporary science—and contemporary philosophy, history, literary criticism and so on—is involved in the struggle. This is a complicated, many-sided matter; it varies from discipline to discipline, and from area to area within a given discipline. One of our chief tasks, therefore, must be that of cultural criticism. We must *test* the spirits, not automatically welcome them because of their great academic prestige. Academic prestige, wide, even nearly unanimous acceptance in academia, declarations of certainty by important scientists—none of these is a guarantee that what is proposed is true, or a genuine deliverance of reason, or plausible from a theistic point of view. Indeed, none is a guarantee that what is proposed is not animated by a spirit wholly antithetical to Christianity. We must discern the religious and ideological connections; we can't automatically take the word of the experts, because their word might be dead wrong from a Christian standpoint.

Finally, in all the areas of academic endeavor, we Christians must think about the matter at hand from a Christian perspective; we need Theistic Science. Perhaps the discipline in question, as ordinarily practiced, involves a methodological naturalism; if so, then what we need, finally, is not answers to our questions from *that* perspective, valuable in some

ways as it may be. What we really need are answers to our questions from the perspective of *all* that we know—what we know about God, and what we know by faith, by way of revelation, as well as what we know in other ways. In many areas, this means that Christians must rework, rethink the area in question from this perspective. This idea may be shocking, but it is not new. Reformed Christians have long recognized that science and scholarship are by no means religiously neutral. In a way this is our distinctive thread in the tapestry of Christianity, our instrument in the great symphony of Christianity. This recognition underlay the establishment of the Free University of Amsterdam in 1880; it also underlay the establishment of Calvin College. Our forebears recognized the need for the sort of work and inquiry I've been mentioning, and tried to do something about it. What we need from our scientists and other academics, then, is both cultural criticism and Christian science.

We must admit, however, that it is our *lack* of real progress that is striking. Of course there are good reasons for this. To carry out this task with the depth, the authority, the competence it requires is, first of all, enormously difficult. However, it is not just the *difficulty* of this enterprise that accounts for our lackluster performance. Just as important is a whole set of historical or sociological conditions. You may have noticed that at present the Western Christian community is located in the twentieth-century Western world. We Christians who go on to become professional scientists and scholars attend twentieth-century graduate schools and universities. And questions about the bearing of Christianity on these disciplines and the questions within them do not enjoy much by way of prestige and esteem in these universities. There are no courses at Harvard entitled "Molecular Biology and the Christian View of Man." At Oxford they don't teach a course called "Origins of Life from a Christian Perspective." One can't write his Ph.D. thesis on these subjects. The National Science Foundation won't look favorably on them. Working on these questions is not a good way to get tenure at a typical university; and if you are job hunting you would be ill-advised to advertise yourself as proposing to specialize in them. The entire structure of contemporary university life is such as to discourage serious work on these questions.

This is therefore a matter of uncommon difficulty. So far as I know, however, no one in authority has promised us a rose garden; and it is also a matter of absolutely crucial importance to the health of the Christian community. It is worthy of the very best we can muster; it demands powerful, patient, unstinting and tireless effort. But its rewards match its demands; it is exciting, absorbing and crucially important. Most of all, however, it needs to be done. I therefore commend it to you.

NOTES

1. Ps. 104 vs. 5.
2. See Stephen Hawking, *A Brief History of Time* (New York: Bantam Books, 1988) pp. 115ff.
3. *Christian Observer* 1832, p. 437.
4. Here the work of Bas van Fraassen is particularly instructive.
5. As with intuitionist and constructivist mathematics, idealistic interpretations of quantum mechanics and Bell theoretical questions about information transfer violating relativity constraints on velocity.
6. According to Anthony Flew, to suggest that there is real doubt about evolution is to corrupt the youth.
7. Richard Dawkins, *The Blind Watchmaker* (London and New York: W. W. Norton and Co., 1986), pp. 6 and 7.
8. Quoted in Richard Dawkins, *The Selfish Gene* (Oxford: Oxford University Press, 1976), p. 1.
9. *Darwinism Defended*, pp. 326–327.
10. "Evolution as Fact and Theory" in *Hen's Teeth and Horse's Toes* (New York: Norton, 1983).
11. Let me refer you to the following books: *The Mystery of Life's Origins*, by Charles Thaxton, Walter Bradley and Roger Olsen; *Origins*, by Robert Shapiro, *Evolution, Thermodynamics, and Information: Extending the Darwinian Program*, by Jeffrey S. Wicken, *Seven Clues to the Origin of Life* and *Genetic Takeover and the Mineral Origins of Life*, by A. G. Cairns-Smith, and *Origins of Life*, by Freeman Dyson; see also the relevant chapters of Michael Denton, *Evolution: A Theory in Crisis*. The authors of the first book believe that God created life specially, the authors of the others do not.
12. "The Theory of Evolution: Recent Successes and Challenges," in *Evolution and Creation*, ed. Ernan McMullin (Notre Dame: University of Notre Dame Press, 1985), p. 60.
13. "Evolution as Fact and Theory" in *Hen's Teeth and Horse's Toes* (New York: W. W. Norton and Company, 1980), pp. 254–55.
14. "Evolutionary Biology and the Study of Human Nature," presented at a consultation on Cosmology and Theology sponsored by the Presbyterian (USA) Church in Dec. 1987.
15. The issues here are complicated and subtle and I can't go into them; instead I should like to recommend my colleague Alfred Freddoso's powerful piece, "Medieval Aristotelianism and the Case against Secondary Causation in Nature," in *Divine and Human Action,* edited by Thomas Morris (Ithaca: Cornell University Press, 1988).
16. *Op. cit.*, p. 257.
17. *Op. cit.*, pp. 258–259.
18. Claude A. Villee, Eldra Pearl Solomon, P. William Davis, *Biology*, Saunders College Publishing 1985, p. 1012. Similarly, Mark Ridley [*The Problems of Evolution* (Oxford: Oxford University Press, 1985)] takes the fact that the genetic code is universal across all forms of life as proof that life originated only once; it would be extremely improbable that life should have stumbled upon the same code more than once.
19. *The Panda's Thumb* (New York: 1980), p. 181. According to George Gaylord Simpson (1953): "Nearly all categories above the level of families appear in the record suddenly and are not led up to by known, gradual, completely continuous transitional sequences."
20. And even so it helps much less than you might think. It does offer an explanation of the absence of fossil forms intermediate with respect to closely related or adjoining species; the real problem, though, is what Simpson refers to in the quote in the previous footnote: the fact that nearly all categories above the level of families appear in the record suddenly without the gradual and continuous sequences we should expect. Punctuated equilibriumism does nothing to explain the nearly complete absence, in the fossil record, of intermediates between such major divisions as, say, reptiles and birds, or fish and reptiles, or reptiles and mammals.
21. Here see Michael Denton, *Evolution: A Theory in Crisis* (London: Burnet Books, 1983) chapter 12.
22. *Op. cit.*, p. 28.
23. "Science must be provisionally atheistic or cease to be itself." Basil Whilley, "Darwin's Place in the History of Thought" in M. Banton, ed., *Darwinism and the Study of Society* (Chicago: Quadrangle Books, 1961).

42

ERNAN McMULLIN

Evolution and Special Creation

How did God bring the ancestral living things to be? Two broadly different sorts of answer have found favor with believers in a Creator. One is to suppose that God brings the universe into existence already containing the potentialities that are required in order that the complexities of the world we know should "naturally" develop within it. The other is to say that for some of these complexities to develop, God had to "supplement" nature in certain respects, to act in a special way, special not only in the sense of being different from God's ordinary sustaining of the order accessible to us through natural science, but also in the sense that the interruption of that order is aimed at bringing about results that could not otherwise come to be. The first answer is the evolutionary one. What precise *theories* of evolution one chooses to defend is another matter. *Evolution* is a generic label for the natural process whereby potentialities already present are actualized. The second alternative has the somewhat clumsy title of *special creation.*

One who defends the hypothesis of special creation to account for the origin of a particular sort of being (like the first living cells or the first humans) may be quite content to allow an evolutionary account in other contexts. And one who argues, in principle, for the sufficiency of evolutionary models may (if a theist) insist that the natural order itself is created, dependent on God for its very existence. What separates the two is not the general admissibility of the notions of evolution and creation, but the need for "special" episodes in the story of cosmic development. According to one account, they were needed; according to the other, they were not. On the face of it, both sides need to exercise logical caution. How can those who invoke special creation to account for a particular cosmic transition exclude the possibility that an as-yet unthought of evolutionary explanation might later be found for it? Short of providing an already-completed evolutionary account, how could defenders of evolution exclude the possibility that special creation might have occurred at some juncture? The evolutionist is not required to hold (and if a theist will not hold) that special creation is in principle *impossible,* only that it is in general unlikely, or unneeded in specific contexts.

The vigorously negative reaction to the claims of "creation science" in recent decades might easily lead one to overlook the logical and epistemological complexities of the underlying disagreement between proponents of evolution and proponents of special creation. What came to be called creation science was an aberrant solution forced on defenders of the special creation alternative by the constraints imposed on public school education due to the accepted interpretation of the Constitution of the United States. Its manifest logical inadequacy led ultimately to the legal findings in the celebrated Overton judgment (Arkansas, 1981) striking down the mandatory teaching of creation science as an alternative to evolution[1] and might easily mislead one into supposing that special creation can at this point be dismissed out of hand in discussions of the origins of life. But creation science is only one of the many variant versions of special creation, and assuredly one of the more vulnerable.

Zygon 28 (September 1993), pp. 299–335. Reprinted by permission.

It seems worthwhile, then, to look closely at a very different and much more sophisticated sort of defense of special creation. Alvin Plantinga is a well-known philosopher of religion whose work in epistemology, metaphysics, and modal logic is widely known and justly respected. In a recent essay, "When Faith and Reason Clash: Evolution and the Bible," he proclaims the merits of special creation in the light of what he perceives as inadequacies in the current evolutionary account of origins, and he proposes the antecedent likelihood, in a general way, of special creation from the theological standpoint of the Christian.[2] His principal targets are those evolutionists who, he believes, covertly rely on an antitheistic premise in order to make inflated claims for the certainty of what he calls the "Grand Evolutionary Scheme." His essay is an extended exercise in the epistemology of scientific theory from the perspective of a religious believer; though I disagree with some of its main conclusions, I shall not, I hope, underrate their force.

THEISTIC SCIENCE

Plantinga's thesis in regard to evolution is that, for the Christian, the claim that God created humankind, as well as many kinds of plants and animals, separately and specially, is more probable than the thesis of common ancestry (TCA) that is central to the theory of evolution (Plantinga 1991a, 22, 28). His larger context is that of an exhortation to Christian intellectuals to join battle against "the forces of unbelief," particularly in academia, instead of always yielding to "the word of the experts." These intellectuals must be brought to "discern the religious and ideological connections; . . . [they must not] automatically take the word of the experts, because their word might be dead wrong from a Christian standpoint" (1991a, 30). The implication many would take from this is that Christian intellectuals should ally themselves with the critics of evolution, that it may somehow be to their *advantage* to find flaws in the case for evolution.

The "science" these Christian intellectuals profess will not be of the usual naturalist sort. Their account of the origin of species, for instance, will be at odds with that given by Darwin, on grounds that are distinctively Christian in content. Despite the fact that claims such as these on the part of the Christian depend on what he or she knows "by faith, by way of revelation," Plantinga believes that they can appropriately be called science, and he suggests as a label for them "theistic science" (1991a, 29). An important function of this broader knowledge would be revisionary; he reminds us that "Scripture can correct current science." His theistic science bears some similarity to the creation science that has commanded the headlines in the United States so often in recent decades. Like the creation scientists, he maintains that in the present state of knowledge the best explanation of the origin of many kinds of plants and animals is an interruption in the ordinary course of natural process, a moment when God treats "what he has created in a way different from the way in which he ordinarily treats it" (1991a, 22). Like them, he relies on a critique of the theory of evolution, pointing to what he regards as fundamental shortcomings in the Darwinian project of explaining new species by means of natural selection and emphasizing recent criticisms of one or other facet of the synthetic theory from within the scientific community itself. Like them, he calls for a struggle against prevailing scientific orthodoxy, one that may pit the teachers of Christian youth against the "experts."

But the differences between Plantinga and the creation scientists are even more basic. Most of the latter believe in a "young earth" dating back only a few thousand years, and they attempt to undermine the many arguments that can be brought against this view. Plantinga allows "the evidence for an old earth to be strong and the warrant for the view that the Lord teaches that the earth is young to be relatively weak" (1991a, 15). The creation scientists argue for a whole series of related cosmological theses (that stars and galaxies do not change, that the history of the earth is dominated by the occurrence of catastrophe, and so forth); Plantinga focuses on the single issue of the origins of living things, especially of humankind. And he is in the end more concerned with combating the claims of certainty made by many evolutionists than he is with arguing that the Christian is irrevocably committed against a full evolutionary account of origins. He allows, as the creation scientists, I suspect, would not, that as evolutionary science advances, his own present estimate that special creation is more likely to account for some of the major transitions in the story of life on earth might have to give way.

In the debates regarding the teaching of creation science in the public schools, its defenders attempted to detach their arguments from any sort of reliance on Scripture, or more generally, from theological considerations, whereas Plantinga appeals explicitly to the scriptural understanding of the manner of God's action in the world. The former make a heroic attempt to qualify their creationism as scientific, in what they take to be the conventional sense of the term *scientific*. Their effort, I think it is fair to say, was hopeless right from the start. They would undoubtedly have preferred to defend a view more explicitly based on Genesis, but the exigencies of the constitutional restrictions on the public school curriculum prevented this. The scientists among them attempted to shore up their case by citing various consonances between the catastrophism of their young-earth account and the geological record. But the inspiration for their account lay, and clearly *had* to lie, in the Bible. Trying to fudge this, though understandable under the circumstances, proved a disastrous strategy.

Plantinga offers a far more consistent theme. True, his "theistic science" will not pass constitutional muster, so it will not serve the purposes for which creation science was originally advanced. But that is not an argument against it; it is merely a consequence of the unique situation of public education in the United States, a situation that imposes losses as well as gains. I do not think, however, that theistic science should be described *as* science. It lacks the universality of science, as that term has been understood in the later Western tradition.[3] It also lacks the sort of warrant that has gradually come to characterize a properly "scientific" knowledge of nature, one that favors systematic observation, generalization, and the testing of explanatory hypothesis. Theistic science appeals to a specifically Christian belief, one that lays no claim to assent from a Hindu or an agnostic. It requires faith, and faith (we are told) is a gift, a grace, from God. To use the term *science* in this context seems dangerously misleading; it encourages expectations that cannot be fulfilled.

Plantinga objects to the sort of methodological naturalism that would deny the label *science* to any explanation of natural process that invokes the special action of God; indeed, he characterizes it, in Basil Willey's phrase, as provisional atheism. "Is there really any compelling or even decent reason for thus restricting our study of nature?" he asks (Plantinga 1991a, 27). But, of course, methodological naturalism does not restrict our study of nature; it just lays down which sort of study qualifies as *scientific*. Calling on the special action of God to explain the origins of the major phyla in the way Plantinga does transcends the boundaries of science.[4] This is not primarily because God is involved (Aristotle's argument for a First Mover, for example, could be counted a broadly naturalistic one), but because the action is a special one inaccessible to any sort of test on our part and because of the sort of evidence that has to be invoked, evidence that does not lend itself to evaluation by the standard techniques of natural science, however loosely these be defined.[5]

If someone wants to pursue another approach to nature—and there are many others—the methodological naturalist has no reason to object. Scientists *have* to proceed in this way; the methodology of natural science gives no purchase on the claim that a particular event or type of event is to be explained by invoking God's "special" action or by calling on the testimony of Scripture. Calling this *methodological* naturalism is simply a way of drawing attention to the fact that it is a way of characterizing a particular *methodology*, no more. In particular, it is not an ontological claim about what sort of agency is or is not possible. Dubbing it *provisional atheism* is objectionable; the scientist who does not include among the alternatives to be tested when attempting to explain some phenomenon an action that would not lend itself to such tests is surely not to be accused of atheism, even of a provisional sort. "What we need," Plantinga tells us, "is a scientific account of life that isn't restricted by methodological naturalism" (1991a, 29). But, of course, if it is not so restricted, it is simply improper to call it *scientific*, in the light of long and unequivocal contrary usage.

Let me make myself clear. I do not object (as the concluding section of this essay makes clear) to the use of theological considerations in the service of a larger and more comprehensive world view in which natural science is only one factor. I would be willing to use the term *knowledge* in an extended sense here, though I am well aware of some old and intricate issues about how faith and knowledge are to be related. (See, for example, Kellenberger 1972, ch. 10.) But I would not be willing to use the term *science* in this context. Nor do I think it necessary to do so in order to convey the respectability of the claim being made: that theology may appropriately modulate other parts of a person's belief-system, including those deriving from science. I would be much more restrictive than Plantinga is, however, in allowing for the situation he describes as "Scripture correcting current science."[6] But before I analyze our differences, it may be useful first to lay out the large areas where we agree.

POINTS OF AGREEMENT

What really galls Plantinga are the views of people like Richard Dawkins and William Provine who not only insist that evolution is a proven "fact," but who suppose that this somehow undercuts the reasonableness of any sort of belief in a Creator. Their argument hinges on the notion of design. The role of the Creator in traditional religious belief (they claim) was that of designer; the success of the theory of evolution has shown that design is unnecessary. Hence, there is no longer any valid reason to be a theist. In a recent review of a history of the creationist debate in the United States, Provine lays out this case, and concludes that Christian belief can be made compatible with evolutionary biology only by supposing that God "works through the laws of nature" instead of actively steering biological process by way of miraculous intervention. But his view of God, he says, is "worthless," and "equivalent to atheism" (Provine 1987). (On this last point, Plantinga and he might not be so far apart.) He chides scientists for publicly denying, presumably on pragmatic grounds, that evolution and Christian belief are incompatible; they *must*, he says, know this to be nonsense.

Plantinga puts his finger on an important point when he notes that for someone who does not believe in God, evolution in some form or other is the only *possible* answer to the question of origins. Prior to the publication of *The Origin of Species* in 1859, the argument from design for a Creator was widely regarded as resting directly on biological science.

The founders of physico-theology two centuries earlier (naturalists like John Ray and William Derham) had shown the pervasive presence in nature of means-end relationships, the apparently purposive adjusting of structure and instinctive behavior to the welfare of each kind of organism. Someone who rejected the idea of a designer, therefore, had to face some awkward problems in explaining some of the most obvious features of the living world; it seemed to many as though science itself testified to the existence of God (McMullin 1988).[7]

Darwin changed all this. By undermining the classical arguments from design, he showed that atheism was not, after all, inconsistent with biological science; from then on, the fortunes of atheism as a form of intellectual belief would, to some extent at least, be perceived as depending upon the fortunes of the theory of evolution. No wonder, then, that evolution became a crucial myth of our secular culture (as Plantinga puts it), replacing for many the Christian myth as "a shared way of understanding ourselves at the deep level of religion" (Plantinga 1991a, 17). No wonder also that an attack on the credentials of evolutionary theory would so often evoke from its defenders a reaction reminiscent in its ferocity of the response to heresy in other days.

Is evolution fact or theory? No other question has divided the two sides in the creation-science controversy as sharply. Plantinga argues that someone who denies the existence of a Creator is left with no other option for explaining the origin of living things than an evolutionary-type account. The account thus becomes "fact" not just because of the strength of the scientific evidence in its favor, but because, for the atheist, no other explanation is available. Plantinga objects to the use of the word *fact* in this context because it seems to exclude in principle the possibility of divine intervention, and hence by implication, the possibility of the existence of a Creator. *Fact* seems to convey not just the assurance of a well-supported theory, but the certainty that no other explanation is open.

The debate may often, therefore, be something other than it seems. Instead of being just a disagreement about the weight to be accorded to a particularly complex scientific theory in the light of the evidence available, the debate may conceal a far more fundamental religious difference, each side appearing to the other to call into question an article of faith. Religious believers point out that calling the thesis of common ancestry a *fact* violates good scientific usage; no matter how well-supported a theory may be (they argue), it remains a theory. To nonbelievers, the phrase *merely a theory* comes as a provocation because it suggests a substantial doubt about a claim that appears to them as being beyond question, a doubt prompted furthermore in their view by an illegitimate intrusion of religious belief.

At one level, then, Plantinga's essay can be read as a plea for a more informed understanding of the real nature of the creation-science debate, and a more sympathetic appreciation of what led the proponents of creation science to take the stand they did. Even their defense of a "young" earth (a major point of disagreement between his view and theirs) ought not (he says) to be regarded as "silly or irrational"; a "sensible person" might well subscribe to it after a careful study of the Scriptures. One need not be "a fanatic, or a Flat Earther, or an ignorant Fundamentalist" to hold such a view (Plantinga 1991a, 15). The claim that the earth is ancient is neither obvious nor inevitable; it has to be argued for, and disagreement may, therefore, legitimately occur.

Plantinga is right, to my mind, to see more in the creation-science debate than evolutionary scientists (or the media) have been wont to allow. And the sort of challenge he offers to the defenders of evolution, though it is not new, could serve the purposes of science in the long run if it forces a clarification and strengthening of argument on the other side, or if it punctures the sometimes troubling smugness that experts tend to display when dealing with outsiders. Plantinga leans too far in the other direction, however. In the first place, those who affirm that "evolution is a fact" are not necessarily committed to a covert denial of God's existence. The affirmation itself is, of course, an ambiguous one. A plausible construal of it in this context might run as follows: The belief that the relationships attested to by the fossil record, by comparative morphology, and by molecular biology are best explained in broadly evolutionary terms is true. Calling a theoretical belief *true* customarily means that the cumulative evidence in its favor is so strong that it is safe to affirm it without qualification, just as a geologist might, for example, affirm that the continents of Africa and South America, once in physical contact, have gradually separated from one another. This ought *not* be taken to mean that the alternative can be logically excluded in a completely conclusive way; nothing more than overwhelming likelihood is what scientists normally intend by this sort of usage. One may *object* to this usage, but one cannot impute an implicitly atheistic premise to those who follow it. Such a premise *may* be playing a covert role, but it is equally possible that it may not.

In the second place, the reading of creation science that he urges is rather too charitable. A claim does not have to be obvious or inevitable for its rejection to connote fanaticism or ignorance. If the indirect evidence for the great age of the earth is overwhelming (Plantinga himself allows that it is "strong"), if its denial would call into question some of the best-supported theoretical findings of an array of natural sciences (cosmology, astrophysics, geology, biology), then one is entitled to issue a severe judgment on the legitimacy of the challenge. Perusal of some of the standard works in creation science would lead one to suspect that no matter *how* strong the scientific case were in favor of an ancient earth, it would make no difference to their authors. Their implicit commitment to a literalist interpretation of *Genesis* is such that (to my mind, at least) it appears to block a genuinely rational assessment of the alternatives.

What bothers Plantinga, I suspect, about the use of terms like *fanaticism* here is that from *his* point of view the creation-scientist's heart is in the right place. Anyone who stands up for the maxim of *sola Scriptura* ("Scripture alone") in the modern world, even in contexts as unpromising as the debate about the age of the earth, ought not (he suggests) simply be dismissed as irrational. Creation-scientists may be wrong in holding that the earth is only a few thousand years old, but their intellectual commitment to Scripture ought to be regarded with sympathy by their fellow Christians. I am much less sympathetic to them, in part because of a deeper disagreement about the merits of the *sola Scriptura* premise as well as of the remaining major theses of creation science. Though I would not be as harsh on creation scientists as leading evolutionists have been, I would, as a Christian, want to register disapproval of creation science at least as strong as the latter's, though for reasons that differ in part from theirs. These reasons will become clear, I hope, in what I have to say about Plantinga's analysis of what happens when "faith and reason clash." . . .

THE ANTECEDENT LIKELIHOOD OF SPECIAL CREATION

The most distinctive feature of Plantinga's argument is that he makes a point of *not* calling explicitly upon the two creation narratives in Genesis. Historically, these narratives have provided the main warrant for the traditional Christian belief that God intervened in a special way in the origins of the living world. Defenders of that belief have tended to rely on Genesis, unless they were prevented from doing so, as the recent advocates of creation science were, by extrinsic constraints. Plantinga is, however, under no such constraints. His reason for eschewing the reference to Genesis that one might have expected to find is, rather, an awareness of the problematic character of the literalist approach to the Genesis story of creation (Plantinga 1991b, 81). Instead, he rests his case not on specific scriptural passages, but on a central defining theme in the biblical account of God's dealings with the people of Israel. In this context, at least, God evidently "intervened" or "interrupted" normal human routines in all sorts of ways. (Words like *intervene* are inadequate to convey the action of a Creator with the created universe, Plantinga reminds us, but we do not have any better ones.) Since the God of Abraham brought about God's ends in "special" ways throughout the long history of Israel, it is to be expected (Plantinga suggests) that the same may very well be true at some moments in the much longer story of the development of life on earth.

The issue, be it noted, is not whether God *could* have intervened in the natural order; it is presumably within the power of the Being who holds the universe at every moment in existence to shape that existence freely. The issue, is, rather, whether it is antecedently *likely* that God would do so, and more specifically whether such intervention would have taken the form of special creation of ancestral living kinds. Attaching a degree of *likelihood* to this requires a reason; despite the avowed intention not to

call on Genesis, there might appear to be some sort of residual linkage here. In the absence of the Genesis narrative, would it appear likely that the God of the salvation story would also act in a special way to bring the ancestral living kinds into existence? It hardly seems to be the case.

Might it be that the supposed likelihood of special creation in given cases (e.g., for the "founders" of the major phyla) derives directly from the unlikelihood of there being a scientific explanation in such cases? If there are only two possible types of explanation, and one can be shown to be highly improbable on present evidence, the other automatically gains in likelihood. In this event, a reference to God's dealings with Israel would not be needed. But Plantinga made it clear that this was not his strategy: "It is a bit more probable, *before we look at the scientific evidence,* that the Lord created life and some of its forms—in particular human life—specially" (1991a, 22, emphasis mine).

It is this casting of special creation and evolution as rivals in the domain of cosmological explanation that I find so troubling. If one assumes that there is a presumption in favor of some sort of special creation at the critical moments in the historical development of life (a presumption whose plausibility wanes in regard to specific transitions as the strength of the evolutionary explanation of those transitions increases) one inevitably transforms the field of prehistory into a battleground where the religious believer is engaged in constant skirmishes with the protagonists of evolutionary-type theories, skirmishes that most often end in forced retreat for the religious believer.

Plantinga claims that the Christian believer "has a freedom not available to the naturalist," because the believer is "free to look at the evidence . . . and follow where it leads" (1991a, 28). This would be more persuasive if he were to hold only that the believer holds an extra alternative that allows him or her to be more critical of the shortcomings of the scientific theory. But he proposes something much stronger than that: There is an antecedent *likelihood*, he says, of "special" intervention of this kind at some points in the cosmic process, and hence where the scientific case is weak, the hypothesis of divine intervention has to be allowed the higher likelihood. I am not sure that this *does* in the end allow the Christian believer more freedom than the naturalist. But whatever one makes of that, it certainly ensures conflict; it is likely to maximize the strain between faith and reason, as the believer searches for the expected gaps in the scientific account.

In his 1991b, Plantinga appears to change ground somewhat. On the one hand, he says: "I remain confident that TCA is relatively unlikely given a Christian or theistic perspective and the empirical evidence" (1991b, 108). But now the warrant for claiming the antecedent likelihood of special creation appears to shift from the salvation story to the "empirical evidence." Quoting Francis Crick and Harold Kein on the difficulty of explaining how the first cells originated, he concludes that "we have every reason to doubt that life arose simply by the workings of the laws of physics" (1991b, 102). He goes on:

> It therefore looks as if God did something special in the creation of life. (Of course, things may change; that is how things look *now*.) And if he did something special in creating life, what would prevent him from doing something special at other points, in creating human life, for example, or other forms of life? . . . I am therefore inclined to maintain my suggestion that the antecedent probability, from a theistic point of view, is somewhat against the idea that all the kinds of plants and animals, as well as humankind, would arise just by the workings of the laws of physics and chemistry. (1991b, 102)

The antecedent probability (no longer strictly antecedent) now seems to depend on the current lack of plausible scientific accounts of how the first cells could have originated. (Crick, who is notably unsympathetic to theistic belief, would surely not agree with the inference being drawn from this!) In his 1991b, Plantinga is more intent on shifting the burden of proof, and on combating claims for the antecedent probability, on theological grounds, of a naturalist account favoring TCA. If TCA were correct, "we should expect much stronger evidence than we actually have. . . . The actual empirical evidence must be allowed to speak more loudly than speculative theological assumptions" (1991b, 102). So much for his original claim that the story of God's dealings with Israel spoke loudly in favor of special creation over TCA! . . .

THE INTEGRITY OF GOD'S NATURAL WORLD

Plantinga's original argument relied on the premise that God's special intervention in the cosmic process is antecedently probable. Here is where he and I really part ways. My view would be that from the theological and philosophical standpoints, such intervention is, if anything, antecedently *improbable*. Plantinga builds his case by recalling that "according to Scripture, [God] has often intervened in the working of his cosmos" (Plantinga 1991a, 22). And the examples he gives are the miracles recounted in Scripture and the life, death, and resurrection of Jesus Christ. I want to recall here a set of old and valuable distinctions between nature and supernature, between the order of nature and the order of grace, between cosmic history and salvation history. The train of events linking Abraham to Christ is not to be considered an analog for God's relationship to creation generally. The Incarnation and what led up to it was unique in its manifestation of God's creative power and a loving concern for the created universe. To overcome the consequences of human freedom, a different sort of action on God's part was required, a transformative action culminating in the promise of resurrection for the children of God, something that (despite the immortality claims of the Greek philosophers) lies altogether outside the bounds of nature.

The story of salvation is a story about men and women, about the burden and the promise of being human. It is about free beings who sinned and who therefore *needed* God's intervention. Dealing with the human predicament "naturally," so to speak, would not have been sufficient on God's part. But no such argument can be used with regard to the origins of the first living cells or of plants and animals. The biblical account of God's dealings with humankind provides no warrant whatever for supposing that God would have brought the ancestors of the various kinds of plants and animals to be outside the ordinary order of nature. The story of salvation *does* bear on the origin of the first humans. If Plantinga were merely to say that God somehow leaned into cosmic history at the advent of the human, Scripture would clearly be on his side. How this "leaning" is to be interpreted is, of course, another matter.[8] But his claim is a much stronger one. . . .

TOO MUCH AUTONOMY?

But what are we to make of Plantinga's objection that having life coming gradually to be according to the normal regularities of natural process is "semi-deistic," i.e., that it attributes too much autonomy to the natural world? He says:

> God could have accomplished this creating in a thousand different ways. It was entirely within his power to create life in a way corresponding to the Grand Evolutionary scenario . . . to create matter . . . together with laws for its behavior, in such a way that the inevitable[9] outcome of matter's working according to these laws would be first, life's coming into existence three or four billion years ago, and then the various higher forms of life, culminating as we like to think, in humankind. This is a semi-deistic view of God and his working. (Plantinga 1991a, 21)

He contrasts this alternative with the one he favors:

> Perhaps these laws are *not* such that given enough time, life would automatically emerge. Perhaps he did something different and special in the creation of life. Perhaps he did something different and special in creating the various kinds of animals and plants (Plantinga 1991a, 22).

Plantinga's characterization of the first alternative as semideistic is intended to validate the second alternative as the appropriate one for the Christian to choose. But why should the first alternative be regarded as semideistic? He allows that it was within God's power to bring about cosmic evolution, but then asserts that to say God *did* in fact fashion the world in this way would be semideistic. This is puzzling. It would be semideistic, perhaps, if we *already* knew that God had intervened in bringing into existence some kinds of plants and animals, in which case the "grand evolutionary scenario" would attribute a greater degree of autonomy to the natural world than would be warranted. But this is exactly what we do *not* know. And to assume that we *do* know it would beg the question.

The problem may lie in the use of the label *semi-deistic*.[10] A semideist, Plantinga remarks, could go so far as to allow that God "starts everything off" and "constantly sustains the world in existence" and could

even maintain that "any given causal transaction in the universe requires specific divine concurrent activity." All this would, apparently, not be enough to make such a view acceptable. What more could be needed? Defining God's relationship with the natural order in terms of creation, conservation, and *concursus*, has been standard, after all, among Christian theologians since the Middle Ages. Perhaps what still needs to be made explicit is that God *could* also, if God so chose, relate to the created world in a different way, either by way of special creation, or in the dramatic mode of a grace that overcomes nature and of wonders that draw attention to the covenant with Israel and ultimately to the person of Jesus. The possibility of such an "intrusion" on God's part into human history, of a mode of action that lies *beyond* nature, must not be excluded in advance, must indeed be affirmed. I take it that the denial that such a mode of action *is* possible on the part of the Being who creates and conserves and concurs is what constitutes semideism, in Plantinga's sense of that term.

But someone who asserts that the evolutionary account of origins is the best-supported *one* is *not* necessarily a semideist in this sense. Some defenders of evolution—notably those who deny the existence of a Creator and are, therefore, not deists of any sort—would, of course, exclude special creation in this way, in principle. But there is no intrinsic connection whatever between the claim that God did, in fact, choose to work through evolutionary means and the far stronger claim that God *could* not have done otherwise. Nor, of course, is there any reason why someone who defends the evolutionary account of origins should go on to deny that God might intervene in the later human story in the way that Christians believe God to have done.

In sum, then, at least *four* alternatives would have to be taken into account here. There are those who defend the evolutionary account of origins, and also rejecting the existence of God, would (if pressed) say that life could not *possibly* have come to be except through evolution. There may be those who maintain that God created, conserves, and concurs in the activity of the universe but *could* not "intervene" in a special way in its history to bring new kinds of animals and plants to be, for example. These (if they exist) are the semideists Plantinga describes. Then there are those who prefer the evolutionary account of origins on the grounds of evidence that this is in fact most probably the way it happened, but who are perfectly willing to allow that it was within the Creator's power to speed up the story by special creation of ancestral kinds of plants and animals, even though (in their view) this was not what God did. This is a view that a great many Christians from Darwin's day to our own have defended; it is the view I am proposing here. It is *not* semideistic. And finally, there is the option of special creation: that God *did*, in fact, intervene by bringing various kinds of living things to be in a "special" way.

When Plantinga presents two alternatives only, the second being that God might "perhaps" have intervened as defenders of special creation believe occurred, he must be supposing that the other alternative, the "grand evolutionary scenario," is one that excludes such a "perhaps"; i.e., that excludes, *in principle*, the possibility that God could have intervened in a special way in the natural order. What I am challenging is this supposition. The Thesis of Common Ancestry can claim, as we have seen, an impressive body of evidence in its own right. It need not rely on, nor does it entail any in-principle claim about what God could or could not do.[11]

CONCLUSION

So, finally, how *should* the Christian regard this thesis? Perhaps better, since there are evidently "distinctive threads in the tapestry of Christianity" in Plantinga's evocative metaphor (1991a, 30), how might someone respond who sees in the Christian doctrine of creation an affirmation of the integrity of the natural order? TCA implies a cousinship extending across the entire living world, the sort of coherence (as Leibniz once argued) that one might expect in the work of an all-powerful and all-wise Creator. The "seeds," in Augustine's happy metaphor, have been there from the beginning; the universe has in itself the capacity to become what God destined it to be from the beginning, as a human abode, and for all we know, much else.

When Augustine proposed a developmental cosmology long ago, there was little in the natural science of his day to support such a venture. Now that has changed. What was speculative and not quite coherent has been transformed, thanks to the labors of countless workers in a variety of different scientific fields. TCA allows the Christian to fill out the metaphysics of creation in a way that (I am persuaded) Augustine and Aquinas would have welcomed. No longer need one suppose that God must have added plants here and animals there. Though God *could* have done so, the evidence is mounting that the resources of the original creation were sufficient for the generation of the successive orders of complexity that make up our world.

Thus, common ancestry gives a meaning to the history of life that it previously lacked. From another perspective, this history now appears as preparation. The uncountable species that flourished and vanished have left a trace of themselves in us. The vast stretches of evolutionary time no longer seem quite so terrifying. Scripture traces the preparation for the coming of Christ back through Abraham to Adam. Is it too fanciful to suggest that natural science now allows us to extend the story indefinitely farther back? When Christ took on human form, the DNA that made him son of Mary may have linked him to a more ancient heritage stretching far beyond Adam to the shallows of unimaginably ancient seas. And so, in the Incarnation, it would not have been just human nature that was joined to the Divine, but in a less direct but no less real sense all those myriad organisms that over the aeons had unknowingly shaped the way for the coming of humanity.[12]

Anthropocentric? But of course: The story of the Incarnation *is* anthropocentric. Reconcilable with the evolutionary story as that is told in terms of chance events and blind alleys? I believe so, but to argue it would require another essay. Unique? Quite possibly not: Other stories may be unfolding in very different ways in other parts of this capacious universe of ours. Terminal? Not necessarily: We have no idea what lies ahead for humankind. The transformations that made us what we are may not yet be ended. Antecedently probable from a Christian perspective? I will have to leave that to the reader.

NOTES

1. For the text of the judgment, "McLean vs. Arkansas," see Gilkey 1985, 266–301. The judgment is not itself without some logical difficulties; see Quinn 1984.

2. Plantinga's essay was featured in a special issue of *Christian Scholar's Review* 21 (1991), 8–32 (here 1991a). The issue carried critical responses by Howard Van Till (33–45), and myself (55–79), as well as a detailed reply by Plantinga (here 1991b). The present essay is a revised and considerably augmented version of my paper in that volume. I am grateful to Dr. Plantinga for our discussions of these issues, and for the characteristic care he took in responding to my original criticisms.

3. In defense of his usage, Plantinga notes that theology at an earlier time was called a science (1991b, 98). But this usage was recognized to be problematic from the Aristotelian viewpoint of that time. To the objection that theology cannot be regarded as a science because it proceeds from premises not admitted by all, Aquinas responds that because these premises are revealed by God, they can be accepted on authority, just as optics takes its principles from geometry (*Summa Theologica*, Vol. 1 q.1, a.2). But this does not really answer the objection adequately, since the revealed character of these premises is not admitted by all. And the Aristotelian distinction between what is better known to us and what is better known "in itself" will not do the work. When the Aristotelian conception of science (deduction from self-evident premises) was gradually abandoned in the seventeenth century, the new conceptions that succeeded it made the extension of the term *science* to theology even more problematic, particularly in the present context of the knowledge of nature.

4. Calling it God's "direct" action would leave matters ambiguous, since it could be said that God's action in sustaining the world in existence is direct action; this sort of action is, of course, not in dispute here. What makes God's "special" action inaccessible to the methods of natural science is that it lies, as medieval philosophers put it, "outside nature," outside the pattern of regularities that afford a foothold for later inquirers. The most that science could do where "special" action is claimed, as in the case of miracles, would be to exclude, as far as possible, alternative "natural" explanations. But when special creation is supposed to have occurred in the early history of life on earth, this (as we shall see) is *very* difficult to do.

5. This argument does not depend on an ability to draw a sharp demarcation between science and nonscience. Scientists often rely on principles of natural order of a broadly metaphysical sort, but these are in principle accessible to all; they are over the long run at least partially adjudicable in terms of the "success" (in a fairly specific sense) of the the-

ories employing them. (See McMullin, 1993.) Reliance on Scripture is another matter entirely.

6. As an illustration of how Scripture could "correct current science," Plantinga remarks: "If, for example, current science were to return to the view that the world has no beginning, and is infinitely old, then current science would be wrong" (Plantinga 1991a, 14). I do not believe that Scripture *does* prescribe that the universe had a beginning in time, in some specific technical sense of the term *time;* the point of the creation narratives is the dependence of the world on God's creative act, to my mind, not that it all began at a finite time in the past. A world that has always existed would still (as Aquinas emphasized) require a creator. As an illustration of how complex the notion of temporal beginnings has become, the Hawking model of cosmic origins mentioned by Plantinga does not imply that the universe is infinitely old (as that phrase would ordinarily be understood), but rather that as we trace time backwards to the Big Bang, the normal concept of time may break down as we approach the initial singularity some 15 billion years ago. The history of "real time" (as Hawking calls it) would still be finite in the same terms as before, as he explicitly points out (Hawking 1988, 138). The question of whether or not the time elapsed in cosmic history is finite or infinite depends, in part, on the choice of physical process on which to base the time scale, particularly on whether it is cyclic or continuous. The question of the finitude or infinity of past time, so much debated by medieval philosophers and theologians, cannot straightforwardly be answered in absolute terms. The notion of time measurement is far more complex and theory-dependent than earlier discussions allowed. But the theological *point* of the biblical account of creation remains untouched by technical developments such as these (McMullin 1981, 35).

7. The exponents of physico-theology were not entirely sure how to classify their arguments from design concerning origins. These could not be directly tested in the normal empirical ways, but it did seem as though "naturalist" explanations could be systematically excluded.

8. "God fashioned Adam from the dust of the earth and breathed into his nostrils the breath of life" (Genesis 2:7). The "fashioning" here could be that of a billion years of evolutionary preparation of that "dust" to form beings that for the first time could freely affirm or freely deny their maker. Pope Pius XII in his encyclical *Humani Generis* (1950) allowed that such an evolutionary origin of the human

body was an acceptable reading of the Genesis text. But he added that the human *soul* could not be so understood; souls must be "immediately created" by God (1950, 181). The Platonic-sounding dualism underlying this distinction requires further scrutiny. The uniqueness of God's covenant with men and women and of the promise of resurrection does not require that there be a naturally immortal soul, distinct in its genesis and history from its "attendant" body. But it is unnecessary to develop this issue here, since Plantinga's challenge extends to the evolutionary account of the plant and animal worlds, not simply of the human alone.

9. *Inevitable* is a word that defenders of evolution, whether theists or not, would be inclined to challenge. It suggests that the evolutionary process is, at least in a general way, deterministic or predictable. But this is just what nearly all theorists of evolution would deny.

10. In the entry under *deism* in *The Encyclopedia of Religion,* Allen Wood remarks that the term *deism* tended over time to become "a vague term of abuse" when used by Christian writers with regard to hypotheses that in their view attributed an undue degree of autonomy to the universe.

11. There is one further perspective on this matter of semideism that I have set aside above. The occasionalists of the fourteenth century maintained that God is the *only* cause, strictly speaking, of what happens in the world. What appears to be causal action within the world is for them no more than temporal succession. Things do not have natures that specify their actions; rather, the fact that they act according to certain norms must be directly attributed to God's intentions. There is no reason in this view why God should not, for example, suddenly make new kinds of plants and animals appear, if God so wishes; since there is no order of *nature,* God is committed only to the reasonable stability of (more or less) regular succession on which human life depends. (The issue that separated the nominalists from the Aristotelian defenders of real causation in nature is brought out very well in the essay by Alfred Freddoso [1988] cited by Plantinga.) In this perspective, the issue of special creation comes to be posed in a quite different way. Any view which affirms the sufficiency of the natural order for bringing about the origins of life might be dubbed by the occasionalist as semideist. When I read the paragraph where Plantinga says that someone who maintains that God creates, conserves, and concurs in the activity of the universe

can still be semideistic, my first reaction was to assume that this committed him to occasionalism, since it would seem that it is *only* from the occasionalist perspective that this view of God's relationship with the natural order would be classed as semideist. But Plantinga is quite evidently not an occasionalist; his treatment of natural science implies that he believes in the operation of secondary causation in nature. Thus, I have assumed in the discussion above that he must have had something else in mind when speaking of semideism, namely, the openness of creation to the supernatural order of grace and miracle. Incidentally, the occasionalist *would* be likely to believe that special creation is antecedently more probable, and (in Berkeley's version, at least) might tend to question a theory, like the theory of evolution, which depends on the reality of such causes as genetic mutation.

12. Though the alert reader will have caught echoes of the theology (not the biology) of Teilhard de Chardin, the affinities with the Christology of Karl Rahner are, perhaps, more immediate. See, for example, Rahner 1961, 30.

REFERENCES

Freddoso, Alfred. 1988. "Medieval Aristotelianism and the Case against Secondary Causation in Nature." In *Divine and Human Action,* ed. T. V. Morris, 74–118. Ithaca, N.Y.: Cornell Univ. Press.

Gilkey, Langdon. 1985. *Creationism on Trial.* Minneapolis: Winston.

Hawking, Stephen. 1988. *A Brief History of Time.* New York: Bantam Books.

Kellenberger, James. 1972. *Religious Discovery, Faith, and Knowledge.* Englewood Cliffs, N.J.: Prentice-Hall.

McMullin, Ernan. 1967. *Galileo, Man of Science.* New York: Basic Books.

———. 1981. "How Should Cosmology Relate to Theology?" In *The Sciences and Theology in the Twentieth Century,* ed. A. R. Peacocke, 17–57. Notre Dame, Ind.: Univ. of Notre Dame Press.

———. 1988. "Natural Science and Belief in a Creator." In *Physics, Philosophy, and Theology,* 49–79, ed. R. J. Russell, W. R. Stoeger, and G. V. Coyne. Notre Dame, Ind.: Univ. of Notre Dame Press and Rome: Vatican Observatory Press.

———. 1993. "Indifference Principle and Anthropic Principle in Cosmology." *Studies in the History and Philosophy of Science* 24 (1993): 359–389.

Pius XII. [1950] 1981. *Humani Generis.* In *The Papal Encyclicals 1939–1958,* ed. Claudia Carlen. Raleigh: McGrath.

Plantinga, Alvin. 1991a. "When Faith and Reason Clash: Evolution and the Bible." *Christian Scholar's Review* 21:8–32.

————. 1991b. "Evolution, Neutrality, and Antecedent Probability: A Reply to Van Till and McMullin." *Christian Scholar's Review* 21:80–109.

Provine, William B. 1987. Review of *Trial and Error. The American Controversy over Creation and Evolution,* by E. J. Larson. In *Academe* 73 (1):50–52.

Quinn, Philip. 1984. "The Philosopher of Science as Expert Witness." In *Science and Reality,* ed. J. T. Cushing, C. F. Delaney, and G. M. Gutting, 32–53. Notre Dame, Ind.: Univ. of Notre Dame Press.

Rahner, Karl. 1961. "Christology within an Evolutionary View of the World." In *Theological Investigations,* 157–92. Vol. 5. Baltimore: Helicon.

43

PETER ATKINS

does science have limitations?

Purposeless People

When confronted with the analysis of any concept, however complex, the only intellectually honest approach is to explore the extent to which an absolutely minimal explanation can account for the reliable evidence. There is no justification for departing from this procedure when the complex concept in question is that of the person, however deeply emotive it may be, and however much we may long for a reassuring outcome. Only if a minimal approach is explicitly demonstrated to be inadequate may there be some justification for indulging in the soft furnishings of additional hypotheses. We should begin our exploration, therefore, by asking whether the concept of person, which I take to be the concept of our individual existence, persistence, and role in this universe, can be explained without the sugar-coating of invented attributes of persons, additions that have been pro-

posed by the under-informed or the wiley perhaps, and have been adhered to generally by the religious.

We should ask whether the concept of personal existence can survive stripped-down explanations and their ramifications. Is there any support for the existence of something beyond the absolutely sparse? Is there life beyond bones? Are the fat and tallow of religious, philosophical, and psychological forms of justification necessary and not merely desirable? And if fat and tallow are found to be unnecessary, is there any justification for an ethical view amid the bones of people's purposelessness?

I will argue that it is intellectually dishonest at this stage of human development to resort to the artifice of supposing man's existence to be justified by recourse to something beyond this world. I will argue that the time is ripe for the faithful to relinquish their prejudices and to examine with an open mind the possibility that the world is a happy accident, that we might be creatures of chance, nothing more than fragments of highly organized matter, and that

---

Persons and Personality, ed. A. Peacocke and G. Gillette (Oxford: Basil Blackwell, 1987) pp. 12–32. Reprinted by permission of the author.

there may be nothing intrinsically special about us apart from the complexity of our responses. I will also argue that the long-term future of everything is oblivion and annihilation.

I challenge anyone who seeks, hopes, and believes in a seductively richer option and a rosier destiny to accompany me to the bedrock of existence and to build their beliefs on it only in so far as their shelter is shown to be essential. I will argue that *all* softenings of my absolutely barren view of the concept of person and of the foundations of this wonderful, extraordinary, delightful world are sentimental wishful thinking.

THE OMNICOMPETENCE OF SCIENCE

The spring of my apparently barren but, as I hope to show, deeply enriching world-view, is my belief that science is all-competent.

I base this belief on the observation that science has never encountered a barrier that it has not surmounted or that it cannot reasonably be expected to surmount eventually. I admit that this is only a belief, but I will argue that it is supportable, and that it is simpler than the alternatives and should therefore be given priority over more elaborate views. In alluding to domains not yet fully conquered, I have in mind the at least arguable—and certainly not yet explicitly denied—view that, in due course, science will be equipped to deal with aesthetic and religious experience, and will be able to account for our perception of ourselves as distinct, responding entities. It will do so, I believe, by showing that the characteristically human capacities which we lump together for convenience of discourse as 'human spirit' or 'soul' are no more than states of the brain, and likewise that extension of the idea of a soul to expectation of eternal persistence is already quite plainly explicable in terms of a deep-seated desire to avoid, and the inability to come to terms with, the prospect of one's own annihilation.

I do not see any evidence to support the claim that there are aspects of the universe closed to science. Given the success of science in encroaching on the territory traditionally regarded as that of religion, I can accept that many people *hope* that its domain of competence will prove bounded, with things of the spirit on one side of the fence, things of the flesh on the other. But *until it is proved otherwise,* there is no reason to suppose that science is incompetent when it brings its razor to bear on belief. Until the day that science is explicitly shown to be incompetent, we should acknowledge that its not-yet-stopped razor is slicing through the fabric of the heavens and leading us towards an extraordinary deep understanding of the composition, organization, and origin of the world.

As science's razor continues to slice, so it is revealed that much that was once inexplicable stems from the workings of laws that are simplicity itself. What grounds are there for assuming that the razor will become blunt or will run against the uncuttable? Pessimism? Fear? Outrage? Surely such emotional cringing is poor reason for not permitting this supreme device of the human intellect, our science, to run its course.

If we are to approach our topic with an open mind, we should prepare ourselves to see sliced from the concept of person many of some people's most cherished beliefs. It would be premature to say that science, which at present is undeniably in full flood, cannot deal with the great questions. Give science time: it is in the midst of its achieving; do not yet merely deny its omnicompetence, and do not resort to traditional explanations unnecessarily. In assessing whether a purely scientific world-view is likely to be complete, do not assume that because religious views have been around longer, they are more likely to be right. In wondering whether a sparse scientific view of the person could suffice, I think it only fair to play the game of reversing history, of envisaging a religious upstart battling against a high ground held for millennia by science. Could anyone seriously take religion's mysteries to be more compelling than science's public achievements? I picture a dog suddenly woken into our intellectual level and presented with the offerings of religion and science. What dog, unfettered by our cultural heritage and free from the iron grip of religion's social, economic, private, and artistic propaganda, would opt for religion's 'explanations'? Surely, any honest dog would accept that science was so well along the road to full explanation, that it should side with science and discard religion, at least until—if ever—science failed to deliver.

THE JOURNEY INTO SIMPLICITY

My attitude is that the omnicompetence of science and, in particular, the simplicity revealed by its insights should be accepted as a working hypothesis until, if ever, it is proved inadequate. This is relevant to our discussion in a multitude of different ways. Among them is the lesson taught by science about the power of the unconstrained and undirected to lead to rich consequences, consequences that are so rich that they can readily be mistaken for purposeful, directed events.

I think it worth exploring this last statement more deeply, for many non-scientists see science as an increasingly complex edifice, with each new discovery, each new theoretical concept, adding one more pimple to an already over-carbuncled and bunioned body. Such a view could not be further from the truth. Each new discovery of fundamental science—and we are not at this stage concerned with its applications—reveals one more facet of an underlying *simplicity*, a simplicity that allows more to be explained by fewer concepts and precepts.

Not all scientists, never mind non-scientists, recognize or acknowledge this simplicity. Some see only the enormous effort and complex equipment needed to make even the seemingly most trivial advance in understanding and confuse the complexity entailed in gathering information with the underlying simplicity revealed by the information so gathered. Some, seeing how new ideas overthrow old familiar ones, as when quantum theory replaced classical mechanics, do not see beyond the loss of the familiar to the sharp insight that comes from discarding approximations and shifting viewpoint. Some, seeing the mathematics required both to express an idea and to relate it to an observation, confuse the complexity with which simple concepts band together to masquerade as complex phenomena with the simplicity of the concepts themselves.

If there is a deep message that a scientist should convey to non-scientists, it is that simplicity can have consequences of extraordinary complexity, a revelation resulting from science's ability to discern the ways in which these simplicities tangle into testable, observable complexities.[1]

THE MOTIVE POWER
OF CHANGE

I must take you to one more level of scientific explanation if you are to appreciate the power of science to attack great problems, and if you are to see what I have in mind when I speak of complexity as the child of simplicity.

The second law of thermodynamics,[2] that child of the steam-engine, has, in the course of the nineteenth and twentieth centuries, grown to become the greatest liberator of the human spirit, and the steam-engines from which it sprang can now be seen to have forged wings for humanity's aspirations. Sadi Carnot—and it is fitting that he was named after a poet—pointed the way to the formulation of this law that acts as signpost in the direction of natural change, a law of such universality that it applies to every kind of change, from the most primitive—such as the cooling of hot metals and the expansion of gases—to the marginally more complex—such as the synthesis of ammonia and the rusting of iron—to the most complex—such as the emergence of new species, the formation of opinions, introspection, hallucination, self-deception, and comprehension.

Carnot's reflections leading to this law, as well as those of his successors Kelvin and Clausius, who formulated it rigorously and explored some of its consequences, were views of matter and its behavior as seen from the outside. Theirs were powerful formulations of descriptions which did not, on their own, lead to insight into the spring of the world. It was left to the short-sighted Boltzmann to see further than all his contemporaries and to formulate an explanation of the second law in terms of the behavior of atoms before their existence was generally accepted. Many of his contemporaries doubted the credibility of Boltzmann's assumptions and feared that his work would dethrone the concept of purpose that they presumed to exist within the world of change, just as Darwin had recently dispossessed its outer manifestations. Suffering from their scorn, Boltzmann, himself committed to the omnicompetence of science, was overcome by instability and unhappiness, and killed himself.

Yet now we know that Boltzmann's interpretation of the second law in terms of atoms was right in broad terms. Today no one seriously doubts the existence of atoms, and no one has reason to suppose that they are driven by a sense of purpose. They are driven by forces, their intrinsic properties—their mass, their charge, and, more deeply, their wave nature—determining their response and their paths through the world, bundled together as an element here and an elephant there. In the last twentieth century, we see that the direction of natural, spontaneous change, the direction taken by events when the universe is left to free-wheel, is that of increasing universal chaos.

Boltzmann's insight into the deep structure of natural change, which accounts for all the events of the world captured by Carnot's, Kelvin's, and Clausius's phenomenological thermodynamics and has now been extended beyond them, is that the universe is sinking, purposelessly, into ever greater chaos. Thus, energy tends to disperse from compact, highly concentrated regions, such as a nucleus, a lump of coal, a sandwich, or the sun. Similarly, the energy of orderly motion tends to become chaotic, as when a bundle of atoms grouped together as a ball hits a wall and the uniform motion of its atoms is rendered chaotic by the collision, and the net motion is randomized into chaotic thermal motion. Likewise, the location of particles becomes chaotic, when they move, collide, and spread, as when a gas expands into a vacuum.

These three kinds of sinking into chaos account for all natural change.[3] That they account for the decay and decomposition of matter may be self-evident. But the richness of these processes, and that of science, is due to the fact that the very same processes also account for the emergence of structures. That is, the second law allows for the abatement of chaos in one region, as long as there is a greater flood elsewhere, so that, overall, the universe becomes more chaotic. This extraordinary, creative, constructive characteristic of the collapse into chaos is a result of the universe being a network of interdependences: pull a string here and a lever moves there. Falling water here drives a lathe there. A chemical reaction here is linked to another reaction there. A new substance produced in one reaction is used in another linked reaction, and together they add to the chaos of the world; but whereas one reaction has resulted in decay, as when a sandwich is eaten, the other has given birth to a protein.

The universe is an astonishingly, but not incredibly, rich, interconnected network of events. It moves forward as a result of its gradual sinking into chaos, but its interconnectedness is so rich that a surge of chaos there may effloresce into a cathedral, a symphony, or a deed here. Structures—in the broadest sense, including raising a stone in a field to build a house—moving a molecule in a brain to form an opinion—are local abatements of chaos, driven by a greater surge of chaotic dispersal elsewhere.

CHAOS, PURPOSE, AND THE EMERGENCE OF PEOPLE

Of course, we cannot yet trace a surge of chaos at some point and identify it with a particular deed. Perhaps we never will be able to, for it is undeniable that the brain is an exceptionally complex test-tube. Yet this criticism is irrelevant. It is enough for our present purposes to expose the bedrock of the scientific explanation of events: that the chaotic dispersal of energy and matter is interconnected in an intricate web, as in a diabolical organ where depressing one key may sound a chord of a billion unpredictable notes.

As yet, little has been said of purpose. At this stage science can perform its elucidation without appealing to the shroud of obscurity of man-made artifice. A block of hot metal will cool, not because its purpose is to cool, but because the spontaneous chaotic dispersal of its energy results in it cooling. A shoot emerges from a seed and grows into a plant, not because the seed's purpose is to grow, but because the intricate network of reactions in its cells are gearboxes that propel its growth as the rest of the world sinks a little further into chaos. The lily is a flag hoisted by collapse into purposeless chaos. All the extraordinary, wonderful richness of the world can be expressed as growth from the dunghill of purposeless, interconnected decay.

People, too, have emerged as the same dunghill has effloresced. One molecule capable of reproducing itself in its own image is all it needs to set the world on the progress that has culminated in it being peopled with persons. All that was needed initially was a supply of suitable molecules that could be linked together into one larger molecule with the capacity to act as a template for its own replication. There are several conjectures as to how the appropriate organic molecules might have accumulated in the warm, wet, storm-ridden, and ultraviolet-soaked conditions prevailing on this planet three and a half billion years ago, and how they might have been combined into molecules that are the ancestors of today's DNA. Science, although far from sure, is certainly not defeated when it comes to suggesting how chance may have transformed the inanimate into the animate, with the potential for humanity.

One such global grandfather molecule could have entered the world through the blind, purposeless action of the second law, and at its inception become king. The blind activity of the second law, leads to replications of the king, and with more than one king, the world is at war and natural selection rules. I do not intend to trace the steps of evolution,[4] except to say that we can see in broad outline, and here and there in detail, how, given the conditions prevailing on this planet, the grand sinking into chaos led to us.

We have inherited the earth, at least temporarily, because our ancestors were equipped by chaos with mobility and brains that could respond ever more adroitly to the pressures of circumstance. We are, in a word, the children of chaos. At root there is only corruption and the unstemmable tide of chaos. That is the bleakness we should accept, as the starting-point for our analysis of the concept of person, for science can account for our emergence without the imposition of any extraneous view; it does not need to smear on to its clear, sinewy explanations any invented concept, particularly that of purpose. Purpose is unnecessary; all that is required to account for the emergence of people is interconnectedness and time.

Yet, when we look around and see beauty, when we look within and experience consciousness, when we participate in the delights of life, we feel in our hearts that the heart of the universe is richer by far. But do not be seduced: that is sentiment, and not what we should know in our minds. Unless it can be explicitly demonstrated otherwise, we should adopt the view that all attributes of persons have grown in response to the pressures for survival, and that any rationalization of them in terms of *additional* hypotheses, such as that of a creating, rewarding, or admonishing god or a teleological sense of purpose, are unjustified, superfluous superimpositions.

The bare bones of the scientific explanation of the emergence, existence, and temporary persistence of persons are that the universe is sinking into chaos. Faith in things beyond is a psychologically motivated fat and tallowing of these bones: God, an afterlife, the concept of purpose, are merely attempts to ameliorate the prospect of death, to unload the burden of guilt, and to soften the hardships of life. There is not one iota of justification for them beyond assertion, wishful thinking, and hallucination.

Although I can see that there is in some sense a justifiable temporally local sense of purpose—as for example, when I attempt to persuade others to my views (for reasons I could state), and more generally to contribute to the cultural heritage of the world; but I know that in the long term such activity is futile. For, after billions of years, even though we may have mastered the galaxies, founded new, young worlds, built our own private utopias next to stars we have learned to ignite and perhaps refresh, and have individually acquired physical and mental immortality, there will come a time when activity in the universe will cease. Then, at the dead end of the world, when all matter has decayed, when the expansion of the universe has distended space so much that it is a perfect void, we and all our achievements will come to naught. In the end there will be only dead flat space-time, our castles will have gone, as well as our libraries, our achievements, our selves. We, who will no longer be, will then listen in vain in the void for the Last Trumpet.

The sound of the Last Trumpet could, I suppose, be that of the Big Crunch, when the expanding universe, gripped by its own gravitation, ceases its current expansion, falls back on itself like a returning high-thrown ball, and crashes together in a final fury. Such a crunch could be the precursor of another Big Bang, a new First Trumpet, another cycle of expansion and contraction, another episode of groping for understanding in the liberty of newly refreshed space-time. But, aside from the fact that I consider it more likely that this is a one-shot universe, with a window briefly and only once opened for comprehension and aspiration, the Big Crunch would erase the tracks of our progress through this episode, and so, just as surely as in a one-shot universe, our achievements would be as though they had never been.

There is no reason to believe, except wishful thinking, that the long-term future of the human race, if it has one, even if it should evolve into superhumans, will be anything other than annihilation and oblivion. The long-term future, and in that sense purpose, of humanity should be regarded as empty. We came from nothing and will return to nothing, leaving behind neither material nor intellectual castles. All traces of our wars, poems, theories, and aspirations, all traces of our existence, and all traces of the existence of all matter will be erased.

NOTES

1. P. W. Atkins, *The Creation* (W. H. Freeman, Oxford, 1981).
2. P. W. Atkins, *The Second Law* (W. H. Freeman, New York, 1984).
3. Ibid.
4. R. Dawkins, *The Selfish Gene* (Oxford University Press, Oxford, 1976).

44

MARTIN GARDNER

Science and the Unknowable

Existence, the preposterous miracle of existence! To whom has the world of opening day never come as an unbelievable sight? And to whom have the stars overhead and the hand and voice nearby never appeared as unutterably wonderful, totally beyond understanding? I know of no great thinker of any land or era who does not regard existence as the mystery of all mysteries.

—John Archibald Wheeler

One of the fundamental conflicts in philosophy, perhaps the most fundamental, is between those who believe that the universe open to our perception and exploration is all there is, and those who regard the universe we know as an extremely small part of an unthinkably vaster reality. These two views were taken by those two giants of ancient Greek philosophy, Plato and Aristotle. Plato, in his famous cave allegory, likened the world we experience to the shadows on the wall of a cave. To turn this into a mathematical metaphor, our universe is like a projection onto three-dimensional space of a much larger realm in a higher space-time.

For Aristotle the universe we see, although parts of it are beyond human comprehension, is everything. It is a steady-state cosmos, self-caused, having no beginning or end. There is no Platonic realm of transcendent realities and deities. Plato succumbed to what Paul Kurtz likes to call the "transcendental temptation." Aristotle managed to avoid it.

In recent years cosmologists have blurred the distinction between the universe we know and tran-

Skeptical Inquirer 22 (November/December 1998), pp. 20–23. Used by permission of the *Skeptical Inquirer*.

scendent regions by positing a "multiverse" in which an infinity of universes are continually exploding into existence, each with a unique set of laws and constants. This is one way to defend the anthropic principle against the argument that the universe's fine tuning is evidence of a Designer. It is known that if any of some dozen constants is altered by a minuscule fraction it would not be possible for suns and planets to form, let alone life to evolve. The counter argument: If there is an infinity of universes, each with an unplanned, random set of constants, then obviously we must exist only in a universe with constants that permit life to evolve.

The multiverse concept, however, is far from a step toward Platonic transcendence. The other universes do not differ from ours in any truly fundamental way. They all spring into being in response to random fluctuations in the same laws of quantum mechanics, varying only in the accidental way their Big Bang creates laws. There is still no need to leap from a godless nature to transcendental regions that somehow lie beyond the multiverse.

A few cosmologists and far-out philosophers have gone much further. They conjecture that all possible universes exist—that is, every universe based on a noncontradictory set of laws. In the many-worlds interpretation of quantum mechanics, the universe is constantly splitting into parallel worlds, but these countless worlds all obey the same laws. The multi-multiverse of all-possible-worlds is a much larger ensemble, obviously infinite because the number of logically consistent possibilities is infinite. Most physicists do not buy this view because it is the utmost imaginable violation of Occam's razor. Leibniz's notion of a Creator who surveyed all logically possible worlds, then selected what She considered the most desirable, is surely a simpler conjecture by many orders of magnitude.

A question now arises. As science steadily advances in its knowledge of nature, never reaching absolute certainty but always getting closer and closer to understanding nature, will it eventually discover everything?

We have to be careful to define what is meant by "everything." There is a trivial sense in which humanity cannot possibly know all there is to know. We will never know how many hairs were on Plato's head when he died, or whether Jesus sneezed while delivering the Sermon on the Mount. We will never know all the decimal digits of pi, or all possible theorems of geometry. We will never know all possible theorems just about triangles. We will never know all possible melodies, or poems, or novels, or paintings, or jokes, or magic tricks because the possible combinations are limitless. Moreover, as Kurt Gödel taught us, every mathematical system complex enough to include arithmetic contains theorems that cannot be proved true or false within the system. Whether Gödelian undecidability may apply to mathematical physics is not yet known.

When physicists talk about TOEs (Theories of Everything) they mean something far less trivial. They mean that all the fundamental laws of physics eventually will become known, perhaps unified by a single equation or a small set of equations. If this happens, and physicists find what John Wheeler calls the Holy Grail, it will of course leave unknown billions and billions of questions about the complexities that emerge from the fundamental laws.

At the moment, cosmologists do not know the nature of "dark matter" that holds together galaxies, or how fast the universe is expanding, and hundreds of other unanswered questions. Biologists do not know how life arose on Earth or whether there is life on planets in other solar systems. Evolution is a fact, but deep mysteries remain about how it operates. No one has any idea how complex organic molecules are able to fold so rapidly into the shapes that allow them to perform their functions in living organisms. No one knows how consciousness emerges from the brain's complicated molecular structure. We do not even know how the brain remembers.

Such a list of unknowns could fill a book, but all of them are potentially knowable if humanity survives long enough. Too often in the past scientists have decided that something is permanently unknowable only to be contradicted a few generations later. On the other hand, many scientists have predicted that physics was near the end of its road only to have enormous new revolutions of knowledge take place a few decades later.

In recent years, just when it was thought that all the basic particles had been found or conjectured, along came superstrings, the most likely candidate at the moment for a TOE. If superstring theory is correct, it means that all fundamental particles are made of incredibly tiny loops of enormous tensile strength. The way they vibrate generates the entire zoo of particles.

What are superstrings made of? As far as anyone knows they are not made of anything. They are pure mathematical constructs. If superstrings are the end of the line, then everything that exists in our universe, including you and me, is a mathematical construction. As a friend once said, the universe seems to be made of nothing, yet somehow it manages to exist.

On the other hand, superstrings may, at some future time, turn out to be composed of still smaller entities. Many famous scientists, notably Arthur Stanley Eddington, David Bohm, Eugene Wigner, Freeman Dyson, and Stanislaw Ulam, believed that the universe has bottomless levels. As soon as one level is penetrated, a trap door opens to a hitherto unsuspected sub-basement. These sub-basements are infinite. As the old joke goes, it's turtles all the way down. Here is how Isaac Asimov expressed this opinion in his autobiography *I, Asimov*: "I believe that scientific knowledge has fractal properties; that no matter how much we learn, whatever is left, however small it may seem, is just as infinitely complex as the whole was to start with. That, I think, is the secret of the Universe."

A similar infinity may go the other way. Our universe may be part of a multiverse, in turn part of a multi-multiverse, and so on without end. As one of H. G. Wells' fantasies has it, our cosmos may be a molecule in a ring on a gigantic hand.

Even if the universe is finite in both directions, and there are no other worlds, are there fundamental questions that can never be answered? The slightest reflection demands a yes.

Suppose that at some future date a TOE will provide all the basic laws and constants. Explanation consists of finding a general law that explains a fact or a less general law. Why does Earth go around the sun? Because it obeys the laws of gravity. Why are there laws of gravity? Because, Einstein revealed, large masses distort space-time, causing objects to move along geodesic paths. Why do objects take geodesic paths? Because they are the shortest paths through space-time. Why do objects take the shortest paths? Now we hit a stone wall. Time, space, and change are given aspects of reality. You can't define any of these concepts without introducing the concept into the definition. They are not mere aspects of human con-

sciousness, as Kant imagined. They are "out there," independent of you and me. They may be unknowable in the sense that there is no way to explain them by embedding them in more general laws.

Imagine that physicists finally discover all the basic waves and their particles, and all the basic laws, and unite everything in one equation. We can then ask "Why that equation?" It is fashionable now to conjecture that the Big Bang was caused by a random quantum fluctuation in a vacuum devoid of space and time. But of course such a vacuum is a far cry from nothing. There had to be quantum laws to fluctuate. And why are there quantum laws?

Even if quantum mechanics becomes "explained" as part of a deeper theory—call it X—as Einstein believed it eventually would be, then we can ask "Why X?" There is no escape from the superultimate questions: Why is there something rather than nothing, and why is the something structured the way it is? As Stephen Hawking recently put it, "Why does the universe go to all the bother of existing?" The question obviously can never be answered, yet it is not emotionally meaningless. Meditating on it can induce what William James called an "ontological wonder-sickness." Jean Paul Sartre called it "nausea." Fortunately such reactions are short-lived or one could go mad by inhaling what James called "the blighting breath of the ultimate why?"

Consider the extremely short time humanity has been evolving on our little planet. It seems unlikely that evolution has stopped with us. Can anyone believe that a million years from now, if humanity lasts, that our brains will not have evolved far beyond their present capacities? Our nearest relatives, the chimpanzees, are incapable of understanding why three times three is nine, or anything else taught in grade school. It is difficult to imagine that a million years from now our brains will not be grasping truths about the universe that are as far beyond what we now can know as our understanding is beyond the mind of a monkey. To suppose that our brains, at this stage of an endless process of evolution, are capable of knowing everything that can be known strikes me as the ultimate in hubris.

If one is a theist, obviously there is a vast unknowable reality, transcending our universe, a "wholly other" realm impossible to contemplate without an emotion of what Rudolph Otto called the *mysterium tremendum.* But even if one is an atheist or agnostic, the Unknowable will not go away. No philosopher has written more persuasively about this than agnostic Herbert Spencer in the opening chapters of his *First Principles* (1894).

On the beginning hundred pages of this book, in a part titled "The Unknowable," Spencer argues that a recognition of the Unknowable is the only way to reconcile science with religion. The emotion behind all religions, aside from their obvious superstitions and gross beliefs, is one of awe toward the impenetrable mysteries of the universe. Here is how Spencer reasoned:

> One other consideration should not be overlooked —a consideration which students of Science more especially need to have pointed out. Occupied as such are with established truths, and accustomed to regard things not already known as things to be hereafter discovered, they are liable to forget that information, however extensive it may become, can never satisfy inquiry. Positive knowledge does not, and never can, fill the whole region of possible thought. At the uttermost reach of discovery there arises, and must ever arise, the question—What lies beyond? As it is impossible to think of a limit to space so as to exclude the idea of space lying outside that limit; so we cannot conceive of any explanation profound enough to exclude the question—What is the explanation of that explanation? Regarding Science as a gradually increasing sphere, we may say that every addition to its surface does but bring it into wider contact with surrounding nescience. There must ever remain therefore two antithetical modes of mental action. Throughout all future time, as now, the human mind may occupy itself, not only with ascertained phenomena and their relations, but also with that unascertained something which phenomena and their relations imply. Hence if knowledge cannot monopolize consciousness— if it must always continue possible for the mind to dwell upon that which transcends knowledge, then there can never cease to be a place for something of the nature of Religion; since Religion under all its forms is distinguished from everything else in this, that its subject matter passes the sphere of the intellect.

By "Religion" Spencer did not mean religion in the usual sense of worshipping God or gods, but only a sense of awe and wonder toward ultimate mysteries. For him Science and Religion were two essential aspects of thought; Science expressing the knowable, Religion the unknowable. The two merge without contradiction; "If Religion and Science are to be reconciled," he writes, "the basis of reconciliation must be this deepest, widest, most certain of all facts—that the Power which the Universe manifests to us is inscrutable."

No matter how many levels of generalization are made in explaining facts and laws, the levels must necessarily reach a limit beyond which science is powerless to penetrate.

> In all directions his investigations eventually bring him face to face with an insoluble enigma; and he ever more clearly perceives it to be an insoluble enigma. He learns at once the greatness and the littleness of the human intellect—its power in dealing with all that comes within the range of experience, its impotence in dealing with all that transcends experience. He, more than any other, truly *knows* that in its ultimate nature nothing can be known.

The rest of Spencer's *First Principles,* titled "The Knowable," is an effort to summarize the science of his day, especially what was then known about evolution.

> But an account of the Transformation of Things, given in the pages which follow, is simply an orderly presentation of facts; and the interpretation of the facts is nothing more than a statement of the ultimate uniformities they present— the laws to which they conform. Is the reader an atheist? The exposition of these facts and these laws will neither yield support to his belief nor destroy it. Is he a pantheist? The phenomena and the inferences as now to be set forth will not force on him any incongruous implication. Does he think that God is immanent throughout all things, from concentrating nebulae to the thoughts of poets? Then the theory to be put before him contains no disproof of that view. Does he believe in a Deity who has "given unchanging laws to the Universe"? Then he will

find nothing at variance with his belief in an exposition of those laws and an account of the results.

Boundaries and Barriers: On the Limits of Scientific Knowledge (Addison-Wesley, 1996), edited by John Casti and Anders Karlqvist, is one of a spate of recent books on the topic. For almost all its authors, the term "limits" is confined to unsolved but potentially solvable questions. Most of the authors agree with what the editors say in their introduction: "Unlike mathematics, there is no knock-down airtight argument to believe that there are questions about the rest of the world that we cannot answer in principle."

Only British astronomer John Barrow has the humility to disagree. He concludes his contribution as follows:

> In this brief survey we have explored some of the ways in which the quest for a Theory of Everything in the third millennium might find itself confronting impassable barriers. We have seen there are limitations imposed by human intellectual capabilities, as well as by the scope of technology. There is no reason why the most fundamental aspects of the laws of nature should be within the grasp of human minds, which evolved for quite different purposes, nor why those laws should have testable consequences at the moderate energies and temperatures that necessarily characterize life-supporting planetary environments. There are further barriers to the questions we may ask of the universe, and the answers that it can provide us with. These are barriers imposed by the nature of knowledge itself, not by human fallibility or technical limitations. As we probe deeper into the intertwined logical structures that underwrite the nature of reality, we can expect to find more of these deep results which limit what can be known. Ultimately, we may even find that their totality characterizes the universe more precisely than the catalogue of those things that we can know.

Barrow later expanded these sentiments in his 1998 book *Impossibility: The Limits of Science and the Science of Limits* (Oxford University Press). Here are some passages from his final courageous chapter:

The idea that some things may be unachievable or unimaginable tends to produce an explosion of knee-jerk reactions amongst scientific (and not so scientific) commentators. Some see it as an affront to the spirit of human inquiry: raising the white flag to the forces of ignorance. Others fear that talk of the impossible plays into the hands of the anti-scientists, airing doubts that should be left unsaid lest they undermine the public perception of science as a never-ending success story. . . .

We live in strange times. We also live in strange places. As we probe deeper into the intertwined logical structures that underwrite the nature of reality, I believe that we can expect to find more of these deep results which limit what can be known. Our knowledge about the Universe has an edge. Ultimately, we may even find that the fractal edge of our knowledge of the Universe defines its character more precisely than its contents; that what cannot be known is more revealing than what can.

George Gamow once described science as an expanding circle, not on a plane but on a sphere. It reaches a maximum size, after which it starts to contract until finally the sphere is covered and no more fundamental knowledge about the universe remains. In recent years numerous physicists, Hawking for instance, have expressed similar hopes. Richard Feynman suggested that although the circle may start to contract, it will become ever more difficult to obtain new knowledge and close the circle completely.

That science will soon discover everything is far from a recent hope. William James, lecturing at Harvard more than a century ago, attacked the hope with these words:

> In this very University . . . I have heard more than one teacher say that all the fundamental conceptions of truth have already been found by science, and that the future has only the details of the picture to fill in. But the slightest reflection . . . will suffice to show how barbaric such notions are. They show such a lack of scientific imagination, that it is hard to see how one who is actively advancing any part of science can make a mistake so crude. . . .

Our science is a drop, our ignorance a sea. Whatever else be certain, this at least is certain—that the world of our present natural knowledge *is* enveloped in a larger world of *some* sort of whose residual properties we at present can frame no positive idea.

Infinite In All Directions, the title of Freeman Dyson's 1988 book, says it all. Near the close of his third chapter he has this to say about a different hope:

It is my hope that we may be able to prove the world of physics as inexhaustible as the world of mathematics. Some of our colleagues in particle physics think that they are coming close to a complete understanding of the basic laws of nature. They have indeed made wonderful progress in the last ten years. But I hope that the notion of a final statement of the laws of physics will prove as illusory as the notion of a formal decision process for all of mathematics. If it should turn out that the whole of physical reality can be described by a finite set of equations, I would be disappointed. I would feel that the Creator had been uncharacteristically lacking in imagination. I would have to say, as Einstein once said in a similar context, "Da könnt' mir halt der liebe Gott leid tun" ("Then I would have been sorry for the dear Lord").

PART 8 SUGGESTIONS FOR FURTHER READING

Carvin, W. P. *Creation and Scientific Explanation.* Edinburgh: Scottish Academic Press, 1988.

Peacocke, A. R. *Creation and the World of Science.* Oxford: Oxford University Press, 1979.

Pennock, Robert T. *Tower of Babel: The Evidence against the New Creationism.* Cambridge, MA: MIT Press, 1998.

Peters, Ted, ed. *Science and Theology: The New Consonance.* Boulder, CO: Westview Press, 1998.

Polkinghorne, John. *Reason and Reality: The Relationship between Science and Theology.* Valley Forge, PA: Trinity Press International, 1991.

Russell, Bertrand. *Religion and Science.* London: Oxford University Press, 1953.

Stanesby, Derek. *Science, Reason, and Religion.* New York: Routledge, 1998.

Swinburne, Richard. *Faith and Reason.* Oxford: Clarendon Press, 1981.

PART 9

CONTEMPORARY ISSUES
AND APPLICATIONS

In addition to the philosophical problems raised by science in general, there are a number of problems raised by particular sciences. For example, quantum mechanics raises special problems about the nature of unobservables, biology raises special problems about the nature of explanation, and parapsychology raises special problems about the nature of confirmation. The selections in this part provide a sampling of the philosophical debate that goes on within these disciplines.

In his article on the Natural Ontological Attitude, Fine argued that quantum mechanics was able to progress without assuming that the entities it dealt with were real. Guy Vandegrift presents one of the reasons that the practitioners of quantum mechanics were reluctant to take a realist stance toward subatomic particles. If it is assumed that subatomic particles exist independently of their being observed, then it seems that they must be in some sort of psychic contact with one another. The traditional way out of this situation (known as the Copenhagen interpretation of quantum mechanics) is to assume that, in the situation Vandegrift discusses, an atom emits not particles but rather a wave. According to Max Born, this is a wave of probability that represents the chances of finding a particle in a particular state if we perform a particular type of measurement. Performing the measurement is said to "collapse the wave function" and to bring one of the probabilities into actuality. The problem with this interpretation is that it implies that we bring the world into existence in the act of measuring it. Einstein could never accept this aspect of quantum mechanics; he couldn't believe that the moon exists only when someone is looking at it.

One can believe in the objective reality of subatomic particles if one is willing to countenance the existence of faster-than-light connections between particles. David Bohm, for example, has shown that if we assume subatomic particles emit pilot waves that function like feelers and travel faster than the speed of light, then we can believe that particles have all of their properties independently of their being observed. Most scientists have been unwilling to follow Bohm in this, however, because the pilot waves cannot be observed and because, according to Einstein's special theory of relativity, nothing can travel faster than the speed of light.

Victor J. Stenger suggests that we can dispel the paradox from this situation by taking the probability wave to be real. The correlation between the properties of distant subatomic particles, then, is no more mysterious than the correlation of distant television sets tuned to the same channel. These TV sets display the same images because they are receiving the same wave. Stenger still owes us an explanation of how the wave turns into a particle. He claims that it is not the act of measurement that collapses the wave function, but the wave's interacting with a macroscopic object. So not only do we no longer need to assume that signals travel faster than the speed of light; we also do not need to assume that reality is mind dependent.

Evolutionary theory has traditionally been used to explain why certain organisms have certain physical characteristics. The explanations usually appeal to the fact that having those characteristics improves the organism's chances of survival. For example, we could explain why humans have an opposable thumb by citing the fact that having an opposable thumb makes it more likely that one will live long enough to reproduce. Because those that are more likely to live long enough to reproduce are more likely to have offspring, and because their offspring are more likely to have opposable thumbs, we would expect that after a suitably long period of time, most members of that population would have an opposable thumb.

Sociobiologists claim that this same line of reasoning can be used to explain why certain organisms engage in certain behaviors. For example, E. O. Wilson, the father of sociobiology, has suggested that religious belief may have a biological basis because being a member of a religious group can confer a survival advantage on its members. Stephen Jay Gould rejects these sorts of explanations, not on the grounds that they aren't true, but on the grounds that they aren't testable. Especially in the case of humans, there is no way we could do the breeding experiments necessary to test such hypotheses. Furthermore, these explanations often have unsavory political implications. So biologists should refrain from offering them.

Arthur L. Caplan points out that Darwin's theory was also criticized on the grounds that it is untestable. When a theory holds the promise of explaining such a wide variety of phenomena, however, objections concerning testability don't carry much weight. In Darwin's case, the objection turned out to be unfounded because the mechanisms underlying natural selection were eventually discovered. Natural selection is simply a label for complex interactions between genes, genotypes, phenotypes, and environments. Even if genes for specific behaviors have not yet been identified, that doesn't mean they won't be identified in the future. In principle, then, the hypothesis that a specific group of genes is responsible for a certain behavior (or behavioral disposition) could be tested by genetic engineering. For example, we could create one clone with the genes in question and another without them. Some may object to this procedure on ethical grounds, but that does not mean the hypothesis is untestable.

Robert H. Thouless claims that psychic phenomena may soon lead to a paradigm shift of the sort discussed by T. S. Kuhn. Telepathy (reading another's mind), clairvoyance (seeing distant objects without using one's senses), precognition (seeing into the future), and psychokinesis (moving objects with one's mind) cannot be explained within the current scientific paradigm. These phenomena, then, are anomalous, and such anomalies have historically led to scientific revolutions.

Daisie and Michael Radner think that Thouless's prediction of an imminent paradigm shift is premature. For a phenomenon to count as an anomaly, there must be general agreement as to when it occurs. But in the case of psychic phenomena, no such agreement exists. In determining whether events are due to random chance or to the existence of psychic phenomena, researchers count scores above chance, scores below chance, and scores that are initially above chance but then decline below chance as evidence for the existence of psychic phenomena. Why not also include scores that are initially below chance but then incline above chance? There simply is no agreement about what counts as evidence for psychic phenomena. In the absence of such agreement, any purported evidence for psychic phenomena cannot be considered true evidence of an anomaly. So there is no reason to believe that parapsychology will lead to a paradigm shift any time soon.

45

GUY VANDEGRIFT

Bell's Theorem and Psychic Phenomena

As a teacher of physics, I try to distil exotic theories so as to be understood and appreciated by the largest possible audience. What I recently learned in an advanced quantum mechanics class was so disturbing that I felt compelled to express the concept as simply as possible, without destroying the correctness of the argument. It appears that elementary particles act as if their behavior were linked by channels of communication that can be best described as 'psychic'.

The philosophical implications of Bell's theorem and the Einstein–Podolsky–Rosen (EPR) paradox have been recognized for several years[1] and are discussed in most modern textbooks on quantum physics.[2] Although quantum mechanics apparently 'resolves' the EPR paradox, we can think about it without using quantum mechanics. In fact, quantum mechanics can so overwhelm a person's common sense that one tends not to think about what is really happening.

It must be understood that we are not talking about a theory, but the results of actual experiments,[3] first performed in 1971. These experiments involve two particles that are simultaneously ejected from a single atom, in opposite directions, as shown in Figure 1. Previous experiments have used either photons or protons, but for the sake of clarity I am going to invent a hypothetical experiment involving neutrons.

It is important to know something about the nature of the measurements made on the particles. In the case of neutrons, the particle might be passed over the north pole of a magnet which is oriented

perpendicularly to the path. For reasons explained by quantum theory, the neutron is deflected either towards or away from the magnet, and always in the same amount, as shown in Figure 2.

The measurement essentially interrogates the neutron, asking 'Are you attracted?'. The response by the neutron is either 'Yes' or 'No'. The measurement can be made in a number of different ways by changing the angle that the magnet makes about the path of the neutron. Figure 3 shows the experiment seen end on, with the different possible orientations for the magnet, the original path of the neutron and its two possible paths after passing the magnet.

After one measurement, subsequent measurements on the same neutron are of no interest to us. We get to ask each neutron one and only one question. However, since there are two neutrons, we get to ask two questions. And, by choosing the magnet orientations, we get to choose from a number of possible pairs of questions. For our purposes, we restrict ourselves to three possible orientations of the magnet: 0°, 45°, and 90°. Thus we ask each neutron one and only one of the following questions:

Are you attracted to a magnet oriented at 0°?

Are you attracted to a magnet oriented at 45°?

Are you attracted to a magnet oriented at 90°?

It is fascinating that these particles are far apart at the time of the interrogation. One experiment with photons puts a distance of twenty meters between interrogation points. It is generally believed that traditional communication between the particles during interrogation is impossible. There is certainly no known physical mechanism for communication. The experimentalists took the trouble to change the ques-

The Philosophical Quarterly 45 (October 1995), pp. 471–476. Reprinted by permission of Blackwell Publishers Ltd. and University of Southern California.

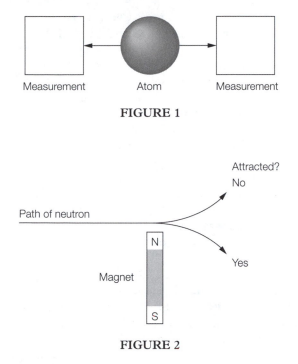

FIGURE 1

FIGURE 2

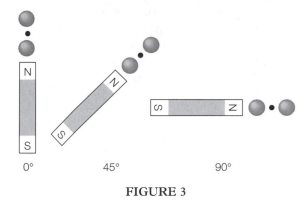

FIGURE 3

tions so rapidly that any such 'communication' between particles would require information to travel faster than the speed of light. Such 'superluminar' communication violates the principle of causality, which says that it is impossible to change the past.[4]

THE STORY OF HARRY AND SALLY

Let us anthropomorphize the situation by labelling the particles with the names of two people, Harry and Sally. We shall also change the questions to something that a human might appreciate. As any logical person will recognize, this re-labelling does not change the fundamental paradox inherent in Bell's theorem. Harry and Sally are being investigated by researchers in psychic phenomena who have constructed two sound-proof isolation chambers, designed so that no information can leave or enter once the doors are shut. Every morning, Harry and Sally enter separate isolation chambers and each is asked one question, randomly selected from the following list:

Are you hungry?

Are you thirsty?

Are you happy?

The experiment is repeated every day for many years, and the results are tabulated. Sally and Harry may or may not be asked the same question on any given day. Before entering the isolation chambers, neither knows which question will be asked. Sally and Harry are unable to communicate with each other during the interrogation.

After their daily interrogation, Sally and Harry leave their isolation chambers and go on with their usual daily routines. It is understood that Harry and Sally are under no obligation to tell the truth during the interrogation. They are allowed to discuss and plan their answers before and after each interrogation, if they wish.

THE RESULTS

After looking at the data from a large number of interrogations, the researchers notice that each question is answered 'Yes' 50% of the time, and 'No' 50% of the time. The researchers also observe that the answers are never in violation of the following two rules:

Rule 1. If Sally and Harry are asked the same question, they always give the same answer.

This implies that they consciously or unconsciously knew the answers to all three questions before they entered the isolation chambers. The justification for this claim is that if Harry changes his mind, there is no mechanism by which Sally would know of this so as to change her mind in order to comply with Rule (1). Since both are prepared to give the same answers, there are eight possible sets of answers that they could have prepared each morning before entering the isolation chambers, as shown in Table 1:

Table 1

	Hungry?	Happy?	Thirsty?
1.	No	No	No
2.	No	No	Yes
3.	No	Yes	No
4.	No	Yes	Yes
5.	Yes	No	No
6.	Yes	No	Yes
7.	Yes	Yes	No
8.	Yes	Yes	Yes

Rule 2. There is a pattern among the answers whenever different questions are asked. For example, if one is hungry, the other one is never thirsty. The pattern is summarized in Table 2, which contains six rules which are obeyed whenever different questions are asked:

Table 2

	Hungry?	Happy?	Thirsty?
a.	Yes	—	No
b.	No	—	Yes
c.	Yes	Yes	—
d.	No	No	—
e.	—	No	No
f.	—	Yes	Yes

In Table 2, line (f) says that one is always happy when the other is thirsty. To be more precise, one claims to be happy whenever the other claims to be thirsty. Note that it is never verified that Harry and Sally are in compliance with Rule (1) and Rule (2) on the same day. If both are asked the same question, Rule (1) is verified, and if they are asked different questions, Rule (2) is verified. It is impossible to verify both rules on the same day because only two questions are asked. Only after the experiment has been repeated many times is it clear that both Rule (1) and Rule (2) are never violated.

However, since Sally and Harry do not know in advance which questions will be asked, they must somehow preselect a set of answers that obeys all the rules. But this is impossible! For example, suppose Harry and Sally select item (1) in Table 1. This is in violation of rule (a) in Table 2. The reader is asked to verify that each possibility listed in Table 1 is inconsistent with one of the rules in Table 2.

There are only two mechanisms by which Harry and Sally could always obey both rules. They could communicate with each other while in the isolation chambers, thus changing their story in the middle of the interrogation. Or they could know in advance which questions would be asked. Either mechanism would require psychic powers on the part of Harry and Sally.

It has been clearly stated by experts that the apparent telepathy displayed by neutrons is limited by the fact that we cannot send signals via the interrogations. In other words, if Sally says she is hungry, we cannot ask her if she chose that answer because Harry said he was hungry, or because Harry said he was happy. Sally cannot even say whether Harry was the first to be interrogated. She apparently does not know herself!

BACK TO ACTUAL NEUTRONS

One significant difference exists between an actual EPR experiment and this story. It turns out that Harry and Sally do not always obey the rules of Table 2. In fact, they break Rule (2a) as often as they obey it. It was Bell who in 1965 pointed out that the correlations predicted by quantum theory are impossible without what might loosely be called 'communication' between the particles. These 'impossible' correlations have been observed in a number of experiments.

To put this talk about correlations into everyday language, we have a situation where Harry and Sally have good days and bad days with respect to their 'psychic powers'. However, the good days happen so often that it would defy the laws of probability for them to obey Rules (1) and (2) as often as they do. Imagine, for example, that Harry and Sally obey the rules on only half the days, with the selection of those days being completely random. On those days when Harry and Sally do not obey the rules, they simply give random answers. However, half the time, they obey both rules, thus doing the 'impossible'. Such 'impossible' behavior could be detected statistically by observing for a large number of days.

CONCLUSION

Bell's theorem is truly astonishing, more astonishing than the rest of quantum mechanics, which makes bizarre predictions about small objects. According to quantum mechanics, large objects also display this bizarre behavior, but to a much lesser extent. For example, the 'uncertainty principle' predicts that it is possible for someone to be suddenly and mysteriously transported to a high mountain in Tibet, only to return just as mysteriously. However, a simple estimate indicates that the probability of anyone's being transported a significant distance is so remote that it will never happen in the age of the universe. In other words, most of the bizarre quantum behavior attributed to particles is due to the fact that they are so tiny.

On the other hand, there seems to be no fundamental reason why two people could not put themselves into what might be called a 'degenerate mixed-energy-state' and reproduce what Harry and Sally have done here. I did not intend to write an essay on psychic phenomena, and made this analogy because it is the most direct description of what the EPR experiment is actually doing. I do not believe in mental telepathy, miracles or any other occult phenomenon. This affair with Bell's theorem has shaken me to the bone.

NOTES

1. See, e.g., Bernard d'Espagnat, 'The Quantum Theory and Reality', *Scientific American,* 241(5), November 1979, pp. 158–81, in which a simple proof of Bell's theorem can be found; also James T. Cushing and Ernan McMullin (eds), *Philosophical Consequences of Quantum Theory* (Univ. of Notre Dame Press, 1989).

2. See Hans C. Ohanian, *Principles of Quantum Mechanics* (New York: Prentice Hall, 1990).

3. Alain Aspect, Jean Dalibard and Gerald Roger, 'Experimental Test of Bell's Inequalities Using Time-Varying Analyzers', *Physical Review Letters,* 49 (1982), p. 1805.

4. See Edwin F. Taylor and John A. Wheeler, *Spacetime Physics* (New York: W. H. Freeman, 1966).

46

VICTOR J. STENGER

The Unconscious Quantum

FASTER–THAN–LIGHT COMMUNICATION?

Let us examine in some detail whether the correlations implied by the apparent empirical violation of Bell's inequality can be used to communicate superluminally, indeed instantaneously, between two separated points in space.

Such a possibility has been proposed in the past by physicists Jack Sarfatti and Nick Herbert, among others.[1] . . . Sarfatti and Bohm had witnessed a demonstration of Uri Geller's alleged psychic abilities in London in 1974. Sarfatti came home from his London stint convinced that the mind is a quantum system, able to affect other quantum systems such as radioactive nuclei, as reported in Geller's London performance. Since then Sarfatti has been involved with Herbert, Fred Alan Wolf, and others in various initiatives that promote the quantum-consciousness connection, though Sarfatti told me he no longer believes in Geller's powers. Gary Zukav acknowledges Sarfatti as a "catalyst" for his book on science mysticism, *The Dancing Wu Li Masters*.[2]

Superluminal communicators based on the EPR experiment of the type proposed by Sarfatti utilize the notion that a decision made at one end of the beam line, such as the orientation of a polarizer, will instantaneously affect what is observed at the other end. This would seem to suggest that signals can be sent from one beam end to the other faster than the speed of light.

One version of such a device has been analyzed by David Mermin.[3] Mermin considered an appara-

tus similar to the Orsay experiments [described by Vandegrift] except that electrons rather than photons are used, as in Bohm's version of the EPR experiment.

The basic experimental setup is [as follows:] At the center is a singlet (total spin zero) source S of pairs of electrons that are emitted along the two opposite beam lines. At the end of each beam line is a "spin meter," an inhomogeneous magnetic field that is used to measure electron spin. Each spin meter can be rotated about its axis so that the spin component of either electron can be measured along any desired axis. For simplicity, the beam lines are taken to be equally distant so the electrons from a given pair arrive simultaneously at each beam end (in the laboratory reference frame).

Also for simplicity, only three possible settings, 120° apart, as illustrated in Figure 1, are considered. Electron detectors in each spin meter measure whether the electron is spinning along or opposite the selected axis. At spin meter A, a red light on the apparatus flashes on in the first case and a green light in the second. This arrangement is reversed at spin meter B so that the same color light will flash at both ends when the axes are in the same position and the electrons from a singlet pair pass through. Let us label the three directions 1, 2, and 3.

Now suppose that Alphonse at the end of beam line A wishes to send a message to Beth at the end of beam line B. He can attempt to do this by coding a message in terms of the three spin meter settings. Suppose he sets his spin meter to 1 and observes the red light flash, meaning that an electron spinning along the 1 axis was detected. At the end of beam line B, Beth sets her spin meter to 1 and detects the other electron in the singlet pair and sees her red light flash.

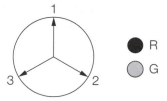

FIGURE 1. Each spin meter in an EPR communicator can be rotated to one of three positions 1, 2, and 3 that are 120 degrees apart. When at a given setting, the spin meter measures whether the electron's spin is along or opposite the chosen axis. Lights flash red or green depending on which of the two results occurs.

Suppose Alphonse next sets his spin meter to 2 and sees a green light. If Beth moves her meter to 2, she will also see a green light, and a message 1-2 will have been transmitted. Thus it would appear that a signal containing the information 1-2 has been transmitted instantaneously from Alphonse to Beth.

However, note that in the above example Beth had to know ahead of time to set her spin meter to 1 and 2 consecutively. Thus she already knew the message and no new information was transmitted. In order for information to be transmitted, Beth must try different settings. Still, one might think that information transfer is still possible since the spin measurements at the ends of the two beam lines are more correlated, according to quantum mechanics, than they would be if the measurements were independent.

Mermin tested this hypothesis by writing a simulation program on a computer and printing out the results. Rather than simply reprinting his results, I have written my own program to check for myself and add a trimming. Mermin's program sends a random message from A, but any message should exhibit the same results, namely, the lack of information transfer. In my simulation, I can allow Alphonse to send any arbitrary message and then see if the message is received.

For purposes of illustration, assume Alphonse tries to send the message 1-2-3, repeating it over and over again by setting his spin meter consecutively in each of the three orientations and detecting an electron in each case. The computer then randomly decides whether the spin of the electron detected by Alphonse is along or opposite the chosen axis, that is, whether Alphonse's light flashes red or green. He has no control over this.

The computer, acting for Beth, randomly chooses her spin meter orientations and determines, from the rules of quantum mechanics, what color light flashes at Beth's detector. Those rules are simple: (1) if spin meters A and B are at the same position, the lights will flash the same color; (2) if the spin meters are at different positions, they will flash colors randomly with the same color as A occurring on average one-quarter of the time ($\cos^2 120° = 1/4$) and the opposite color three-quarters of the time.

SAMPLE RESULTS

Following is a typical output. Each run will be different in detail if started with a different random number. The first sixty trials are listed below, and the totals for the runs of six thousand trials are given at the end.

The three spin meter positions are indicated by 1, 2, and 3. R and G specify the two colors that represent electron spins being measured along or opposite the spin meter axes. Thus "1 G 3 R" means spin meter setting 1 for the message transmitted at A with a green light flash, and spin meter setting 3 at the receiver B seeing the red light flash. The message 1-2-3 is repeatedly transmitted, so each line below corresponds to two attempts to send this message. For example, in the first line the message being sent is 1-2-3-1-2-3 and the message received is 3-3-2-1-2-1.

I have indicated in boldface type those cases where the spin meter settings are the same. Note that whenever this occurs, the same colors flash. But as we see by the totals at the end, no information is transmitted.

1G3R	2G3G	3G2R	**1G1G**	**2G2G**	**3R1R**
1R1R	**2G2G**	3R2R	1R3R	2G1R	**3G3G**
1R2R	2R3R	3R2R	**1G1G**	**2G2G**	3G1G
1G2R	2R3R	**3R3R**	1G2R	**2R2R**	3G1R
1G3R	**2R2R**	3G2R	1G2R	2R1R	3R2R
1R2R	2G3R	3G1R	**1R1R**	**2R2R**	**3R3R**
1R2R	2R3R	3R1R	**1R1R**	2G1R	3R1R
1G2G	2G3R	3R2R	1R3R	2G3R	3R2R
1G1G	2G1R	3R2R	1R2R	**2G2G**	**3R3R**
1R2R	2G3R	**3G3G**	1G3G	2R3R	3G1R

Summary of results from six thousand trials:

Fraction of light flashes of the same color = 0.499

Fraction of light flashes of different color = 0.501

Expected statistical fluctuation: ±0.01

We see that the lights flash the same color half the time and different colors the other half. . . . You will find the same results for any message: equal numbers of red and green light flashes, within statistical fluctuations. . . . No information is transmitted from the end of one beam line to the other, despite the apparent correlation that exists when the two spin meter settings are identical. We can conclude from this that no signals move at superluminal speeds in our EPR experiment.

Further, the application of conventional quantum mechanics to the experiment did not involve any unexpected statistical results. The $\cos^2 \theta$ rule used to compute the probability that an electron will be detected with a spin axis making an angle θ with the spin axis of the other electron in the pair, 120° in our case, is exactly the same as given by Malus's law for polarized light, discovered by Etienne Louis Malus in 1809. This result is trivially derived in today's physics classes by simply projecting the field vector along the new axis and squaring to get the intensity. It applies equally well for electrons in this case.

CORRELATION WITHOUT COMMUNICATION

Correlation over spacelike separations does not require superluminal communication or quantum mechanics. . . . Detectors located at separated points along a wave front can simultaneously receive signals from a transmitting station and obtain whatever information is carried by that signal. For example, a television station sends out modulated electromagnetic waves that carry information that is reconstructed by thousands of television sets located all over the region. All the sets along a circumference of equal distance from the station receive the same picture simultaneously. Note that the television sets have to be tuned to the same channel to get the correlated message. They could be miles from one an-

other, but this does not imply that the sets are communicating with each other at infinite speeds, telling each other what channels are selected.

We can imagine a wave function (thinking of it as some kind of real wave) being emitted from the singlet source in an EPR experiment analogous to the electromagnetic wave from the television station. It informs the spin meter detectors that the electrons they are receiving have a certain chance of being observed to deflect one way or the other as they pass though the magnetic fields of the meters, for various orientations of their field axes.

The orientations at both ends must be known in order to compute the correct probability. This is analogous to having to know when two television sets are tuned to the same channel to predict what is being viewed. However, a spooky action at a distance is implied when we interpret the setting at one end to instantaneously affect what is known at the other. In the television metaphor, it would be as if changing channels on my set immediately affected what you see on your set across town.

There are correlations between the individual measurements in our simulation of the EPR experiment. The table above, which lists simulated individual events, shows that the lights always flash the same color when the orientations are the same. Of course they should, or else angular momentum would not be conserved and we would have a severe violation of a long-established principle of physics. This principle alone, without quantum mechanics, requires at least some correlation between the results observed at the two beam lines.

Still, the experimental violation of Bell's inequality implies an additional correlation beyond this. My computer program uses a knowledge of the orientation of A in determining the outcome of the measurement at B. The relative angle between the two was assumed to be known in calculating the probability for the light to flash a given color.

The program simulates individual events in a statistical fashion that is consistent with the Born probability postulate $P = |\psi|^2$. The probability follows from quantum mechanics.

$P = \cos^2 \theta$, really just Malus's law again, where θ is the relative orientation of the spin meters at the ends of the beam lines.

But, you might ask, doesn't a superluminal correlation exist in the way I implemented the quantum mechanical theory to simulate the outcome of individual particle detections? Does this not demonstrate that nonlocality is required after all? Actually, it nicely illustrates the point I have been trying to make that nonlocality can exist in theory, or, as in this case, computer simulation, without being required in reality.

A computer simulation, like a mathematical theory, can contain nonlocal correlations among the abstract entities it uses to represent an experiment. It matters not how we choose to manufacture our theoretical constructs, so long as they successfully describe the observed data and so long as people do not assume these constructs correspond to "real" elements of nature.

In the computer program, we simulate the experiment event by event. Each experiment has a given relative orientation of spin axes that must be assumed in simulating the outcome under those conditions. This is the contextual nature of quantum physics.

In the process we generate individual outcomes that are not specifically predicted by quantum mechanics. Quantum mechanics was only used to give the probability for a given outcome; the rest was determined by the computer's random number generator. We have done nothing different from what we might have done in simulating a coin toss by calling the random number generator and printing H when it returns a number between 0 and 0.499999 and T when it returns a number between 0.5 and 0.999999. The fact that we are simulating individual outcomes does not mean that we have programmed a theory that determines the outcome of an individual coin toss in the real world. In fact, we are probably simulating something close to what actually happens in a nondeterministic universe. Perhaps, as some have suggested, the computer program is a better metaphor for physical systems than mathematical solutions to the equations of physics.

Both our simulated experiment and conventional quantum theory agree on the observed outcome of an attempt at superluminal communication by means of an EPR apparatus. It would not work, and so Einstein's speed limit would remain inviolate.

47

STEPHEN JAY GOULD

Sociobiology and the Theory of Natural Selection

NATURAL SELECTION AS STORYTELLING

Ludwig von Bertalanffy, a founder of general systems theory and a holdout against the neo-Darwinian tide, often argued that natural selection must fail as a comprehensive theory because it explains *too* much—a paradoxical, but perceptive statement. He wrote (1969: 24, 11):

> If selection is taken as an axiomatic and *a priori* principle, it is always possible to imagine auxiliary hypotheses—unproved and by nature unprovable—to make it work in any special case. . . . Some adaptive value . . . can always be construed or imagined.
>
> I think the fact that a theory so vague, so insufficiently verifiable and so far from the criteria otherwise applied in "hard" science, has become a dogma, can only be explained on sociological grounds. Society and science have been so steeped in the ideas of mechanism, utilitarianism, and the economic concept of free competition, that instead of God, Selection was enthroned as ultimate reality.

Similarly, the arguments of Christian fundamentalism used to frustrate me until I realized that there are, in principle, no counter cases and that, on this ground alone, literal bibliolatry is bankrupt. The theory of natural selection is, fortunately, in much better straits. It could be invalidated as a general cause of

evolutionary change. (If, for example, Lamarckian inheritance were true and general, then adaptation would arise so rapidly in the Lamarckian mode that natural selection would be powerless to create and would operate only to eliminate.) Moreover, its action and efficacy have been demonstrated experimentally by 60 years of manipulation within *Drosophila* bottles—not to mention several thousand years of success by plant and animal breeders.

Yet in one area, unfortunately a very large part of evolutionary theory and practice, natural selection has operated like the fundamentalist's God—he who maketh *all* things. Rudyard Kipling asked how the leopard got its spots, the rhino its wrinkled skin. He called his answers "just-so-stories." When evolutionists try to explain form and behavior, they also tell just-so stories—and the agent is natural selection. Virtuosity in invention replaces testability as the criterion for acceptance. This is the procedure that inspired von Bertalanffy's complaint. It is also the practice that has given evolutionary biology a bad name among many experimental scientists in other disciplines. We should heed their disquiet, not dismiss it with a claim that they understand neither natural selection nor the special procedures of historical science.

This style of storytelling might yield acceptable answers if we could be sure of two things: 1) that all bits of morphology and behavior arise as direct results of natural selection, and 2) that only one selective explanation exists for each bit. But, as Darwin insisted vociferously, and contrary to the mythology about him, there is much more to evolution than natural selection. (Darwin was a consistent pluralist who viewed natural selection as the most important agent of evolutionary change, but who accepted a range of other agents and specified

Sociobiology: Beyond Nature/Nurture? G.W. Barlow and J. Silverberg, eds. (Boulder: Westview Press, 1980) pp. 257–269. Reprinted by permission of the publisher.

the conditions of their presumed effectiveness. In Chapter 7 of the *Origin* (6th ed.), for example, he attributed the cryptic coloration of a flat fish's upper surface to natural selection and the migration of its eyes to inheritance of acquired characters. He continually insisted that he wrote his 2-volume *Variation of Animals and Plants under Domestication* (1868), with its Lamarckian hypothesis of pangenesis, primarily to illustrate the effect of evolutionary factors other than natural selection. In a letter to *Nature* in 1880, he used the sharpest and most waspish language of his life to castigate Sir Wyville Thomson for caricaturing his theory by ascribing all evolutionary change to natural selection.)

Since God can be bent to support all theories, and since Darwin ranks closest to deification among evolutionary biologists, panselectionists of the modern synthesis tended to remake Darwin in their image. But we now reject this rigid version of natural selection and grant a major role to other evolutionary agents—genetic drift, fixation of neutral mutations, for example. We must also recognize that many features arise indirectly as developmental consequences of other features subject to natural selection—see classic (Huxley 1932) and modern (Gould 1966 and 1975; Cock 1966) work on allometry and the developmental consequences of size increase. Moreover, and perhaps most importantly, there are a multitude of potential selective explanations for each feature. There is no such thing in nature as a self-evident and unambiguous story.

When we examine the history of favored stories for any particular adaptation, we do not trace a tale of increasing truth as one story replaces the last, but rather a chronicle of shifting fads and fashions. When Newtonian mechanical explanations were riding high, G.G. Simpson wrote (1961: 1686):

> The problem of the pelycosaur dorsal fin . . . seems essentially solved by Romer's demonstration that the regression relationship of fin area to body volume is appropriate to the functioning of the fin as a temperature regulating mechanism.

Simpson's firmness seems almost amusing since now—a mere 15 years later with behavioral stories in vogue—most paleontologists feel equally sure that the sail was primarily a device for sexual display. (Yes, I know the litany: It might have performed both functions. But this too is a story.)

On the other side of the same shift in fashion, a recent article on functional endothermy in some large beetles had this to say about the why of it all (Bartholomew and Casey 1977: 883):

> It is possible that the increased power and speed of terrestrial locomotion associated with a modest elevation of body temperatures may offer reproductive advantages by increasing the effectiveness of intraspecific aggressive behavior, particularly between males.

This conjecture reflects no evidence drawn from the beetles themselves, only the current fashion in selective stories. We may be confident that the same data, collected 15 years ago, would have inspired a speculation about improved design and mechanical advantage.

SOCIOBIOLOGICAL STORIES

Most work in sociobiology has been done in the mode of adaptive storytelling based upon the optimizing character and pervasive power of natural selection. As such, its weaknesses of methodology are those that have plagued so much of evolutionary theory for more than a century. Sociobiologists have anchored their stories in the basic Darwinian notion of selection as individual reproductive success. Though previously underemphasized by students of behavior, this insistence on selection as individual success is fundamental to Darwinism. It arises directly from Darwin's construction of natural selection as a conscious analog to the laissez-faire economics of Adam Smith with its central notion that order and harmony arise from the natural interaction of individuals pursuing their own advantages (see Schweber 1977).

Sociobiologists have broadened their range of selective stories by invoking concepts of inclusive fitness and kin selection to solve (successfully I think) the vexatious problem of altruism—previously the greatest stumbling block to a Darwinian theory of social behavior. Altruistic acts are the cement of stable societies. Until we could explain apparent acts of self-sacrifice as potentially beneficial to the genetic fitness of sacrificers themselves—propagation of genes through enhanced survival of kin, for example—the prevalence of altruism blocked any Darwinian theory of social behavior.

Thus, kin selection has broadened the range of permissible stories, but it has not alleviated any methodological difficulties in the process of storytelling itself. Von Bertalanffy's objections still apply, if anything with greater force because behavior is generally more plastic and more difficult to specify and homologize than morphology. Sociobiologists are still telling speculative stories, still hitching without evidence to one potential star among many, still using mere consistency with natural selection as a criterion of acceptance.

David Barash (1976), for example, tells the following story about mountain bluebirds. (It is, by the way, a perfectly plausible story that may well be true. I only wish to criticize its assertion without evidence or test, using consistency with natural selection as the sole criterion for useful speculation.) Barash reasoned that a male bird might be more sensitive to intrusion of other males before eggs are laid than after (when he can be certain that his genes are inside). So Barash studied two nests, making three observations at 10-day intervals, the first before the eggs were laid, the last two after. For each period of observation, he mounted a stuffed male near the nest while the male occupant was out foraging. When the male returned he counted aggressive encounters with both model and female. At time one, males in both nests were aggressive toward the model and less, but still substantially, aggressive toward the female as well. At time two, after eggs had been laid, males were less aggressive to models and scarcely aggressive to females at all. At time three, males were still less aggressive toward models, and not aggressive at all toward females. Barash concludes that he has established consistency with natural selection and need do no more (1976: 1099–1100):

These results are consistent with the expectations of evolutionary theory. Thus aggression toward an intruding male (the model) would clearly be especially advantageous early in the breeding season, when territories and nests are normally defended. . . . The initial, aggressive response to the mated female is also adaptive in that, given a situation suggesting a high probability of adultery (i.e., the presence of the model near the female) and assuming that replacement females are available, obtaining a new mate would enhance the fitness of males. . . . The decline in male-female aggressiveness during incubation and fledgling stages could

be attributed to the impossibility of being cuckolded after the eggs have been laid. . . . The results are consistent with an evolutionary interpretation. In addition, the term "adultery" is unblushingly employed in this letter without quotation marks, as I believe it reflects a true analogy to the human concept, in the sense of Lorenz. It may also be prophesied that continued application of a similar evolutionary approach will eventually shed considerable light on various human foibles as well.

Consistent, yes. But what about the obvious alternative, dismissed without test in a line by Barash: male returns at times two and three, approaches the model a few times, encounters no reaction, mutters to himself the avian equivalent of "it's that damned stuffed bird again," and ceases to bother. And why not the evident test: expose a male to the model for the *first* time *after* the eggs are laid.

We have been deluged in recent years with sociobiological stories. Some, like Barash's, are plausible, if unsupported. For many others, I can only confess my intuition of extreme unlikeliness, to say the least—for adaptive and genetic arguments about why fellatio and cunnilingus are more common among the upper classes (Weinrich 1977), or why male panhandlers are more successful with females and people who are eating than with males and people who are not eating (Lockard et al. 1976).

Not all of sociobiology proceeds in the mode of storytelling for individual cases. It rests on firmer methodological ground when it seeks broad correlations across taxonomic lines, as between reproductive strategy and distribution of resources, for example (Wilson 1975), or when it can make testable, quantitative predictions as in Trivers and Hare's work on haplodiploidy and eusociality in Hymenoptera (Trivers and Hare 1976). Here sociobiology has had and will continue to have success. And here I wish it well. For it represents an extension of basic Darwinism to a realm where it should apply.

SPECIAL PROBLEMS FOR HUMAN SOCIOBIOLOGY

Sociobiological explanations of human behavior encounter two major difficulties, suggesting that a Darwinian model may be generally inapplicable in this case.

Limited Evidence and Political Clout

We have little direct evidence about the genetics of behavior in humans; and we do not know how to obtain it for the specific behaviors that figure most prominently in sociobiological speculation—aggression, conformity, etc. With our long generations, it is difficult to amass much data on heritability. More important, we cannot (ethically, that is) perform the kind of breeding experiments, in standardized environments, that would yield the required information. Thus, in dealing with humans, sociobiologists rely even more heavily than usual upon speculative storytelling.

At this point, the political debate engendered by sociobiology comes appropriately to the fore. For these speculative stories about human behavior have broad implications and proscriptions for social policy—and this is true quite apart from the intent or personal politics of the storyteller. Intent and usage are different things; the latter marks political and social influence, the former is gossip or, at best, sociology.

The common political character and effect of these stories lies in the direction historically taken by innatist arguments about human behavior and capabilities—a defense of existing social arrangements as part of our biology.

In raising this point, I do not act to suppress truth for fear of its political consequences. Truth, as we understand it, must always be our primary criterion. We live, because we must, with all manner of unpleasant biological truth—death being the most pervasive and ineluctable. I complain because sociobiological stories are not truth but unsupported speculations with political clout (again, I must emphasize, quite apart from the intent of the storyteller). All science is embedded in cultural contexts, and the lower the ratio of data to social importance, the more science reflects the context.

In stating that there is politics in sociobiology, I do not criticize the scientists involved in it by claiming that unconscious politics has intruded into a supposedly objective enterprise. For they are behaving like all good scientists—as human beings in a cultural context. I only ask for a more explicit recognition of the context and, specifically, for more attention to the evident impact of speculative sociobiological stories. For example, when the *New York Times* runs a weeklong front page series on women and their rising achievements and expectations, spends the first four days documenting progress toward social equality, devotes the last day to potential limits upon this progress, and advances sociobiological stories as the only argument for potential limits—then we know that these are stories with consequences:

> Sociologists believe that women will continue for some years to achieve greater parity with men, both in the work place and in the home. But an uneasy sense of frustration and pessimism is growing among some advocates of full female equality in the face of mounting conservative opposition. Moreover, even some staunch feminists are reluctantly reaching the conclusion that women's aspirations may ultimately be limited by inherent biological differences that will forever leave men the dominant sex (*New York Times*, Nov. 30, 1977).

The article then quotes two social scientists, each with a story.

> If you define dominance as who occupies formal roles of responsibility, then there is no society where males are not dominant. When something is so universal, the probability is—as reluctant as I am to say it—that there is some quality of the organism that leads to this condition.

> It may mean that there never will be full parity in jobs, that women will always predominate in the caring tasks like teaching and social work and in the life sciences, while men will prevail in those requiring more aggression—business and politics, for example—and in the 'dead' sciences like physics.

Adaptation in Humans Need Not Be Genetic and Darwinian

The standard foundation of Darwinian just-so stories does not apply to humans. That foundation is the implication: if adaptive, then genetic—for the inference of adaptation is usually the only basis of a genetic story, and Darwinism is a theory of genetic change and variation in populations.

Much of human behavior is clearly adaptive, but the problem for sociobiology is that humans have so far surpassed all other species in developing an alternative, non-genetic system to support and transmit adaptive behavior—cultural evolution. An adaptive behavior does not require genetic input and Darwinian selection for its origin and maintenance in humans; it may arise by trial and error in a few individuals that do not differ genetically from their groupmates, spread by learning and imitation, and stabilize across generations by value, custom and tradition. Moreover, cultural transmission is far more powerful in potential speed and spread than natural selection—for cultural evolution operates in the "Lamarckian" mode by inheritance through custom, writing and technology of characteristics acquired by human activity in each generation.

Thus, the existence of adaptive behavior in humans says nothing about the probability of a genetic basis for it, or about the operation of natural selection. Take, for example, Trivers' (1971) concept of "reciprocal altruism." The phenomenon exists, to be sure, and it is clearly adaptive. In honest moments, we all acknowledge that many of our altruistic acts are performed in the hope and expectation of future reward. Can anyone imagine a stable society without bonds of reciprocal obligation? But structural necessities do not imply direct genetic coding. (All human behaviors are, of course, part of the potential range permitted by our genotype—but sociobiological speculations posit direct natural selection for specific behavioral traits.) As Benjamin Franklin said: "We must all hang together, or assuredly we shall all hang separately."

FAILURE OF THE RESEARCH PROGRAM FOR HUMAN SOCIOBIOLOGY

The grandest goal—I do not say the only goal—of human sociobiology must fail in the face of these difficulties. That goal is no less than the reduction of the behavioral (indeed most of the social) sciences to Darwinian theory. Wilson (1975) presents a vision of the human sciences shrinking in their independent domain, absorbed on one side by neurobiology and on the other by sociobiology.

But this vision cannot be fulfilled, for the reason cited above. Although we can identify adaptive behavior in humans, we cannot tell thereby if it is genetically

based (while much of it must arise by fairly pure cultural evolution). Yet the reduction of the human sciences to Darwinism requires the genetic argument, for Darwinism is a theory about genetic change in populations. All else is analogy and metaphor.

My crystal ball shows the human sociobiologists retreating to a fallback position—indeed it is happening already. They will argue that this fallback is as powerful as their original position, though it actually represents the unravelling of their fondest hopes. They will argue: yes, indeed, we cannot tell whether an adaptive behavior is genetically coded or not. But it doesn't matter. The same adaptive constraints apply whether the behavior evolved by cultural or Darwinian routes, and biologists have identified and explicated the adaptive constraints. (Steve Emlen [1980] reports, for example, that some Indian peoples gather food in accordance with predictions of optimal foraging strategy, a theory developed by ecologists. This is an exciting and promising result within an anthropological domain—for it establishes a fruitful path of analogical illumination between biological theory and non-genetic cultural adaptation. But it prevents the assimilation of one discipline by the other and frustrates any hope of incorporating the human sciences under the Darwinian paradigm.)

But it does matter. It makes all the difference in the world whether human behaviors develop and stabilize by cultural evolution or by direct Darwinian selection for genes influencing specific adaptive actions. Cultural and Darwinian evolution differ profoundly in the three major areas that embody what evolution, at least as a quantitative science, is all about:

1. Rate. Cultural evolution, as a "Lamarckian" process, can proceed orders of magnitude more rapidly than Darwinian evolution. Natural selection continues its work within *Homo sapiens,* probably at characteristic rates for change in large, fairly stable populations, but the power of cultural evolution has dwarfed its influence (alteration in frequency of the sickling gene vs. changes in modes of communication and transportation). Consider what we have done with ourselves in the past 3000 years, all without the slightest evidence for any biological change in the size or power of the human brain.

2. Modifiability. Complex traits of cultural evolution can be altered profoundly all at once (social revolution, for example). Darwinian change is much slower and more piecemeal.

3. Diffusibility. Since traits of cultural evolution can be transmitted by imitation and inculcation, evolutionary patterns include frequent and complex anastomosis among branches. Darwinian evolution in sexually reproducing animals is a process of continuous divergence and ramification with few opportunities for coming together (hybridization or parallel modification of the same genes in independent groups).

I believe that the future will bring mutual illumination between two vigorous, independent disciplines—Darwinian theory and cultural history. This is a good thing, joyously to be welcomed. But there will be no reduction of the human sciences to Darwinian theory, and the research program of human sociobiology will fail. The name, of course, may survive. It is an irony of history that movements are judged successful if their label sticks, even though the emerging content of a discipline may lie closer to what opponents originally advocated. Modern geology, for example, is an even blend of Lyell's strict uniformitarianism and the claims of catastrophists (Rudwick 1972; Gould 1977). But we call the hybrid doctrine by Lyell's name, and he has become the conventional hero of geology.

I welcome the coming failure of reductionistic hopes because it will lead us to recognize human complexity at its proper level. For consumption by *Time*'s millions, my colleague Bob Trivers maintained: "Sooner or later, political science, law, economics, psychology, psychiatry, and anthropology will all be branches of sociobiology" (*Time*, Aug. 1, 1977: 54). It is one thing to conjecture, as I would allow, that common features among independently developed legal systems might reflect adaptive constraints and might be explicated usefully with some biological analogies. It is quite another to state, as Trivers did, that the mores of the entire legal profession will be subsumed, along with a motley group of other disciplines, as mere epiphenomena of Darwinian processes.

I read Trivers' statement the day after I had sung in a full production of Berlioz' *Requiem*. And I re-membered the visceral reaction I had experienced upon hearing the 4 brass choirs, finally amalgamated with the 10 tympani in the massive din preceding the great *Tuba mirum*—the spine tingling and the involuntary tears that almost prevented me from singing. I tried to analyze it in the terms of Wilson's conjecture—reduction of behavior to neurobiology on the one hand and sociobiology on the other. And I realized that this conjecture might apply to my experience. My reaction had been physiological and, as a good mechanist, I do not doubt that its neurological foundation can be ascertained. I will also not be surprised to learn that the reaction has something to do with adaptation (emotional overwhelming to cement group coherence in the face of danger, to tell a story). But I also realized that these explanations, however "true," could never capture anything of importance about the meaning of that experience.

And I say this not to espouse mysticism or incomprehensibility, but merely to assert that the world of human behavior is too complex and multifarious to be unlocked by any simple key. I say this to maintain that this richness—if anything—is both our hope and our essence.

SUMMARY

Even since Darwin proposed it, the theory of natural selection has been marred by an uncritical style of speculative application to the study of individual adaptations: one simply constructs a story to explain how a shape, function, or behavior might benefit its possessor. Virtuosity in invention replaces testability, and mere consistency with evolutionary theory becomes the primary criterion of acceptance. Although this dubious procedure has been used throughout evolutionary biology, it has recently become the primary style of explanation in sociobiology.

Human sociobiology presents two major problems related to this tradition. First, evidence is so poor or lacking that speculative storytelling assumes even greater importance than usual. Secondly, the existence of behavioral adaptation does not imply the operation of Darwinian processes at all—for non-genetic cultural evolution, working in the Lamarckian mode, dwarfs by its rapidity the importance of slower Darwinian change. The sociobiological vision of a reduction of the human sciences to biology via Darwinism and natural selection will fail. Instead, I anticipate fruitful, mutual illumination by analogy between independent theories of the human and biological sciences.

REFERENCES

Barash, D. 1976. Male response to apparent female adultery in the mountain bluebird (*Sialia currucoides*): An evolutionary interpretation. *American Naturalist* 110:1097–1101.

Bartholomew, G.A., and T.M. Casey. 1977. Endothermy during terrestrial activity in large beetles. *Science* 195:882–883.

Bertalanffy, L. von. 1969. Chance or law. *In* A. Koestler (ed.). *Beyond reductionism*. Hutchinson, London.

Cock, A.G. 1966. Genetical aspects of metrical growth and form in animals. *Quarterly Review of Biology* 41:131–190.

Darwin, C. 1868. *The variation of animals and plants under domestication*. John Murray, London.

———. 1880. Sir Wyville Thomson and natural selection. *Nature* 23:32.

Emlen, S. 1980. In G.W. Barlow and J. Silverberg, eds. *Sociobiology: Beyond Nature/Nurture?* Westview, Boulder, Colorado.

Gould, S.J. 1966. Allometry and size in ontogeny and phylogeny. *Biological Reviews* 41:587–640.

———. 1975. Allometry in primates, with emphasis on scaling and the evolution of the brain. *In* approaches to primate paleobiology. *Contributions to Primatology* 5: 244–292.

———. 1977. Eternal metaphors of paleontology. *In* A. Hallam (ed.). *Patterns of evolution*. Elsevier, Amsterdam, pp. 1–26.

Huxley, J. 1932. *Problems of relative growth*. MacVeagh, London.

Lockard, J.S., L.L. McDonald, D.A. Clifford, and R. Martinez. 1976. Panhandling: Sharing of resources. *Science* 191:406–408.

Rudwick, M.J.S. 1972. *The meaning of fossils*. Macdonald, London.

Schweber, S.S. 1977. The origin of the *Origin* revisited. *Journal of the History of Biology* 10:229–316.

Simpson, G.G. 1961. Some problems of vertebrate paleontology. *Science* 133:1679–1689.

Trivers, R. 1971. The evolution of reciprocal altruism. *Quarterly Review of Biology*. 46:35–57.

Trivers, R., and H. Hare. 1976. Haplodiploidy and the evolution of the social insects. *Science* 191:249–263.

Weinrich, J.D. 1977. Human sociobiology: Pair-bonding and resource predictability (effects of social class and race). *Behavioral Ecology and Sociobiology* 2:91–118.

Wilson, E.O. 1975. *Sociobiology: The New Synthesis*. Harvard University Press, Cambridge, Massachusetts.

48

ARTHUR L. CAPLAN

Say It Just Ain't So: Adaptational Stories and Sociobiological Explanations of Social Behavior

I

Whatever else sociobiology has achieved since the appearance of E. O. Wilson's *magnum opus* on the subject, *Sociobiology: The New Synthesis,* in 1975, it has been a boon to the field of publishing. In the six years since the appearance of Wilson's book, over fifty texts and anthologies have appeared in print devoted solely to the topic of sociobiology.[1] And many other volumes are in press or on the drawing boards. Philosophers, anthropologists, political scientists, biologists, ethnologists, sociologists, and theologians are among the ranks of scholars to be found mulling over the methods and messages of sociobiology at book length. The number of articles published in scholarly journals, magazines, and newspapers on sociobiology is in the hundreds, if not the thousands. One does not have to subscribe to a scientometric approach to the sociology of science to realize that this sudden voluminous burst of publications represents a relatively rare phenomenon in the recent history of science.

Not only is the number of publications devoted to sociobiology in recent years notably large, but these publications have, for the most part, been remarkably supportive of the sociobiological approach to the study of animal and human behavior. Even subtracting those texts jostling with Wilson's for primacy of place as *the* founding document of sociobiology, there still remain an overwhelming number of books and articles touting the merits and

virtues of kin selection, reciprocal altruism, parental investment, and stable strategy theory for understanding the customs and mores of mites, monkeys, and men.[2] Sociobiologists have been quite open about their cannibalistic intentions toward those in the social sciences and the humanities who ply their trades in any area pertaining to social behavior. Eliminative reduction at best, total paradigmatic replacement at worst, are to be the ultimate fates of large areas of the social sciences and humanities in the sociobiological research program.[3] Ironically, despite these dire prospects, the competition has been quite fierce among many scholars in the targeted fields of inquiry to see who can be swallowed first in the maw of biological reductionism.[4]

Edibility aside, it is interesting to note that the first round of the debate about the adequacy and import of sociobiology seems to have been won by the pro-sociobiological camp. The various ethical and methodological criticisms leveled against sociobiology have either been absorbed or ignored.

Sociobiologists have checked some of their early tendencies toward brazen scientistic excess. Genuine second-thoughts concerning the influences of ideology, racism, sexism, sweeping generalizations, and crude hereditarianism are present in most current discussions of human sociobiology.

Methodological worries about the reification of behavioral traits, dubious comparative methodology, and lack of concrete factual support for sociobiological theorizing have, for the most part, however, been studiously ignored. Sociobiologists and their admirers in the social sciences have been so enamored of the powerful models sociobiology has presented that worries

The Philosophical Forum 13 (Winter–Spring 1981–1982), pp. 144–160. Reprinted with permission from the publisher.

about method seem like so much irrelevant whistling in the Darwinian wind. Most biologists, anthropologists, political scientists, sociologists, economists, and psychologists rallying to the sociobiological banner have not been sympathetic to "in principle" methodological concerns. Rather, these scientists, particularly those concerned with human behavior, have been eager to test the applicability of sociobiological models to particular instances of behavior within their own areas of interest.[5]

Some critics[6] of sociobiology have been inclined to explain this receptivity to sociobiology on ideological grounds—social scientists, like their biological brethren, being products of a capitalistic, bourgeois culture, find it easy to see the egoistic invisible hand of the free market in the tiniest bit of social activity. However, such explanations fail to account for the fact that sociobiology has not captured the hearts and minds of all social scientists, and the equally important fact that the critics of sociobiology have been just as vulnerable to the siren song of capitalism as their supposedly benighted peers. Crude ideological explanations of the positive reception accorded sociobiology seem as suspect as the crude hereditarianism and reductionism that so irritate the critics of this field.

Another somewhat less simplistic, ideological explanation of sociobiology's initial success in the scientific lists is that scientists with personal, professional, and institutional prestige can keep almost any idea or theory, no matter how wild or wacky, alive and respectable. A moment's reflection on the amenability of alchemy, phrenology, parapsychology, mesmerism, social Darwinism, eugenics, creationism, and astrology to serious methodological and/or moral critiques[7] should give any critic of sociobiology pause about the efficacy of such efforts. If certain denizens of Harvard, Oxford, Michigan, and Washington take a fancy to a particular theory, then rational critique may not serve as the quickest means to theoretical refutation.

The positive response to sociobiology in diverse academic circles raises other interesting questions for students of conceptual change and theoretical evolution. If prescriptive methodological admonitions have thus far lacked efficacy in derailing sociobiology, this may be a result of the inadequacy of the prescriptions or of their untimeliness. Whether one subscribes to a traditional falsificationist view of the-

ory evolution or to some variant of the paradigm, research program, themata, or problem-solving approaches for understanding conceptual change, the recent history of the origin and dissemination of sociobiology may require certain refinements to be made in applying these views to the analysis of the initial success of specific theories. It is, for example, highly unlikely that demands for critical experiments, falsification of key models, or the rigorous explication of central concepts will find receptive ears among the originators and initial proponents of theories. Before an anthropologist can critically test the applicability of kin selection for understanding the marriage customs of a particular group of people, the anthropologist must be enthusiastic enough and sanguine enough about the utility of the model to make the effort to test it. Proponents of sociobiological models are aware, albeit perhaps tacitly, that this is so, and, thus, pitch their approach accordingly. War, infanticide, sexual behavior, gender roles, inheritance patterns, etc., have all served as lures to whet the appetites of the anthropological world for "biologization." In the initial stages of theory dissemination, cries for tests, falsification, or confirmation are likely to fall on deaf ears, while the scientists concerned attempt to establish for themselves whether there is enough meat present in the theory to merit an effort at satisfying normal methodological strictures.

It should also be noted that the broader the scope and explanatory power claimed by a scientific theory, the less likely it is that any single scientist will feel competent to undertake its falsification. Once sociobiology established itself as possessing models worth considering, the very scope of the theory made it highly unlikely that any single scientist or research group would have the expertise required to falsify it. If sociobiology is a kind of biological analogue to the long-sought unified field theory in physics, then its failure or success is not likely to hinge solely on its adequacy for explaining the flight behavior of the red-tailed deer or the marriage customs of the Yanomamo Indians.

Prescriptive methodological strictures in the philosophy of science which advise the scientist to test, falsify, confirm, or corroborate are often imparted by philosophers as if there existed an audience of omnipotent experts. But this rugged frontiersman image of the practicing scientist fails to account for the limited effect single cases of empirical refutation have in contexts in which theories are in their initial stages of development and encompass many types of subject matters. It seems petty, at best, to demand immediate testability from those involved in the earliest stages of theory formation and proselytization.[8] It is merely naive to insist upon immediate falsifiability as a hallmark of legitimacy to potential converts who suffer from serious self-doubts about the adequacy of the theoretical foundations of their own disciplines, as is the case in vast stretches of the social sciences. In a situation in which a theoretical vacuum is perceived to exist, it will take more than a few worries about morals and method to divert the eye of practitioners from the alluring glint of a prospective comprehensive theory.

In the initial stages of theory development, scope, coherence, and need seem to count far more heavily in understanding theory acceptance than the criteria of testability and confirmation, which seem more appropriate for assessing the validity and utility of mature theories. It is interesting to note that of all the objections that have been raised against sociobiology, those that appear to have had the greatest impact among the theory's proponents and those scientists who have not committed themselves one way or another are not specific to sociobiology *per se*. Rather, the two methodological criticisms that have generated the most attention are really worries about the adequacy of a particular style of theorizing utilized in many evolutionary accounts.

Stephen Jay Gould and Richard Lewontin have attacked sociobiology on the grounds that its explanations are no more than clever stories woven to highlight the role of sociobiological models in the genesis of social behavior, and that these stories are riddled with a Panglossian faith in the optimizing power of natural selection.[9] However, as both Gould and Lewontin acknowledge, these criticisms really apply, in their view, to many evolutionary accounts in biology. To the degree that such methodological worries have been effective, they seem to call into question the adequacy of a rather common explanatory strategy in theoretical analyses of all aspects of organic evolution, and, only indirectly, the particular and distinctive claims of sociobiology. But perhaps this should not be considered surprising, for, as I have tried to suggest, it is particularly difficult to mount methodological criticisms of new theories having both great scope and few, if any, theoretical competitors. Theories in their initial stages of development may not be able to withstand very much methodological heat. It may be that those methodological criticisms which extend in their scope to more established theories have a better chance of being effective *qua* methodology. Indirection may be a better critical tack to take with regard to nascent theories than direct confrontation. However, it is not clear that the particular criticisms leveled by Gould and Lewontin are valid against either explanations of non-behavioral traits based upon the modern synthetic theory of evolution or the analyses of social behavior proffered by sociobiologists.

II

A key problem with many sociobiological accounts, according to Gould and Lewontin, is that they amount to no more than the weaving of plausible stories. Gould alleges that many sociobiologists commit the fallacy of what he terms—borrowing a phrase from Rudyard Kipling—the "just-so story." He cites approvingly Ludwig von Bertalanffy's indictment of the modern synthetic theory of evolution as just so much fantastic myth-making: "If selection is taken as an axiomatic and *a priori* principle, it is always possible to imagine auxiliary hypotheses—unproved and by nature unprovable—to make it work in any special case. . . . Some adaptive value . . . can always be construed or imagined."[10] Gould notes that von Bertalanffy's objection applies with a vengeance to sociobiological explanations of behavior: "von Bertalanffy's objections still apply, if anything with greater force, because behavior is generally more plastic and more difficult to specify and homologize than morphology. Sociobiologists are still telling speculative stories, still hitching without evidence to one potential star among many, still using mere consistency with natural selection as a criterion of acceptance."[11]

A strange sense of *déjà vu* surrounds the just-so story complaint against sociobiology. For, as Gould rightly observes, it is precisely this complaint that has been levelled against the adequacy of Darwinian accounts of evolution for much of the twentieth century.

Evolutionary theory has been charged time and again with the sins of excessive malleability, *ad hocness,* and unfalsifiability.[12]

It is precisely these concerns which led Karl Popper and various other philosophers[13] to relegate Darwinism to the dustbin of metaphysics. The charge lingers on to the present day, particularly in paleontological and taxonomic circles. Biologists in these fields have shown an astounding capacity for self-flagellation as they bemoan the theoretical docility of many of their peers who are unable to see that the modern synthetic theory of evolution rests upon sandy mythological foundations.[14]

The problem with Gould's complaint, as with von Bertalanffy's before him, is that it throws the scientific baby of evolutionary science out with a vast amount of naive historicist bathwater. Of course it is true that myth-making and deliberate *ad hocness* have no place in evolutionary biology—sociobiological or otherwise. But the appeal to remove storytelling from evolutionary biology conflates important differences between science, history, stories, and myths.

Popper and those critics of evolutionary theory such as von Bertalanffy, who were inspired by Popper's criticisms of historicist laws, fallaciously view evolutionary theory as a maxim in search of biological facts to explain. Natural selection, on the Popperian account, is a principle invoked to explain each and every twist and turn in the history of organic life on this planet. If giraffes have long necks, it is, presumably, as a result of natural selection. If a sub-group of giraffes is found possessing short necks, this too is presumably a consequence of natural selection. The accordion-like ability of natural selection in explaining disparate facts reminds Popper and other kindred spirits of two other allegedly similar and suspect explanatory outlooks—Marxism and Freudianism.[15] Popper smells historicism in Darwinism and argues, since the events of human and animal history are unique, they cannot be subsumed under a principle such as natural selection which purports to explain all of organic history.[16]

The primary problem with this line of attack against Darwinism and its descendant, the modern synthetic theory of evolution, is that it totally and utterly misconstrues the status of natural selection. Natural selection is not a maxim of any sort, *a priori* or otherwise. It is, as Darwin stated repeatedly, a metaphor for describing both the processes and outcomes of biological evolution. Natural selection *per se* explains nothing. Recent misguided efforts by a variety of authors[17] to the contrary notwithstanding, natural selection cannot and should not be reified to the status of a nomological principle. Rather, natural selection is a useful label for referring to an extraordinarily complex array of causal interactions occurring at the level of genes, genotypes, phenotypes, and environments. It is the laws and generalizations of genetics, development, ecology, and demography which ultimately are invoked by biologists to explain change and descent in the history of life. Natural selection is simply a covering term or place-holder for describing the various processes involved in producing evolutionary change, or the products of such processes.[18]

Perhaps the easiest way of seeing the emptiness of the charge that natural selection is utilized in all evolutionary accounts as an *a priori* explanatory law is by reflecting upon a rather peculiar episode in the recent history of the philosophy of biology. During the 1950s and early 1960s a number of articles were written by some of the leading advocates of the modern synthetic theory denying the existence of laws in evolutionary biology. For example, Ernst Mayr wrote: "Uniqueness is particularly characteristic for evolutionary biology. It is quite impossible to have, for unique phenomena, general laws like those existing in classical mechanics."[19] George G. Simpson argued that: "History does not correspond with possible mechanistic models, such as some in the physical sciences. That history is not simple and tidy is unfortunate, perhaps, but it is true. . . . The human desire for neat and unequivocal conclusions explains the long and necessarily futile search for simple, absolute, deterministic laws of evolution."[20]

Ironically, numerous philosophers in the 1970s have argued quite strenuously that evolutionary biology does indeed have laws. For example, Michael Ruse has argued that population genetics surely has identifiable laws, e.g., the Hardy-Weinberg law, and since population genetics is at the core of evolutionary theory, then evolutionary theory must surely have laws.[21]

This juxtaposition of opposing views about the existence of laws in evolutionary theory appears, at first glance, to be utterly bizarre. Defenders of the scientific status of contemporary evolutionary theory, such as Ruse, claim to find laws in nearly all evolutionary explanations, while leading proponents of the theory, such as Mayr and Simpson, argue that the theory has no laws, can never have laws, and that this state of affairs confers a distinctive status upon evolutionary theorizing in comparison with other branches of natural science!

However, the contradictory nature of these points of view concerning laws in evolutionary theory evaporates once it is seen that the biologists and philosophers involved were really talking at cross-purposes. Surely Mayr and Simpson are as aware as Ruse and other philosophers of biology of the vast edifice of laws, theorems, generalizations, models, and principles erected in the past seventy-five years in the fields of population genetics, ecology, and demography. Their aim in arguing against the existence of laws or maxims in evolutionary theory was to debunk the existence of a particular type of law—historical, purposive, or directional laws.[22] Mayr, Simpson, and other architects of the modern synthetic theory were concerned to refute the views of Berg, Schindewolf, Teilhard de Chardin, and others who argued for various orthogenetic explanations of evolution—that evolution can only proceed through fixed, predetermined cycles or stages of development.[23]

The arguments of biologists such as Mayr and Simpson against the existence of laws in evolutionary biology are best understood as arguments against historicist interpretations of the history of life—the very point of such concern to Popper, von Bertalanffy and, most recently, Gould. Darwinian evolutionary biologists, from Darwin himself to contemporary exponents of Neo-Darwinism, have been adamant opponents of all forms of historicism in explaining the history of life. At the heart of the Darwinian analysis of evolution is the belief that historical phenomena in the organic world can only be explained by ahistorical, mechanistic laws. Thus, philosophers of biology, such as Ruse, are not, as they have often thought, really at odds with contemporary evolutionists over the nature of laws in evolutionary biology. Vagaries of language have simply obscured the fact that Darwinians, Popperians, and devotees of the received view of scientific theories, such as Ruse, are all in agreement about the nomological character of laws in evolutionary theory. There are no distinctive laws of evolution in the sense of historicist or directional laws. There

are no evolutionary laws which can subsume distinct events in the history of life in order to explain such events. Rather, non-historical laws can be applied to particular events or occurrences in the history of life in order to explain subsequent changes and developments. Evolutionists only have available the mechanistic, nomological generalizations and models of population genetics, demography, ecology, molecular biology, and now sociobiology for explaining events in the history of life.[24]

If my analysis of the misunderstanding that has arisen over the character of laws in evolutionary explanations is correct, then the inappropriateness of the charge that evolutionary accounts are based upon a blind adherence to the law of natural selection becomes apparent. Darwin and the Neo-Darwinists who follow in his scientific footsteps have no tolerance for the type of metaphysical invocation of explanatory principle so anathematized by Popper, von Bertalanffy, and Gould. While it is indeed possible to challenge the belief that historical or directional laws of evolution do not exist or that such laws are conceptually incoherent,[25] the fact is that the modern synthetic theory has no truck with this type of law. The essence of the Darwinian approach to the explanation of biological evolution over time is that such changes can only be explained by means of laws, principles, and models that make no essential reference to time or history as subsumed variables.

Sociobiologists, if they can be fairly characterized as anything, can surely be characterized as Darwinian in explanatory outlook. Their models of kin selection, parental investment, reciprocal altruism, and the like, are meant to explain the evolution of social behavior by means of the interactions that obtain between genotypes, phenotypes, and environments. Natural selection for sociobiologists, as for any evolutionary biologist committed to a Darwinian understanding of evolution, can never serve as an *a priori* maxim, unfalsifiable nomic principle, or *ad hoc* explanatory device. It is simply a phrase that acts as a capsule summary for the complex set of causal interactions that, acting over time, eventuate in the myriad forms of life, traits, and behaviors we refer to as the end-products of evolution. Whatever the sins of sociobiologists may be, and they may, as Gould and others have noted, be numerous, reification of natural selection into an *a priori* law or principle is not one of them. Sociobiologists may believe that natural selection should be at the heart of any explanation of evolutionary change in animals and humans, past or present, living or dead. But this belief is merely a belief in the power and scope of evolutionary theory to explain all aspects of organic change in every species. While this belief may be (and probably is) false, its falseness does not arise as a consequence of the invocation of natural selection as some sort of all-powerful, untestable law of nature. This red herring derives its fishy smell from a failure to perceive the single-minded devotion to anti-historicist views of history that permeates the Darwinian, Neo-Darwinian, and sociobiological view of life.

III

The criticism that sociobiologists adopt an explanatory strategy toward all social behavior, such that they view it solely as the result of natural selection and thereby make their accounts untestable or *a priori*, seriously misconstrues the meaning of natural selection. I know of no single instance where any sociobiologist has argued for a metaphysical interpretation of natural selection *sensu* Gould or von Bertalanffy. What sociobiologists do is construct hypotheses about the evolution of various social behaviors based upon an emended version of Neo-Darwinism. The emendations they make to Darwinian theory involve (1) the recognition that similar environmental forces act similarly on similar genotypes, (2) that certain forms of behavior can be mutually beneficial to interacting organisms, and (3) that parents are in direct competition with their own offspring for environmental resources.[26] These emendations are still part and parcel of the modern synthetic theory which posits various laws governing the interaction of genes, phenotypes, environments, and isolating factors. The request that sociobiologists cease spinning selectionist stories about social behavior is equivalent to the request that they not extend the scope of the modern synthetic theory to social behavior—which of course is the central aim of the sociobiological enterprise. Sociobiological accounts are not, as Gould suggests, "consistent" with evolutionary theory—they are *derived* directly from an emended version of that theory.

Numerous philosophers of history[27] have claimed that there are a number of additional criteria that distinguish stories and myths from history. For example, there is near unanimity of opinion about the claim that among the properties possessed by history, as opposed to stories and myths, are internal consistency, the avowed intention to produce a "factual" account of past events, and the willingness of historians to test their accounts against publicly available forms of evidence. Stories normally lack all of these characteristics.

If such criteria can be utilized to distinguish history from stories, myths, and fables, then surely sociobiological accounts count as history, not stories. Sociobiological explanations of the incest taboo, homosexuality, panhandling among humans, and inheritance patterns among persons in various cultural settings are constructed so as to be grounded in an established theory (an emended version of the modern synthetic theory of evolution), to be "factual," and to be testable by publicly available evidence. Indeed, the evidence for the adequacy of sociobiological accounts regarding these phenomena seems to refute many of these hypotheses. But the real point at issue is that many sociobiological accounts do approximate the classificatory standards for being understood as history (perhaps false history but still history) operative in the social sciences and human history, which is probably all that can reasonably be asked of sociobiological hypotheses on methodological grounds at this point in time. While sociobiological accounts of the origins of social behavior may indeed be slap-dash or false, they are patently not fictions or fables.

The metaphysical status of natural selection is not the only methodological criticism that has had some impact in the current debate about sociobiology. Linked to the charge of story-telling is the claim that sociobiologists are Panglossian in their approach to the analysis of social behavior. Gould and R. C. Lewontin have criticized what they term the "adaptationist program" both in sociobiology and in evolutionary theorizing in general. They describe this program as follows:

Studies under the adaptationist program generally proceed in two steps:
1. An organism is atomized into "traits" and these traits are explained as structures optimally designed by natural selection for their functions. . . .

2. After the failure of part-by-part optimization, interaction is acknowledged via the dictum that an organism can't optimize each part without imposing expenses on others. The notion of "trade-off" is introduced and organisms are interpreted as best compromises among competing demands.[28]

It is not just story-telling that disturbs Gould and Lewontin, but the telling of stories which always portray organisms as optimally designed compromises for solving environmental challenges.

I believe Gould and Lewontin have accurately characterized the specific algorithm followed by most evolutionary biologists in attempting to formulate explanations for the existence and persistence of various traits and behaviors in organisms. They are surely correct in noting that the parsing of organisms into composites of individual traits and behaviors introduces an air of artificiality into explanations in evolutionary biology.[29] They are also correct in noting that while evolutionary biologists (and sociobiologists) make *pro forma* obeisance to the causal efficacy of such factors as drift, mutation, recombination, and allometry in textbook expositions of the modern synthetic theory, in practice these factors are forgotten in the rush to invoke environmental forces as the primary and sole type of causal factors guiding the evolutionary process. As Gould notes, Panglossian adaptationalism has been particularly prevalent in sociobiological circles: "Most work in sociobiology has been done in the mode of adaptive storytelling based upon the optimizing character and pervasive power of natural selection."[30] Sociobiologists are prone to positing explanations of any and all behaviors as optimal solutions to fitness problems posed by environmental and conspecific forces.

The suggestion by Gould and Lewontin that evolutionists should remember that there are other forces at work in producing evolutionary change is, it must be noted, of no special relevance to sociobiological theorizing. It is a criticism that can be offered against current evolutionary theorizing in general.

Gould and Lewontin do not provide in their criticism a philosophical explication of the concept of adaptation itself. They offer their criticism of the adaptationalist program as part of their overall concern with storytelling in evolutionary biology, and, in passing, to sociobiology. However, as I have tried to show, their concern about storytelling in evolutionary theorizing seems to me to be misplaced. Evolutionary biologists operating within a Darwinian framework are in no way committed to a metaphysical belief in the explanatory power of a reified form of natural selection. Once the critique of the adaptationalist program is decoupled from the misguided critique of storytelling, however, an interesting problem does emerge—perhaps evolutionary biologists and their sociobiological brethren have misunderstood the concept of adaptation.

Adaptation is almost always defined by evolutionary biologists in terms of some advantage conferred upon an organism by a particular trait or behavior. Usually the advantage is understood in terms of increased reproductive fitness. The definition proferred by E. O. Wilson in *Sociobiology: The New Synthesis* is typical. Wilson defines adaptation as ". . . any structure, physiological process or behavioral pattern that makes an organism more fit to survive and to reproduce in comparison with other members of the same species."[31]

The problem with this and other such definitions is that they conflate the notions of advantage and adaptation. There are, as Gould and Lewontin correctly observe, many ways in which advantageous traits and behaviors come to exist in organisms. Some beneficial traits arise as a result of the processes involved in natural selection. Others exist simply as a result of mutation, drift, allometry, or contingent accidents.

Consider an example. A dark coat color may be quite beneficial to certain members of a particular species of rat if the coloring makes it harder for birds and other predators to find the rats. Rats may acquire dark coats through the process of natural selection, or, as a consequence of moving about through muddy terrain. The process by which the dark coat color appears is irrelevant in terms of the survival and reproductive advantages such coloration may provide. But on the standard definition of the sort given by Wilson, *any* feature that confers an advantage is, by definition, an adaptation.

The obvious difficulty with such a definition is that it forces biologists to view every beneficial or advantageous property of organisms as adaptations. Surely what concerns the biologist is not only whether a property, trait, or behavior is beneficial, but also how it came to exist in an organism or species. Adaptation, like such terms as "crater" and "hybrid," is an historical concept.[32] In order to ascertain whether something is a crater, a hybrid, or an adaptation, we need to know something about its past, its etiology.

Adaptations can only result from the process of natural selection. But advantages in terms of survival, fitness, or efficiency can result from a variety of processes and causes. Thus, it would appear that advantage is a poor index of adaptation, and vice versa.

Once advantageousness is dropped from the definition of adaptation, it becomes easier to utilize this term in contexts where its use may have seemed somewhat awkward. There are numerous instances in the history of life where well-adapted species have become extinct. If adaptation is viewed solely in terms of benefits to survival and reproduction, it is rather difficult to explain the common phenomenon of extinction. But, if adaptation is defined as an historical concept, denoting characters, traits, or behaviors produced as a result of the process of natural selection, then the extinction of adapted species is not quite as peculiar a phenomenon.

Similarly, if adaptation is defined in terms of survival and reproductive advantages which accrue to organisms relative to other organisms, then only those organisms possessing the greatest advantages can be considered adapted. The fascination with optimality noted by Gould and Lewontin in most evolutionary accounts is actually a by-product of the conflation of the notions of adaptation and advantage. If adaptation is defined as any trait which makes an organism fitter than its peers then, by definition, only optimal traits can count as adaptations. If, however, the process of natural selection produces a range of traits in the members of a species, those traits that result are no less adaptations for their lack of optimality.

If adaptation is defined solely in terms of etiology, then comparisons can be made as to which adaptations are more or less advantageous without biasing evolutionary accounts toward an erroneous belief in the optimizing power of natural selection. Nor need it be assumed that whatever advantages an organism possesses for survival and reproduction exist as a consequence of natural selection—an assumption which is particularly suspect in explana-tions of social behavior where advantages may be conferred by learning, mimicry, or culture.[33]

Gould and Lewontin's criticism of Panglossian adaptationalism is valid, but they fail correctly to diagnose the cause of the problem. Evolutionary biologists and sociobiologists are certainly entitled to try to explain every property of an organism as an adaptation—there is no *a priori* reason why such a research strategy is untenable. But it is not tenable to try to link the notions of adaptation and advantage by definitional fiat. Some adaptations are optimally advantageous, but others are not. Natural selection could conceivably be the cause of every organic property known to man, but, if the fossil record is to be believed, there is no guarantee that this process necessarily produces benefits or advantages as a matter of course. The percentage of organisms that have both survived and reproduced over the course of time belies the comprehensive optimizing power of evolution.[34]

V

The sociobiology debate reveals a number of important facts about theory acceptance in science. Methodological concerns are unlikely to be effective antidotes to theory acceptance during the initial phases of the dissemination of a theory, particularly if the theory has a broad scope and few theoretical competitors. It appears easier to capture the attention of a new theory's proponents and potential converts by pointing out methodological flaws in more established theories which may bear upon or support the new theory. The methodological criticisms which have had the most telling impact against sociobiology highlight these features nicely since the most effective criticisms to-date have been directed against evolutionary theory in general and not against the specific models and claims of sociobiology. Ironically, the normative criticisms of sociobiology which commanded so much attention in the earliest phases of the sociobiology debate appear to have had little impact in hindering the acceptance of the theory.[35]

Upon examination, however, the two most powerful methodological objections to sociobiology—the "just-so story" complaint and the charge of "Panglossian adaptationalism"—do not appear to withstand critical scrutiny. The former objection misconstrues the meaning of natural selection in evolutionary accounts; the latter objection fails to locate the real source of confusion concerning adaptation: the conflation of advantage with adaptation in definitions of the concept. Neither objection appears valid as a criticism of the soundness of the modern synthetic theory of evolution and, thus, neither objection is likely to fatally damage the intellectual prospects of sociobiology. However, by focusing attention on the concepts of natural selection and adaptation, Gould and Lewontin may have succeeded in highlighting some of the explanatory excesses of sociobiology. For while current evolutionary theory may not be liable for the sins of sociobiology, it may, if properly understood, set strict limits on the adequacy of purely biological explanations of social behavior.

NOTES

1. Useful bibliographies are included in M. Ruse, *Sociobiology: Sense or Nonsense?* (Dordrecht: D. Reidel, 1979), and N. A. Chagnon and W. I. Irons, eds., *Evolutionary Biology and Human Social Behavior* (North Scituate, Mass.: Duxbury, 1979).

2. Recent examples include P. Singer, *The Expanding Circle* (New York: Farrar, Straus & Giroux, 1981), W. J. Ong, *Fighting For Life* (Ithaca: Cornell, 1981), and A. Rosenberg, *Sociobiology and the Preemption of Social Science* (Baltimore: Johns Hopkins, 1980).

3. A. Rosenberg, *Ibid.* Also, E. O. Wilson. "Biology and the Social Sciences," *Daedalus*, 106, 127–140.

4. See for example any of the papers in R. D. Alexander and D. W. Tinkle, eds., *Natural Selection and Social Behavior* (New York: Chiron Press, 1981), or in the Chagnon and Irons volume cited in Note 1.

5. See A. Somit, ed., *Biology and Politics* (Paris: Mouton, 1976), D. Barash, *Sociobiology and Behavior* (New York: Elsevier, 1976), and J. T. Bonner, *The Evolution of Culture* (Princeton: Princeton University Press, 1980).

6. M. Sahlins, *The Use and Abuse of Biology* (Ann Arbor: University of Michigan, 1976); S. Rose, "It's Only Human Nature: The Sociobiologists Fairyland," *Race and Class*, 20, No. 3 (1979), 158–170; the articles by E. Allen, et al., and S. J. Gould reprinted in A. Caplan, ed., *The Sociobiology Debate* (New York: Harper & Row, 1978), and the

introduction by Ashley Montagu in A. Montagu, ed., *Sociobiology Examined* (Oxford: Oxford University Press, 1980).

7. See M. D. Hanen, M. J. Ostler, R. G. Wyant, eds., *Science, Pseudo-Science and Society* (Waterloo: Wilfred Laurier Press, 1980.

8. David Hull stresses this point in his "Scientific Bandwagon or Traveling Medicine Show?," in M. S. Gregory, A. Silvers, and D. Sutch, eds., *Sociobiology and Human Nature* (San Francisco: Jossey-Bass, 1978), pp. 136–163.

9. S. J. Gould and R. Lewontin, "The Spandrels of San Marco and the Panglossian Paradigm: A Critique of the Adaptationist Programme," *Proceedings of the Royal Society of London*, B, 205 (1979), 581–598. Also, R. Lewontin, "Adaptation," *Scientific American*, 239 (1978), 156–169; S. J. Gould, "Sociobiology: The Art of Storytelling," *New Scientist*, 80 (1978), 530–533; and S. J. Gould "The Evolutionary Biology of Constraint," *Daedalus*, 109 (1980), 39–52.

10. S. J. Gould, "Sociobiology and the Theory of Natural Selection," in G. Barlow and J. Silverberg, eds., *Sociobiology: Beyond Nature/Nurture?* (Boulder: Westview, 1980), p. 257.

11. *Ibid.*, p. 260.

12. For an excellent review of different versions of these criticisms, see M. Ruse, *The Philosophy of Biology* (London: Hutchinson, 1973), Chapters II and III. Also, see A. Caplan "Darwinism and Deductivist Models of Theory Structure," *Studies in History and Philosophy of Science*, 10 (1979), 341–353.

13. See especially K. R. Popper, *The Poverty of Historicism*, 3rd ed. (London: Routledge & Kegan Paul, 1961), Chapter IV.

14. See the review of Ernst Mayr's *Evolution and the Diversity of Life* by N. Platnick in *Systematic Zoology*, 26 (1977), 224–228; D. E. Rosen and D. G. Buth, "Empirical Evolutionary Research Versus Neo-Darwinian Speculation," *Systematic Zoology*, 29 (1980), 300–308; and Niles Eldredge and Joel Cracraft, *Phylogenetic Patterns and the Evolutionary Process* (New York: Columbia, 1980).

15. Popper, *The Poverty of Historicism*, pp. 73–83. Also, see K. Popper, *Objective Knowledge* (Oxford: Oxford University Press, 1972).

16. An excellent critique of Popper's views can be found in M. Ruse, "Karl Popper's Philosophy of Biology," *Philosophy of Science*, 44 (1977), 638–661.

17. M. B. Williams, "Deducing the Consequences of Evolution," *Journal of Theoretical Biology*, 29

(1970), 342–385, and E. S. Reed, "The Lawfulness of Natural Selection," *The American Naturalist,* 118 (1981) 61–71.

18. A. L. Caplan, "Testability, Disreputability and the Structure of the Modern Synthetic Theory of Evolution," *Erkenntnis,* 13 (1978), 261–278.

19. E. Mayr, "Cause and Effect in Biology," rpt. in R. Munson, ed., *Man and Nature* (New York: Dell, 1971), p. 114.

20. G. G. Simpson, "Evolutionary Determinism," rpt. in R. Munson, ed., *Man and Nature* (New York: Dell, 1971), p. 210.

21. See Ruse, *The Philosophy of Biology,* and Chapter II of his *Sociobiology: Sense or Nonsense?* (Dordrecht: Reidel, 1979).

22. A clear explanation of the difference between functional or mechanistic laws, and historical or directional laws has been given by Maurice Mandelbaum:

    [A functional law] would only enable us to predict immediately subsequent events, and each further prediction would have to rest upon knowledge of the initial and boundary conditions obtaining at that time. The second type of law [a directional law] would not demand a knowledge of subsequent initial conditions. . . . For if there were a law of directional change which could be discovered in any segment of history, we could extrapolate to the past and to the future without needing to gather knowledge of the initial conditions obtaining at each successive point in the historical process.

    M. Mandelbaum, "Societal Laws," in W. H. Dray, ed., *Philosophical Analysis and History* (New York: Harper and Row, 1966), p. 234.

    Karl Popper denies the possibility of this type of directional law as abject historicism in the *Poverty of Historicism, op. cit.,* Note 15. See also Chapter 7 of Mandelbaum's *History, Man and Reason* (Baltimore: Johns Hopkins, 1971), for a superb discussion of the problems facing proponents of directional laws in history and the biological sciences.

23. For a defense of the orthogenetic approach to evolution, see M. Grene, "Two Evolutionary Theories," *British Journal for the Philosophy of Science,* 9 (1959), 110–127.

24. However, Stephen J. Gould has done some plumping for historicism in evolutionary theory. See his "The Promise of Paleobiology as a Nomothetic Evolutionary Discipline," *Paleobiology,* 6 (1980), 96–118.

25. P. Urbach, "Is Any of Popper's Arguments Against Historicism Valid?," *British Journal for the Philosophy of Science,* 29 (1978), 117–130, and A. Olding, "A Defense of Evolutionary Laws," *British Journal for the Philosophy of Science,* 29 (1978), 131–143.

26. See R. Dawkins, *The Selfish Gene* (Oxford: Oxford University Press, 1976).

27. See, for example, W. B. Gallie's discussion of stories and histories in *Philosophy and the Historical Understanding,* 2nd ed. (New York: Shocken, 1968), Chapters II and III.

28. S. J. Gould and R. C. Lewontin, "The Spandrels of San Marco and the Panglossian Paradigm: A Critique of the Adaptationist Programme," p. 585.

29. See also R. C. Lewontin, "Fitness, Survival and Optimality," in D. J. Horn, G. R. Stairs, and R. D. Mitchell, eds., *Analysis of Ecological Systems* (Columbus: Ohio State, 1979), pp. 3–21.

30. S. J. Gould, "Sociobiology and the Theory of Natural Selection," p. 259.

31. E. O. Wilson, *Sociobiology: The New Synthesis* (Cambridge: Harvard, 1975), p. 577.

32. Morton Beckner distinguishes among three classes of concepts that appear to him to be characteristic of and unique to biological theory—"polytypic," "historical," and "functional." He notes that:

    if we describe a contemporary system by means of a historical concept . . . we are presupposing that the system has actually had such and such a history. To call a plant "hybrid corn," for example, is to presuppose that the plant is a first filial descendant of a cross between two distinct strains of corn, in the sense that if it is in fact not such a descendant, it is logically impossible for it to be hybrid corn.

    M. Beckner, *The Biological Way of Thought* (Berkeley: University of California, 1968), p. 25.

    It is just this sense of "historical" that seems to me best to characterize the concept of adaptation. If a trait or behavior is not the end-result of the process of natural selection, it cannot be an adaptation regardless of the advantages it may confer on the organism or organisms which possess it.

33. S. C. Washburn, "Human Behavior and the Behavior of Other Animals," *American Psychologist,* 33 (1978), 12–24.

34. See the discussion of rates of extinction in Chapter X of T. Dobzhansky, F. Ayala, G. Slebbins and J. Valentine, *Evolution* (San Francisco: W. H. Freeman, 1977).

35. W. L. Albury, "Politics and Rhetoric in the Sociobiology Debate," *Social Studies of Science,* 10 (1980), 519–536.

49

ROBERT H. THOULESS

Parapsychology during the Last Quarter of a Century

There are two principal interlinked differences between the directions of interest of parapsychology today and psychical research as it was in its early years; one is a new way of using experimentation, the other is a shift of interest from the problem of whether psi phenomena really occur to that of finding out what can be known about their nature and properties.

It is true that early psychical researchers such as Myers and Gurney did experiments, but they seem to have regarded them primarily as means of confirming the real occurrence of psi phenomena. On the other hand, they expected to elucidate the nature of psi phenomena by more and more careful examination of spontaneous cases. They did not seem to have any idea of the usefulness of the experiment as a method of finding an answer to a question about the nature of the thing investigated.

Perhaps the most important early contribution made by the Parapsychology Laboratory at Duke University was that it led the way in giving experiment this new role. On rereading J. B. Rhine's first book, *Extra-Sensory Perception,* I am amazed at the number of characteristics of psi that were correctly identified during the early years at Duke. The important point is that parapsychology was now based on experiment as its basic method and that experiment was being used, as in other branches of science, as a means of finding out about what was being investigated. This is a direction of research that was started well before the period covered by the present survey, but it has continued to influence

Progress in Parapsychology (Durham, NC: Parapsychology Press, 1971), pp. 221–235. Reprinted with permission from the publisher.

research during our period, both in the way of confirming the early indications of the nature of psi and in finding out new facts about it. . . .

SCIENTIFIC REVOLUTIONS

The conviction that what is important now is to find out more about the nature of ESP has been strengthened by T. S. Kuhn's book, *The Structure of Scientific Revolutions,* published in 1962. Although the author does not mention parapsychology in this book, he gives an illuminating account of the kind of situation in scientific development of which the emergence of the psi phenomena is an example. In an address I gave to the Royal Institution nineteen years ago, I said: "In the fact that we have experimental results that are unexpected and inexplicable, we have in parapsychology a situation favorable to a profitable advance in theory." Kuhn argues that this is characteristic of all the advances in scientific theory which he calls "scientific revolutions."

These revolutions are such turning points in scientific development as the passage from the Ptolemaic cosmology to that of Copernicus, from creationism to Darwin's evolutionary theory of organic development, from Newtonian to relativity dynamics, and so on. In all of these and in many similar cases, expectations based on a generally accepted system of concepts, laws, and experimental techniques have been found in some respects not to be fulfilled. These nonfulfilments of expectation, such as the failure of the Micheson–Morley experiment to detect motion of the earth through the ether, are called by Kuhn "anomalies." He notes that the first reaction of scientists engaged in normal research in their subject is to ignore anomalies in the hope that they will in the end be found to be explicable in terms of the accepted system of concepts. Only when further research confirms the reality of the anomaly and the failure of the accepted system of theoretical explanation to accommodate it, a state of tension arises which leads to a scientific revolution in which the old explanatory system is abandoned and a new one adopted; in Kuhn's terminology the old "paradigm" is replaced by a new one.

All of this is illuminating for the present position of parapsychology and for the recognition of its immediate tasks. The demonstration of the reality of extrasensory perception and of psychokinesis is a demonstration of the presence of a series of anomalies. How little they are expected is shown by the violent reaction of such psychologists as Hansel against them. If the history of science is a reliable guide in this matter, we should expect them to be rejected until a new explanatory system is put forward which will accommodate them; it is not to be expected that rejection will be overcome merely by the accumulation of stronger evidence in favor of their reality.

THE PRESENT TASK

It would, however, be a misunderstanding of the implications of Kuhn's ideas to infer that our task now is to think out a new paradigm. It is not thus that scientific revolutions have taken place in the past. The call is rather to more detailed and more precise research. As we know more about the psi phenomena and as our knowledge becomes more exact, the shape of the future paradigm will gradually become clear. Then an individual like Darwin, Newton, or Einstein will put forward a new explanatory system in terms of which the phenomena of psi will not merely be explained, but will be shown to be such as we should have expected.

Kuhn then seems to give us additional reason for thinking that the last twenty-five years in parapsychology must be judged, not by the amount of additional evidence that has been produced for the real occurrence of psi, but rather by how much more has been found out about psi. If this is agreed, the answer must be, I think, that real progress has been made but that all that has been done remains small in comparison with what remains to be found out before we can claim to have an adequate understanding of how psi fits into our total picture of how things are.

I think it is reasonable to guess that the time for the emergence of a new paradigm that will accommodate psi is likely to be a long way ahead. An obvious difference between parapsychology and the physical sciences is that, in the physical sciences, expectations can be more easily formulated precisely in quantitative terms, and one can more easily make the exact measurements that are necessary to test experimental expectations.

One of the immediate tasks of psi research would seem to be that of finding out methods of getting more reliable psi measurements. This might be by developing more fruitful experimental designs, by devising better ways of selecting experimental subjects, or by discovering ways of training experimental subjects. All of these have, of course, been tried during the last twenty-five years, but without conspicuous success. I think the search should be continued. . . .

50

DAISIE RADNER AND MICHAEL RADNER

Parapsychology: Pre-Paradigm Science?

THE IMPORTANCE OF A PARADIGM

In *The Structure of Scientific Revolutions,* historian and philosopher of science Thomas S. Kuhn stresses the dependence of fact gathering on what he calls a *paradigm.* Other philosophers have noted vagueness and ambiguity in Kuhn's notion of a paradigm, but such problems need not detain us here.[1] All we need is a rough characterization. Roughly speaking, a paradigm is a theoretical framework that guides and unifies research in a given area. It dictates how the phenomena are to be assembled and classified, poses the problems—both conceptual and experimental—for the scientist to work on, and provides the techniques for solving them.

According to Kuhn, "one of the things a scientific community acquires with a paradigm is a criterion for choosing problems that, while the paradigm is taken for granted, can be assumed to have solutions" (p. 37). Problem solving within the context of a paradigm is *puzzle solving.* What is a puzzle? It is a problem for which a solution is assured. There are rules for what counts as a solution and for how the solution is to be obtained (think of a crossword puzzle). The rules for scientific puzzles are implicit in the paradigm. An example of a puzzle generated by a paradigm is how to explain the shape of the earth, particularly the flattening at the poles, in terms of classical mechanics. Classical mechanics requires that there should be an explanation, and it provides

the mathematical principles in terms of which the explanation is to be given. In calling this sort of problem a "puzzle," Kuhn does not mean to imply that it is trivial. A great deal of effort and ingenuity is needed to solve the puzzles of science.

Prior to the development of a paradigm, there may be a time when scientists have to do their research without a commonly accepted body of beliefs or set of standard procedures to go by. Kuhn emphasizes the haphazard and floundering character of this preparadigm stage. "In the absence of a paradigm or some candidate for paradigm, all of the facts that could possibly pertain to the development of a given science are likely to seem equally relevant. As a result, early fact-gathering is a far more nearly random activity than the one that subsequent scientific development makes familiar" (p. 15). Fact gathering during this stage is likely to yield a morass of data. Spurious similarities get noted while important differences are overlooked. Phenomena are described in great detail, but often the descriptions omit just those factors that subsequently turn out to be the most illuminating. Some phenomena are reported that later investigators cannot repeat or confirm.

The early history of paleontology provides a good illustration of fact gathering during a preparadigm stage. Nowadays we take it for granted that fossils are petrified remains of organisms that perished a long time ago. But this was not at all obvious to early investigators. The term "fossil" was originally used to refer to anything dug up from the earth. Fossil objects included gems (for example, garnet), metals (copper, gold), rocks (marble), hardened fluids (amber) and stones (gypsum, loadstone). Fossils in the modern sense were classified as stones. When they were first discovered, it was far from evident that they were organic remains. Some of them bore no resemblance to any organism now living, either because they represented extinct groups such as belemnites and crinoids, or because they were disguised by their mode of preservation; for example, the shell had dissolved and the rest of the organism was petrified without it.

Others did resemble living organisms, but it was difficult to see how they could have been the *remains* of living organisms, given the positions in which they were found. Fossils of marine organisms were discovered on hilltops far from the sea. If they were the remains of marine life, how did they get there? Some early naturalists tried to use Noah's Flood to account for the anomaly; but the Flood, at least as described in the Bible, was not long enough or violent enough to have carried the shells to their present position. Without a satisfactory theory of geological change, the organic interpretation of fossils was difficult to defend. Until a paradigm emerged, the fossil situation remained chaotic.

Without a paradigm to guide research, there was ample opportunity for individuals to go off in all directions and suggest interpretations having no evidential support. One of the most notorious instances was Scheuchzer's "Deluge man." Johann Scheuchzer (1672–1733) was a Swiss physician and naturalist and an advocate of the theory that fossils were relics of the Flood. In 1715, he found a fossil skeleton that he confidently identified as a specimen of *Homo diluvii testis*, or "man, a witness of the Deluge." A century later, the great French naturalist Georges Cuvier (1769–1832) established that Scheuchzer's "Deluge man" was nothing but a giant salamander.

PARAPSYCHOLOGY: PREPARADIGM SCIENCE?

Parapsychologists cite Kuhn to convince themselves and others that they are at a legitimate stage of scientific development. In an article entitled "ESP Research at Three Levels of Method," published in the *Journal of Parapsychology*, September 1966, R. A. McConnell asserts that Kuhn's "description of what he calls this 'pre-paradigmatic period of 'science' fits the situation precisely as we find it in parapsychology today." Does it? The author goes on to note, "In the absence of theory, facts are gathered indiscriminately and confusion is the result." Nevertheless, he insists that parapsychology has made progress in spite of its lack of a proper theoretical foundation. Even though there is no paradigm, there is still consensus on methodology, on concepts (psi-missing, decline effect), and on what count as acceptable results.

Now according to Kuhn's analysis, either research is guided by a paradigm, or else it is undertaken in an atmosphere of confusion and chaos. The parapsychologists try to evade both alternatives. They have no paradigm to guide them, yet there is no disagreement about how to approach, describe, or interpret phenomena. Parapsychologist Robert H. Thouless gives away the game when he suggests that we think of psychical research as puzzle-solving activity in the Kuhnian sense.[2] Puzzle solving is working in the context of a paradigm. The parapsychologists admit that they have no paradigm, yet they go about their work as though they did.

Will parapsychology ever come up with a paradigm and so gain a place among the sciences? Parapsychologists would have us believe that they are on their way to a paradigm, but in fact the nature of their enterprise ensures that they will never arrive at one. Parapsychology is in a terminal fact-gathering stage.

When we compare parapsychology to other disciplines in which fact gathering gave way to paradigm-directed research, we see an important difference. Let us return to the history of paleontology. Questions arose in the early days about the nature and origin of fossils. Through discovery and debate, the answers slowly emerged. Classification schemes gradually improved; the line between fossils and other objects dug up from the earth grew more and more distinct. The organic theory won out over other attempts to explain the origin of fossils in terms of "plastic forces" and "seminal breezes" that supposedly molded the stones into resemblances of living organisms. Fossils were associated with geological strata, which in turn suggested a time sequence. The Flood theory gave way to a more satisfactory explanation of the position of fossil remains. Without the Deluge theory, there was no more need for the "Deluge man"; he could go back to being a salamander.

The paleontologists used their growing understanding of fossils to sift our faulty data and bad reports. Their fact gathering went hand in hand with their search for a theory that could make sense out of the data.

The situation in parapsychology is quite different. It is true that psychical researchers do on occasion give up data. In 1928, J. B. Rhine examined Lady Wonder, a horse that answered questions by arranging numbered and lettered blocks with her nose. He concluded that she was telepathic. A magician later found that Lady's owner was signaling the horse. No respectable parapsychologist today believes in the psychic powers of Lady Wonder. In 1974, Walter J. Levy, then director of the Institute for Parapsychology, was caught cheating in his experiments on psychokinesis in rats. Levy was promptly ousted from the Institute and all his previously published work was called into question.[3] Note that in both these cases the rejection of the data was prompted by something other than the researchers' increasing knowledge of their subject matter. In both cases, somebody was actually caught in the act of cheating. When parapsychologists give up data, it is not because they understand psi better than before. It is not even because they have proved methods of detecting psi; for . . . they continue to stand by the results of earlier experiments even though serious criticisms have been raised concerning the design of those experiments. They go about their fact gathering unhindered by the constraint of having to make sense out of the data.

The parapsychologists would probably reply that they are indeed trying to make sense out of psi as they gather their evidence, and that they have in fact made some progress along these lines. They have found that the psi process operates in a lawlike manner. The lawfulness manifests itself in *decline effects* and *position effects*. Decline effects include: (1) decline in average score (number of hits) during the course of a run or a series of runs, and (2) decline in score *variance* during the course of a series of runs. To illustrate the second type of decline, suppose that the subject is put through a series of 6 runs, each consisting of 25 trials, where the mean chance expectation is 1 in 5; or 5 correct guesses in each run. Suppose further that the subject scores 1

in the first run, 0 in the second, 12 in the third, and 4, 6, 5 in the remaining runs. The scores in the first three runs fluctuate more widely from the theoretical mean than the scores in the last three. Though the scores themselves do not decline in the course of the experiment—in fact, the lowest scores are achieved at the beginning—there is a decline in score variance.

Position effects are patterns of distribution of hits over the record page. They include: (1) vertical decline, which means that there are more hits in the upper half of the page than in the bottom half, (2) horizontal decline, more hits in the left half than in the right, (3) diagonal decline, the most hits in the upper left quarter and the fewest in the lower right, and (4) U-curve, with the beginning higher than the end.

Without a plausible theory, however, there is no reason to connect any of these scoring patterns with *psi*, which has been defined by parapsychologists as "extrasensorimotor exchange with the environment." Suppose that analysis of experimental results at the Radner Institute for Pseudoscientific Research revealed that subjects tended to score most of their hits at the end of a run, and that when the record pages were examined, there tended to be more hits in the lower half of the page, more in the right half, and more in the lower right quarter than in the upper left. What would the parapsychologists say about such findings? One possible way of responding would be to say: "Because this scoring pattern is precisely the opposite of the decline curve that is characteristic of psi activity, there must have been something else going on in your experiment that accounts for the result." But this response is not open to them, for their argument that the decline effect is a "sign of psi" is a statistical argument: It must be psi because it isn't chance. Either the statistical argument supports our "incline effect" *as well as* their decline effect—our effect is as much beyond chance as theirs—or else it supports neither. The only avenue open to them would be to include the "incline effect" as another "sign of psi." Sometimes scores go up and sometimes they go down. Psi is like that.

Imagine the pioneers in the field of electricity taking a similar line. Sometimes copper is a conductor and sometimes it is an insulator. Sometimes a resisting material decreases the current and sometimes it doesn't. Electrical research would have gone the way of alchemy.

For all their decline effects and position effects, psychical researchers have failed to bring order into the chaos of psi activity. All they have succeeded in doing is to introduce another level of chaos. Now, in addition to the hodgepodge of alleged psi phenomena, we have a hodgepodge of alleged statistical regularities. Neither helps to make sense of the other.

It is interesting that the so-called signs of psi are used to let data in but virtually never to throw data out. Parapsychologists are quite willing to accept the decline and position effects as *sufficient conditions* of psi. They believe that whenever there is a decline in average score or in variance, there is psi. Thus they are able to rescue many experiments that provide no evidence of psi when the total score alone is taken into account. But they are not so willing to allow the "signs of psi" to count as *necessary conditions* of psi. They are reluctant to concede that whenever there is psi there is a decline in average score or in variance. That would commit them to throwing out some experimental results that would otherwise come out positive. Searching through the articles in the *Journal of Parapsychology*, we found only one case in which there was any willingness to use the "signs of psi" as both necessary and sufficient conditions. That was the case of Levy, the researcher who was caught cheating. It was suggested that a search for decline effects in his earlier work might show some

of it to be genuine. The data were already under suspicion. Since they would have been thrown out anyway, there was nothing to fear in checking them over for "signs of psi."

NOTES

1. Kuhn's paradigm notion contains a number of valuable components, although Kuhn does not provide rigorous explications of them in *The Structure of Scientific Revolutions*, 2nd ed. (Chicago: University of Chicago Press, 1970). Wolfgang Stegmüller follows up some implications for theory structure in *The Structure and Dynamics of Theories* (New York: Springer-Verlag, 1976). Bas van Fraassen's book *The Scientific Image* (Oxford: Clarendon Press, 1980) explores the issue of why-questions and sheds light on the presuppositional nature of paradigms.
2. Thouless, "Parapsychology," p. 298.
3. For the parapsychology community's reaction to the Levy incident, see J. B. Rhine, "A New Case of Experimenter Unreliability," *Journal of Parapsychology* 38 (1974):215–225; "Second Report on a Case of Experimenter Fraud," *Journal of Parapsychology* 39 (1974):306–325.

REFERENCE

Kuhn, Thomas S. *The Structure of Scientific Revolutions*, 2nd ed. Chicago: University of Chicago Press, 1970.

PART 9 SUGGESTIONS FOR FURTHER READING

Philosophy of Physics

Auyang, Sunny Y. *How Is Quantum Field Theory Possible?* New York: Oxford University Press, 1995.

Cushing, J. T., and E. McMullin. *Philosophical Consequences of Quantum Theory: Reflections on Bell's Theorem.* Notre Dame, IN: University of Notre Dame Press, 1989.

Davies, P. C. W., and John Gribbin. *The Matter Myth: Dramatic Discoveries that Challenge Our Understanding of Physical Reality.* New York: Simon & Schuster, 1992.

d'Espagnat, Bernard. *Veiled Reality: An Analysis of Present-Day Quantum Mechanical Concepts.* Reading, MA: Addison-Wesley, 1995.

Kosso, Peter. *Appearance and Reality: An Introduction to the Philosophy of Physics.* New York: Oxford University Press, 1998.

Stapp, Henry P. *Mind, Matter, and Quantum Mechanics.* New York: Springer-Verlag, 1993.

Philosophy of Biology

Hull, David L., and Michael Ruse, eds. *The Philosophy of Biology.* Oxford: Oxford University Press, 1998.

Mayr, Ernst. *Toward a New Philosophy of Biology: Observations of an Evolutionist.* Cambridge, MA: Harvard University Press, 1988.

Sober, Elliott. *Conceptual Issues in Evolutionary Biology.* Cambridge, MA: MIT Press, 1994.

Sterelny, Kim, and Paul E. Griffiths. *Sex and Death: An Introduction to the Philosophy of Biology.* Chicago: University of Chicago Press, 1999.

Wilson, Fred. *Empiricism and Darwin's Science.* Dordrecht: Kluwer Academic Publishers, 1991.

Philosophy of Parapsychology

Broad, C. D. *Religion, Philosophy, and Psychical Research.* London: Routledge and Kegan Paul, 1953.

Broughton, Ricard S. *Parapsychology: The Controversial Science.* New York: Ballantine, 1991.

Flew, Antony, ed. *Readings in the Philosophical Problems of Parapsychology.* Buffalo, NY: Prometheus Books, 1995.

Jahn, R. G., and B. J. Dunne. *Margins of Reality.* New York: Harcourt, Brace, Jovanovich, 1987.

Kurtz, Paul, ed. *A Skeptic's Handbook of Parapsychology.* Buffalo, NY: Prometheus Books, 1985.

Radin, Dean. *The Conscious Universe: The Scientific Truth of Psychic Phenomena.* San Francisco: HarperCollins, 1997.

Sagan, Carl. *The Demon Haunted World: Science as a Candle in the Dark.* New York: Random House, 1995.